Electromagnetics and Network Theory and their Microwave Technology Applications

Stefan Lindenmeier • Robert Weigel
Editors

Electromagnetics and Network Theory and their Microwave Technology Applications

A Tribute to Peter Russer

Springer

Editors

Prof. Dr.-Ing. habil. Stefan Lindenmeier
Universität der Bundeswehr München
Institut für Hoch- und
Höchstfrequenztechnik
Werner-Heisenberg-Weg 39
85577 Neubiberg
Germany
stefan.lindenmeier@unibw.de

Prof. Dr.-Ing. Dr.-Ing. habil. Robert Weigel
Universität Erlangen-Nürnberg
Lehrstuhl für Technische Elektronik
Cauerstr. 9
91058 Erlangen
Germany
r.weigel@ieee.org

ISBN 978-3-642-43223-1 ISBN 978-3-642-18375-1 (eBook)
DOI 10.1007/978-3-642-18375-1
Springer Heidelberg Dordrecht London New York

Cover design: eStudio Calamar S.L.

Printed on acid-free paper

Springer is part of Springer Science+Business Media (www.springer.com)

Preface

On October 8–9, 2008, we organized an IEEE MTT-S International Mini-Symposium on Electromagnetics and Network Theory and its Microwave Applications at Munich University of Technology (TUM), Munich, Germany. This symposium was dedicated to Peter Russer on the occasion of his 65th birthday and his retirement. During his career as researcher in the field of Electromagnetics and Network Theory Peter Russer achieved not only a multitude of outstanding scientific results but he also had the special gift to bring researchers together and to build up an international network of scientists in this field. This network was base of the successful symposium which provided an international forum for the discussion of the challenges and perspectives of electromagnetics and network theory and their microwave applications in various aspects. Invited presentations have been given by Josef A. Nossek of TUM, President of Association for Electrical, Electronic & Information Technologies, VDE, Franz X. Kärtner of Massachusetts Institute of Technology, MIT, and of course by Peter Russer, TUM. In oral sessions and an interactive forum 48 reviewed scientific contributions were presented. Half of those contributions have been further extended now to be combined in this book in order to give a compact overview about actual research in the field of Electromagnetics and Network Theory and its Microwave Applications.

The book is subdivided into basic topics of applications and theory in this field as there are antennas and wave propagation, microwave- and communication-systems and methods for the numerical modelling of components, networks and structures being part of these systems. In a first section an actual state of research in antennas and propagation is given since the description of antennas as well as wave propagation in RF-lines and electric networks is crucial for the investigation of microwave systems like radar-, radio-location- and communication-systems.

Especially in mobile applications, radar-, radio-location- and navigation-systems as well as microwave sensors are more and more in use. An actual state of research in this field is given in the second section. Actual results of research on such systems are shown for automotive radar, a high precision radio-location-system, RF-sensors and RF-measurement technologies. The wide field of communication systems is discussed in the third section where an overview about further progress in mobile communication and wireless data transmission is given and results of actual research are shown.

In the fourth section actual numerical methods are discussed which are essential for the estimation of electromagnetic effects in all the applications shown previously. In the scope between the treatment of very tiny structures and very large structures new ways are shown for the numerical modelling of the electromagnetic field in nanostructures as well as in macrostructures and large periodic structures. In the last chapter we present the autobiography of Peter Russer which shows in a very good example, that the combined treatment of all the aspects mentioned above leads to achievements which may seem almost impossible. But, speaking with his words, the impossible just takes longer.

At this point we would like to take the occasion to give a brief summary on the very successful scientific work history and Peter Russer's extraordinary achievements – both as an outstanding researcher and as a distinguished educator. Peter Russer was born in Vienna, Austria in 1943, during World War II. After finishing elementary school and gymnasium in Vienna, he studied Electrical Engineering at the Vienna University of Technology where he received the Dipl.-Ing. degree in 1967. He continued at his Alma Mata and became a young research assistant working towards the doctoral degree under the supervision of the late Professor Hans Pötzl on "Josephson electronics", for which he received the Dr. techn. degree in 1971. Shortly after (1971) he joined the AEG-Telefunken Research Institute in Ulm, Germany, where, for ten years, he worked on fibre optic communication, solid-state electronic circuits, noise analysis, laser modulation and fibre optic gyroscopes. At the young age of 38 (in 1981), he was offered a Full Professorship at the TUM and to become Director of the Institute of High Frequency Engineering, where he has been since. His service to TUM was only briefly interrupted from 1992 to 1995 when he was selected the Founding Director of the Ferdinand Braun Institute in Berlin, Germany, a position which was also associated with a Guest Professorship at the Technical University of Berlin. In September 1995 he returned to TUM, and from 1997 to 1999 he served as Dean of the Faculty of Electrical and Information Engineering.

Peter Russer is a renowned scholar and highly respected teacher who is devoted to his students. He has developed and taught a large variety of courses in RF techniques, microwaves, quantum electronics and optical communications. His scripts and monographs are superb teaching tools and have served as basis for a couple of distinguished textbooks. Peter Russer was also the mastermind behind the international Master of Science in the Microwave Engineering curriculum at the TUM which is running very successfully since eight years. His fine teaching skills have attracted a great number of young talents to become his Master and Ph.D. students. Over the years he has graduated a total of nearly 500 students of which about 70 received their doctoral degree. Many of his students have started successful careers in industry and academia and continue to keep close ties with their mentor and 'Doktorvater'. Quite a high number of his Ph.D. students like Erwin Biebl, Franz X. Kärtner, Gerhard Fischerauer, Gerd Scholl, Josef Hausner, Sebastian Sattler and ourselves have become University Professors, respectively at TUM, Massachusetts University of Technology, University of Bayreuth, Hamburg University of federal armed forces, University of Bochum, Munich University of federal armed forces, and University of Erlangen-Nuremberg; and these so-called "Peter's Boys" have greatly contributed to a special journal issue published in summer of 2008 (Peter's Boys – Making Frequencies Think, Frequenz – Journal of RF-Engineering and Telecommunications, vol. 62, no. 7–8, July/August 2008, pp. 153–207).

Peter Russer is well known internationally for his many innovative and significant contributions to Josephson electronics, fibre optic communication and gyroscopes, laser modulation, solid-state electronics, noise analysis techniques, Bragg cell-based spectrum analyzers, integrated optics, surface acoustic waves, hyperthermia, microwave superconductivity, linear/nonlinear circuit design methods, design of integrated microwave and millimetre-wave circuits, numerical techniques in computational electromagnetics, and lately also to electromagnetic compatibility (EMC). In most of these fields, Peter Russer has clearly pioneered the research from numerous points of view. Let us give just three examples: (1) The publication *H. Hillbrand, P. Russer, "An Efficient Method for Computer Aided Noise Analysis of Linear Amplifier Networks", IEEE Transactions on Circuits and Systems, vol. 23, no. 4, April 1976, pp. 235–238* laid the basis for the theoretical foundation for the noise analysis of two-ports using correlation matrices, a technique which meanwhile is being used in nearly all network analysis computer codes. (2) On December 21, 1978, Erich Kasper and Peter Russer, who in those days were colleagues at AEG-Telefunken in Ulm published their patent (Germany, no. DE2719464) entitled *Verfahren zur Herstellung von Hochfrequenztransistoren* which describes the invention of the SiGe heterobipolar transistor (HBT), a semiconductor device which is crucial for the implementation of silicon integrated millimetre-wave circuits (SIMMWICs) which nowadays are very successfully applied in communications, sensing and radar at millimetre-wave frequencies. (3) Peter Russer's pioneering work on the foundations of the Transmission Line Matrix (TLM) modelling of electromagnetic fields has been widely acclaimed as the most rigorous approach to put this technique on solid ground. In all his research areas, Peter Russer's work demonstrates an exceptional quality, originality, and technical impact. Many times he has been able to transfer his scientific results into innovative application beneficial for the economy and for the society. To this date, Peter Russer has authored and co-authored more than 140 refereed journal publications, more than 500 conference papers, 6 books and 20 book chapters. The impact of his academic work is complemented by the numerous novel ideas and approaches he developed for industry as evidenced by the more than 50 patents he holds or has applied for. Reflecting on all these merits, it is no surprise that Peter Russer has received several high-ranking awards and honours including the 1979 NTG award for his seminal paper "Electronic circuits for high-bit rate digital fibre optic communication systems". In 1994 he was elected IEEE Fellow for his fundamental contributions to noise analysis and low-noise optimization of linear electronic circuits with general topology. In 2006, he received the IEEE Microwave Theory and Techniques-Society Distinguished Educator Award, also in 2006 the Fellowship of the Council for Technical Sciences in Germany (ACATECH), and in 2007 the Honorary Doctoral degree from the Moscow State University of Aviation.

During his professional career, Peter Russer was not only very active in research and teaching, he has also greatly contributed time and talent to the well-being of the scientific community. He is a member of IEEE, EuMA, URSI, ITG, DPG and ÖPG. Besides serving as chairman, organizer, member of technical program and steering committees of numerous conferences, workshops, society chapters, sessions etc., he also serves the scientific community as reviewer for national and international journals, conferences and research foundations. Just to note a few of these activities: Peter Russer organized and chaired the European Microwave Conference in Munich in 1999, has been chair of the German IEEE MTT/AP Joint Chapter, has been chair of URSI's commission D – Electronics and Photonics, has been a member of the German Science Foundation's (DFG) senate board for collaborative research centres, has been Associate Editor of the IEEE Transactions on Microwave Theory and Techniques, has been chair of the IEEE MTT-Society's Technical Committee on Field Theory, and has been a member of the EuMA board of directors.

It always was and still is an honour to know Peter Russer personally and for so many years. He has now moved into his status of an Emeritus of Excellence which has been awarded to him by his university TUM and which shows, that his university still counts on his very valuable contributions. We are sure he will go on in continuing his service to the scientific society and we are looking forward to it.

We cordially thank Dr. Daniel Brenk of the University of Erlangen-Nuremberg and Carmen Wolf of Springer who wisely supervised the edition of this book.

Munich
Erlangen
July 2011

Stefan Lindenmeier
Robert Weigel

Contents

Part IV Numerical Methods for Electromagnetic Field Modeling

Part I
Antennas and Propagation

Chapter 1
A Hybrid MoM/UTD Method for the Analysis of a Monopole Antenna in an Aperture

Christoph Ullrich and Peter Russer

Automotive antennas are usually realized as conformal antennas that are placed on the car glazing. Therefore they reside in the apertures of the metallic car body. In a simplified representation the passenger cabin is an absorbing cavity which features one or more apertures.

The coupling of an electromagnetic wave through an aperture into a cavity is a well-known problem in electromagnetic compatibility as it describes the shielding effectiveness of a metal encasing. This problem has already been successfully solved with the Method of Moments (MoM) [2, 12, 23]. A modified version of this problem are apertures that are penetrated by a wire [4]. However, in these cases from literature the aperture is excited by an incident wave whereas the model of an automotive antenna has to be excited by a source in the aperture plane. A source model for this excitation in the aperture plane is given in this work.

First a simple model of an automotive antenna is created: the outer shell of the car body with the window opening is represented by a metal screen with an aperture. The passenger compartment with lossy interior materials is simplified to an absorber-clad cavity. A representation of this model is given in Fig. 1.1.

1.1 Method of Moments with Magnetic Current Density

1.1.1 Integral Equations with Magnetic Charge

In order to calculate the field distribution in the aperture we introduce a fictitious magnetic charge $\underline{\mathcal{Q}}_m$ in addition to the electric charge $\underline{\mathcal{Q}}_e$ and electric current $\underline{\mathcal{I}}_e$.

C. Ullrich (✉)
AUDI AG, 85045 Ingolstadt, Germany
e-mail: christoph.ullrich@audi.de

P. Russer
Technische Universität München, Arcisstraße 21, 80333 Munich, Germany
e-mail: russer@tum.de

S. Lindenmeier and R. Weigel (eds.), *Electromagnetics and Network Theory and their Microwave Technology Applications*, DOI 10.1007/978-3-642-18375-1_1,

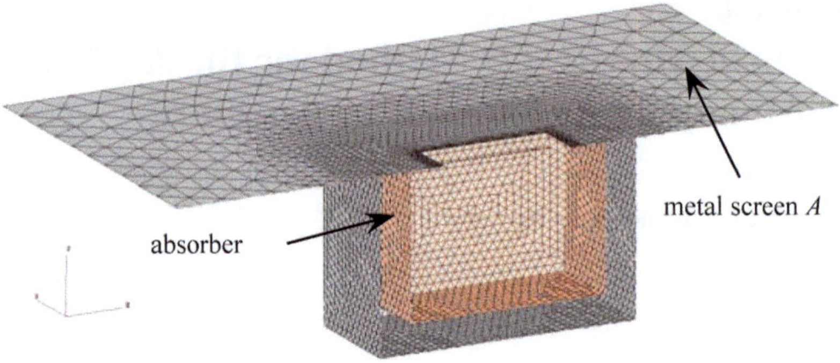

Fig. 1.1 Mesh of an aperture antenna backed by an absorbing body

Moving magnetic charges lead to a magnetic current $\underline{\mathcal{J}}_m$. Then we can write the Maxwell's equations [18] in the following form

$$\mathrm{d}\underline{\mathcal{E}} = -j\omega\underline{\mathcal{B}} + \underline{\mathcal{J}}_m \tag{1.1a}$$

$$\mathrm{d}\underline{\mathcal{H}} = j\omega\underline{\mathcal{D}} + \underline{\mathcal{J}}_e \tag{1.1b}$$

$$\mathrm{d}\underline{\mathcal{D}} = \underline{\mathcal{Q}}_e \tag{1.1c}$$

$$\mathrm{d}\underline{\mathcal{B}} = \underline{\mathcal{Q}}_m \; . \tag{1.1d}$$

In this form Maxwell's equations show almost perfect symmetry. Therefore solutions that were developed to calculate the electric current density on electric conductors can directly be applied to solve magnetic currents in an aperture. At this point it should be noted that magnetic charge and magnetic current do not necessarily exist physically but are solely used as a means to simplify the solution.

Let the metal screen A be of infinite extension. Then we cover the aperture with a perfect magnetic conductor (PMC). We use the equivalence principle [6] to replace the electromagnetic sources that cause the radiation from the aperture by an equivalent magnetic current on both sides of the PMC. This impressed magnetic current must cause the same field distribution in the both half spaces that are separated by the metal screen as if no PMC were present.

From the magnetic current $\underline{\mathcal{J}}_m$ we can derive the magnetic surface current density on the PMC $\underline{\mathcal{J}}_m^{PMC}$ and from this the desired value of the electric field in the aperture $\underline{\mathcal{E}}^{Aperture}$ [3]:

$$\mathsf{n} \wedge \underline{\mathcal{E}}^{Apertur} = -\mathsf{n} \wedge \underline{\mathcal{J}}_m^{PMC} \tag{1.2}$$

The electromagnetic field which is excited by $\underline{\mathcal{J}}_m$ can only be derived from a scalar potential if the field is irrotational in the considered domain. By introducing a potential partitioning surface (PPS) the space surrounding the conductor is separated in such a way that all possible integration paths encircling the conductor are cut

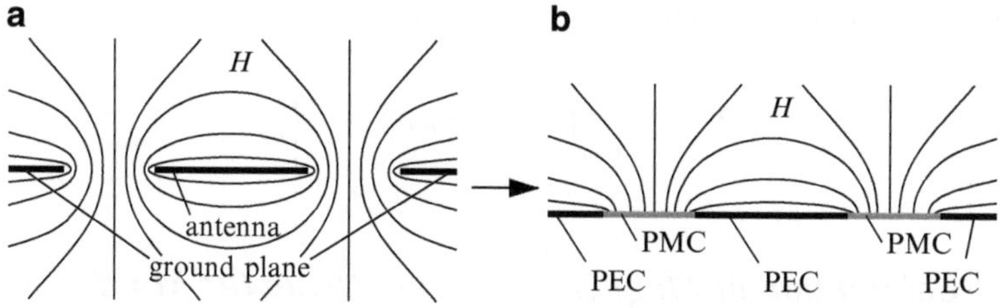

Fig. 1.2 Separation of possible integration paths by a PPS

by a PPS [13]. The PMC covering the aperture divides the problem space in two subspaces each with an irrotational magnetic field. Therefore the PMC also acts as PPS as shown in Fig. 1.2. A solution of the problem can be found by first solving the subproblems in the subspaces and then matching the field at the PMC boundary.

Furthermore, on the surface of the PMC we have $\underline{\mathscr{J}} = 0$ and $\underline{\mathscr{Q}}_e = 0$. Therefore the electrical current density is divergence free. Hence in analogy to the derivation of the EFIE [24] we can develop the magnetic field $\underline{\mathscr{H}}$ in dependance to the magnetic current $\underline{\mathscr{J}}_m$ [21]:

$$\underline{\mathscr{H}} = j\omega\varepsilon \int_V \underline{\mathscr{G}}_m \wedge \underline{\mathscr{J}}_m \tag{1.3}$$

$\underline{\mathscr{G}}_m$ itself is defined as the Green's dyad

$$\underline{\mathscr{G}}_m = (1 - \frac{1}{k^2}\mathrm{d} \star \mathrm{d}\star) G_{m0} \mathscr{I}. \tag{1.4}$$

Here it should be mentioned that in comparison to the Green's dyad in [24] we have a sign change of the second term which can be traced back to the remaining asymmetry of the signs in the Maxwell's equations as given in (1.1).

G_{m0} is given by

$$G_{m0} = \frac{e^{-jk|\mathbf{r}-\mathbf{r}'|}}{4\pi\mu|\mathbf{r}-\mathbf{r}'|} \tag{1.5}$$

and $\mathscr{I}$ denotes the unit double one-form [18]. With (1.4) and (1.5) we can write (1.3) as

$$\underline{\mathscr{H}} = j\omega\varepsilon\mu \int G_{m0}\mathscr{I} \wedge \underline{\mathscr{J}}_m - \frac{j}{\omega}\mathrm{d} \star \mathrm{d} \star \int G_{m0}\mathscr{I} \wedge \underline{\mathscr{J}}_m. \tag{1.6}$$

Considering the Lorenz gauge $\star\,\mathrm{d} \star \underline{\mathscr{A}}^e = j\omega\varepsilon\mu\Phi^e$ this can be shortened to

$$\underline{\mathscr{H}} = j\omega\underline{\mathscr{A}}^e + \mathrm{d}\Phi^e, \tag{1.7}$$

where $\underline{\mathscr{A}}^e$ and Φ^e denote the *electric* vector potential

$$\underline{\mathscr{A}}^e = \varepsilon\mu \int G_{m0}\mathscr{I} \wedge \underline{\mathscr{J}}_m \tag{1.8}$$

and the *magnetic* scalar potential

$$\Phi^e = -\frac{j}{\omega}\int \star \mathrm{d} \star G_{m0}\mathscr{I} \wedge \underline{\mathscr{J}}_m. \tag{1.9}$$

1.1.2 Calculation of Magnetic Currents Using the MoM

In order to solve (1.6) we apply the Method of Moments [6, 22]. To this end we expand the unknown magnetic current $\underline{\mathscr{J}}_m$ in the aperture into

$$\underline{\mathscr{J}}_m(\mathbf{r}) = \sum_n^N V_n \mathsf{f}_n(\mathbf{r}), \tag{1.10}$$

where V_n denote generalized voltage amplitudes and $\mathsf{f}_n(\mathbf{r})$ denotes a suitable basis one-form [1, 22]. We insert (1.10) into (1.6) and test the resulting equation with f_m in an application of the Galerkin method. After partial integration we yield:

$$\begin{aligned}\int_S \mathsf{f}_m \wedge \underline{\mathscr{H}} = \quad & j\omega\varepsilon\mu \textstyle\int_S \mathsf{f}_m \wedge \int_S' G_{m0}\mathscr{I} \wedge \sum_n^N V_n \mathsf{f}_n \\ & -\tfrac{j}{\omega} \textstyle\int_S \mathrm{d}\mathsf{f}_m \wedge \int_S' \star \mathrm{d} \star G_{m0}\mathscr{I} \wedge \sum_n^N V_n \mathsf{f}_n\end{aligned} \tag{1.11}$$

Consideration of the boundary conditions for the tangential magnetic field on the surface of the PMC $\underline{\mathscr{H}}^{tan}$ for an incident magnetic field $\underline{\mathscr{H}}^{in}$ yields

$$\mathsf{n} \wedge \underline{\mathscr{H}}^{in} = -\mathsf{n} \wedge \underline{\mathscr{H}}^{tan}. \tag{1.12}$$

When the basis one-forms f_n are defined on the surface of the PMC, we can write the left side of (1.11) as

$$-\int_S \mathsf{f}_m \wedge \underline{\mathscr{H}}^{tan}. \tag{1.13}$$

The resulting equation can be written as a system of linear equations of dimension $N \times N$:

$$\mathbf{I} = \mathbf{YV} \tag{1.14}$$

The N-dimensional vectors $\mathbf{I}$ and $\mathbf{V}$ aggregate the generalized excitation currents I_n and generalized voltage amplitudes V_n that have been introduced in (1.10). The coefficients of $\mathbf{Y}$ can be calculated very similar to the coefficients of $\mathbf{Z}$ in [17]. With (1.14) and (1.11) we can write

$$Y_{mn} = j\omega\mu\varepsilon \int_S \mathsf{f}_m \wedge \int_S' G_{m0}\mathscr{I} \wedge \mathsf{f}_n - \frac{j}{\omega} l_m \int_S' \star \mathrm{d} \star G_{m0}\mathscr{I} \wedge \mathsf{f}_n. \tag{1.15}$$

Both $\mathfrak{f}_n$ and $\mathfrak{f}_m$ are each integrated over 2 triangles. Thus, in order to calculate both terms on the right side of (1.15) we have to solve 4 separate integrals. Therefore we write:

$$j\omega\mu\varepsilon \int_S \mathfrak{f}_m \wedge \int_S' G_{m0}\mathscr{I} \wedge \mathfrak{f}_n = \mathscr{A}^e_{m+,n+} + \mathscr{A}^e_{m+,n-} + \mathscr{A}^e_{m-,n+} + \mathscr{A}^e_{m-,n-} \quad (1.16)$$

$$\frac{j}{\omega} l_m \int_S' \star \mathrm{d} \star G_{m0}\mathscr{I} \wedge \mathfrak{f}_n = -\Phi^e_{m+,n+} + \Phi^e_{m+,n-} + \Phi^e_{m-,n+} - \Phi^e_{m-,n-} \quad (1.17)$$

The potentials introduced in (1.7) are the same as the magnetic vector potential $\mathscr{A}$ and the electric scalar potential Φ as defined in [17] save the definition of the Green's function G_{m0}. Accordingly, the coefficients of (1.16) and (1.17) can directly be derived from the coefficients in [17]:

$$\mathscr{A}^e_{m\pm,n\pm} = \frac{\varepsilon}{\mu} \cdot \mathscr{A}_{m\pm,n\pm} \quad (1.18)$$

$$\Phi^e_{m\pm,n\pm} = \frac{\varepsilon}{\mu} \cdot \Phi_{m\pm,n\pm} \quad (1.19)$$

With these coefficients the magnetic current distribution in the aperture can be calculated by

$$\mathbf{V} = \mathbf{Y}^{-1}\mathbf{I}. \quad (1.20)$$

1.1.3 Magnetic Excitation

The excitation of the antenna is dual to the excitation as explained by Makarov [14]. It is modeled by a delta gap source across the infinitesimal gap between the two triangles on both sides of the feeding edge. Therefore the excitation vector **I** contains the element

$$I_m = l_m I_0, \quad (1.21)$$

which describes the excitation across the gap. All other elements disappear, so that

$$I_n = 0 \quad \text{for all } n \neq m. \quad (1.22)$$

If more than one edge is fed, (1.21) holds true for all edges.

Figure 1.3 shows the basic principle of an automotive glass antenna: a frame surrounding an aperture. In the plane of the aperture an antenna structure – in this case a monopole – is inlaid. The excitation occurs at the edge of the aperture in the excitation region. When calculating the electric surface current density the excitation is realized as an impressed current across the feed region in positive x-direction.

If the method described here is applied, the feed is realized through an impressed magnetic current. In order to compare the results achieved with this method with those obtained by the usual approach using electric surface current densities, the electric feed current of the antenna has to be transformed into an equivalent

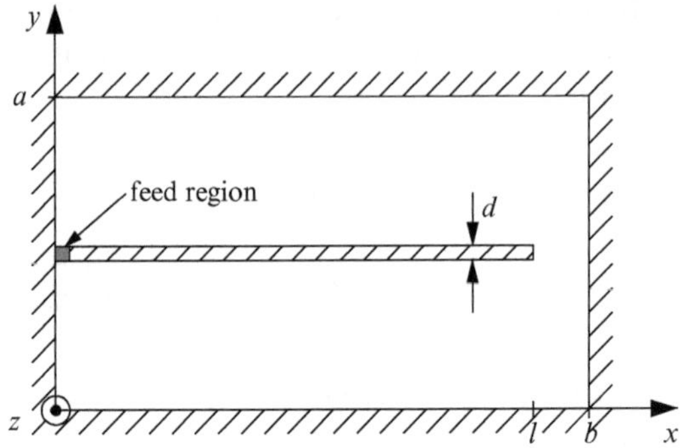

Fig. 1.3 Aperture with monopole antenna

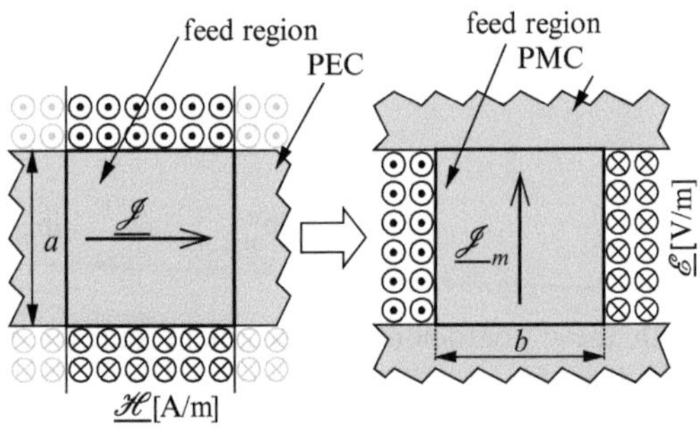

Fig. 1.4 Transformation of the feed into an equivalent magnetic current

magnetic current. Figure 1.4 illustrates how an x-directed, electric current density $\underline{\mathscr{J}}$ can be transformed into a y-directed equivalent magnetic surface current density $\underline{\mathscr{J}}_m$. According to Babinet's principle the field caused by these two current densities is the same [18]. For the application of the method using electric surface currents, the whole PEC area of the structure surrounding the aperture has to be meshed whereas with magnetic surface currents only the aperture region respectively the area of the introduced PMC has to be meshed. Solely the feed region is present in the mesh for both methods (cf. also Fig. 1.4).

In order to directly compare the results using magnetic currents with those achieved with a feed with the voltage V_0 first an arbitrary current density I_0 is chosen. After the calculation of the coefficients $\mathbf{V}$ of $\underline{\mathscr{J}}_m$ by means of (1.20) the result has to be normalized as follows:

$$\mathbf{V}_{norm} = \frac{\mathbf{V} \cdot V_0}{V_m / l_m}, \tag{1.23}$$

Here $l_m = b$ is the length of the feeding edge, V_0 is the equivalent feeding voltage that was used during the calculation using electric surface densities and V_m is the magnetic current on the feeding edge m. If $\mathbf{V}_{norm}$ is used to calculate the electromagnetic fields in the problem space, the results are directly comparable to those achieved with electric surface current densities and a feed of V_0.

1.1.4 Radiation by an Aperture with Magnetic Currents

The field radiated from the aperture with the magnetic current distribution $\underline{\mathcal{J}}_m$ can be calculated by inserting the Lorenz gauge $\star d \star \underline{\mathcal{A}}^e = j\omega\varepsilon\mu\Phi^e$ [15] in (1.7) which yields

$$\underline{\mathcal{H}} = \frac{1}{j\omega\mu\varepsilon} \mathrm{d} \star \mathrm{d} \star \underline{\mathcal{A}}^e + j\omega\underline{\mathcal{A}}^e. \tag{1.24}$$

Using (1.10) and (1.8) we can directly calculate $\underline{\mathcal{A}}^e$ from the coefficients V_n. In order to calculate the electric field we insert the constitutive equation $\underline{\mathcal{D}} = \varepsilon \star \underline{\mathcal{E}}$ [18] into (1.1b) and yield

$$j\omega\varepsilon \star \underline{\mathcal{E}} = \mathrm{d}\underline{\mathcal{H}} - \underline{\mathcal{J}}_e. \tag{1.25}$$

As there is no electric conductor in the aperture the electric current density in the aperture is $\underline{\mathcal{J}}_e = 0$. If we insert (1.24) into (1.25) and consider $\mathrm{d}\,\mathrm{d}\mathcal{V} = 0$ we obtain [18]

$$\underline{\mathcal{E}} = \frac{1}{\varepsilon} \mathrm{d}\underline{\mathcal{A}}^e. \tag{1.26}$$

When calculating $\underline{\mathcal{A}}^e$ one should keep in mind that the magnetic current density is present on both sides of the introduced PMC. Hence for the calculation of field quantities the current density $\underline{\mathcal{J}}_m^{ff} = 2\underline{\mathcal{J}}_m$ has to be inserted into (1.8). Then the electric field in the aperture is given by

$$\mathrm{d}\underline{\mathcal{E}} = 2\underline{\mathcal{J}}_m. \tag{1.27}$$

A numerically efficient way of calculating the far field can be achieved if every basis one-form is considered as a small magnetic dipole [14]. This dipole spans the distance between the center points $\mathbf{r}_m^{c\pm}$ of the two triangles of each basis one-form and has the length

$$h = |\mathbf{h}| = \left|\mathbf{r}_m^{c-} - \mathbf{r}_m^{c+}\right|. \tag{1.28}$$

A constant magnetic current of amplitude $2V_m l_m$ is impressed on the whole length of the dipole. The field radiated by this small magnetic dipole is given by [18]

$$\underline{\mathcal{E}}_\phi = \frac{hV_m l_m}{2\pi}\left(\frac{1}{r^2} + \frac{jk}{r}\right) e^{-jkr} \sin\theta\, r \sin\theta\, \mathrm{d}\phi. \tag{1.29}$$

In case of the far-field $kr >> 0$ only the part with $1/r$ remains and (1.29) can be simplified. For the numerical computation it is generally advisable to use cartesian coordinates. With the transformation into cartesian coordinates and (1.29) the far field of a single magnetic dipole at the point $\mathbf{r} = [r_x r_y r_z]^T$ becomes

$$\underline{\mathcal{E}}^{MoM}(\mathbf{r}) = \frac{jkV_m l_m}{2\pi r} e^{-jkr} \left[(h_y r_z - h_z r_y)\,\mathrm{d}x + (h_z r_x - h_y r_z)\,\mathrm{d}y + (h_x r_y - h_y r_x)\,\mathrm{d}z\right], \tag{1.30}$$

with h_x, h_y and h_z from (1.28). Then a simple sum over all dipoles yields the sought-after total electric field at the point $\mathbf{r}$.

1.1.5 Comparison to Results Achieved with Electric Surface Currents

In order to verify the method introduced in the above section the structure as shown in Fig. 1.3 was analyzed. The aperture of length $a = 1,14$ m and width $b = 0,73$ m is placed in a perfectly conducting screen of infinite extension. A monopole antenna of length $l = 0,99$ m and width $d = 0,01$ m is fed against the left edge of the aperture. With the MoM and magnetic currents only the aperture has to be discretized which leads to the mesh with 1856 triangles which is shown in Fig. 1.5. The structure was fed by a magnetic current of frequency $f = 200$ MHz in positive y-direction. The feed region is highlighted in Fig. 1.3.

For comparison the structure was also analyzed using the commercially available software EMCStudio [5]. The amplitude of the electric field in the aperture was read out with field probes. For the calculation in EMCStudio the finite metal screen had to be considered in the model. The size of the metal screen was limited to $6\,\text{m} \times 6\,\text{m}$ as at this size the field in the aperture did not change significantly.

Notwithstanding the fact that the model for electrical surface currents was discretized much coarser, this model needed 3466 triangles. The finer mesh for calculation with magnetic currents as shown in Fig. 1.5 only needed 1850 elements.

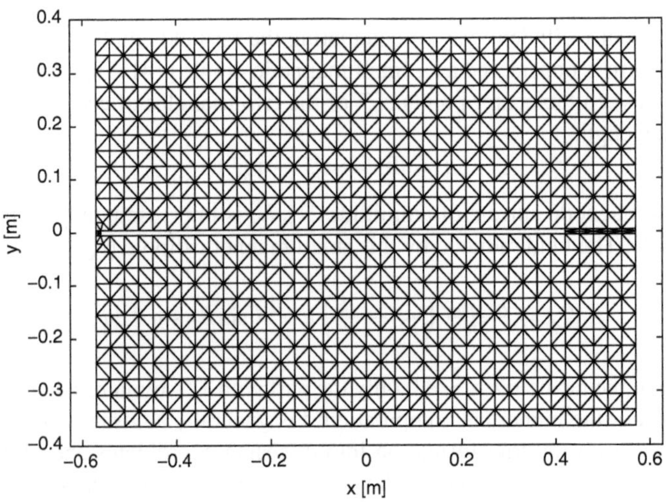

Fig. 1.5 Mesh of the aperture

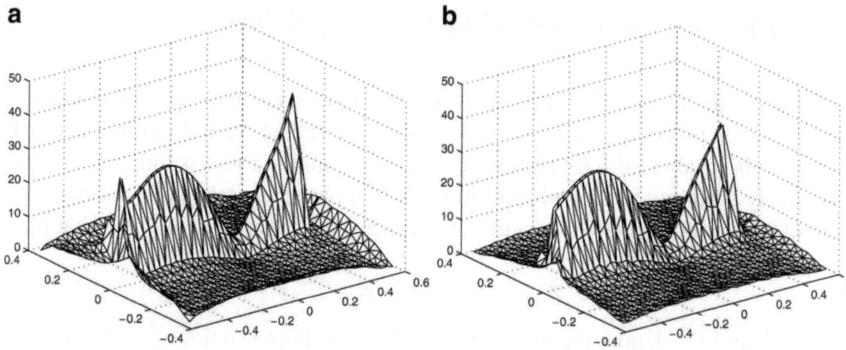

Fig. 1.6 Electric field E in (V/m) calculated with (**a**) magnetic currents an (**b**) commercial software EMCStudio

Figure 1.6a shows the electrical field which was calculated with the method using magnetic currents explained above. For comparison Fig. 1.6b shows the result of the field computation given by EMCStudio. The results are almost identical. The differences at the feed point can be traced back to the fact that in the model using magnetic currents the magnetic current impressed at the feed point is evaluated at a very large electrical field at this point. The surface current density decays with $1/r$ away from the feed point. In EMCStudio this impressed current only occurs on the PEC and therefore the singularity at the feed point cannot be seen in the aperture. However, with magnetic currents this singularity is in the discretized region are and therefore is clearly visible in the results.

1.2 Finite Extension of Screen

In the method given above an aperture in a PEC screen of infinite extension was assumed. In order to account for the necessarily finite extension of this screen the Uniform Geometric Theory of Diffraction (UTD) is applied. The UTD is an extension to Geometrical Optics which overcomes some of its limitations.

Already in the seventeenth century Francesco Maria Grimaldi observed that a ray of light impeding on a sharp edge is seen as a bright line at the edge from a viewpoint in the shadow [16]. The explanation of this problem was not possible at the time as the transversal property of electromagnetic waves was still unknown. Thus only in 1896 Arnold Sommerfeld could physically correctly explain the diffraction effects at a straight edge through a seminal analytic derivation. However, the application potentials thereof where only opened up after the development of radar technology in World War II. Thus the UTD was developed into a cohesive theory only after 1950 by Keller [7].

1.2.1 The Uniform Geometrical Theory of Diffraction

In Geometrical Optics (GO) one assumes that electromagnetic waves of extremely high frequency propagate along rays. The effects of reflection and refraction which are easily observable for light rays are correctly accounted for. Assuming a metallic respectively well-conducting object, no refraction of the ray takes place. Therefore the field at a point is given by a sum with proper phase overlay of all rays that pass through this point:

$$\underline{\mathscr{E}}_{GO} = \underline{\mathscr{E}}^{in}\mathrm{u}^{in} + \underline{\mathscr{E}}^{r}\mathrm{u}^{r}, \tag{1.31}$$

where $\underline{\mathscr{E}}^{in}$ and $\underline{\mathscr{E}}^{r}$ denote the incident respectively reflected field. $\underline{\mathscr{E}}^{r}$ can easily be calculated from the incident field with the according reflection coefficient [18]. As both fields are only present in the illuminated region, they have to be multiplied with the step function u^{in} and u^{r}. These functions each are 0 in the shadow region and 1 in the illuminated region. In reference to Fig. 1.7 the rays $\mathbf{S}_g$ in the shadow region are neglected.

The UTD in addition accounts for the effect of diffraction of the electromagnetic radiation at edges [7] so that field quantities in the geometrical optics shadow region can be determined, too. Therefore (1.31) is corrected by the magnitude of the edge diffracted field $\underline{\mathscr{E}}^{d}$ [11]

$$\underline{\mathscr{E}}_{UTD} = \underline{\mathscr{E}}^{in}\mathrm{u}^{in} + \underline{\mathscr{E}}^{r}\mathrm{u}^{r} + \underline{\mathscr{E}}^{d}. \tag{1.32}$$

It should be noted that $\underline{\mathscr{E}}_d$ is not limited to the shadow region. Diffracted rays $\mathbf{S}_g$ are also present in the other regions of Fig. 1.7. From basic physical laws it is obvious that the complete field at the shadow boundaries has to be continuous. Therefore the diffracted field $\underline{\mathscr{E}}_d$ must be discontinuous at the shadow boundary as the field calculated with the GO has a step at this boundary. The sum of the two fields must not have a discontinuity. The same hold true for the reflection boundary [11].

In order to determine the diffracted field $\underline{\mathscr{E}}^{d}$ we first develop the incident field in a Luneberg-Kline-series [8]

$$\underline{\mathscr{E}}^{in}(\mathbf{r}) \approx e^{-jk\psi(\mathbf{r})} \sum_{m=0}^{\infty} \frac{\underline{\mathscr{E}}_m(\mathbf{r})}{(j\omega)^m}, \tag{1.33}$$

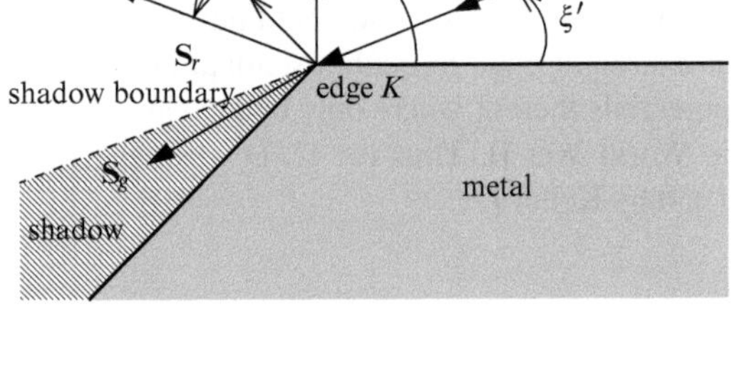

Fig. 1.7 Incident, reflected, and diffracted ray

where $\psi(\mathbf{r})$ describes the form of the wavefront at the point $\mathbf{r}$. Equation (1.33) is an exact description of the present field and must also satisfy the wave equation for a source free space

$$\left(\nabla + k^2\right) \underline{\mathscr{E}} = 0. \tag{1.34}$$

From the solenoidality follows directly the auxiliary constraint

$$\mathrm{d} \star \underline{\mathscr{E}} = 0. \tag{1.35}$$

As the UTD is an asymptotic approximation for very high frequencies, the series expansion can be truncated after the first element of (1.33) because all further elements vanish for large ω due to the factor $(1/\omega)^m$. Inserting the first element of (1.33) in (1.34) we obtain

$$\frac{\partial \underline{\mathscr{E}_0}}{\partial s} + \frac{1}{2}\left(\nabla^2 \psi\right) \underline{\mathscr{E}_0} = 0. \tag{1.36}$$

Here $\nabla \psi = \mathbf{s}$ defines the direction of the ray which is defined normal to the wavefront $\psi(\mathbf{r}) = \mathit{const}$. Because of this definition it is sufficient to henceforth only consider the scalar quantity s which denotes the distance along the ray path [11]. With (1.33) in (1.35) and setting the coefficients of the series to zero we yield the eikonal equation for a medium with $\varepsilon = 1$:

$$|\nabla \psi| = 1 \tag{1.37}$$

Assuming very high frequencies the series expansion from (1.33) can be truncated after the first element. Therefore the approximation of $\underline{\mathscr{E}}$ along the ray $\mathbf{s}$ can be simplified to

$$\underline{\mathscr{E}}(s) \approx e^{-jk\psi(s)} \underline{\mathscr{E}_0}(s). \tag{1.38}$$

Equation (1.36) can be integrated directly. This more elaborate integration is given by Kouyoumjian in [9]. This yields

$$\underline{\mathscr{E}_0}(s) = \underline{\mathscr{E}_0}(0) \sqrt{\frac{\rho_1 \rho_2}{(\rho_1 + s)(\rho_2 + s)}}. \tag{1.39}$$

Here ρ_1 and ρ_2 are the main radii of curvature of the wave front at the point $s = 0$. This distance is also illustrated in Fig. 1.8: incident rays are diffracted at the edge of the illuminated object. Therefore all rays emanate from this edge. Thus this edge is the first caustic. The second caustic is created by the bundling of the rays at the distance $\rho_2 - \rho_1$ from the edge. The radius of curvature ρ_2 depends on the curvature of the edge. Taking into consideration (1.37) and the fact that $\mathbf{s}$ is by definition normal to the wavefront defined by ψ we can deduct [11]

$$\psi(s) = \psi(0) + s. \tag{1.40}$$

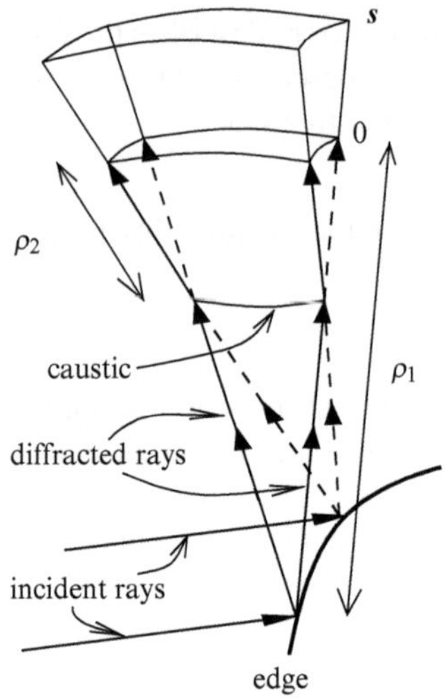

Fig. 1.8 Pair of incident rays which are diffracted into an astigmated tube of rays [7]

With (1.40), (1.39) and (1.38) follows for the field along the diffracted ray

$$\underline{\mathscr{E}}(s) = \underline{\mathscr{E}}(0)e^{-jk\psi(0)}e^{-jks}\sqrt{\frac{\rho_1\rho_2}{(\rho_1+s)(\rho_2+s)}}\,. \tag{1.41}$$

Due to the assumptions for the high frequency approximation the solution depends primarily on the immediate surrounding of the diffraction point Q_b. Therefore the incident wave at this point can be considered as a locally plane wave and the factor $\psi(0)$ in (1.41) can be neglected. So the only remaining unknown is $\underline{\mathscr{E}}(0)$ at the diffraction point. If the diffracted field is denoted by $\underline{\mathscr{E}}^d$ we yield

$$\underline{\mathscr{E}}^d(s) = \underline{\mathscr{E}}^d(0)e^{-jks}\sqrt{\frac{\rho_1\rho_2}{(\rho_1+s)(\rho_2+s)}}\,. \tag{1.42}$$

A suitable choice for the origin of the diffracted ray is the diffraction point Q_b on the edge. However, the edge is a caustic at which (1.42) is singular. On the other hand it is obvious that $\underline{\mathscr{E}}^d(s)$ in (1.42) has to exist independently of the choice of origin. Therefore

$$\lim_{\rho_2\to 0}\underline{\mathscr{E}}^d(0)\sqrt{\rho_2} \tag{1.43}$$

must exist. Furthermore $\underline{\mathscr{E}}_d$ must be proportional to the incident field $\underline{\mathscr{E}}^{in}$ at the point Q_b. So we can write:

$$\lim_{\rho_2\to 0}\underline{\mathscr{E}}^d(0)\sqrt{\rho_2} = \underline{\mathscr{E}}^{in}(Q_b)\cdot\mathbf{D}, \tag{1.44}$$

where **D** is a still undefined diffraction matrix which links incident and diffracted ray. Inserting (1.44) into (1.42) we obtain

$$\underline{\mathscr{E}}^{d} \approx \underline{\mathscr{E}}^{in} \cdot \mathbf{D} \cdot \sqrt{\frac{\rho}{s(\rho+s)}}\, e^{-jks}. \tag{1.45}$$

$\underline{\mathscr{E}}^{in}$ denotes the incident field at the diffraction point and s is the distance along the diffracted ray from the diffraction point. In the case given here ρ is the distance between origin and the diffraction point.

1.2.1.1 Determination of Diffraction Matrix D

Assuming a z-directed edge the Dirichlet and Neumann boundary conditions for the electric respectively magnetic fields have to be satisfied:

$$\underline{E}_z = 0 \quad \text{and} \quad \frac{\partial \underline{H}_z}{\partial n} = 0, \tag{1.46}$$

where n is the edge normal. Considering (1.46) and (1.45) we obtain for the z-component of the electric and magnetic field:

$$\underline{E}_z^d \approx \underline{E}_z^i \cdot D_s \cdot \sqrt{\frac{\rho}{s(\rho+s)}}\, e^{-jks} \quad \text{and} \quad \underline{H}_z^d \approx \underline{H}_z^i \cdot D_h \cdot \sqrt{\frac{\rho}{s(\rho+s)}}\, e^{-jks} \tag{1.47}$$

D_s and D_h denote the scalar diffraction coefficients which ensure the satisfaction of the Dirichlet respectively Neumann boundary conditions.

For the satisfaction of the Neumann boundary conditions only the components of the ray $\mathbf{s}'$ and $\mathbf{s}$ which are normal to the edge have to be considered. Therefore we define $\overline{\xi}'$ and $\hat{\xi}$ as the projection of the incident ray $\mathbf{s}'$ respectively diffracted $\mathbf{s}$ into the xy-plane which is normal to the edge (cf. Fig. 1.7).

In analogy to this we define the vectors $\hat{\beta}'$ and $\hat{\beta}$ which indicate the direction of the incident ray $\mathbf{s}'$ respectively diffracted $\mathbf{s}$ in a plane which is spanned by the diffracting edge and the direction of the incident respectively diffracted ray. The diffracted ray $\mathbf{s}$ does not have to lie in the same plane as the incident ray $\mathbf{s}'$. As shown in Fig. 1.9 it may lie in a cone which is created by the rotation of the outgoing ray around the diffracting edge. Therefore $\hat{\beta}'$ and $\hat{\beta}$ may lie in two different planes.

The unit vectors $\hat{\xi}'$ and $\hat{\xi}$ are always perpendicular to the diffraction edge. For $\hat{\beta}'$ and $\hat{\beta}$ applies

$$\hat{\beta}' = \mathbf{s}' \times \hat{\xi}' \quad \text{and} \quad \hat{\beta} = \mathbf{s} \times \hat{\xi}. \tag{1.48}$$

From simple geometrical considerations (cf. Fig. 1.9) we obtain for the relation between transversal field along the rays $\mathbf{s}'$ and $\mathbf{s}$ and the z-directed component of the

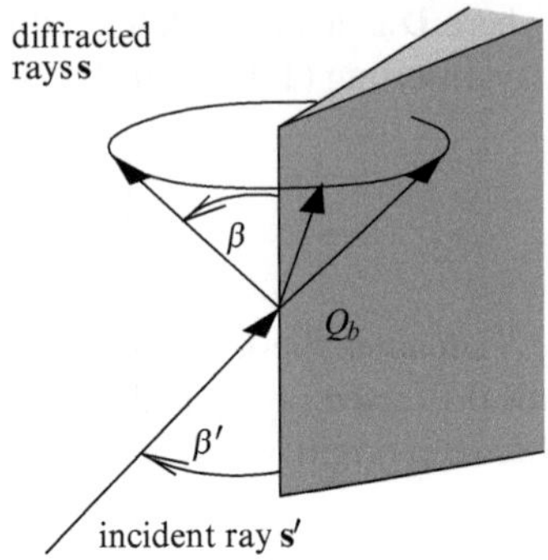

Fig. 1.9 Angle of incidence and emergence of the rays

electric respectively magnetic field at the edge

$$\underline{E}^i_z = \underline{E}^i_{\beta'} \sin\beta \tag{1.49a}$$

$$\underline{H}^i_z = \frac{1}{\eta}\underline{E}^i_{\phi'} \sin\beta \tag{1.49b}$$

$$\underline{E}^d_z = -\underline{E}^d_{\beta} \sin\beta \tag{1.49c}$$

$$\underline{H}^d_z = -\frac{1}{\eta}\underline{E}^d_{\phi} \sin\beta, \tag{1.49d}$$

with $\eta = \sqrt{\frac{\mu}{\varepsilon}}$. From (1.47) and (1.49) follows

$$\underline{E}^d_{\beta} \approx -\underline{E}^i_{\beta'} \cdot D_s \cdot \sqrt{\frac{\rho}{s(\rho+s)}}\, e^{-jks} \quad \text{and} \tag{1.50a}$$

$$\underline{H}^d_{\phi} \approx -\underline{E}^i_{\phi'} \cdot D_h \cdot \sqrt{\frac{\rho}{s(\rho+s)}}\, e^{-jks}. \tag{1.50b}$$

If this derivation is also carried out for x- and y-directed edges after merging the results one obtains the diffraction matrix [10]

$$\mathbf{D} = -\hat{\beta}'\hat{\beta} D_s - \hat{\xi}'\hat{\xi} D_h, \tag{1.51}$$

where $\hat{\beta}'\hat{\beta}$ respectively $\hat{\xi}'\hat{\xi}$ denote the dyadic product [18] of the two vectors $\hat{\beta}'$ and $\hat{\beta}$ respectively $\hat{\xi}'$ and $\hat{\xi}$.

The diffraction coefficients D_h and D_s for a screen could be derived for the first time by Arnold Sommerfeld in 1896 [19]. Sommerfeld solved the problem by setting up a double Riemann space in which the boundary value integral could be solved analytically. Following this he developed the field in a series expansion with Bessel functions in order to approximately calculate the diffraction coefficients. The calculation of the diffraction coefficients of a wedge is basically an extension of this problem to n-fold Riemann spaces as is descriptively shown in the commented translation of Sommerfelds work by Nagem et al. [20].

With this derivation the scalar diffraction coefficients $D_{s,h}$ for a wedge are [10]

$$D_{s,h}\left(\xi,\xi';\beta\right) = -\frac{e^{-j(\pi/4)}}{2\sqrt{2\pi k}\sin\beta}\cdot\left\{\frac{F\left[2kL\cos^2\{(\xi-\xi')/2\}\right]}{\cos\left[(\xi-\xi')/2\right]}\right. \mp \left.\frac{F\left[2kL\cos^2\{(\xi+\xi')/2\}\right]}{\cos\left[(\xi+\xi')/2\right]}\right\} \quad (1.52)$$

with the distance parameter for a spherical wave

$$L = \frac{r_p\rho_p}{r+\rho_p}\sin^2\beta. \quad (1.53)$$

The distance parameter is necessary to account for the influence of the form of the wave front at the diffraction point which was assumed to be plane in (1.42). Here ρ_p denotes the distance between the origin of the incident ray and the diffraction point P and r_p denotes the distance between the diffraction point P and the observation point. That means that (1.53) defines the curvature of the wavefront which so far has been defined by ρ_1 and ρ_2.

In the analytic derivation, in order to integrate the series expansion with Bessel functions one has to solve a transition function at the shadow and reflection boundaries. This transition function $F(X)$ from (1.52) is given by [10, 19]

$$F(X) = 2j\sqrt{X}\,e^{jX}\int_{\sqrt{X}}^{\infty} e^{-j\tau}\,\mathrm{d}\tau. \quad (1.54)$$

As Sommerfeld could not analytically solve the Fresnel integral contained in this formula he used an approximation of a truncated series expansion. Due to this his diffraction coefficients were not valid in the transition regions. However, Sommerfeld could already define a region in which his approximation was valid with an error of $\Delta < \varepsilon$. When developing the Geometrical Theory of Diffraction (GTD) Keller used the same approximation. Thus fields close to the shadow and reflection boundary could not be calculated.

Today Fresnel integrals can be computed numerically in an efficient way and thus this approximation is not needed any more. Therefore (1.52) is still valid at the caustics. This method then is called Uniform Theory of Diffraction (UTD) whose sole difference to the GTD is its validity at the caustics.

1.3 Calculation of Total Electric Field

Considering the structure given in Fig. 1.1 as a simple model of a vehicle one can assume that the absorber clad cavity behind the aperture absorbs all radiation in its direction. If the metal screen A is in the xy-plane with $z = 0$, we have

$$\underline{\mathscr{E}}(z < 0) = 0. \tag{1.55}$$

In the model analyzed here the diffraction edge is always in the same plane as the origin of the wave. Thus we have $\xi' = 0$ for all points on the edge. Therefore we yield $D_s = 0$ and the diffraction matrix can be simplified considerably. The diffraction coefficient D_h becomes for $\xi' = 0$ [21]:

$$D_h(\xi;\beta) = \frac{-e^{-j\pi/4}}{\sqrt{2\pi k}\sin\beta}\left\{\frac{F[2\,kL\cos^2(\xi/2)]}{\cos(\xi/2)}\right\} \tag{1.56}$$

For an angle of incidence of $\xi' = 0$ incident and reflected field are superimposed onto each other at the point P of the edge. Therefore the magnitude of the *incident* field $\underline{\mathscr{E}}^{in}$ at the Point P is only have of the present field $\underline{\mathscr{E}}^{MoM}$. So we have for the field diffracted at point P of the edge

$$\underline{\mathscr{E}}^{d}_{\theta P} \approx \frac{1}{2}\underline{\mathscr{E}}^{MoM}_{\theta}\left(\rho_P, \theta, \phi\right)\; D_h(\xi;\beta)\cdot\sqrt{\frac{\rho_P}{r(r+\rho_P}}\cdot e^{-jkr_P}, \tag{1.57}$$

where $\underline{\mathscr{E}}^{MoM}_{\theta}$ was calculated with (1.30). r is the distance between observation point and origin.

For every observation angle ϕ in the far field we only have to consider two edge points $A(\rho_{cA}, \pi/2, \phi)$ and $B(\rho_{cB}, \pi/2, \phi + \pi)$ which are on opposite sides of the metal screen A. This is due to the fact that the absolute value of the angle β of the incident ray must be equal to the absolute value of the angle β'. Therefore the total electric field is given by

$$\underline{\mathscr{E}}^{tot}_{\theta} = \underline{\mathscr{E}}^{MoM}_{\theta} + \underline{\mathscr{E}}^{d}_{\theta A} + \underline{\mathscr{E}}^{d}_{\theta B}, \tag{1.58}$$

where $\underline{\mathscr{E}}^{d}_{\theta A}$ and $\underline{\mathscr{E}}^{d}_{\theta B}$ were calculated with (1.57).

The combination of the calculation of the electric field in an infinite PEC screen with the MoM and magnetic surface currents and the subsequent correction of this field with the UTD results in a hybrid MoM/UTD method with magnetic surface currents.

1.4 Results

The structure as shown in Fig. 1.5 was analyzed with the hybrid method. The aperture has the dimensions as given in Sect. 1.1.5. The feed is realized by a magnetic current in positive y-direction.

The metal screen A has a size of 6×6 m. The structure was also analyzed with EMCStudio [5] where the cavity of size $2.3 \times 1.85 \times 1.5$ m was clad with absorbing material and included in the simulation model.

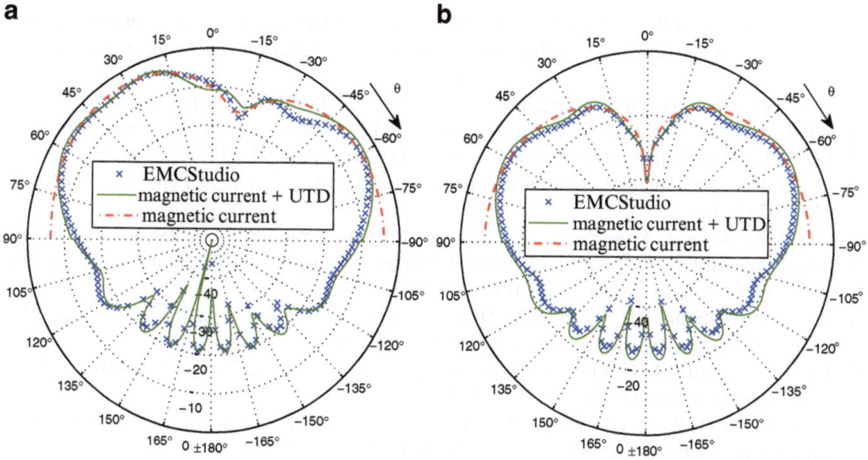

Fig. 1.10 E_θ for $f = 200\,\text{MHz}$ (dBV/m) in (**a**) the xz-plane and (**b**) the yz-plane

The graphs in Figs. 1.10a and Fig. 1.10b show the calculated field in the xz- respectively yz-plane for $f = 200\,\text{MHz}$. For comparison the far field calculated with magnetic currents without the UTD correction is also shown. This illustrates the effect of the hybrid method. Very good agreement between the MoM/UTD result and the analysis with the commercial tool EMCStudio is achieved [21].

The MoM/UTD hybrid requires a calculation time of 65 s on a notebook with 1.7 GHz and 1.25 GB RAM whereas the solution with electric surface currents needed 38 min on a Cluster with 24 parallel 3-GHz-CPUs with a total of 72 GB RAM. Therefore the calculation time of the hybrid method is smaller by orders of magnitude. For the simulation of antenna structures which are placed in an aperture backed on one side by an absorbing body it is very advantageous to use the hybrid method shown here. The calculation time drops to a fractional amount and the results are comparable to the analysis with alternative methods.

Especially for antennas operating at higher frequencies, such as GPS and SDARS antennas, which can also be designed as slot antennas, the method shown here can efficiently account for the influence of electrically large parts such as the roof or trunk lid in automotive applications.

References

1. M.J. Bluck, S.P. Walker, High-order discrete helmholtz decompositions for the electric field integral equation. IEEE Trans. Antennas Propag. **55**, 1338–1347 (2007)
2. C.M. Butler, Y. Rahmat-Samii, R. Mittra, Electromagnetic penetration through apertures in conducting surfaces. IEEE Trans. Antennas Propag. **26**, 82–93 (1978)
3. C.M. Butler, K.R. Umashankar, Electromagnetic excitation of a wire through an aperture-perforated conducting screen. IEEE Trans. Antennas Propag. **24**, 456–462 (1976)

4. V. Daniele, M. Gilli, S. Pignari, EMC prediction model of a single wire transmission line crossing a circular aperture in a planar screen. IEEE Trans. Electromagn. Compat. **38**, 117–126 (1996)
5. EMCoS Consulting and Software: EMCStudio v4.0 (2008), http://www.emcos.com
6. R.F. Harrington, *Time-Harmonic Electromagnetic Fields*. (McGraw-Hill, New York, 1961)
7. J.B. Keller, Geometrical theory of diffraction. J. Opt. Soc. Am. **52**, 116–130 (1962)
8. M. Kline, An asymptotic solution of Maxwell's equations. Commun. Pure Appl. Math. **4**, 225–262 (1951)
9. R.G. Kouyoumjian, Asymptotic high-frequency methods. Proc. IEEE **53**, 864–876 (1965)
10. R.G. Kouyoumjian, P.H. Pathak, A uniform geometrical theory of diffraction for an edge in a perfectly conducting surface. Proc. IEEE **62**, 1448–1461 (1974)
11. R.G. Kouyoumjian, P.H. Pathak, *The Dyadic Diffraction Coefficient for a Curved Edge*. The Ohio State University Electroscience Laboratory (1974)
12. J. Lin, W.L. Curtis, M.C. Vincent, Electromagnetic coupling to a cable through apertures. IEEE Trans. Antennas Propag. **24**, 198–203 (1976)
13. S. Lindenmeier, P. Russer, Design of planar circuit structures with an efficient magnetostatic-field solver. IEEE Trans. Microw. Theory Tech. **45**, 2468–2475 (1997)
14. S. Makarov, MoM antenna simulations with matlab: RWG basis functions. IEEE Antennas Propag. Mag. **43**, 100–107 (2001)
15. R. Nevels, C. Shin, Lorenz, Lorentz, and the Gauge. IEEE Antennas Propag. Mag. **43**, 70–71 (2001)
16. J. Priestley, *The History and Present State of Discoveries relating to Vision, Light, and Colours*. (J. Johnson, London, 1772)
17. S.M. Rao, D.R. Wilton, A.W. Glisson, Electromagnetic scattering by surfaces of arbitrary shape. IEEE Trans. Antennas Propag. **30**, 409–418 (1982)
18. P. Russer, *Electromagnetics, Microwave Circuit and Antenna Design for Communications Engineering*. (Artech House Publishers, London, 2006)
19. A. Sommerfeld, Mathematische theorie der diffraction. Math. Ann. **47**, 317–374 (1896)
20. A. Sommerfeld, R.J. Nagem, M. Zampolli, G. Sandri, *Mathematical Theory of Diffraction*. (Birkhäuser, Boston, 2004)
21. C. Ullrich, K.F. Warnick, P. Russer, Radiation from a monopole antenna backed by an absorbing body using a hybrid MoM/UTD approach. In *Proceedings of the International Symposium on Antennas and Propagation, IEEE* (2008)
22. C. Ullrich, Efficiente Simulations methoden für die Optimierung von komplexen Fahceugantennensystemen. (Curillier, Gōttingen, 2009)
23. T. Wang, R.F. Harrington, J.R. Mautz, Electromagnetic scattering from and transmission through arbitrary apertures in conducting bodies. IEEE Trans. Antennas Propag. **38**, 1805–1814 (1990)
24. K.F. Warnick, D.V. Arnold, Electromagnetic green functions using differential forms. J. Electromagnet. Wave **10**, 427–438 (1996)

Chapter 2
Electromagnetic and Network Theory of Waveguide Radiation by Spherical Modes Expansions

Cristiano Tomassoni, Mauro Mongiardo, Peter Russer, and Roberto Sorrentino

2.1 Introduction

In recent years, modal techniques have been successfully improved and are increasingly used for dealing with design of waveguide discontinuities and passive components [1–8], due to their efficiency and also because they provide rigorous and useful network representations. One distinguished characteristic of modal techniques is to separate the transverse field behavior from the longitudinal one; this decoupling makes it feasible to consider electromagnetic wave propagation inside a waveguide as a superposition of transmission lines (each pertaining to a mode) which couple only at discontinuities. Electromagnetic field representation inside a waveguide with finite cross-section, is therefore achieved by a discrete summation of the relevant waveguide modes.

A similar procedure can be followed when considering free-space as a waveguide, by means of a spherical mode expansion [9, pp. 445–450], [10–14]. Free-space field expansion in terms of spherical modes presents several advantages:

- Straightforward extension of the modal techniques used for waveguide problems also to radiation problems
- Derivation of rigorous equivalent networks

C. Tomassoni (✉)
Università di Perugia, via G. Duranti, 93, 06125 Perugia, Italy
e-mail: tomassoni@diei.unipg.it

M. Mongiardo
Università di Perugia, via G. Duranti, 93, 06125 Perugia, Italy
e-mail: mongiardo@diei.unipg.it

P. Russer
Technische Universität München, Arcisstr. 21, 80333 Munich, Germany
e-mail: russer@tum.de

R. Sorrentino
Università di Perugia, via G. Duranti, 93, 06125 Perugia, Italy
e-mail: sorrentino@diei.unipg.it

S. Lindenmeier and R. Weigel (eds.), *Electromagnetics and Network Theory and their Microwave Technology Applications*, DOI 10.1007/978-3-642-18375-1_2,

- Derivation of boundary conditions for terminating numerical methods based on space discretization
- Treatment of conformal boundaries, as e.g. finite circular flange, semi-spherical dielectric covering, etc.

While the use of spherical modes for dealing with conformal geometries is quite natural and has received a considerable interest in the past, it is noted here that the first two points are of particular relevance for computer-aided design and have received modest attention. In fact they allows to extend the very successful modal techniques also to free-space radiation problems. In [15–18] a general procedure has been described for the systematic partitioning of complex problems into subdomains and their rigorous description in terms of networks. By using spherical transmission lines for describing propagation in free-space, one can obtain network representations similar for closed and open problems. In addition it is also feasible to describe the interaction between distant radiators without the necessity of discretizing the region of space between them. The value of the spherical wave expansion for solving antenna problems has been already noted in [19–21]. In these works this approach has been used for coupling the analysis of cavity-backed microstrip antennas performed by a three-dimensional finite element method, to other antennas of the same type but placed far away.

In this paper we describe the electromagnetics and network theory of waveguide radiation by considering as an example an array of waveguides radiating on a *finite* circular flange plane. We rigorously solve the above problem, which to the best of our knowledge has only be solved by discretization methods so far, and we also provide an equivalent network for this type of structures. The problem of flange-mounted array radiation has been considered in the literature, quite some time ago [22]. Fundamental progress has been made in the excellent work of Trevor Bird, of which we cite just a few contributions [24–26] Further refinements have been proposed in [27] by taking into account the field singularities and by considering strategies for the efficient design with the adjoint network method in [28]. The case of radiation from elliptical horns has been considered in [29] and an efficient scheme for parallel computation has been provided in [30]. In all the above cases the presence of an infinite flange plane has always been assumed. The finite flange problem has been considered in the past in [31] by using a geometrical theory of diffraction (GTD) approximation for dealing with a rectangular flange. In [32] an hybrid technique combining moment method, FEM and GTD has been used to attack the finite flange problem. Notably, the case of a circular flange backed by a metallic semi-sphere (see Fig. 2.1 for a sketch) is amenable of considerable analytical progress when a spherical mode expansion is used. As a result, it is possible to investigate finite flange effects with a modest numerical effort and to provide a rigorous network for this structure.

The paper is structured as follows: in Sect. 2.2 the spherical mode expansion is introduced and is applied to the problem of radiation from an array of flange-mounted rectangular waveguides and its equivalent network is also established. In particular, in this section we start from the statement of the problem (in Sect. 2.2.1), illustrate the analogy with modal techniques for closed waveguides (in Sect. 2.2.2),

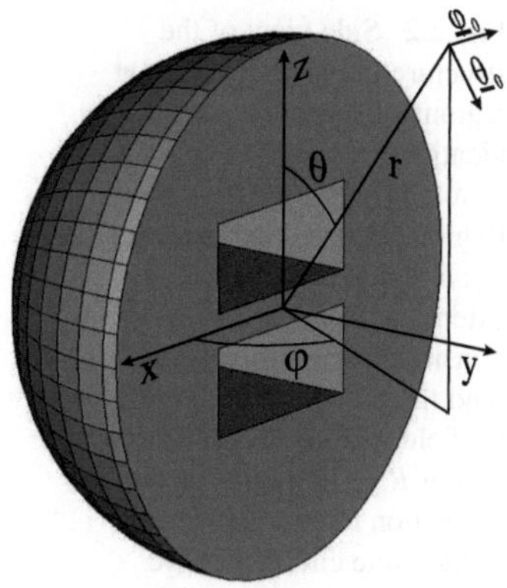

Fig. 2.1 Three-dimensional view of a flange-mounted small array of rectangular waveguides radiating into FREE SPACE. Note the finite circular flange; for modeling purposes the backside of the array is considered as a perfectly conducting half-sphere

and we recall the spherical mode expansion (in Sect. 2.2.3) and the generalized transformer (in Sect. 2.2.5). Characterization of the transition region has been outlined in [14] and is not repeated here; furthermore, since description of waveguide regions is well known it has not been considered in the following. Finally, in Sect. 2.3 the rigorous analysis of a small array of rectangular waveguides radiating into a *finite* circular flange plane is considered and some numerical results are provided.

2.2 Generalized Network of Radiating Waveguides in Terms of Spherical Modes

2.2.1 Statement of the Problem

We consider, with reference to Fig. 2.1, an array of rectangular waveguides, mounted on a *finite* circular flange, radiating into free-space. For modeling purposes, it is assumed that the backside of the circular flange is a metallic half sphere (hemisphere).

In order to illustrate the methodology it is sufficient to refer to an array of just two waveguides. In Fig. 2.1 the three-dimensional view of the structure is shown and in Fig. 2.2 is sketched its side-view. From the latter figure we see that different regions of space have been introduced:

- Waveguide regions, denoted as region R_g^i, for the ith waveguide
- Transition Region, denoted by region R_t, i.e. a semi-spherical region of space of the same diameter of the circular flange
- A region of space R_r extending from the end of the transition region up to infinity (and therefore comprising the far-field region)

The apertures, i.e. the boundary between regions R_g^i and R_t, are denoted by S_a. Note that in Fig. 2.2 we have considered a surface S_t separating the transition region

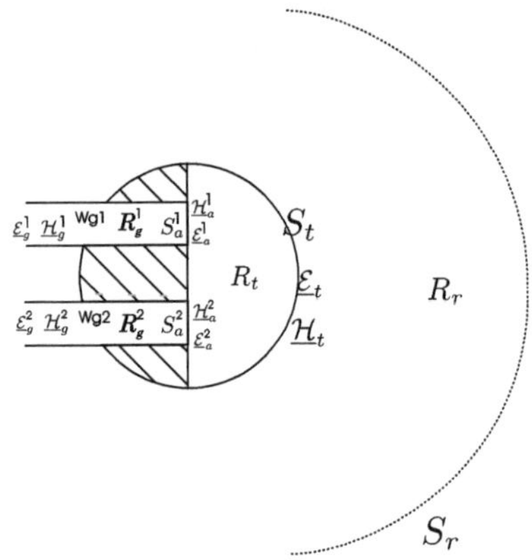

Fig. 2.2 Side view of the structure in Fig. 2.1; different regions of space have been identified: waveguide regions, denoted by R_g^i; a "Transition Region" denoted as region R_t; a region of space extending from the end of the transition region up to infinity (and therefore comprising the far-field region) denoted as region R_r. The radius of the Transition region is the same of the finite circular flange

from region R_r and a surface S_r, in region R_t, where the far-field may be evaluated. The Transition Region is bounded on one side by the flange plane on which the surface S_a lies; the surface S_t provides the remaining part of the boundary.

2.2.2 *Modal Analysis and Equivalent Networks*

As noted before, one significant advantage of using the spherical mode expansion is the fact that it allows to extend modal techniques also to free-space problems. Let us refer to Fig. 2.3, where in the lower right corner is sketched a waveguide problem representing two waveguides (denoted with 1 and 2 respectively), a resonator (region 3), and another waveguide (region 4) attached to the resonator via an iris. On the lower left side of Fig. 2.3 is sketched our problem: the two waveguides radiates into the semi-spherical transition region (region 3) and then into free-space. In the upper part of Fig. 2.3 is sketched the equivalent network representing *both* problems. Note the presence of a generalized transformer between regions 3 and 4. In the waveguiding problem (lower right sketch) the field on the iris between region 3 and 4 is coupled to the waveguide modes of region 4; the coupling matrix corresponds to a generalized transformer. In the case of finite circular flange the field on the semi-spherical surface of the Transition region is coupled to the spherical mode expansion of free-space, taking into account the presence of the metallic semi-sphere.

2.2.3 *Spherical Modes*

The EM-field in spherical coordinate may be written as superposition of spherical modes as

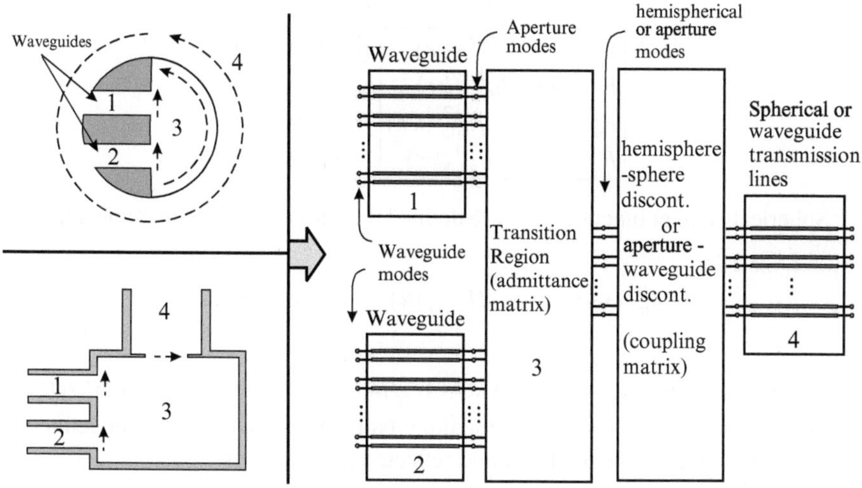

Fig. 2.3 Equivalent network of the two problems sketched in the *left part* of the figure; on the *higher left side* is shown the problem of two waveguides radiating on a finite circular flange; on the *right side* is shown the equivalent waveguiding problem

$$\mathbf{E}_t(r,\theta,\varphi) = \sum_{n=1}^{\infty}\sum_{m=0}^{n}\sum_{j}^{\substack{TEe,\,TMe\\TEo,\,TMo}} \frac{V_{m,n}^{(j)}(r)}{r}\mathbf{e}_{m,n}^{(j)}(\theta,\varphi) \tag{2.1}$$

$$\mathbf{E}_r(r,\theta,\varphi) = \sum_{n=1}^{\infty}\sum_{m=0}^{n}\sum_{j}^{\substack{TMe\\TMo}} \frac{n(n+1)}{-j\omega\varepsilon}\frac{I_{m,n}^{(j)}(r)}{r^2}T_{m,n}^{(j)}(\theta,\varphi)\underline{r}_0 \tag{2.2}$$

$$\mathbf{H}_t(r,\theta,\varphi) = \sum_{n=1}^{\infty}\sum_{m=0}^{n}\sum_{j}^{\substack{TEe,\,TMe\\TEo,\,TMo}} \frac{I_{m,n}^{(j)}(r)}{r}\mathbf{h}_{m,n}^{(j)}(\theta,\varphi) \tag{2.3}$$

$$\mathbf{H}_r(r,\theta,\varphi) = \sum_{n=1}^{\infty}\sum_{m=0}^{n}\sum_{j}^{\substack{TEe\\TEo}} \frac{n(n+1)}{j\omega\mu}\frac{V_{m,n}^{(j)}(r)}{r^2}T_{m,n}^{(j)}(\theta,\varphi)\underline{r}_0 \tag{2.4}$$

with equivalent voltages and currents given by:

$$\begin{cases} V_{m,n}^{(j)}(r) = V_{0_{m,n}}^{(j)+}F_n^{V+}(kr) + V_{0_{m,n}}^{(j)-}F_n^{V-}(kr) \\ I_{m,n}^{(j)}(r) = Y_0\left[V_{0_{m,n}}^{(j)+}F_n^{I+}(kr) - V_{0_{m,n}}^{(j)-}F_n^{I-}(kr).\right] \end{cases} \tag{2.5}$$

In the above equation V_o^+ and V_o^- represent the incident (toward ∞) and reflected (from ∞) spherical waves amplitude, while F functions are defined in terms of spherical Hankel function as:

	TM	TE
$F_n^{V+}(kr)$	$h_n^{(2)\prime}(kr)$	$h_n^{(2)}(kr)$
$F_n^{V-}(kr)$	$h_n^{(1)\prime}(kr)$	$h_n^{(1)}(kr)$
$F_n^{I+}(kr)$	$-jh_n^{(2)}(kr)$	$jh_n^{(2)\prime}(kr)$
$F_n^{I-}(kr)$	$jh_n^{(1)}(kr)$	$-jh_n^{(1)\prime}(kr).$

Spherical Hankel functions $h_n^{(\ell)}$ are defined in terms of Hankel functions $H_n^{(\ell)}$ as:

$$h_n^{(\ell)}(x) = \sqrt{\frac{\pi x}{2}} H_{n+\frac{1}{2}}^{(\ell)}(x) \qquad \text{where } \ell = 1, 2 \tag{2.6}$$

and we denote with $Y_0 = \sqrt{\frac{\varepsilon}{\mu}}$ and $k = \omega\sqrt{\mu\epsilon}$ the admittance and the wavenumber, respectively, of the considered medium. Note that from (2.5) we can define the modal admittance for incident (+) and reflected (−) waves as:

$$Y_c^+ = Y_0 \frac{F_n^{I+}(kr)}{F_n^{V+}(kr)} \qquad Y_c^- = -Y_0 \frac{F_n^{I-}(kr)}{F_n^{V-}(kr).} \tag{2.7}$$

The tesseral harmonics $T_{m,n}^{(j)}(\theta,\varphi)$, for both TE and TM modes, are defined as:

$$T_{m,n}^{\left\{\begin{array}{c} even \\ odd \end{array}\right\}} = A \left\{\begin{array}{c} \cos(m\varphi) \\ \sin(m\varphi) \end{array}\right\} P_n^m(\cos\theta). \tag{2.8}$$

Finally, the way to evaluate the eigenfunction of electric type $\mathbf{e}_{m,n}(\theta,\varphi)$ depends on the mode type (TM_e, TM_o, TE_e, TE_o)

$$\mathbf{e}_{m,n}^{\left\{\begin{array}{c} TM_e \\ TM_o \end{array}\right\}} = A\left[\left\{\begin{array}{c} \cos(m\varphi) \\ \sin(m\varphi) \end{array}\right\} \frac{d}{d\theta} P_n^m(\cos\theta)\underline{\theta}_0 + \frac{m}{\sin(\theta)} \left\{\begin{array}{c} -\sin(m\varphi) \\ \cos(m\varphi) \end{array}\right\} P_n^m(\cos\theta)\underline{\varphi}_0\right] \tag{2.9}$$

$$\mathbf{e}_{m,n}^{\left\{\begin{array}{c} TE_e \\ TE_o \end{array}\right\}} = A\left[\frac{m}{\sin(\theta)} \left\{\begin{array}{c} -\sin(m\varphi) \\ \cos(m\varphi) \end{array}\right\} P_n^m(\cos\theta)\underline{\theta}_0 + -\left\{\begin{array}{c} \cos(m\varphi) \\ \sin(m\varphi) \end{array}\right\} \frac{d}{d\theta} P_n^m(\cos\theta)\underline{\varphi}_0\right] \tag{2.10}$$

where m ranges from 0 to n for even modes and from 1 to n for odd modes. The relevant eigenfunction of magnetic type is:

$$\mathbf{h}_{m,n}(\theta,\varphi) = \underline{r}_0 \times \mathbf{e}_{m,n}(\theta,\varphi), \tag{2.11}$$

and the normalization constant A is defined as:

$$\int_0^{2\pi} \int_0^{\pi} |\mathbf{e}_{m,n}(\theta,\varphi)|^2 \sin\theta d\theta d\varphi = 1. \tag{2.12}$$

The above formulation is referred to free-space modes and the modal eigenfunctions $\mathbf{e}_{m,n}(\theta, \varphi)$ are defined on a spherical surface (spherical modes). In this paper we need also to use modes defined on an hemispherical surface (hemispherical modes). With reference to Figs. 2.2 and 2.3, the free-space region 4 (spherical modes) is connected to the transition region 3 (hemispherical modes).

Hemispherical modes can be considered as modes of the half-space. In fact, by taking a free-space region and dividing it into two parts by using an infinite metal plane passing trough the origin of the reference system, we obtain two half-spaces and in each half-space propagate hemispherical modes. Hemispherical modes can be then obtained starting from the spherical modal set by imposing the boundary conditions due to the presence of the electric plane.

In particular if we consider the metal plane lying on the plane xz (the same plane of our circular flange, see Fig. 2.1), the hemispherical modal set is obtained by taking *TEe* and *TMo* modes only (discarding *TEo* and *TMe*) from the spherical modal set and by multiplying them by $\sqrt{2}$ to re-normalize.

2.2.4 Transition Region

In our problem, according to Fig. 2.4, the transition region is a hemispherical portion of space where a hemispherical port and some rectangular ports are present. For such a region it is possible to find an admittance matrix:

$$\begin{bmatrix} \mathbf{I}_r \\ \mathbf{I}_h \end{bmatrix} = \begin{bmatrix} [Y]_{r,r} & [Y]_{r,h} \\ [Y]_{h,r} & [Y]_{h,h} \end{bmatrix} \begin{bmatrix} \mathbf{V}_r \\ \mathbf{V}_h \end{bmatrix} \tag{2.13}$$

where subscripts r and h stand for rectangular and hemispherical ports, respectively. $\mathbf{V}_r$ ($\mathbf{I}_r$) is a vector containing modal voltages (currents) of all rectangular apertures, while $\mathbf{V}_r$ ($\mathbf{I}_r$) is a vector containing modal voltages (currents) of the hemispherical port. Submatrices $[Y]_{r,r}$ $[Y]_{r,h}$ $[Y]_{h,r}$ $[Y]_{h,h}$ can be found taking advantages from some properties of hemispherical modes, as detailed in [14].

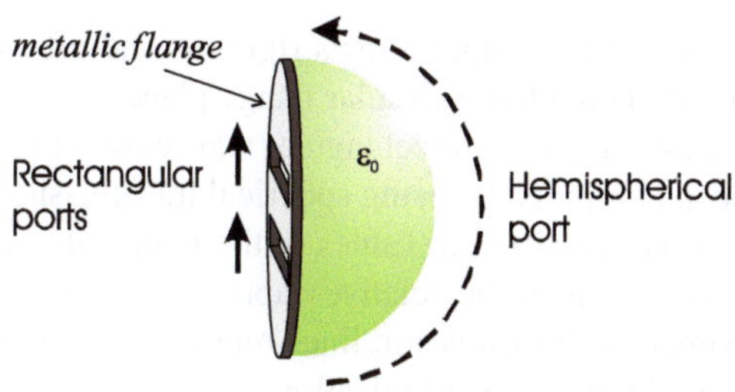

Fig. 2.4 The transition region is a hemispherical portion of space bounded by metal and ports. In particular, the hemispherical surface of the hemisphere is the hemispherical port while the flat surface is composed by a metallic flange with rectangular apertures. Rectangular apertures correspond to rectangular ports

2.2.5 Coupling Between Spherical and Hemispherical Modes

The surface separating region 3 and region 4 on Fig. 2.3 represents the discontinuity between the hemispherical port of region 3 and the spherical port of region 4. Such a discontinuity is a region of zero volume and, similarly to a discontinuity between two waveguides, can be represented by an equivalent network composed solely by transformers [15, 18] and can be studied by applying the mode-matching techniquc and by evaluating the coupling matrix $[M]$. The coupling matrix representation is:

$$\mathbf{V}_s = [M]\,\mathbf{V}_h \tag{2.14}$$

$$\mathbf{I}_h = -\,[M]^T\,\mathbf{I}_s \tag{2.15}$$

where $\mathbf{V}_s$ ($\mathbf{I}_s$) and $\mathbf{V}_h$ ($\mathbf{I}_h$) are vectors containing equivalent voltages (currents) relating to the spherical modes and hemispherical modes, respectively.

The jth element of the ith row of the coupling matrix ($g_{i,j}$) can be evaluated by the coupling integral:

$$g_{i,j} = \int_0^{\pi}\int_0^{\pi} \mathbf{e}_i^{(s)}(\theta,\varphi)\cdot\mathbf{e}_j^{(h)}(\theta,\varphi)\sin\theta d\theta d\varphi \tag{2.16}$$

where the index i (combination of TE_e, TE_o, TM_e, TM_o, m, n) indicate the i-th spherical modes while the index j (combination of TE_e, TM_o, m, n) indicate the j-th hemispherical modes. Obviously, superscript s and h refers to spherical modes and hemispherical modes, respectively.

By inserting (2.9–2.10) into (2.16), it can be noted that the coupling integrals can be conveniently written as product of integrals depending on the variable φ only and integrals depending on the variable θ only. The evaluation of coupling integrals is detailed in the appendix.

2.3 Discussion and Numerical Results

The approach presented in this paper allows rigorous and efficient modeling of a waveguide array mounted on a finite circular flange plane.

In Fig. 2.5 the S_{11} and S_{12} of two rectangular waveguides mounted on a circular flange plane have been computed by using spherical transmission lines and verified against CST simulations, for different values of the flange diameter. Details on the structure geometry are given in the relative caption. The very efficient modeling achieved by using spherical transmission lines represent a considerable advantage when the structure should undergo optimization.

In Fig. 2.6 we have plotted the scattering parameters relative to the case of two rectangular waveguides, placed as shown in the inset, when changing the dimensions of the circular flange. It is apparent that the flange dimension has an effect both on the magnitude and phase of the scattering parameters. It can be seen that

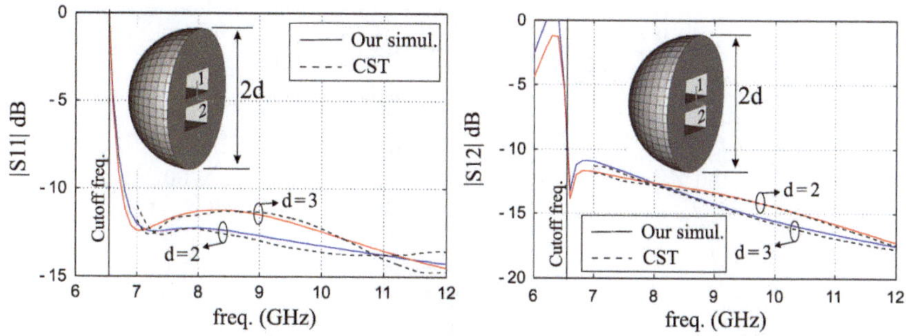

Fig. 2.5 S_{11} and S_{12} of two WR90 rectangular waveguides mounted on a circular flange plane. The distance between waveguide center is 14 mm. The radius 'd' of the spherical flange is expressed in cm. The results have been computed by using spherical transmission lines (*continuous line*) and verified against CST simulation (*dashed line*)

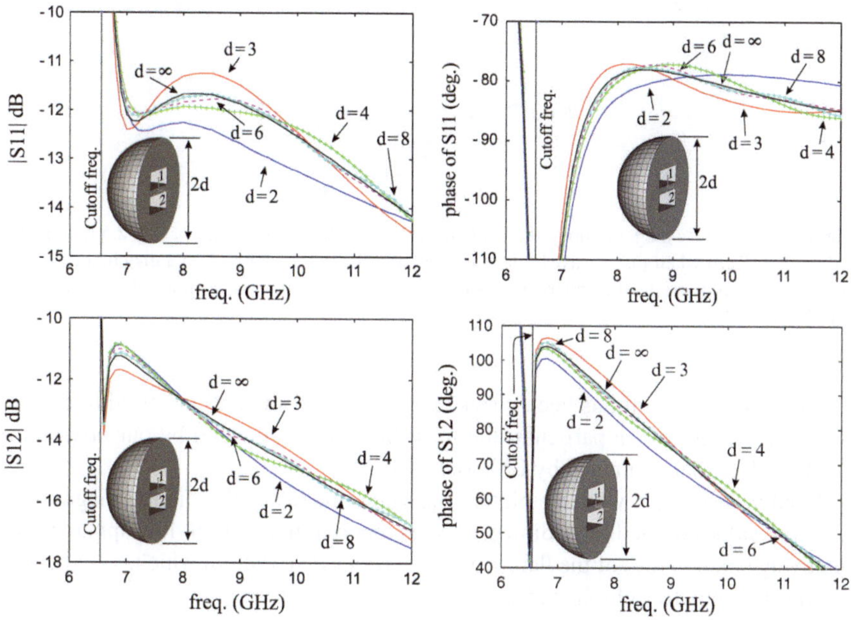

Fig. 2.6 Variation of the scattering parameters with the flange dimensions for the same structure of Fig. 2.5

results obtained by using a finite flange oscillate around the results related to the case of infinite flange and, as expected, increasing the radius of the circular flange, the amplitude of the oscillation decrease and the results tend to those of the infinite flange.

The effect of the finite flange plane on radiation has been investigated in Fig. 2.7 where we have plotted the directivity on the E-plane at 9.5 GHz for flange radius $d = 2$ cm. Since only the upper waveguide has been fed there is a certain

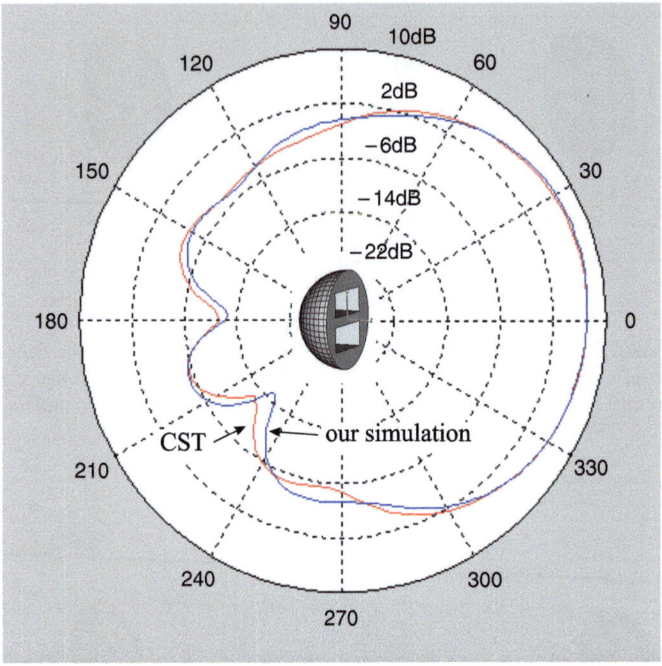

Fig. 2.7 Directivity diagram along the E-plane at 9.5 GHz for the structure of Fig. 2.5 with flange radius $d = 2$ cm when just a waveguide is excited and the other is closed on a matched load. The *continuous line* refers to the spherical mode expansion while the *dashed line* refers to the CST simulation

asymmetry. The result has been checked against CST simulations providing a satisfactory agreement. In particular the max directivity estimated by our program is 7.5 dB, while that extimated by CST is 7.3 dB.

Finally, in Fig. 2.8 we have plotted the three-dimensional radiation diagram for different values of the flange dimensions, as reported in the inset. It is apparent that relatively small values of the flange permit a backside radiation which, as expected, is almost completely eliminated with larger flanges.

2.4 Conclusions

We have considered the problem of an array of rectangular waveguides mounted on a *finite* circular flange plane and radiating into free-space. The use of spherical transmission lines allows a systematic description of the radiation problem and a rigorous network representation. The effect of the *finite* circular flange plane has been rigorously investigated.

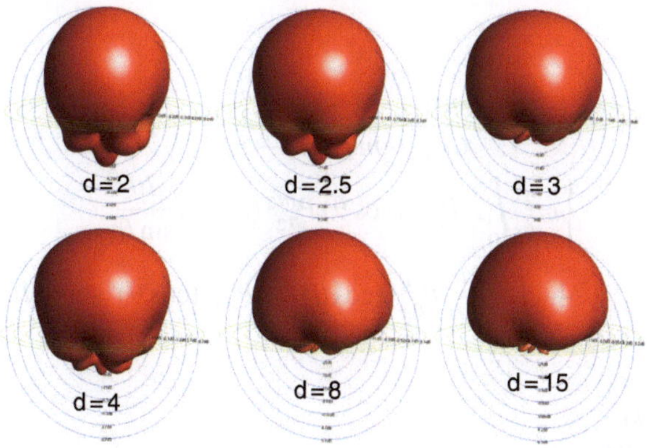

Fig. 2.8 Radiation diagram for the structure of Fig. 2.5 at 9.5 GHz obtained by exciting both waveguides with the same amplitude and phase. The figure refers to a circular flange plane with different radii 'd' (expressed in cm) and shows how the radiated field changes with the flange dimensions

Appendix

Considering spherical and hemispherical modes in (2.9)–(2.10), by recalling that superscript (s) refers to spherical modes while superscript (h) refers to hemispherical modes, the coupling integral (2.16) can be conveniently written as:

$\boldsymbol{TE}^{(s)}_{e_{m_1,n_1}}$ - $\boldsymbol{TE}^{(h)}_{e_{m_2,n_2}}$

$$g_{i,j} = \begin{cases} \frac{1}{\sqrt{2}} \text{ for } m_1 = m_2 \neq 0 \text{ and } n_1 = n_2 \\ 0 \quad \text{Otherwise} \end{cases} \tag{2.17}$$

$\boldsymbol{TE}^{(s)}_{e_{m_1,n_1}}$ - $\boldsymbol{TM}^{(h)}_{o_{m_2,n_2}}$

$$g_{i,j} = 0 \tag{2.18}$$

$\boldsymbol{TE}^{(s)}_{o_{m_1,n_1}}$ - $\boldsymbol{TE}^{(h)}_{e_{m_2,n_2}}$

$$g_{i,j} = 0 \quad \begin{matrix} \text{for } m_1 + m_2 \text{ even} \\ \text{for } n_1 + n_2 \text{ even} \end{matrix} \tag{2.19}$$

otherwise:

$$g_{i,j} = A^{(s)} A^{(h)} \frac{m_1[1-(-1)^{m_1+m_2}]}{(m_1+m_2)(m_1-m_2)} \left\{ m_2^2 \int_0^{\pi} P_{n_1}^{m_1}(\cos\theta) P_{n_2}^{m_2}(\cos\theta) \frac{d\theta}{\sin\theta} + \int_0^{\pi} \frac{d}{d\theta} P_{n_1}^{m_1}(\cos\theta) \frac{d}{d\theta} P_{n_2}^{m_2}(\cos\theta)\ \sin\theta\ d\theta \right\} \tag{2.20}$$

$TE^{(s)}_{o_{m_1,n_1}}$ - $TM^{(h)}_{o_{m_2,n_2}}$

$$g_{i,j} = 0 \tag{2.21}$$

$TM^{(s)}_{e_{m_1,n_1}}$ - $TE^{(h)}_{e_{m_2,n_2}}$

$$g_{i,j} = 0 \quad \begin{array}{l} \text{for } m_1+m_2 \text{ even} \\ \text{for } n_1+n_2 \text{ odd} \end{array} \tag{2.22}$$

otherwise:

$$g_{i,j} = A^{(s)} A^{(h)} \frac{[1-(-1)^{m_1+m_2}]}{(m_1+m_2)(m_1-m_2)} \left\{ m_2^2 \int_0^{\pi} \frac{d}{d\theta} P_{n_1}^{m_1}(\cos\theta) P_{n_2}^{m_2}(\cos\theta) d\theta + m_1^2 \int_0^{\pi} P_{n_1}^{m_1}(\cos\theta) \frac{d}{d\theta} P_{n_2}^{m_2}(\cos\theta)\ d\theta \right\} \tag{2.23}$$

$TM^{(s)}_{e_{m_1,n_1}}$ - $TM^{(h)}_{o_{m_2,n_2}}$

$$g_{i,j} = 0 \quad \begin{array}{l} \text{for } m_1+m_2 \text{ even} \\ \text{for } n_1+n_2 \text{ even} \end{array} \tag{2.24}$$

otherwise:

$$g_{i,j} = A^{(s)} A^{(h)} \frac{m_2[1-(-1)^{m_1+m_2}]}{(m_1+m_2)(m_2-m_1)} \left\{ \int_0^{\pi} \frac{d}{d\theta} P_{n_1}^{m_1}(\cos\theta) \frac{d}{d\theta} P_{n_2}^{m_2}(\cos\theta)\ \sin(\theta)\ d\theta + m_1^2 \int_0^{\pi} P_{n_1}^{m_1}(\cos\theta) P_{n_2}^{m_2}(\cos\theta)\ \frac{d\theta}{\sin(\theta)} \right\} \tag{2.25}$$

$\boldsymbol{TM^{(s)}_{o_{m_1,n_1}}}$ - $\boldsymbol{TE^{(h)}_{e_{m_2,n_2}}}$

$$g_{i,j} = 0 \tag{2.26}$$

$\boldsymbol{TM^{(s)}_{o_{m_1,n_1}}}$ - $\boldsymbol{TM^{(h)}_{o_{m_2,n_2}}}$

$$g_{i,j} = \begin{cases} \frac{1}{\sqrt{2}} \text{ for } m_1 = m_2 \text{ and } n_1 = n_2 \\ 0 \quad \text{Otherwise} \end{cases} \tag{2.27}$$

References

1. R. Sorrentino, M. Mongiardo, F. Alessandri, G. Schiavon, An investigation on the numerical properties of the mode-matching technique. Int. J. Numer. Model. **4**, 19–43 (1991)
2. T. Rozzi, M. Mongiardo, E-plane steps in rectangular waveguide. IEEE Trans. Microw. Theory Tech. **39**, 1279–1288 (1991)
3. F. Alessandri, G. Baini, M. Mongiardo, R. Sorrentino, A 3-D mode matching technique for the efficient analysis of coplanar MMIC discontinuities with finite metallization thickness. IEEE Trans. Microw. Theory Tech. **41**, 1625–1629 (1993)
4. M. Mongiardo, R. Sorrentino, Efficient and versatile analysis of microwave structures by combined mode matching and finite difference methods. IEEE Microw. Guid. Wave Lett., **3**, 241–243, (1993)
5. F. Alessandri, M. Mongiardo, R. Sorrentino, Rigorous mode matching analysis of mitered E-plane bends in rectangular waveguide. IEEE Microw. Guid. Wave Lett. **4**, 408–410 (1994)
6. M. Mongiardo, C. Tomassoni, Modal analysis of discontinuities between elliptical waveguides. IEEE Trans. Microw. Theory Tech. **48**, 597–605 (2000)
7. L. Accatino, M. Mongiardo, Hybrid circuit-fullwave computer-aided design of a manifold multiplexers without tuning elements. IEEE Trans. Microw. Theory Tech. **50**, 2044–2048 (2002)
8. G. Bertin, B. Piovano, L. Accatino, M. Mongiardo, Full-wave design and optimization of circular waveguide polarizers with elliptical irises. IEEE Trans. Microw. Theory Tech. **50**, 1077–1083 (2002)
9. P. Russer, *Electromagnetics, Microwave Circuit and Antenna Design for Communications Engineering*, 2nd edn. (Artech House, Boston, London, 2006)
10. P. Russer, Network-oriented modeling of radiating electromagnetic structures. Elektrik Turk. J. Elec. Engin. **10**(2), 147–162 (2002)
11. M. Mongiardo, P. Russer, Field computations and network representations for open electromagnetic structures. *Elektrotechnik und Informationstechnik*, no. 1. (Springer, Wien New York, 2004), pp. 2–5
12. M. Mongiardo, C. Tomassoni, P. Russer, Generalized network formulation: Application to flange–mounted radiating waveguides. IEEE Trans. Antennas Propag. **55**, 1–12 (2007)
13. M. Mongiardo, P. Russer, R. Sorrentino, C. Tomassoni, Spherical modal expansion for arrays of flange–mounted rectangular waveguides. *37th European Microwave Conference*, Sept. 2007
14. M. Mongiardo, P. Russer, R. Sorrentino, C. Tomassoni, Spherical mode expansions for flange–mounted waveguide apertures. *ICEAA*, Sept. 2007
15. M. Mongiardo, P. Russer, C. Tomassoni, L.B. Felsen, Analysis of n-furcation in elliptical waveguides via the generalized network formulation. IEEE Trans. Microw. Theory Tech. **47** (1999)

16. L. Felsen, M. Mongiardo, P. Russer, Electromagnetic field representations and computations in complex structures I: Complexity architecture and generalized network formulation. Int. J. Numer. Model. **15**, 93–107 (2002)
17. L. Felsen, M. Mongiardo, P. Russer, Electromagnetic field representations and computations in complex structures II: Alternative Green's functions. Int. J. Numer. Model. **15**, 109–125 (2002)
18. P. Russer, M. Mongiardo, L. Felsen, Electromagnetic field representations and computations in complex structures III: Network representations of the connection and subdomain circuits. Int. J. Numer. Model. **15**, 127–145 (2002)
19. J. Rubio, M.A. González, J. Zapata, Analysis of cavity-backed microstrip antennas by a 3-D finite element/segmentation method and a matrix Lanczos-Pad algorithm (SFELP). IEEE Antennas Wireless Propag. Lett. **1**, 193–195 (2002)
20. J. Rubio, M.A. González, J. Zapata, Efficient full-wave analysis of mutual coupling between cavity-backed microstrip patch antennas. IEEE Antennas Wireless Propag. Lett. **2**, 155–158 (2003)
21. J. Rubio, M.A. González, J. Zapata, Generalized-scattering-matrix analysis of a class of finite arrays of coupled antennas by using 3-D FEM and spherical mode expansion. IEEE Trans. Antennas Propag. **53**, 1133–1144 (2005)
22. R.J. Mailloux, Radiation and near field coupling between two collinear open ended waveguides. IEEE Trans. Antennas Propag. **AP-17**, 49–55 (1969)
23. R.J. Mailloux, First-order solutions for mutual coupling between waveguides which propagate two orthogonal modes. IEEE Trans. Antennas Propag. **AP-17**, 740–746 (1969)
24. T.S. Bird, Mode coupling in a planar circular waveguide array. IEE J. Microw. Opt. Acoust. **3**, 172–180 (1979)
25. T.S. Bird, Analysis of mutual coupling in finite arrays of different sized waveguides. IEEE Trans. Antennas Propag. *AP-38*, 166–172 (1990)
26. T.S. Bird, Behavior of multiple elliptical waveguides opening into a ground plane. IEE Proc. **137**, 121–126 (1990)
27. M. Mongiardo, T. Rozzi, Singular integral equation analysis of flange-mounted rectangular waveguide radiators. IEEE Trans. Antennas Propag. 556–565 (1993)
28. M. Mongiardo, R. Ravanelli, Automated design of corrugated feeds by the adjoint network method. Special Issue on automated circuit design using electromagnetic simulators, IEEE Trans. Microw. Theory Tech. **45**, 787–793 (1997)
29. M. Mongiardo, C. Tomassoni, Mutual coupling evaluation for arrays of flange-mounted elliptical waveguides. IEEE Trans. Antennas Propag. **49**, 763–770 (2001)
30. L. Tarricone, C. Tomassoni, M. Mongiardo, A parallel framework for the analysis of metal flanged rectangular aperture arrays. IEEE Trans. Antennas Propag. **49**, 1479–1484 (2001)
31. M. Bailey, Mutual coupling between circular waveguide-fed apertures in a rectangular ground plane. IEEE Trans. Antennas Propag. **22**(4), 597–599 (1974)
32. C.J. Reddy, M.D. Deshpande, C.R. Cockrell, F.B. Beck, Radiation characteristics of cavity backed aperture antennas infinite ground plane using the hybrid FEM/MoM technique and geometrical theory of diffraction. IEEE Trans. Antennas Propag. **44**(10), 1327–1333 (1996)

Chapter 3
Circuit Representation and Performance Analysis of Phased Array Antennas Including Mutual Coupling Effects

Liang Han and Ke Wu

3.1 Introduction

Phased array antennas have gained a prominent position in the design of microwave and millimetre-wave radio and radar systems due to their beam steering capability. In most cases, such phased array structures are large-scaled and may involve a very large number of radiating elements that are interrelated to each other through certain signal routing, feeding mechanism and geometric arrangement. On the basis of the well-established array theory, the array pattern is calculated by the product of an isolated element pattern and related isotropic array factor. This scheme assumes that voltage (current) excitation for each element is uniform (constant) in amplitude but progressively in phase over the entire array. This assumption is valid only for an infinitely extended array. For a finite array, this assumption is very much questionable because it doesn't account for array edge effects as well as non-uniform current distribution that depend on the geometry, frequency, and scan angle. This complicated parameter dependence results from mutual coupling effects observable among all elements in the array. It is usually difficult to explain and formulate the mutual coupling phenomenon, which is generally related to the re-radiation of power through neighbouring elements, and/or electromagnetic interaction and surface-wave propagation within the substrate as well as the influence of feeding network.

If only one element in the array is connected to the excitation point and all other elements are terminated by matched loads, we can obtain an extremely important radiation pattern called active element pattern [1] or scan element pattern [2, 3], which is able to take all the mutual coupling effects into account. In this case, the array pattern can be expressed in the multiplication form of the active element pattern and the isotropic array factor, provided that the array is large enough to

K. Wu (✉) and L. Han
Poly-Grames Research Center, Center for Radiofrequency Electronics Research of Quebec, Department of Electrical Engineering, Ecole Polytechnique (University of Montreal)
e-mail: ke.wu@polymtl.ca

S. Lindenmeier and R. Weigel (eds.), *Electromagnetics and Network Theory and their Microwave Technology Applications*, DOI 10.1007/978-3-642-18375-1_3,

approximate the active element pattern of each element as equal. The importance of active element pattern lies in that it is able to predict the scan blindness for the entire array. In the design of a large array antenna, this excitation scheme is therefore often used because the active element pattern can easily be measured without the need of a complicated power dividing network and phase shifting network required in the case of a forced excitation array.

Regarding the calculation procedure of an active element pattern, the coupling coefficient between each pair of elements in the presence of other elements is needed. These coefficients are often obtained from the mutual impedance or admittance matrix which can be calculated by two main methods, namely, the spatial domain method (element-by-element method) and the spectral domain method (periodic cell method) [2]. The spatial domain method is more suitable for modelling small and medium-sized arrays because it needs to calculate the mutual impedance or mutual admittance of each pair of elements [4–7]. When a large array is concerned, it could consume a large amount of computational resources and time. The spectral domain method has been considered to be more efficient in this case. To include all the mutual coupling effects, periodic boundary conditions are imposed on a single element, which implies that the excitations are the same for all elements except for a progressive exponential multiplier. Therefore, this technique ignores edge or border effects as well as non-uniform current distributions. This is because any large but finite array system has a limited boundary so that the periodic array theorem (Floquet's theorem) is no longer valid, particularly for elements close to the array edges. Of course, there are a number of alternative algorithms, which were developed for reducing the computational requirements by combining these two methods [8–10].

Recently, a new technique was proposed for building up an equivalent circuit network of the antenna array of arbitrary size [11, 12]. It is based on circuit parameter extraction and equivalent model establishment for modeling mutual coupling of arbitrary order. The proposed scheme consists of two main steps. First of all, an equivalent circuit model describing low-order mutual coupling (or adjacent coupling) is characterized and formulated, of which each parametric value is accurately extracted by making use of a numerical calibration procedure with full-wave electromagnetic modeling technique [13, 14]. Then, the circuit model for high-order mutual coupling (or crossover/crosstalk coupling) can be obtained from the lower-order models through a network segmentation procedure, and it can further be used for the modeling of mutual coupling of any higher order. This modeling procedure opens up the possibility of building up equivalent circuit networks of the entire antenna array of arbitrary size in a very accurate and intuitive manner because both the equivalent circuit models of the radiating element and the mutual coupling are available. As a result, the array performance can be accurately analyzed through simple and time-saving circuit simulation.

3.2 Description of Mutual Coupling Decomposition

Let us consider an N-port microwave network consisting of N mutually coupled elements (Fig. 3.1). These elements are considered generally dissimilar; however, they can be identical, such as the elements of a finite periodic structure [15].

Let us assume that element i $(i = 1, 2, \ldots, N)$ in Fig. 3.1 is excited by current I_i, the resulting voltage vector V is related to the excitation current vector I through the following expression.

$$I = YV \tag{3.1}$$

where $Y = [Y_{mn}]_{N\times N}$ is the admittance matrix of this N-port network.

We introduce a matrix $Y^{iso} = [Y_{mn}{}^{iso}]_{N\times N}$ whose diagonal elements are the input admittance of isolated element i as Y_i^{iso},

$$Y_{mn}^{iso} = \begin{cases} Y_i^{iso} & (m = n) \\ 0 & (m \neq n) \end{cases} \tag{3.2}$$

Then, the admittance matrix could be found as a sum of Y^{iso} and M, which reflects such mutual coupling between the elements as,

$$M = \begin{bmatrix} Y_{11} - Y_1^{iso} & Y_{12} & \ldots & Y_{1N} \\ Y_{21} & Y_{22} - Y_2^{iso} & \ldots & Y_{2N} \\ \ldots & \ldots & \ldots & \ldots \\ Y_{N1} & Y_{N2} & \ldots & Y_{NN} - Y_N^{iso} \end{bmatrix} \tag{3.3}$$

Consequently, (3.1) can be rewritten as,

$$V = ZI = (Y^{iso} + M)^{-1} I \tag{3.4}$$

In (3.4), matrix Z is the impedance matrix of the N-port network, and it can be expanded as in [16],

$$Z = (Y^{iso} + M)^{-1} = \sum_{k=0}^{\infty} \left[-(Y^{iso})^{-1} M\right]^k (Y^{iso})^{-1} \tag{3.5}$$

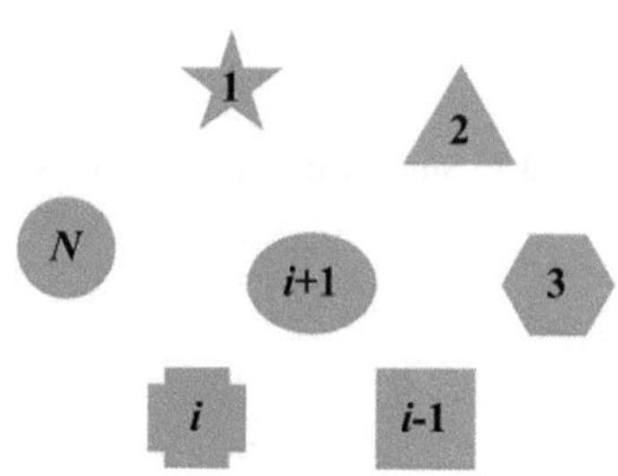

Fig. 3.1 Mutually coupled elements numerated from 1 to N

Studies and practical measurements have already shown that the mutual impedance between elements is generally much smaller than the input impedance of an isolated element in case that the wavelength-normalized distances between elements are larger than half-wavelength. Therefore, matrix M can be seen as a small perturbation of the isolated impedance matrix Y^{iso}, and the sum series is convergent. Moreover, (3.5) suggests the possibility of coupling decomposition, which allows extracting the mutual coupling of arbitrary-order in a consecutive manner.

In our implementation, equivalent circuit networks are built up in order to model mutual coupling of arbitrary-order. It is assumed that the equivalent circuit networks of low-order mutual coupling do not change when used for the extraction of equivalent circuit networks of high-order mutual coupling. In the following sections, this technique will be described in detail with the demonstration of practical examples.

3.3 Modeling of Arbitrary-Order Mutual Coupling

3.3.1 Array Circuit Element Design

A microstrip inset-fed patch antenna resonating at 10 GHz is chosen as an array element (Fig. 3.2). This antenna is designed on substrate RO3003 with its thickness of 0.508 mm. Simulated return loss from 8 to 12 GHz is shown below.

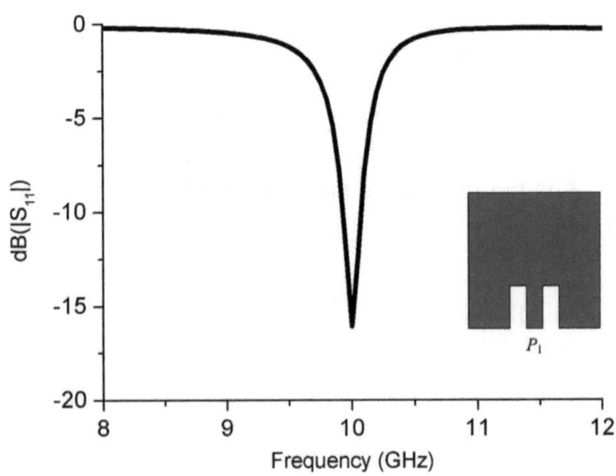

Fig. 3.2 Simulated return loss of a single microstrip inset-fed patch antenna

3.3.2 Modeling of First-Order Mutual Coupling

In order to extract the first-order mutual coupling (immediate adjacent), two coupled microstrip inset-fed patch antennas are placed with an orientation angle of ψ and a distance of r, as shown in Fig. 3.3a. We will model the first-order mutual coupling between two patches in the following three cases: broadside ($\psi = 0°$), echelon ($\psi = 45°$), and collinear ($\psi = 90°$).

Let us designate the admittance matrix of the two coupled patch antennas as $Y^{(1)}$ as below,

$$Y^{(1)} = \begin{bmatrix} Y_{11}^{(1)} & Y_{12}^{(1)} \\ Y_{12}^{(1)} & Y_{11}^{(1)} \end{bmatrix} \tag{3.6}$$

where the superscript indicates that each element in the above matrix is related to the first-order mutual coupling. All the matrix elements are normalized to reference admittance Y_0. ${Y_{11}}^{(1)}$ is the normalized self-admittance of both elements 1 and 2 because two identical elements are used. ${Y_{12}}^{(1)}$ is the normalized mutual admittance reflecting the strength of mutual coupling between the two elements.

The corresponding equivalent circuit model of the above Y-matrix can be sketched as shown in Fig. 3.3b, in which Y_{iso} is the normalized input admittance of an isolated patch and Y_d is defined as the difference between ${Y_{11}}^{(1)}$ and Y_{iso} in order to manifest the effect of element 2 on the self-admittance of element 1 through the first-order mutual coupling.

$$Y_d = Y_{11}^{(1)} - Y_{iso} \tag{3.7}$$

Figures 3.4 and 3.5 respectively plot the variation of extracted Y_d and the variation of mutual admittance ${Y_{12}}^{(1)}$ with respect to the distance between two elements in case of three different orientations. Through Figs. 3.4 and 3.5, we can have the following observations. Firstly, it can be observed in Fig. 3.5 that element 2 has an influence not only on the radiation characteristic of element 1 which is related to

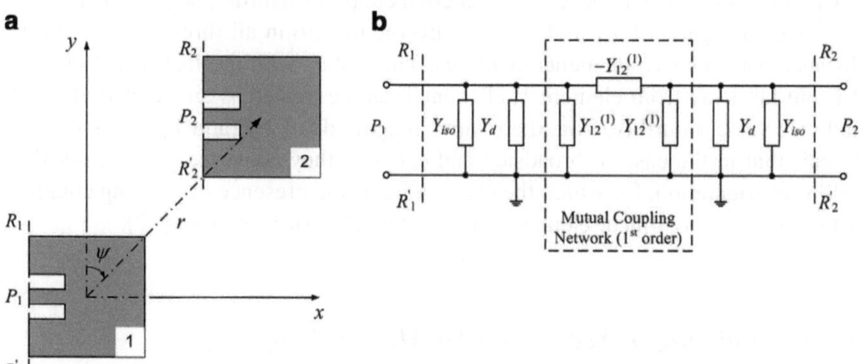

Fig. 3.3 Two coupled microstrip patch antennas: (**a**) geometrical configuration and (**b**) equivalent circuit topology

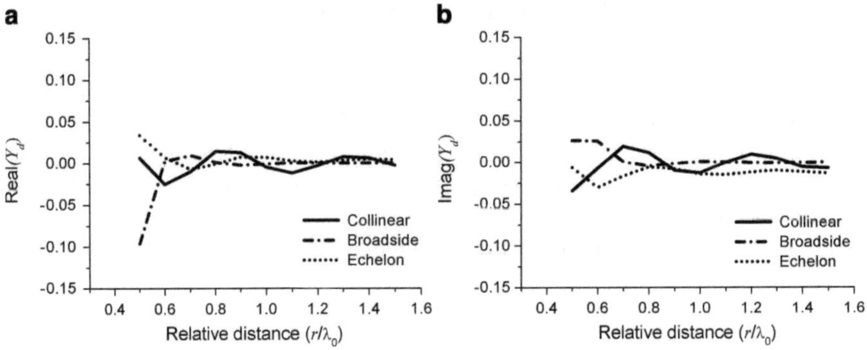

Fig. 3.4 Extracted Y_d: (**a**) real part and (**b**) imaginary part

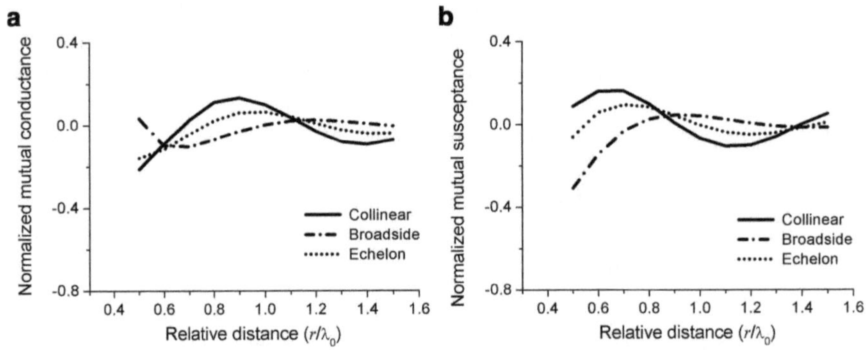

Fig. 3.5 Extracted normalized mutual admittance $Y_{12}^{(1)}$: (**a**) real part and (**b**) imaginary part

self-conductance, but also on its energy storage capability which is related to self-susceptance. Secondly, the variations of Y_d in all three cases are different which verify the geometrical dependence of mutual coupling. Thirdly, both of the real and the imaginary parts of Y_d and $Y_{12}^{(1)}$ converge to zero in all three cases when the distance between two elements increases. This tallies with the fact that if element 2 is put far away from element 1, element 1 can be treated as an isolated element, and vice versa. Fourthly, from the relative magnitude of Y_d (and $Y_{12}^{(1)}$ as well), it reveals that in the case of broadside and echelon, they converge much faster than collinear orientation, for which the explanation is the presence of a strong coupling between the two collinear elements through the TM_0 surface wave [17].

3.3.3 Modeling of Second-Order Mutual Coupling

In this subsection, three coupled elements are used for extracting the equivalent circuit model of second-order mutual coupling. Figure 3.6 shows three coupled

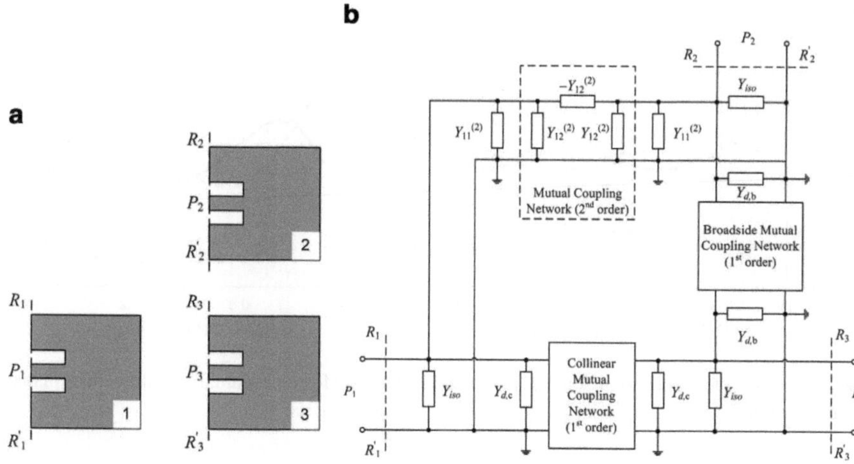

Fig. 3.6 Three coupled microstrip patch antennas: (**a**) geometrical configuration and (**b**) equivalent circuit network

elements in a planar configuration and the corresponding equivalent network topology is depicted in Fig. 3.8. $Y_{d,c}$ and $Y_{d,b}$ correspond to the admittance Y_d in Fig. 3.4 for collinear coupling and broadside coupling, respectively. Y_{iso} is the input admittance of an isolated element. Based on equivalent circuit models of both collinear and broadside first-order mutual coupling, we can obtain an equivalent network topology without the effect of second-order mutual coupling. Therefore, with the calibrated results of these three coupled elements, the equivalent circuit network of the second-order mutual coupling can be calculated through standard network theorem.

From the extracted results of three configurations which are plotted through Figs. 3.7–3.9, we can see that the extracted admittances nearly have the same variation tendency as the first-order mutual coupling such as the convergence towards zero, and a slower decay in the collinear case.

3.3.4 *Modeling of Higher-Order Mutual Coupling*

The entire extraction procedure can be summarized in a flowchart illustrated in Fig. 3.10. The nth-order mutual coupling ($Y_{mc}{}^{(n)}$) between two coupled elements is obtained by the difference of calibrated simulation results ($Y_c^{(n+1)}$) including the nth-order mutual coupling and calculated results ($Y'^{(n+1)}$) based on the cascaded equivalent circuit models of lower-orders mutual coupling which excludes the mutual coupling of the nth-order. Usually, it is suggested to choose strong mutual coupling as low-order for reducing the modeling error. Additionally, a criterion needs to be used for setting the highest order of mutual coupling that should

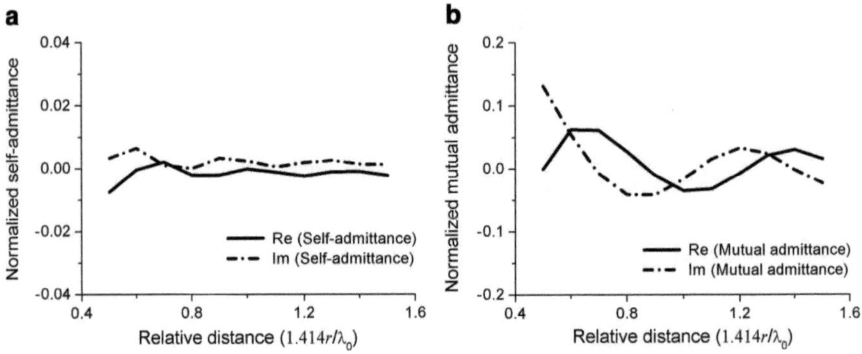

Fig. 3.7 Extracted equivalent circuit parameters of second-order mutual coupling in a planar configuration

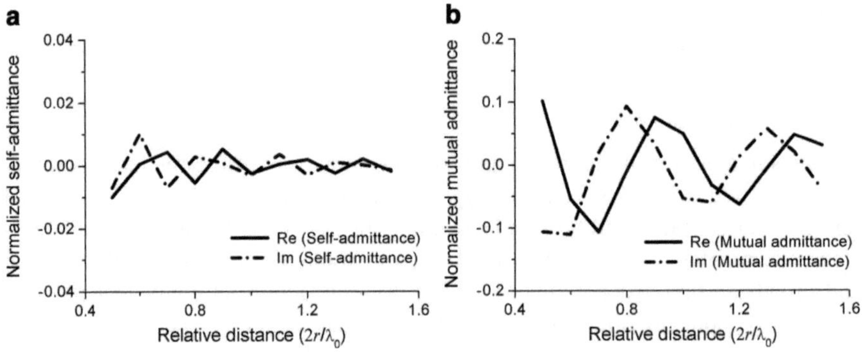

Fig. 3.8 Extracted equivalent circuit parameters of second-order mutual coupling in a collinear configuration

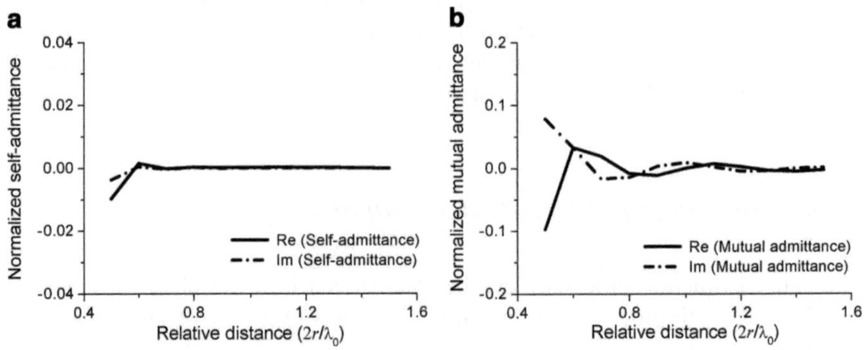

Fig. 3.9 Extracted equivalent circuit parameters of second-order mutual coupling in a broadside configuration

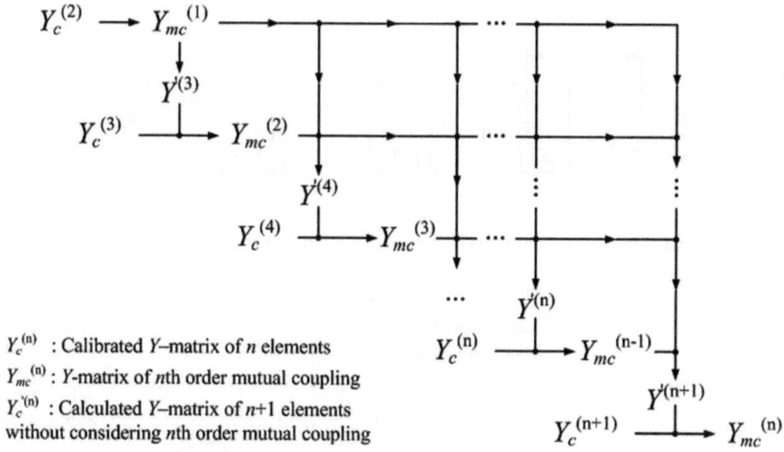

Fig. 3.10 Flowchart of mutual coupling extraction

be taken into account according to specified performance modeling and required design accuracy.

3.4 Design Examples

3.4.1 A 1 × 19 Linear Array

First of all, a linear phased array antenna composed of 19 half-wavelength-spaced elements with a beam direction of $\theta = -30°$ will be modeled with the help of the proposed method to demonstrate its accuracy and efficiency (Fig. 3.11). This array antenna is placed along the x-axis in the xoy-plane.

To begin with, the equivalent circuit models of mutual coupling from the first-order to the fourth-order are extracted and the fourth-order mutual coupling is found to be too weak to be considered in this case. In Fig. 3.12, the normalized self admittances of different elements (elements 1–5) in the array are plotted during the process of establishing the equivalent circuit network of mutual coupling.

From Fig. 3.12, we can make the following observations. First, the influence of neighboring elements on the host element is shown. When there is no neighboring element present, the self admittance is equal to the input impedance of an isolated element. If we consider for example the leftmost element (element 1) with three neighboring elements (elements 2–4, $N = 3$), there is no influence of the next neighboring element (element 5) on it, and this can be seen from the convergence of its admittance. This observation concludes that we only need to consider the mutual coupling of up to the third-order in this case study. Second, Fig. 3.12 also shows the "edge effect". The edge elements (elements 1 and 2) behave differently in the array

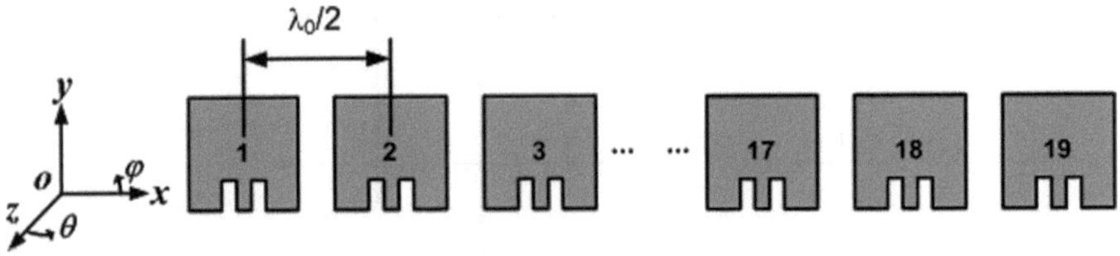

Fig. 3.11 A linear phased array antenna composed of 19 half-wavelength-spaced elements

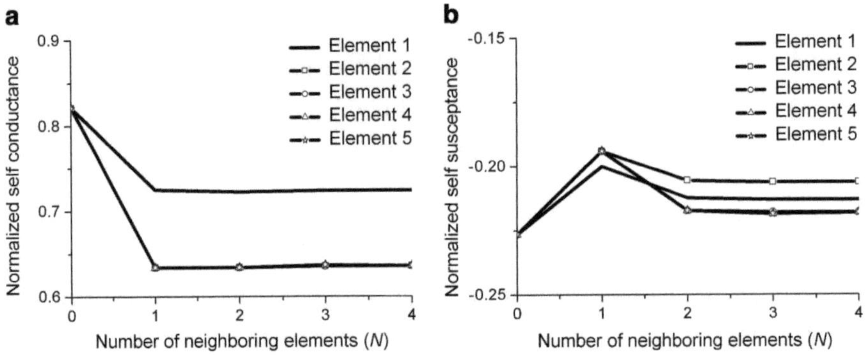

Fig. 3.12 Influences of neighboring elements on the normalized self admittance of the host element

environment from the inner elements (i.e., elements 3–5) in terms of the number of neighboring elements and the order of mutual coupling which should be considered. It is also interesting to find out that the normalized input conductance (Fig. 3.12a) of elements 2 and 3–5 are almost equal, while the susceptance (Fig. 3.12b) is different. As a result, the return losses for these elements are still different. Therefore, in the final array design, elements 1 and 2 as well as their symmetrical counterparts (elements 19 and 18, respectively) should be treated differently from the interior or inner elements (elements 3–17).

Then, the scattering matrix of this 1×19 array is obtained by cascading and simulating the equivalent circuit networks of the patches themselves and their mutual coupling in a commercial circuit simulator (Agilent's ADS).

Finally, we can calculate the array pattern from the simulation results of scattering matrix of the antenna array as it is known that the pattern of a phased array can be expressed by the product of the active element pattern (or "scan element pattern") and the array factor [1, 18]. The array pattern is calculated from four different methods in our investigation for comparison (Fig. 3.13).

The first method (Circuit model) calculates the active element pattern with the help of the proposed circuit modeling technique and multiplies it by the array factor, whereas the second method (Full-wave model) determines the active element pattern by means of the S-matrix obtained from the full-wave simulation in a commercial MoM package and multiplies it by the array factor. In the third method

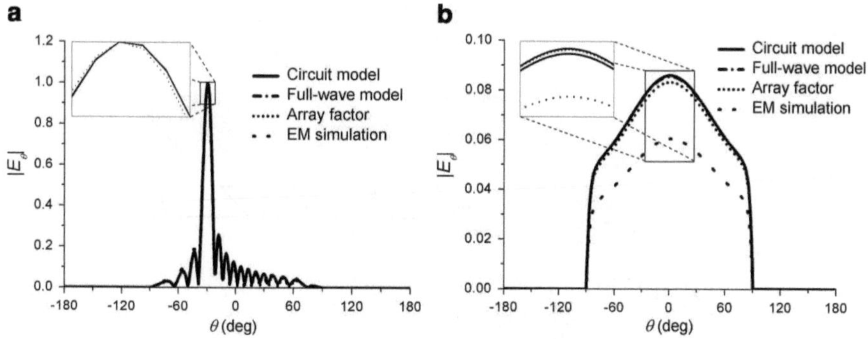

Fig. 3.13 Array pattern comparison: (**a**) *xoy* plane and (**b**) *yoz* plane

Table 3.1 Comparison of array gains calculated by four methods

Method	Array gain
Circuit model	16.6001
Full-wave model	16.6066
Array factor	16.6142
Direct EM simulation	16.6770

(Array factor), the array pattern is obtained from the product of a single element pattern and the array factor. The fourth method (EM simulation) uses a direct full-wave simulation in a commercial MoM package.

From this systematic comparison among the results of these four methods in Fig. 3.13, we can conclude that, for the *xoz*-plane pattern, the calculated results from the proposed circuit model, the full-wave model, and the direct EM simulation are almost the same. However, we can find that there is a deviation among these results and the array pattern obtained from the product of a single element pattern and the array factor. The reason for this behavior is that in the latter calculation, the mutual coupling between elements is not included. On the other hand, for the *yoz*-plane, the array pattern calculated by means of the proposed circuit model is nearly the same as that calculated with the S-matrix obtained from the full-wave results, while there is little difference between these two methods and the array pattern calculated with the help of array factor and the direct EM simulation. It should also be mentioned that a good agreement of the array gains is achieved with these four methods, which are listed in Table 3.1.

3.4.2 A 3 x 3 *Planar Array*

Our proposed modeling technique can also be applied to a planar phased array. Figure 3.14 shows a planar array, of which the elements are half-wavelength spaced on a square lattice ($dx = dy = \lambda_0/2$) with 3 elements along the x-axis and 3 elements along the y-axis.

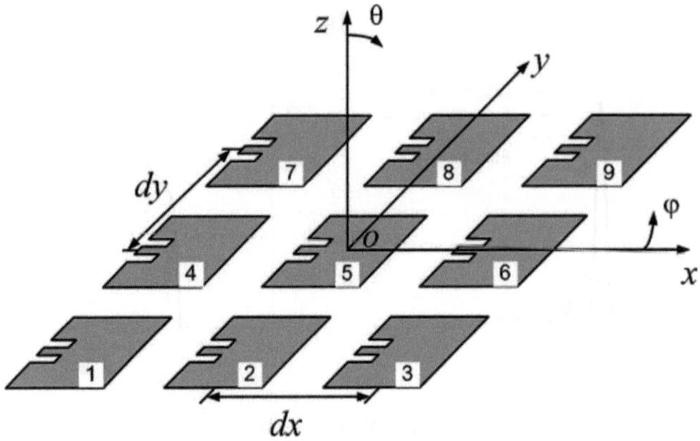

Fig. 3.14 A 3 × 3 planar phased array antenna

Based on the extracted equivalent circuit model of elements and those of mutual coupling up to third-order, a circuit simulation can be performed to analyze this planar array. The simulated S-matrix is then used to calculate the active element pattern of each element. After [18], the active element pattern of mth element of the planar array is given by,

$$E_m^e(r,\theta,\varphi) = F(\theta,\varphi)\frac{e^{-jkr}}{r}V_0\left[e^{j[(i_m-1)u+(j_m-1)v]}+\sum_{n=1}^{K}S_{nm}e^{j[(i_n-1)u+(j_n-1)v]}\right]$$
$$m = 1,2,3\ldots K \tag{3.8}$$

with

$$\begin{cases} u = kdx\sin(\theta)\cos(\varphi) \\ v = kdy\sin(\theta)\sin(\varphi) \end{cases} \tag{3.9}$$

where V_0 is the terminal voltage, $F(\theta,\varphi)$ represents the dominant polarization of the element pattern and S_{nm} is the S-parameter of elements n and m. i_m is the x index of element m and j_m is the y index of element m.

The calculation results are drawn for comparison in Fig. 3.15 at two different planes.

(1) *xoz*-Plane ($\varphi = 0°$)

In this case, $u = kdx\sin(\theta)$ and $v = 0$. The active element pattern of the mth element is simplified as

$$E_m^e(r,\theta,0) = F(\theta,0)\frac{e^{-jkr}}{r}V_0\left[e^{j(i_m-1)u}+\sum_{n=1}^{K}S_{nm}e^{j(i_n-1)u}\right] m = 1,2,3\ldots K \tag{3.10}$$

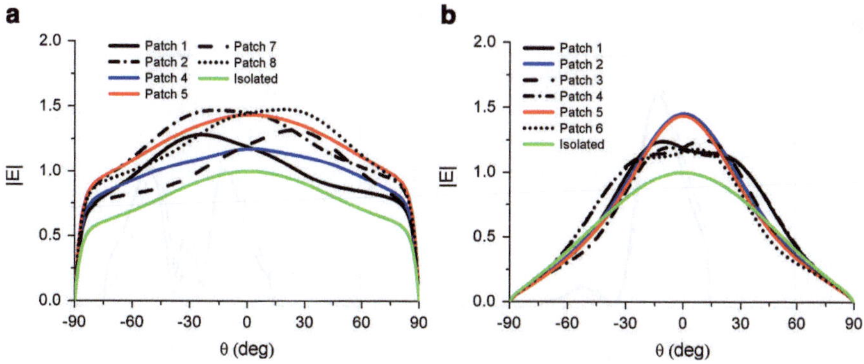

Fig. 3.15 Calculated active element patterns of different elements at both *xoz*-plane and *yoz*-plane

From the above expression, we can see that elements having the same i_m and S-parameters such as elements 1 and 3, 4 and 6 as well as 7 and 9 have the same active element patterns. Therefore, the calculated active element patterns of elements 3, 6 and 9 are omitted in Fig. 3.15a. On the other hand, elements which are placed symmetrically about the $\varphi = 0°$ plane such as elements 1 and 7, and 2 and 8, will have symmetrical active element pattern. This conclusion can be verified by our calculation results in Fig. 3.15a.

(2) *yoz*-Plane ($\varphi = 90°$)

In this case, $u = 0$ and $v = kdy\sin(\theta)$. The active element pattern of the mth element is simplified as

$$E_m^e(r,\theta,90) = F(\theta,90)\frac{e^{-jkr}}{r}V_0\left[e^{j(j_m-1)v} + \sum_{n=1}^{K} S_{nm}e^{j(j_n-1)v}\right] m = 1,2,3\ldots K \tag{3.11}$$

We can come up with the following conclusion in this case. Elements having the same j_m and S-parameters such as elements 1 and 7, 2 and 8, as well as 3 and 9 have the same active element patterns while elements symmetrically placed about the $\varphi = 90°$ plane such as elements 1 and 3, and 4 and 6, should have symmetrical active element pattern. Figure 3.15b verifies this concluding remark.

The array pattern can be obtained by the summation of active element patterns of all array elements. The calculated results shown in Fig. 3.16 agree well with the direct full-wave simulation.

3.5 Conclusions

A novel method is presented for modeling and analyzing antenna array of finite size through circuit network representation. Equivalent circuit models of mutual coupling are extracted successively from low-order to high-order based on network

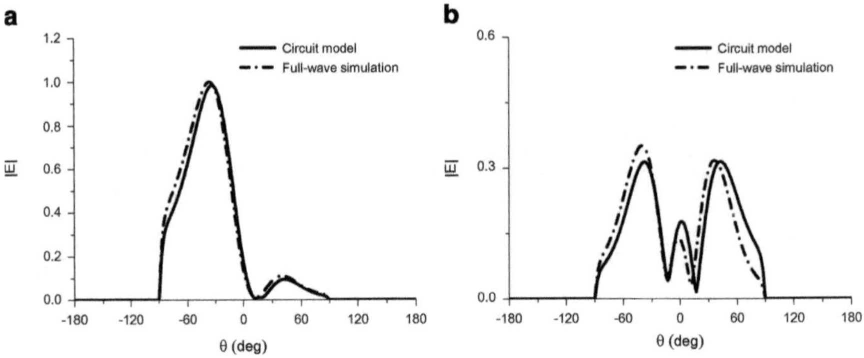

Fig. 3.16 Calculated array pattern

segmentation method and electromagnetic modeling technique. With these equivalent circuit models, the performance of both linear and planar phased array antenna can be calculated in circuit simulator. This technique provides an interesting possibility of predicting the scan blindness phenomenon for phased array of arbitrary-size through a simple circuit simulation. In the end, this would also bridge the gap of design between circuits and antennas through network theory and electromagnetic modeling.

References

1. D. M. Pozar, The active element pattern. *IEEE Trans. Antennas Propag.* **42**, 1176–1178 (1994)
2. R.C. Hansen, *Phased Array Antennas* (Wiley, New York, 1998)
3. R.J. Mailloux, *Phased Array Antenna Handbook* (Artech House, Norwood, MA, 1994)
4. D.M. Pozar, Input impedance and mutual coupling of rectangular microstrip antennas. IEEE Trans. Antennas Propag. **30**, 1191–1196 (1982)
5. D.M. Pozar, Finite phased arrays of rectangular microstrip patches. IEEE Trans. Antennas Propag. **34**, 658–665 (1986)
6. F.J. Demuynck, G.A.E. Vandenbosch, A.R. Van de Capelle, The expansion wave concept–Part I: Efficient calculation of spatial Green's functions in a stratified dielectric medium. IEEE Trans. Antennas Propag. **46**, 397–406 (1998)
7. G.A.E. Vandenbosch, F.J. Demuynck, The expansion wave concept–Part II: A new way to model mutual coupling in microstrip arrays. IEEE Trans. Antennas Propag. **46**, 407–413 (1998)
8. A. Ishimaru, R. Coe, G. Miller, W. Geren, Finite periodic structure approach to large scanning array problems. IEEE Trans. Antennas Propag. **33**(11), 1213–1220 (1985)
9. A. Skrivervik, J. Mosig, Analysis of finite phase arrays of microstrip patches. IEEE Trans. Antennas Propag. **41**(8), 1105–1114 (1993)
10. D. Kelley, W. Stutzman, Array antenna pattern modeling methods that include mutual coupling effects. IEEE Trans. Antennas Propag. **41**(12), 1625–1632 (1993)
11. L. Han, K. Wu, Modeling of arbitrary-order mutual coupling. in *IEEE MTT-S International Microwave Symposium Digest*, 2008, pp. 1389–1392
12. L. Han, K. Wu, Circuit representation and performance analysis of planar phased array antenna including mutual coupling effects. in *IEEE International Mini-Symposium on EMNT*, 2008

13. L. Zhu, K. Wu, Short-open calibration technique for field theory-based parameter extraction of lumped elements of planar integrated circuits. IEEE Trans. Microw. Theory Tech. **50**(8), 1861–1869 (2002)
14. L. Han, K. Wu, W. Hong, L. Li, X.-P. Chen, Embedding of short-open calibration technique in commercial MoM simulators for parameter extraction of planar integrated circuits. in *Proceedings of the Asia-Pacific Microwave Conference*, vol. 3, Yokohama, Japan, Dec. 2006, pp. 1956–1959
15. K.-C. Lee, T.-H. Chu, A circuit model for mutual coupling analysis of a finite antenna array. IEEE Trans. Electromagn. Compat. **38**(3), 483–489 (1996)
16. C.D. Meyer, *Matrix Analysis and Applied Linear Algebra.* (SIAM, Philadelphia, 2000), ch. 3, pp. 126
17. P. Katehi, A generalized method for the evaluation of mutual coupling in microstrip arrays. IEEE Trans. Antennas Propag. **35**, 125–133 (1987)
18. D. Pozar, A relation between the active input impedance and the active element pattern of a phased array. IEEE Trans. Antennas Propag. **51**(9), 2486–2489 (2003)

Chapter 4
Time-Domain Modelling of Group-Delay and Amplitude Characteristics in Ultra-Wideband Printed-Circuit Antennas

Hung-Jui Lam, Yinying Lu, Huilian Du, Poman P.M. So, and Jens Bornemann

4.1 Introduction

With the release of the 3.1–10.6 GHz band for ultra-wideband (UWB) operation, a variety of typical UWB applications evolved; examples are indoor/outdoor communication systems, ground-penetrating and vehicular radars, wall and through-wall imaging, medical imaging and surveillance, e.g. [1, 2].Many future systems will utilize handheld devices for such short-range and high bandwidth applications. Therefore, the realization of UWB antennas in printed-circuit technologies within relatively small substrate areas is of primary importance. And a number of such antennas with either microstrip, e.g. [3–10] or coplanar waveguide feeds, e.g. [11–23], and in combined technologies, e.g. [24, 25], have been presented recently, mostly for the 3.1–10.6 GHz band, but also for higher frequency ranges, e.g. [26].

Since UWB systems involve the transmission and reception of short pulses, the variations of radiated amplitudes and phases over frequency contribute to the distortion of the pulse. While the amplitude variation is usually indicated by changes in the peak gain or radiation patterns, the frequency-dependent phase variation is often omitted, and related data is published only sporadically, e.g., [5, 7, 17, 26]. In order to quantify this behavior, one of two methods is usually applied.

First, in the frequency domain, the spherical wave front in the far field is detected for each frequency, from which the apparent phase center along the antenna surface or axis can be calculated. Alternatively, the phase variation in the near field over the main beam is computed for different phase center points moved from a reference point on the surface of the antenna. Then a valid phase center location is detected if the phase variation over the main beam is within a few degrees. These methods are complicated and time-consuming [26].

H.-J. Lam (✉), Y. Lu, H. Du, P.P.M. So, and J. Bornemann
Department of Electrical and Computer Engineering, University of Victoria, Victoria, BC, Canada V8W 3P6
e-mail: harrylam@uvic.ca, yinlu@rim.com, hliandu@gmail.com, pso@ece.uvic.ca, jbornema@ece.uvic.ca

S. Lindenmeier and R. Weigel (eds.), *Electromagnetics and Network Theory and their Microwave Technology Applications*, DOI 10.1007/978-3-642-18375-1_4,

Secondly, in the time domain, a transient analysis is performed which leads to the group delay. A pulse, whose frequency spectrum covers the bandwidth of the antenna, is generated, applied at the antenna input and its radiated pulse detected. Both pulses are Fourier transformed and their phase response recorded. The group delay is obtained from the derivative of the phase variation with respect to angular frequency [7].

In this paper, the Transmission-Line Matrix (TLM) method in the time domain is utilized to determine the group delay of two printed circuit UWB antennas. The first one is a recently developed, new coplanar-waveguide antenna [27], the second a published microstrip antenna [9,10] with so far no information about phase variations.

4.2 Coplanar UWB Antenna

Figure 4.1 shows the layout and the superimposed coordinate system of the UWB antenna in coplanar technology. It uses an FR4 substrate of 1 mm thickness, an area of 30 × 40 mm (W × L), a permittivity of $\varepsilon_r = 4.7$ and a loss tangent of $\tan\delta = 0.018$. It appears to be a stepped version of a similar antenna presented in [20]. However, there are two fundamental differences. First of all, the antenna in [20] is a slot radiator, which maintains metallic strips at the left and right edges

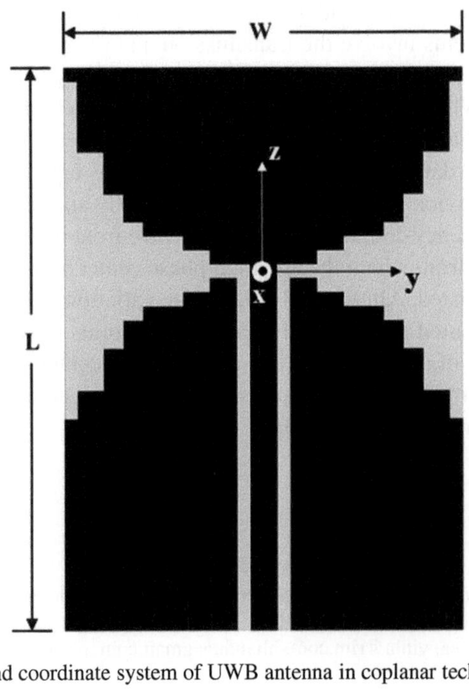

Fig. 4.1 A Layout and coordinate system of UWB antenna in coplanar technology

of the substrate. Such metallic strips are missing in Fig. 4.1 and thus result in a somewhat conical shape of the radiating profile – similar to a tapered slot antenna. Secondly, the stepping is chosen such that the smallest dimension is 0.5 mm. This contributes to low manufacturing sensitivity. However, it also influences the characteristic impedance of the feeding coplanar waveguide, which is significantly higher than the 50Ω coaxial line to be connected at the input. (Note that the coaxial line is also used to physically connect the two ground planes.) As we will show later, this mismatch is not to the detriment of the antenna performance.

The coplanar UWB antenna was designed using the finite-element software HFSS®. For the evaluation of the group-delay characteristics, the antenna was also analyzed by the TLM time-domain field solver MEFiSTo-3D®. Figure 4.2 shows a comparison between the input reflection coefficients obtained with both methods. Note that the connection of the input of the antenna to a coaxial cable is included in both methods. Good agreement is observed, thus verifying the antenna's performance at its input terminal. The input return loss as computer by HFSS between 3.1 and 10.6 GHz is better than 9.4 dB.

The peak gain, computed using HFSS at the dots and spline interpolated, is shown in Fig. 4.3. Its variation versus frequency is comparable to other UWB printed-circuit antennas found in the literature. Note that the direction of the peak gain varies with frequency and, therefore, is not an indication of the amplitude variation in a specific direction.

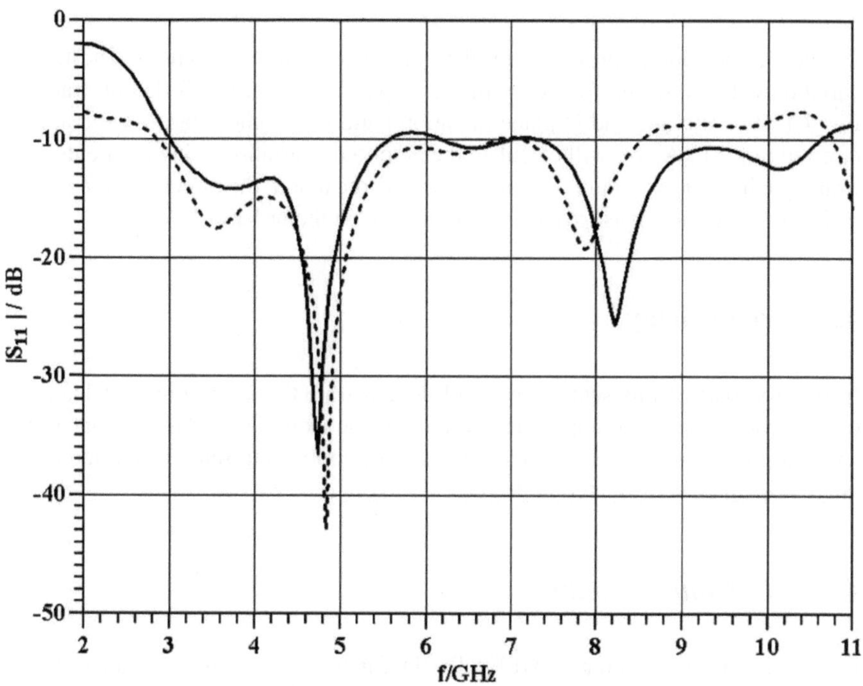

Fig. 4.2 Comparison of input reflection performance between HFSS (*solid line*) and MEFiSTo-3D (*dashed line*)

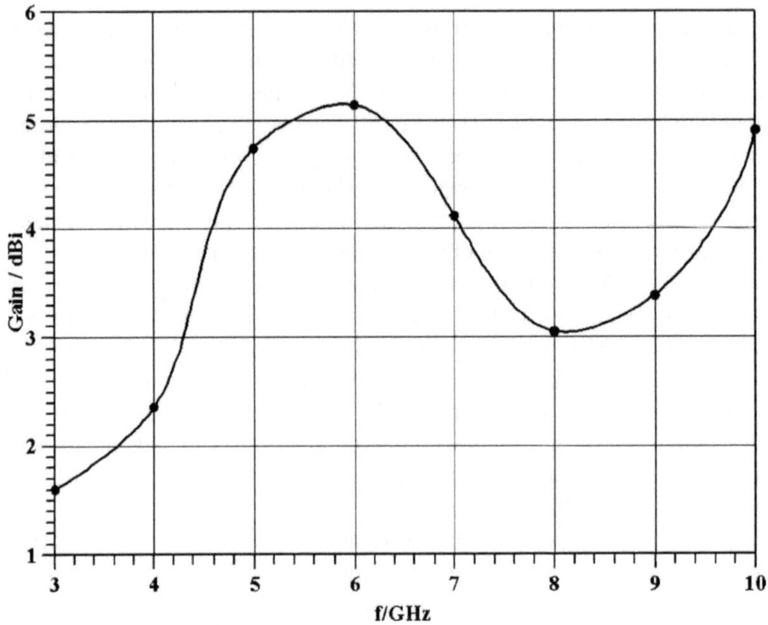

Fig. 4.3 Peak gain of the UWB antenna in CPW technology computed by HFSS (*dots*) and spline interpolated (*solid line*)

Such a variation is presented by the normalized radiation pattern. The E-field variation with angle and frequency in the yz-plane (cf. Fig. 4.1) is demonstrated in Fig. 4.4. (For E-plane and H-plane radiation patterns in other planes, the reader is referred to [27].) As we will calculate the amplitude variation using a time-domain technique in the next section, it is important to note that in the direction of $\theta = \phi = \pi/2$, the variation versus frequency in Fig. 4.4 is in the order of 8–9 dB.

4.3 Group Delay

In the first part of this section, we will demonstrate the time-domain calculation of the group delay and amplitude variation at the example of the coplanar UWB antenna presented in Sect. 4.2. The second part applies the same technique to the microstrip antenna presented in [9, 10].

4.3.1 Coplanar Antenna

Figure 4.5 shows the setup in MEFiSTo-3D. Since the problem is symmetric with respect to a magnetic wall in the xz-plane (all other walls are absorbing boundaries), only half of the computational space is required. The input of the antenna is excited

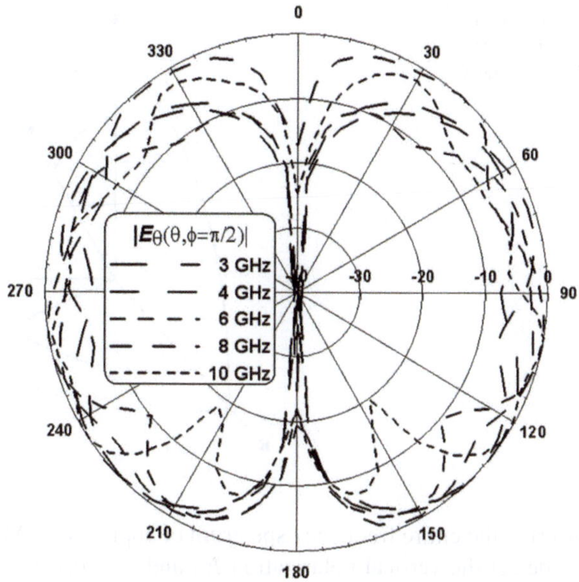

Fig. 4.4 Normalized E-plane radiation pattern (computed with HFSS) in the yz-plane (cf. Fig. 4.1) at various frequencies between 3 and 10 GHz

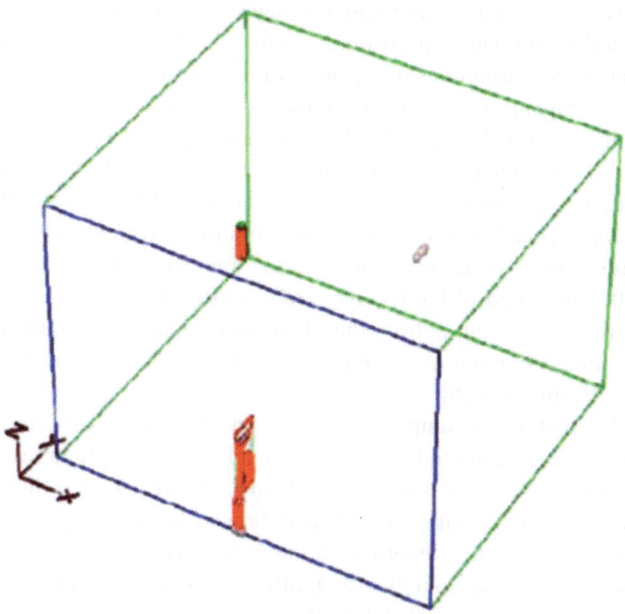

Fig. 4.5 Setup of one half of the coplanar UWB antenna in MEFiSTo-3D including coaxial input port, probes and coaxial reference port

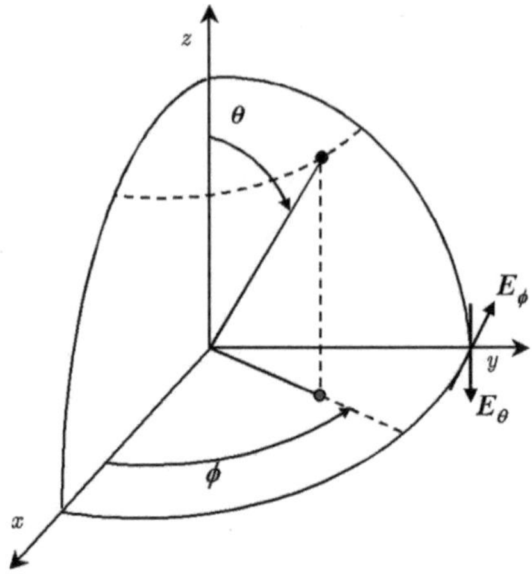

Fig. 4.6 Setup orientation of field components received by probes in Fig. 4.5 with respect to Fig. 4.1

with a pulse covering the entire frequency spectrum of application. At a point in the far field, probes detect the vertical polarization E_θ and the horizontal polarization E_ϕ. Their orientation with respect to Fig. 4.1 are depicted in Fig. 4.6. Note that the coaxial input port and a reference port are included.

Input and detected signals are Fourier transformed to obtain amplitude and phase responses. The group delay is obtained from the derivative of the phase response. Figure 4.7 shows the input time-domain signal together with its corresponding amplitude (in dB) and phase spectrum. Note that the duration of the pulse is about 0.4 ns and the phase variation is in the order of hundreds of degrees.

The radiated signals E_θ (solid lines) and E_ϕ (dashed lines) as detected by the probes in Fig. 4.5 and their amplitude and phase spectra are shown in Fig. 4.8. Figure 4.8a, b confirm that the main polarization is vertical (E_θ) since the detected signal in horizontal polarization (E_ϕ) is at least more than 20 dB below that its vertical component. Figure 4.8c shows the phase variation now in thousands of degrees, which is a result of the ringing of the detected time signal in Fig. 4.8a. Moreover, notice that the main part of the received pulse in Fig. 4.8a looks similar to a negative derivative of the input pulse rather than the original input signal in Fig. 4.7a. Such behaviour is common in antennas that radiate pulses covering a significant frequency spectrum, e.g. [28].

Figure 4.9a, b show the amplitude and group-delay responses, respectively, of the coplanar UWB antenna fed by a coaxial cable. The amplitude response in the main polarization (solid line) is between −40 and −50 dB which is due to the small effective area of the receiving probes. Since the variations in amplitude and phase (group delay) determine the distortion of the pulse transmitted by the antenna, the respective values – as read from the data plotted in Fig. 4.9 – are summarized below for both vertical (VP) and horizontal (HP) polarizations.

Frequency range:	3.1–10.6 GHz
Amplitude variation:	<8.7 db (VP); <23 dB (HP)
Group-delay variation:	<163 ps (VP); <620 ps (HP)

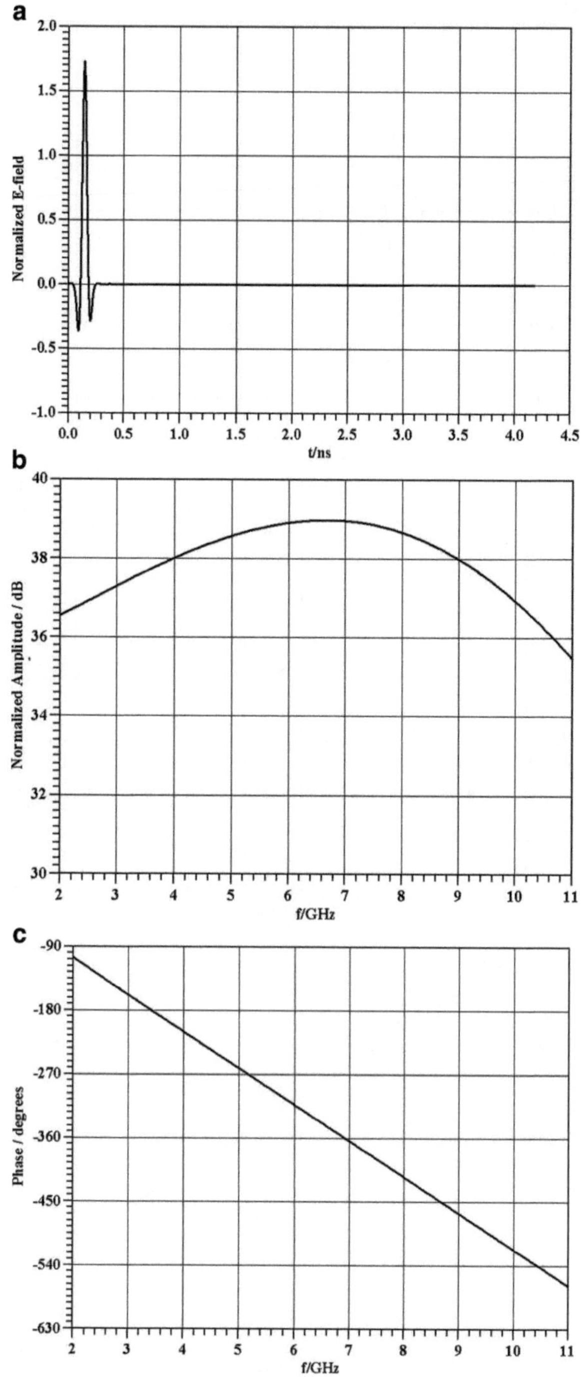

Fig. 4.7 Time-domain signal (**a**), amplitude spectrum (**b**) and phase spectrum (**c**) at the input of the coaxial cable feeding the coplanar antenna (cf. Fig. 4.5)

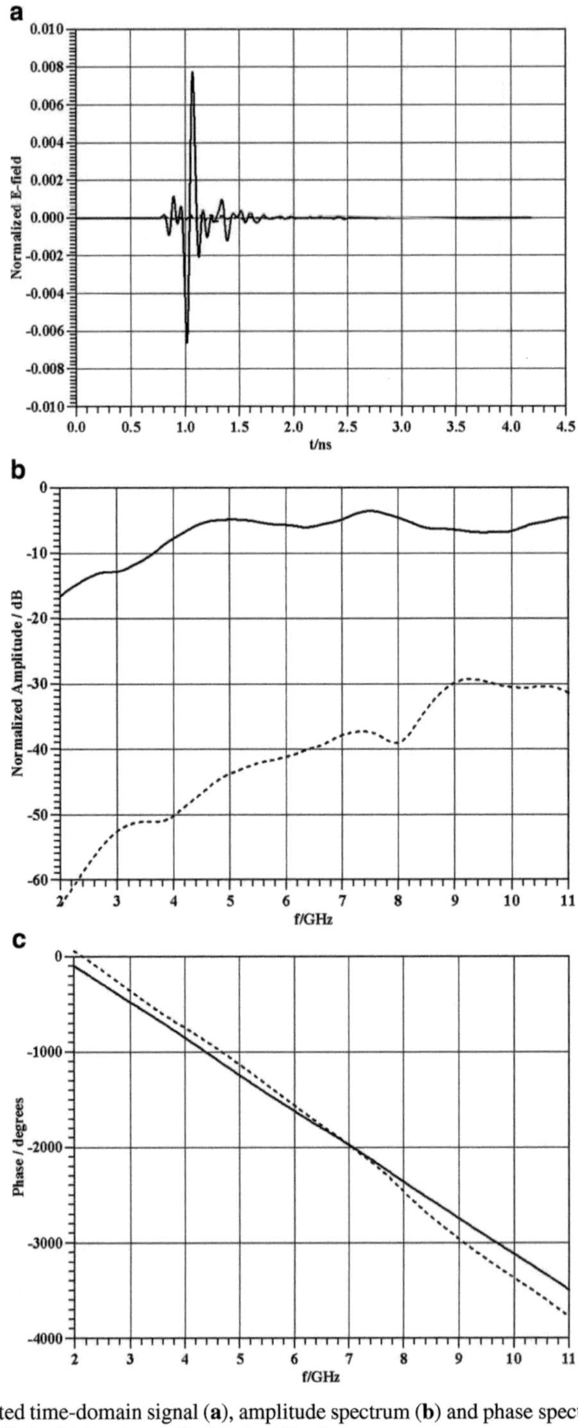

Fig. 4.8 Radiated time-domain signal (**a**), amplitude spectrum (**b**) and phase spectrum (**c**) detected by the probes; E_θ (*solid lines*) and E_ϕ (*dashed lines*)

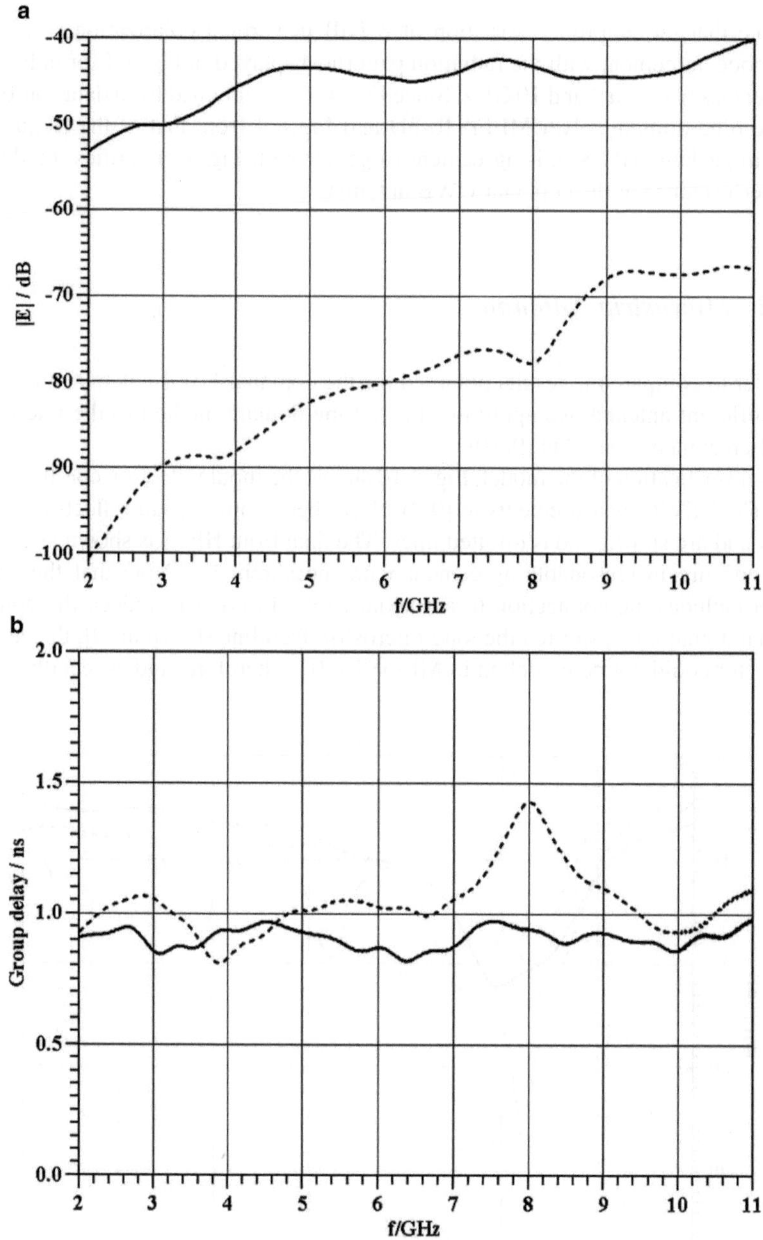

Fig. 4.9 Amplitude response (**a**) and group-delay characteristic (**b**) of coplanar UWB antenna; vertical polarization E_θ (*solid lines*), and horizontal polarization E_ϕ (*dashed lines*)

Note that the amplitude variation of 8.7 dB in vertical polarization (E_θ) is in very good agreement with the radiation patterns displayed in Fig. 4.4 for individual frequencies between 3 and 10 GHz. Since Fig. 4.9 was obtained from data computed by the time-domain solver MEFiSTo-3D and Fig. 4.4 from that of the frequency-domain package HFSS, this agreement (together with Fig. 4.2) verifies the design and performance of the coplanar UWB antenna.

4.3.2 Microstrip Antenna

In order to compare the results obtained for the coplanar UWB antenna with those of a different antenna, we apply the above time-domain method to the microstrip UWB antenna presented in [9, 10].

As a verification of the model, Fig. 4.10 shows the input reflection coefficient (in dB). The VSWR measurements in [9, 10] have been converted to reflection coefficients and are shown as dash-dotted lines. The data from HFSS is shown as dashed lines and are in reasonable agreement with measurements. Note that the HFSS model includes the connection to a coaxial cable. In order to reduce the computational domain, i.e., shorten the long microstrip feed line shown in [9], the coaxial connector could not be modelled in MEFiSTo-3D. Therefore, and especially in the

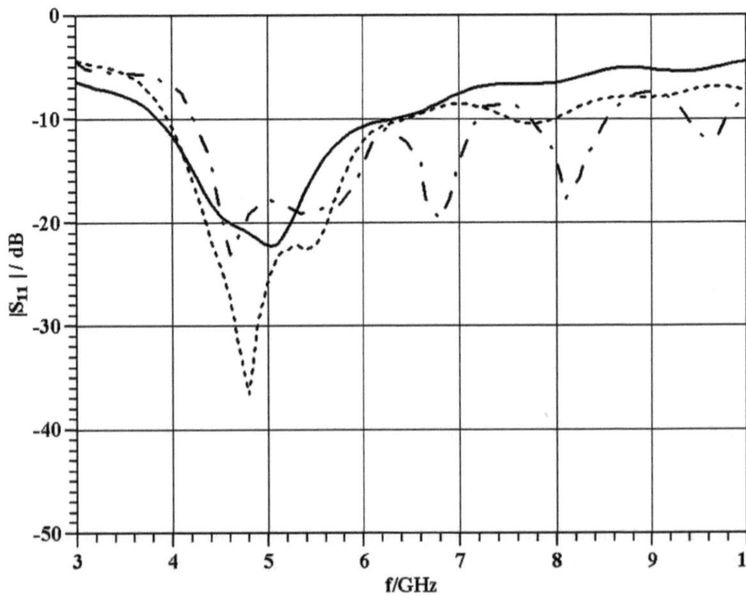

Fig. 4.10 Input reflection coefficient in dB of the microstrip UWB antenna of [9, 10]; calculated values from VSWR measurements in [9,10] (*dash-dotted line*), HFSS (*dashed line*) and MEFiSTo-3D (*solid line*)

higher frequency range, the agreement between measurements and iMEFiSTo-3D is not as good as that with HFSS. However, the basic shape and the reasonably small discrepancies validate the numerical computations.

After exciting the microstrip antenna with a pulse shown in Fig. 4.7, detecting the radiated signal and calculating amplitude and phase responses, the data presented in Fig. 4.11 is obtained. Between 3 and 10 GHz, the amplitude variation in vertical polarization is similar to that of the coplanar UWB antenna in Fig. 4.9a. The signal level difference between horizontal and vertical polarizations in Fig. 4.11a is smaller than that in Fig. 4.9a. This due to the fact that the x-component of the electric field represents the main polarization in a microstrip line if the antenna is oriented in the same way as the coplanar one in Fig. 4.1.

The group delay performances of the microstrip antenna are inferior to those of the coplanar antenna in both polarizations. The following values are obtained:

Frequency range:	3.0–10.0 GHz	
Amplitude variation:	<8.8 db (VP);	<31 dB (HP)
Group-delay variation:	<239 ps (VP);	<1.9 ns (HP)

4.3.3 *Comparison*

Both the coplanar and the microstrip antenna display nearly omnidirectional radiation patterns with characteristics slightly distorting towards 10 GHz (cf. [9, 10] and [27] for details). Over the 3.1–10.6 GHz range, the input reflection coefficient of the coplanar antenna is superior to that of the microstrip antenna. The amplitude variations in vertical polarization are comparable; in horizontal polarization, however, it is 8 dB in favour of the coplanar antenna. The group-delay variations of the coplanar antenna are much smaller than those of the microstrip antenna and, therefore, the coplanar structure of Fig. 4.1 is better suited for UWB applications.

It is noted that a smaller group-delay variation (<100 ps) is reported in [7] for a microstrip UWB antenna with two slots in the radiating patch. However, the gain of than antenna is lower than the one reported in Fig. 4.3 and even drops below 0 dB above 9.8 GHz [7].

4.4 Conclusion

Time-domain techniques, applied here in form of the TLM solver MEFiSTo-3D, present a viable option for the analysis and modelling of UWB printed-circuit antennas. Amplitude characteristics extracted from the time-domain solution agree well with frequency-main methods, which are used for the design of UWB antennas. The computation of group-delay data in an actual application of pulsed transmission is one of the clear advantages of time-domain over frequency-domain techniques.

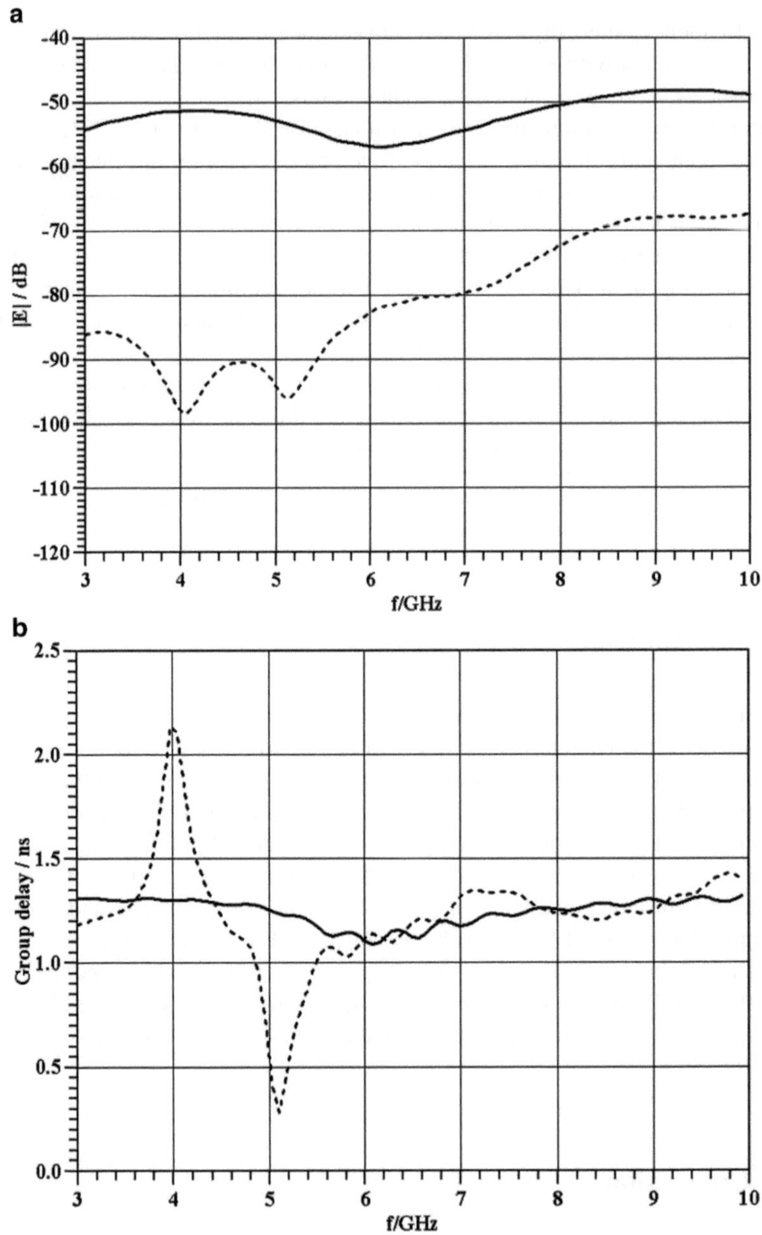

Fig. 4.11 Amplitude response (**a**) and group-delay characteristic (**b**) of the microstrip UWB antenna in [9, 10]; vertical polarization E_θ (*solid lines*), and horizontal polarization E_ϕ (*dashed lines*)

The time-domain modelling procedure presented here is applied to two different printed-circuit UWB antennas, and agreement with frequency-domain computations and measurements is demonstrated.

Acknowledgements The authors gratefully acknowledge financial support for this project through the TELUS Research Grant in Wireless Communications.

References

1. L. Yang, G.B. Giannakis, Ultra-wideband communications: An idea whose time has come. IEEE Signal Proc. Mag. **21**, 26–54 (2004)
2. International Telecommunication Union, Radiocommunication Study Groups, Framework for the introduction of devices using ultra-wideband technology. Document 1/85(Rev.1)-E, 09 Nov 2005
3. K. Kiminami, A. Hirata, T. Shiozawa, Double-sided printed bow-tie antenna for UWB communications. IEEE Antennas Wireless Propag. Lett. **3**, 152–153 (2004)
4. J. Liang, C.C. Chiau, X. Chen, C.G. Parini, Printed circular disc monopole antenna for ultra-wideband applications. IEE Electron. Lett. **40**(20), 1246–1247 (2004)
5. S.H. Choi, J.K. Park, S.K. Kim, J.Y. Park, A new ultra-wideband antenna for UWB applications. Microw. Opt. Technol. Lett. **40**(5), 399–401 (2004)
6. J. Liang, C.C. Chiau, X. Chen, C.G. Parini, Study of a printed circular disc monopole antenna for UWB systems. IEEE Trans. Antennas Propag. **53**, 3500–3504 (2005)
7. Z.N. Low, J.H. Cheong, C.L. Law, Low-cost PCB antenna for UWB applications. IEEE Antennas Wireless Propag. Lett. **4**, 237–239 (2005)
8. J. Liang, C.C. Chiau, X. Chen, C.G. Parini, Printed circular ring monopole antennas. Microw. Opt. Technol. Lett. **45**(5), 372–375 (2005)
9. C.-C. Lin, Y.-C. Kan, L.-C. Kuo, H.-R. Chuang, A planar triangular monopole antenna for UWB communication. IEEE Microw. Wireless Comput. Lett. **15**, 624–626 (2005)
10. H.R. Chuang, C.C. Lin, Y.C. Kan, A printed UWB triangular monopole antenna. Microw. J. **49**, 108–120 (2006)
11. N. Fortino, G. Kossiavas, J.Y. Dauvignac, R. Staraj, Novel antennas for ultrawideband communications. Microw. Opt. Technol. Lett. **41**(3), 166–169 (2004)
12. W. Wang, S.S. Zhong, S.-B. Chen, A novel wideband coplanar-fed monopole antenna. Microw. Opt. Technol. Lett. **43**(1), 50–52 (2004)
13. A.M. Abbosh, M.E. Bialkowski, M.V. Jacob, J. Mazierska, Investigations into an LTCC based ultra wideband antenna. in *Proceedings Asia-Pacific Microwave Conference*, Suzhou, China, Dec 2005. 4 p.
14. C.T.H. Lim, A GCPW-fed printed antenna for UWB applications. in *Proceedings Asia-Pacific Microwave Conference*, Suzhou, China, Dec. 2005, 3 p.
15. X. Chen, J. Liang, P. Li, L. Guo, C.C. Chiau, C.G. Parini, Planar UWB monopole antennas. in *Proceedings Asia-Pacific Microwave Conference*, Suzhou, China, Dec 2005, 4 p.
16. H.K. Lee, J.K. Park, J.N. Lee, Design of a planar half-circle shaped UWB notch antenna. Microw. Opt. Technol. Lett. **47**(1), 9–11 (2005)
17. T.-G. Ma, C.-H. Tseng, An ultrawideband coplanar waveguide-fed tapered ring slot antenna. IEEE Trans. Antennas Propag. **54**, 1105–1110 (2006)
18. Y.-C. Lee, S.-C. Lin, J.-S. Sun, CPW-fed UWB slot antenna. in *Proceedings Asia-Pacific Microwave Conference*, Yokohama, Japan, Dec. 2006, 4 p.
19. S. Nikolaou, D.E. Anagnostou, G.E. Ponchak, M.M. Tentzeris, J. Papapolymerou, Compact ultra wide-band (UWB) CPW-fed elliptical monopole on liquid crystal polymer (LCP). in *IEEE AP-S International Symposium Digest*, Albuquerque, USA, July 2006, pp. 4657–4660

20. E.S. Angelopoulos, A.Z. Anastopoulos, D.I. Kaklamani, Ultra-wideband bow-tie slot antenna fed by a cpw-to-cpw transition loaded with inductively coupled slots. Microw. Opt. Technol. Lett. **48**(9), 1816–1820 (2006)
21. X.-L. Liang, S.-S. Zhong, W. Wang, UWB printed circular monopole antenna. Microw. Opt. Technol. Lett. **48**(8), 1532–1534 (2006)
22. J.-S. Sun, Y.-C. Lee, S.-C. Lin, New design of a CPW-fed ultrawideband slot antenna. Microw. Opt. Technol. Lett. **49**(3), 561–564 (2007)
23. D.-B. Lin, I.-T. Tang, M.-Y. Tsou, A compact UWB antenna with CPW-feed. Microw. Opt. Technol. Lett. **49**(3), 564–567 (2007)
24. Z.N. Chen, X. Qing, Research and development of planar UWB antennas. Suzhou, China, Dec 2005
25. B.L. Ooi, G. Zhao, M.S. Leong, K.M. Chua, C.W.L. Albert, Wideband LTCC CPW-fed two-layered monopole antenna. IEE Electron. Lett. **41**(16), 9–10 (2005)
26. K. Rambabu, H.A. Thiart, J. Bornemann, S.Y. Yu, Ultrawideband printed-circuit antenna. IEEE Trans. Antennas Propag. **54**, 3908–3911 (2006)
27. H.-J. Lam, J. Bornemann, Ultra-wideband printed-circuit antenna in coplanar technology. in *2007 IEEE EMC-S International Symposium Digest, TU-PM-1–7*, Honolulu, USA, July 2007. 4 p.
28. D. Ghosh, A. De, M.C. Taylor, T.K. Sarkar, M.C. Wicks, E.L. Mokole, Transmission and reception by ultra-wideband (UWB) antennas. IEEE Trans. Antennas Propag. Mag. **48**, 67–99 (2006)

Chapter 5
Diffraction of Acoustic and Electromagnetic Waves by Impedance Cones

Jean-Michel L. Bernard, Mikhail A. Lyalinov, and Ning Yan Zhu

5.1 Introduction

One of the chief tasks in advancing the theory of diffraction consists in solving the so-called canonical problems, that is, diffraction of acoustic or electromagnetic waves by bodies of simple shapes like wedges and cylinders. Obviously, cones or conical surfaces of circular cross section are additional examples of these bodies.

Of both theoretical and practical importance are such cones or conical surfaces on whose boundary conditions of impedance type hold for the acoustic or the electromagnetic waves. In spite of their simple shape, the impedance cones have been studied since only around 14 years.

The procedure put forward in [1–6] makes use of the Kontorovich–Lebedev integrals for the representation of the unknown fields, enabling in this way a partial separation of variables. Inverting the boundary conditions expressed in terms of the Kontorovich–Lebedev integrals, non-local difference equations for the spectra follow. Expanding the spectra in Fourier series leads to functional difference equations of the second order for the Fourier coefficients of the spectra. The equivalence between a functional difference equation of the second order and a Fredholm integral equation of the second kind allows the Fourier coefficients to be fast and accurately determined in a numerical way. The knowledge of the Fourier coefficients and hence of the spectra leads to the sought-for solution to the canonical

J.-M.L. Bernard (✉)
Département de Physique Théorique et Appliquée, CEA/DIF-Bruyères le Châtel, 91297 Arpajon cedex, France
e-mail: jean-michel.bernard@cea.fr

M.A. Lyalinov
Department of Mathematical Physics, St. Petersburg University, Petrodvarets, St. Petersburg 198504, Russia
e-mail: lyalinov@yandex.ru

N.Y. Zhu
Institut für Hochfrequenztechnik, Universität Stuttgart, D-70569 Stuttgart, Germany
e-mail: zhu@ihf.uni-stuttgart.de

S. Lindenmeier and R. Weigel (eds.), *Electromagnetics and Network Theory and their Microwave Technology Applications*, DOI 10.1007/978-3-642-18375-1_5,

problem in terms of the Kontorovich–Lebedev integrals. Evaluating these integrals for $kr \to \infty$, where k stands for the wave number and r the distance from the tip of the cone, yields the diffraction coefficient in the oasis.

This procedure has been applied with success to diffraction of acoustic and electromagnetic waves by impedance cones and impedance conical surfaces [7–10]. To demonstrate the prowess of this procedure, therefore, the present paper reports the application of the well-proven procedure to diffraction of an electromagnetic plane wave by an opaque cone whose face is characterised electrically by a diagonal impedance tensor.

The problem under study is formulated in Sect. 5.2, and the procedure of solution is briefly outlined in Sect. 5.3. The numerical results are discussed in Sect. 5.4, while Sect. 5.5 concludes this paper with a discussion of future work.

5.2 Statement of the Problem

The canonical body under study is depicted in Fig. 5.1.

An electromagnetic plane wave impinges upon the cone (the time-dependence used here is $e^{-i\omega t}$):

$$\mathbf{E}^0 = \left(\mathbf{e}_{\vartheta_0} \sin\beta + \mathbf{e}_{\varphi_0} \cos\beta\right) e^{-ikr\cos\hat{\vartheta}(\omega,\omega_0)}, \tag{5.1}$$

$$Z_0\mathbf{H}^0 = \left(\mathbf{e}_{\vartheta_0} \cos\beta - \mathbf{e}_{\varphi_0} \sin\beta\right) e^{-ikr\cos\hat{\vartheta}(\omega,\omega_0)}. \tag{5.2}$$

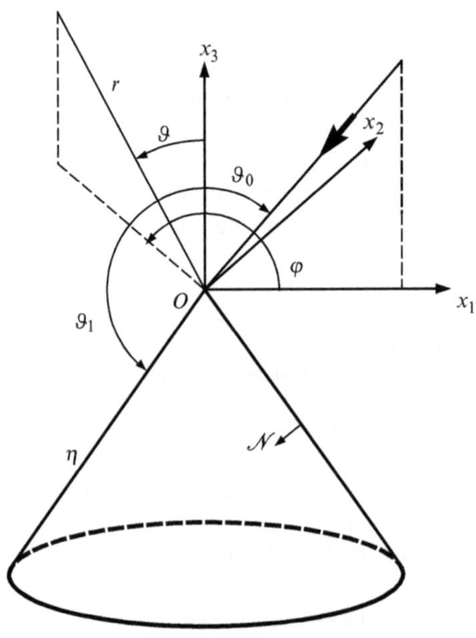

Fig. 5.1 A right circular impedance cone

On the surface of the cone, the electric and magnetic fields are related to each other according to

$$\mathbf{E} - (\mathbf{E} \cdot \mathscr{N})\,\mathscr{N} = -Z_0 \overline{\overline{\eta}}\,\mathscr{N} \times \mathbf{H}, \tag{5.3}$$

where the (with respect to the intrinsic impedance Z_0 of the ambient medium) normalised impedance matrix is diagonal and given by $\overline{\overline{\eta}} = \operatorname{diag}\left[\eta_r \ \eta_\varphi\right]$.

At the tip of the cone, the condition of Meixner and Jones must hold:

$$\int_{S_\delta} \mathbf{e}_r \cdot \left(\mathbf{E} \times \mathbf{H}^* + \mathbf{H} \times \mathbf{E}^*\right) \mathrm{d}S \to 0, \text{ as } \delta \to 0, \tag{5.4}$$

where S_δ means the surface of a sphere of radius δ centred at the tip of the cone.

In addition, radiation conditions in the form of conditions on the far-field asymptotics as discussed in [6] must be met.

For convenience, the electromagnetic fields are expressed in terms of the Debye potentials u and v:

$$\mathbf{E} = \operatorname{curl}\operatorname{curl}(r\mathbf{e}_r u) + ik\operatorname{curl}\,(r\mathbf{e}_r v)\,, \tag{5.5}$$

$$Z_0\mathbf{H} = \operatorname{curl}\operatorname{curl}(r\mathbf{e}_r v) - ik\operatorname{curl}\,(r\mathbf{e}_r u)\,, \tag{5.6}$$

and the Debye potentials depend upon each other on the cone's surface according to

$$\partial_r^2\,(ru) + k^2\,(ru) = -\eta_r\left[\frac{1}{r\sin\vartheta}\partial_{r\varphi}^2\,(rv) + \frac{ik}{r}\partial_\vartheta\,(ru)\right],$$

$$\partial_r^2\,(rv) + k^2\,(rv) = \eta_\varphi^{-1}\left[\frac{1}{r\sin\vartheta}\partial_{r\varphi}^2\,(ru) - \frac{ik}{r}\partial_\vartheta\,(rv)\right]. \tag{5.7}$$

5.3 Procedure of Solution

The first step of the solution procedure lies in decomposing the Debye potentials in an incident part and a scattered part, that is, $u = u_0 + u_s, v = v_0 + v_s$, and representing them in terms of the Kontorovich–Lebedev integrals:

$$(u_s, v_s) = \frac{-2}{k}\sqrt{\frac{2}{\pi}}\left[\int_{-i\infty}^{i\infty} \frac{\nu\sin(\pi\nu)}{\nu^2 - 1/4}\frac{K_\nu(-ikr)}{\sqrt{-ikr}}\right.$$

$$\left.\cdot g_{u,v}(\omega, \omega_0, \nu)d\nu + i\pi\frac{K_{1/2}(-ikr)}{\sqrt{-ikr}}g_{u,v}(\omega, \omega_0, 1/2)\right]. \tag{5.8}$$

Similar expressions hold for u_0 and v_0, with $g_{u,v}$ replaced by $g_{u,v}^0$ which are known.

In the above relations, $K_\nu(z)$ denotes the Macdonald function and $g_{u,v}(\omega, \omega_0, \nu)$ the spectra which are even.

The above integrals converge for real-valued k for

$$\hat{\vartheta}'(\boldsymbol{\omega}, \boldsymbol{\omega}_0) > \pi, \tag{5.9}$$

which defines the oasis, because there are no other wave ingredients in this domain than the incident and tip-diffracted parts. The definition for $\hat{\vartheta}'(\boldsymbol{\omega}, \boldsymbol{\omega}_0)$ is given in [6].

For complex-valued k, when suitably chosen, the above integrals converge outside the oasis, especially also for points on the surface of the cone. Inverting the integral representation of the boundary conditions for the Debye potentials u and v, one gets the following relations for the spectra:

$$g_u^t(\nu+1) - g_u^t(\nu-1) = -\eta_r \left[\frac{\partial_\varphi g_v^t(\nu-1)}{\sin\vartheta_1(\nu-1/2)} \right.$$
$$\left. + \frac{\partial_\varphi g_v^t(\nu+1)}{\sin\vartheta_1(\nu+1/2)} - \frac{2\nu}{\nu^2-1/4} \partial_{\vartheta_1} g_u^t(\nu) \right], \tag{5.10}$$
$$g_v^t(\nu+1) - g_v^t(\nu-1) = +\eta_\varphi^{-1} \left[\frac{\partial_\varphi g_u^t(\nu-1)}{\sin\vartheta_1(\nu-1/2)} \right.$$
$$\left. + \frac{\partial_\varphi g_u^t(\nu+1)}{\sin\vartheta_1(\nu+1/2)} + \frac{2\nu}{\nu^2-1/4} \partial_{\vartheta_1} g_v^t(\nu) \right]. \tag{5.11}$$

The above equations employ the short-hand notations $g_{u,v}^t = g_{u,v}^0 + g_{u,v}$.

It is worth emphasising that the above equations depend neither on the wave number k, nor on the distance to the tip of the cone r.

Thanks to the rotational symmetry of the cone's geometry, the spectra can be expanded into Fourier series of the form:

$$g_{u,v}(\boldsymbol{\omega}, \boldsymbol{\omega}_0, \nu) = \sum_{n=-\infty}^{+\infty} i^n e^{-in\varphi} R_{u,v}(\nu, n, \boldsymbol{\omega}_0) \frac{P_{\nu-1/2}^{-|n|}(\cos\vartheta)}{P_{\nu-1/2}^{-|n|}(\cos\vartheta_1)}, \tag{5.12}$$

$$g_{u,v}^0(\boldsymbol{\omega}, \boldsymbol{\omega}_0, \nu) = \sum_{n=-\infty}^{+\infty} i^n e^{-in\varphi} R_{u,v}^0(\nu, n, \boldsymbol{\omega}_0) \frac{P_{\nu-1/2}^{-|n|}(-\cos\vartheta)}{P_{\nu-1/2}^{-|n|}(-\cos\vartheta_1)}. \tag{5.13}$$

Like the spectra, their Fourier coefficients $R_{u,v}^0$ and $R_{u,v}$ are also even in ν.

It has been taken into account above that the spectra must meet the equation

$$\left[\Delta_\omega + \left(\nu^2 - 1/4\right)\right] g_{u,v}(\boldsymbol{\omega}, \boldsymbol{\omega}_0, \nu) = 0, \tag{5.14}$$

where Δ_ω signifies the Laplace–Beltrami operator and is defined as

$$\Delta_\omega = (\sin\vartheta)^{-1}\partial_\vartheta(\sin\vartheta\,\partial_\vartheta) + (\sin\vartheta)^{-2}\partial_\varphi^2. \tag{5.15}$$

Only then can the Debye potentials, now represented in terms of the Kontorovich–Lebedev integrals, satisfy the scalar wave equation.

Inserting the Fourier series into the relations for the spectra leads to a system of functional difference equations for $R_{u,v}$:

$$\frac{i}{2}\left[R_u(\nu+1)-R_u(\nu-1)\right]=\eta_r w(\nu)R_u(\nu) \\ -\frac{n\eta_r}{2\sin\vartheta_1}\left[\frac{R_v(\nu-1)}{\nu-1/2}+\frac{R_v(\nu+1)}{\nu+1/2}\right]-G_u^0(\nu), \tag{5.16}$$

$$\frac{i}{2}\left[R_v(\nu+1)-R_v(\nu-1)\right]=\eta_\varphi^{-1}w(\nu)R_v(\nu) \\ +\frac{n\eta_\varphi^{-1}}{2\sin\vartheta_1}\left[\frac{R_u(\nu-1)}{\nu-1/2}+\frac{R_u(\nu+1)}{\nu+1/2}\right]-G_v^0(\nu). \tag{5.17}$$

Use new functions defined as

$$\rho_\pm(\nu)=R_u(\nu)\pm(\eta_\mathrm{a}/i)\,R_v(\nu)$$

with $\eta_\mathrm{a}=\sqrt{\eta_r\eta_\varphi}$, the above equations can be brought into a more compact form:

$$\frac{i}{2}\left[a(\nu)\rho(\nu+1)-a(-\nu)\rho(\nu-1)\right]=W(\nu)\rho(\nu)-F^0(\nu), \tag{5.18}$$

with

$$a(\nu)=\begin{pmatrix}1&0\\0&1\end{pmatrix}+\frac{n\eta_\mathrm{a}}{\eta_\varphi\sin\vartheta_1}\frac{1}{\nu+1/2}\begin{pmatrix}1&0\\0&-1\end{pmatrix}, \tag{5.19}$$

$$W(\nu)=\frac{1}{2}\begin{pmatrix}w(\nu)\left(\eta_r+\eta_\varphi^{-1}\right)&w(\nu)\left(\eta_r-\eta_\varphi^{-1}\right)\\w(\nu)\left(\eta_r-\eta_\varphi^{-1}\right)&w(\nu)\left(\eta_r+\eta_\varphi^{-1}\right)\end{pmatrix}; \tag{5.20}$$

the column vectors $\rho(\nu)$ and $F^0(\nu)$ are given by

$$\rho(\nu)=\begin{pmatrix}\rho_+(\nu)\\\rho_-(\nu)\end{pmatrix},\quad F^0(\nu)=\begin{pmatrix}G_u^0(\nu)-i\eta_\mathrm{a}G_v^0(\nu)\\G_u^0(\nu)+i\eta_\mathrm{a}G_v^0(\nu)\end{pmatrix}. \tag{5.21}$$

which are even functions in ν. To fulfil at the tip of the cone the condition of Meixner and Jones, there must be

$$\rho_+(1/2)=\frac{1+\eta_\mathrm{a}n/|n|}{1-\eta_\mathrm{a}n/|n|}\rho_-(1/2). \tag{5.22}$$

The functional difference matrix equation of the second order (5.18) can be written in an equivalent integral form valid inside the strip $|\mathrm{Re}\nu|<1$:

$$\rho(\nu)=B\rho(\nu)+S^1(\nu)+\rho_-(1/2)S^2(\nu)+\rho_+(1/2)S^3(\nu), \tag{5.23}$$

with the operator $B\rho(\nu)$ given by

$$B\rho(\nu) = \frac{1}{2}\mathrm{diag}\,[\Psi(\nu, \Lambda_+),\ \Psi(\nu, \Lambda_-)]$$
$$\cdot \int_{-i\infty}^{i\infty} \mathrm{diag}\,[\Psi(\tau, -r(\Lambda_+), \Lambda_-),\ \Psi(\tau, -r(\Lambda_-), \Lambda_+)]$$
$$\cdot W(\tau) \cdot \rho(\tau) \frac{\sin(\pi\tau)d\tau}{\cos(\pi\nu) + \cos(\pi\tau)}, \tag{5.24}$$

and the remaining terms at the right-hand side by

$$S^1(\nu) = -\frac{1}{2}\mathrm{diag}\,[\Psi(\nu, \Lambda_+),\ \Psi(\nu, \Lambda_-)]$$
$$\cdot \int_{-i\infty}^{i\infty} \mathrm{diag}\,[\Psi(\tau, -r(\Lambda_+), \Lambda_-),\ \Psi(\tau, -r(\Lambda_-), \Lambda_+)]$$
$$\cdot F^0(\tau) \frac{\sin(\pi\tau)d\tau}{\cos(\pi\nu) + \cos(\pi\tau)}, \tag{5.25}$$
$$S^2(\nu) = [0,\ S_-(\nu)]^T \quad \text{if } r(\Lambda_-) < 0,\ = 0 \text{ if not} \tag{5.26}$$
$$S^3(\nu) = [S_+(\nu),\ 0]^T \quad \text{if } r(\Lambda_+) < 0,\ = 0 \text{ if not} \tag{5.27}$$

with

$$r(\Lambda) = \mathrm{sign}\left(\frac{1}{2} + \mathrm{Re}\Lambda\right), \tag{5.28}$$
$$S_\pm(\nu) = \Psi(\nu, \Lambda_\pm)\frac{-\pi}{\cos(\pi\nu)}\mathrm{res}_{1/2}\frac{1}{\Psi(\nu, \Lambda_\pm)}, \tag{5.29}$$
$$\Lambda_\pm = \pm\frac{n\eta_{\mathrm{a}}}{\eta_\varphi \sin\vartheta_1}. \tag{5.30}$$

The special function $\Psi(\nu, \Lambda)$ used above is defined in Appendix 2 of [9].

For points on the imaginary axis of the ν plane, (5.23) becomes a Fredholm integral equation of the second kind which admits a fast and efficient numerical solution.

To determine $\rho(\nu)$, solve at first the following three equations

$$(I - B)\rho^j(\nu) = S^j(\nu), \quad j = 1, 2, 3. \tag{5.31}$$

To find the values $\rho_\pm(1/2)$, one needs another relation between them beside (5.22). Such a relation can be obtained from (5.23) by setting $\nu = 1/2$ either in $\rho_+(\nu)$ for $r(\Lambda_-) < 0$ or in $\rho_-(\nu)$ for $r(\Lambda_+) < 0$.

Thus, for points on the imaginary axis of the ν plane, $\rho(\nu)$ is given by

$$\rho(\nu) = \rho^1(\nu) + \begin{cases} \rho_-(1/2)\rho^2(\nu), & r(\Lambda_-) < 0, \\ 0, & n = 0, \\ \rho_+(1/2)\rho^3(\nu), & r(\Lambda_+) < 0. \end{cases} \tag{5.32}$$

Employing the asymptotic expression for the Macdonald functions contained in the Kontorovich–Lebedev integrals, one arrives with ease at the tip-diffracted field in the far field (inside the oasis $\hat{\vartheta}'(\boldsymbol{\omega}, \boldsymbol{\omega}_0) > \pi$):

$$\mathbf{E}^d = -2\pi \frac{e^{ikr}}{kr} \overline{\overline{D}} \cdot \mathbf{E}^0_{r=0} \left(1 + O\left(\frac{1}{kr}\right)\right), \tag{5.33}$$

$$Z_0 \mathbf{H}^d = \mathbf{e}_r \times \mathbf{E}^d. \tag{5.34}$$

Here, $\overline{\overline{D}}$ denotes the diffraction coefficients in matrix form:

$$\overline{\overline{D}} = \begin{pmatrix} D_{\vartheta\vartheta_0} & D_{\vartheta\varphi_0} \\ D_{\varphi\vartheta_0} & D_{\varphi\varphi_0} \end{pmatrix}. \tag{5.35}$$

The entries of the matrix are related to the integrals of the spectra functions derived above.

5.4 Numerical Results

The above solution procedure has been implemented, the correctness of this procedure and its implementation has been verified through comparison with the results obtained for scalar impedance cones [9], as well as verifying certain known properties of scattering by a rotational symmetric body, among them, the zero-backscattering at axial incidence from such a body with the balanced hybrid condition $\eta_r \eta_\varphi = 1$ [11, 12].

As an example, the normalised radar cross-sections of φ- and ϑ-polarisation for a cone under non-axial incidence are depicted in Figs. 5.2 and 5.3, as a function of the azimuth and co-latitude angles. To illustrate the impact of the anisotropy upon the diffraction behaviour, all parameters for this example take the same values as for one example from [9] (Figs. 3 and 4 there) with the sole exception that now the surface impedance along the azimuth η_φ differs from that in the radial direction. As is conspicuous from Figs. 5.2 and 5.3, the radar cross-sections of different polarisation are of different order of magnitude, whereas they are of the same order of magnitude in the case of a scalar impedance cone (cf. Figs. 3 and 4 of [9]).

5.5 Summary

This paper reported the application of an analytical-numerical procedure to diffraction of an electromagnetic plane wave by an axially anisotropic impedance cone. The key steps consist in incomplete separation of variables with the aid of the Kontorovich–Lebedev integrals, derivation of non-local relations for the spectra,

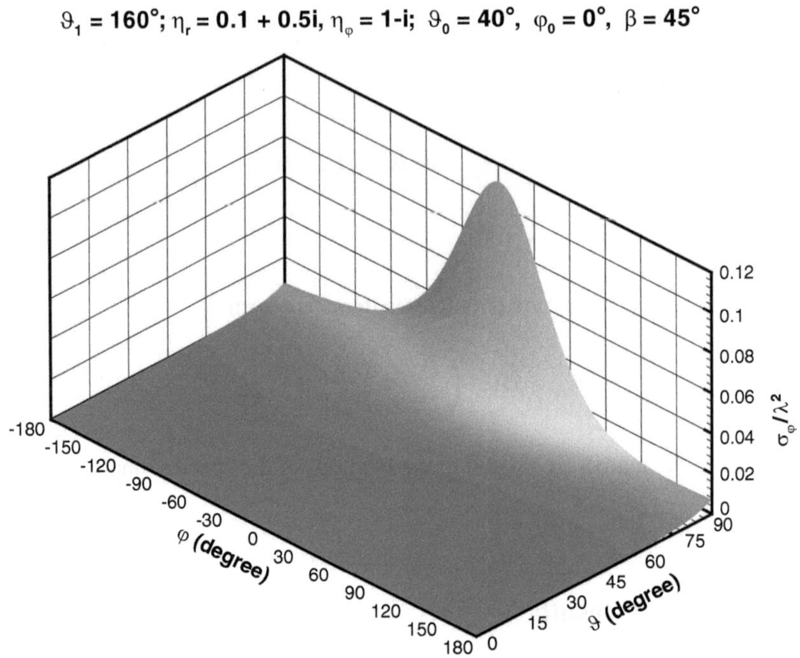

Fig. 5.2 Normalised radar cross-section of φ-polarisation at non-axial incidence as a function of the azimuth and co-latitude angles

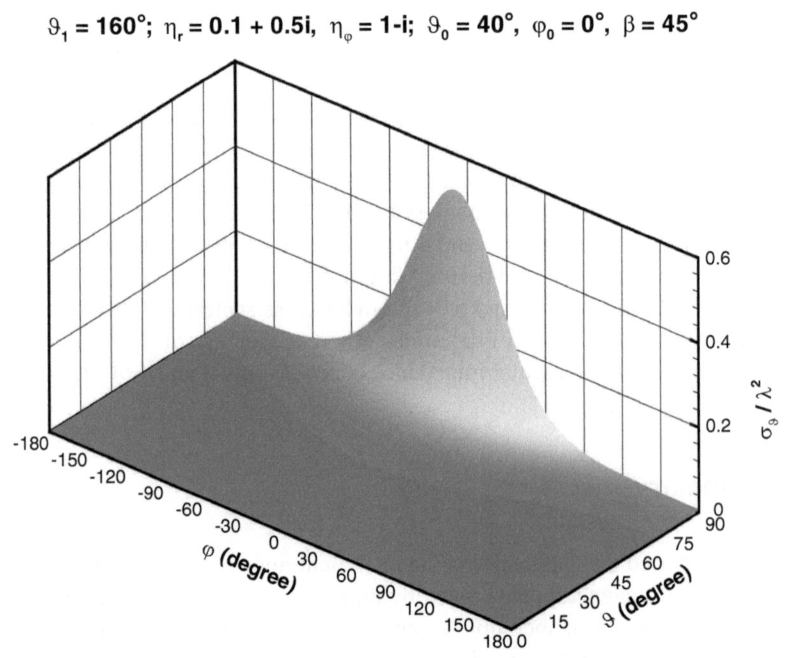

Fig. 5.3 Normalised radar cross-section of ϑ-polarisation at non-axial incidence as a function of the azimuth and co-latitude angles

Fourier-expansion of the spectra, deduction of functional difference equations for the Fourier coefficients, establishment of the equivalence between the functional difference equations of the second order with the Fredholm integral equations of the second kind, numerical solution of the integral equations, and lastly, deduction of the diffraction coefficient in the oasis. Typical behaviour of diffraction has been shown with the aid of an example.

To the authors' knowledge, all studies dealing with impedance cones [1–10] concern diffracted waves in the oasis. Hence, the future work will focus on diffracted waves outside the oasis with the final goal of deriving a uniform formula for the diffraction coefficients.

In addition, it should be beneficial to apply the well-proven techniques for tackling functional difference equations of the second order to other and more demanding problems encountered in the diffraction theory.

References

1. J.M.L. Bernard, Méthode analytique et transformées fonctionnelles pour la diffraction d'ondes par une singularité conique: équation intégrale de noyau non oscillant pour le cas d'impédance constante. rapport CEA-R-5764, Editions Dist-Saclay (1997) [erratum in J. Phys. A, vol.32, p.L45]
2. J.M.L. Bernard, M.A. Lyalinov, The leading asymptotic term for the scattering diagram in the problem of diffraction by a narrow circular impedance cone. J. Phys. A Math. Gen. **32**, L43–L48 (1999) [Replace $\psi(\nu - 1/2)$ by $\psi(\nu + 1/2)$ in (12)]
3. J.M.L. Bernard, M.A. Lyalinov, Spectral domain solution and asymptotics for the diffraction by an impedance cone. IEEE Trans. Antennas Propag. **49**(12), 1633–1637 (2001)
4. J.M.L. Bernard, M.A. Lyalinov, Diffraction of acoustic waves by an impedance cone of an arbitrary cross-section. Wave Motion **33**, 155–181 (2001)
5. Y.A. Antipov, Diffraction of a plane wave by a circular cone with an impedance boundary condition. SIAM J. Appl. Math. **62**(4), 1122–1152 (2002)
6. J.M.L. Bernard, M.A. Lyalinov, Electromagnetic scattering by a smooth convex impedance cone. IMA J. Appl. Math. **69**(3), 285–333 (2004)
7. M.A. Lyalinov, N.Y. Zhu, Acoustic scattering by a circular semi-transparent conical surface. J. Eng. Math. **59**(4), 385–398 (2007)
8. N.Y. Zhu, M.A. Lyalinov, Diffraction by a wedge or by a cone with impedance-type boundary conditions and second-order functional difference equations. PIER B **59**(6), 239–256 (2008)
9. J.M.L. Bernard, M.A. Lyalinov, N.Y. Zhu, Analytical-numerical calculation of diffraction coefficients for a circular impedance cone. IEEE Trans. Antennas Propag. **56**(6), 1616–1623 (2008)
10. M.A. Lyalinov, N.Y. Zhu, V.P. Smyshlyaev, Scattering of a plane electromagnetic wave by a circular hollow cone with thin semi-transparent walls. submitted to IMA J. Appl. Math. **75**(5), 676–719 (2010)
11. V.H. Weston, Theory of Absorbers in Scattering. IEEE Trans. Antennas Propag. **11**(5), 578–584 (1963)
12. K.S. Yee, A.H. Chang, Scattering theorem with anisotropic surface boundary conditions for bodies of revolution. IEEE Trans. Antennas Propag. **39**(7), 1041–1043 (1991)

Part II
Microwave Systems

Chapter 6
Pattern Design and DBF Analysis of a Dielectric Lens Antenna for 77 GHz Automotive Long Range Radar

Peter Wenig and Robert Weigel

6.1 Introduction

Automotive radar sensors are a key component for driver assistance and safety systems. The current trend in automotive long range radar (LRR) is to provide sensors that are capable to meet the accuracy requirements for safety functions such as collision warning and automatic emergency brake. These requirements comprise an improved direction of arrival (DOA) estimation accuracy in comparison to current sensors and the capability to resolve multiple targets even in the same range-velocity cell [6].

A natural solution is an antenna frontend with a uniform linear array (ULA) of antennas that permits the usage of high-resolution subspace based direction-of-arrival (DOA) estimation methods such as MUSIC or ESPRIT in conjunction with digital beam forming (DBF) for target detection [10]. Moreover car manufacturers demand low cost sensors for the automotive mass market with compact dimensions for unobtrusive integration into the car front. Current industrially available sensors use either a rotational symmetric lens antenna fed by a small array of patch radiators [1] or a large twodimensional patch array [2,5,9]. The former concept is not suitable for the application of high-resolution methods or flexible DBF, whereas in the latter concept a linear array of antennas can be used. But the elevation beam shaping with planar antennas acquires a large RF substrate area which is an essential cost factor.

In this work a novel antenna concept is presented that uses a cylindrical dielectric lens. For automotive LRR, the 76 – 77 GHz frequency band is allocated. A uniform linear array (ULA) of microstrip patch subarrays optimized for 76.5 GHz is placed along the focus line of the cylindrical lens. With this arrangement the antenna pattern in the elevation plane of the roadside scenario can be synthesized independently of the azimuth plane by jointly designing the feeding patch antennas along with the lens shape. Moreover it allows the application of DBF and high resolution methods

P. Wenig and R. Weigel (✉)
University of Erlangen-Nuremberg, Cauerstr. 9, 91058 Erlangen, Germany
e-mail: wenig@lfte.de

S. Lindenmeier and R. Weigel (eds.), *Electromagnetics and Network Theory and their Microwave Technology Applications*, DOI 10.1007/978-3-642-18375-1_6,

as mentioned above. The lens antenna is compared to a conventional patch array solution. Furthermore the effects of the lens edges are studied with respect to their impact on the beamformed two-way pattern response.

6.2 System Concept

Figure 6.1 depicts the block diagram of the proposed FMCW frontend concept with multiple parallel receive (Rx) channels. The sensor uses a bistatic antenna configuration in order to reduce transmit signal leakage. Both the transmitting and the receiving antenna are positioned beneath the lens [11].

The lens antenna configuration is shown in more detail in Fig. 6.2. A cylindrical lens is illuminated by a column of serial fed microstrip patches. The orientation of these columns is depicted in the figure, where the y-axis is parallel to the vertical axis of the car. In the azimuth plane (x-z-plane), the receiving antennas form a ULA whose elements are positioned on the focal line of the lens.

The individual receiving array elements are comprised of patch subarrays with a $2 \times N_{\mathrm{E}}$ arrangement. The size of the subarrays controls the spillover radiation and the feed illumination taper for the lens in both elevation and azimuth. The lens-patch-combination is responsible for the elevation-plane beam shaping. Since a projecting dielectric cover for the sensor is indispensable anyway, the lens concept bears no extra assembly step during production.

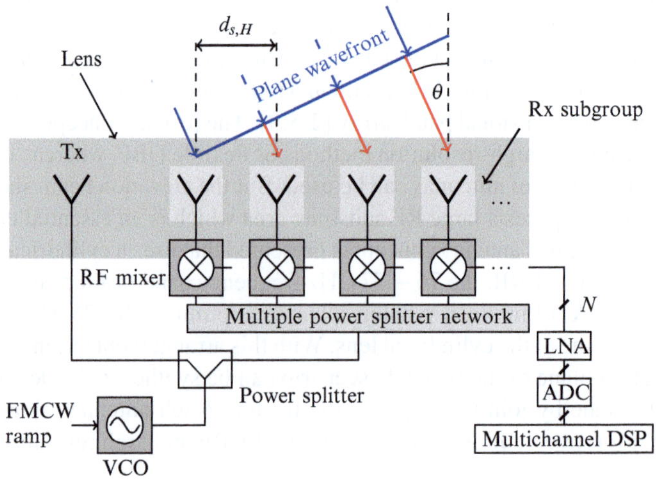

Fig. 6.1 Block diagram of the proposed FMCW radar frontend concept with multiple parallel Rx channels

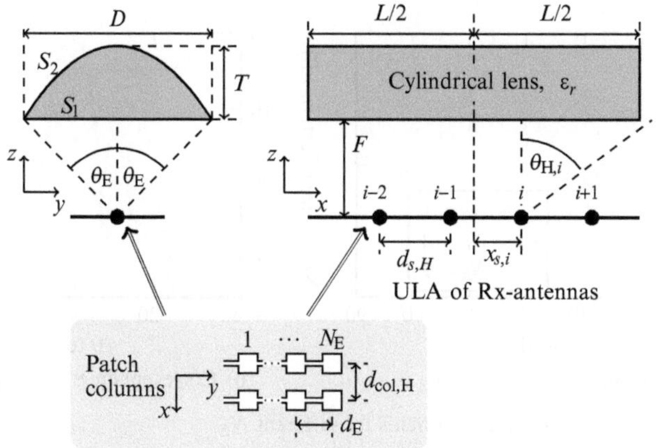

Fig. 6.2 Lens antenna configuration in the elevation-plane (*left*) and azimuth-plane (*right*). The *black dots* mark the phase center positions of the feeding microstrip antennas

6.3 Elevation Pattern Design

In order to arrive at a desired farfield pattern in the elevation plane, multiple factors have to be considered. In the present design, a beamwidth of approx. 4.25° for the combined Tx-Rx pattern response was specified.[1] The farfield pattern can be computed by evaluating the diffraction integral of the field distribution on a virtual aperture layer behind the lens body. Under ideal assumptions, where spillover radiation is neglected, the diameter D and the focal length F affect the size of the virtual aperture, the edge illumination on the aperture plane and the thickness of the lens T. Moreover, the choice of the lens shape affects the transfer function for the field distribution at the illumination side of the lens to its aperture side. The design of the feed antenna pattern controls the spillover radiation and also affects the aperture field distribution. For the current design, an unzoned plano-convex lens shape was chosen.

Regarding the feeding patch columns, N_E can be varied, whereas the patch spacing d_E is largely determined by the equal-phase condition for broadside radiation. Therefore the elevation-plane feed-pattern can be adjusted only in relatively coarse steps. The square patches were fed by a microstrip line connecting at the radiating edges, since the limitation on a single RF-layer is favorable for meeting automotive low cost design requirements. This feed concept produces an asymmetric lens illumination due to spurious feedline radiation. In the following the effect of the patch column length N_E on the farfield patterns of the whole lens antenna system is studied. All simulated radiation patterns were generated by full wave analysis

[1] The transmitting antenna is responsible for producing a focused illumination of the roadside scenario in the azimuth plane. Since it is also positioned under the lens, its elevation pattern largely equals that of the receiving antennas.

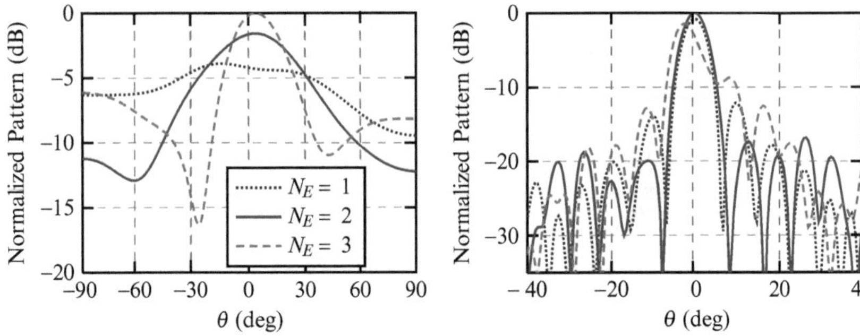

(a) Simulated feed farfields (b) Corresponding lens farfields

Fig. 6.3 Simulated elevation plane farfields for different N_E

with CST Microwave Studio. Figure 6.3a compares the feed farfield patterns for $N_E \in \{1, 2, 3\}$ and Fig. 6.3b shows the corresponding lens farfield patterns. The lens focal distance was fixed to $F = 15$ mm while the diameters D were optimized for the desired beamwidth in each case. Note that all patterns are normalized to the maximum occurring value. With only a single patch radiator ($N_E = 1$) the very low amplitude taper and the quite substantial spillover loss lead to a very poor sidelobe attenuation and a slightly lower gain compared to $N_E = 2$. For the latter choice the simulation shows a sidelobe attenuation of about 16.5 dB. Even though the feed antenna with $N_E = 3$ yield the highest gain, in combination with the lens pattern there is a significant pattern deformation, along with an increase in sidelobe levels and a slight gain loss. This can be attributed to the feed pattern minimum at approx. $-28°$ which is within the lens' field of view. Therefore, only the $N_E = 2$ case is considered below.

6.4 Comparison with Planar Column Antenna

In order to provide an equivalent beamwidth without a lens and using the same substrate parameters as before, a patch antenna column with $N_E = 16$ is needed, if equals spacings d_E are used along the column. Since the presented lens antenna configuration is thought as a replacement for an antenna layout comprised of such longer patch columns (in the following addressed as "column" antenna), the performance of these two concepts will be compared. Figure 6.4 shows the simulated and measured farfield elevation patterns, respectively, of the feed antenna with $N_E = 2$, the corresponding lens antenna pattern and the column antenna pattern. The plots are normalized to the maximum gain of the lens antenna. The gain increase of the lens compared to the feed antenna is about 9 dB. The maximum gain exceeds the gain of the column antenna by 1.5 dB. Sidelobes are attenuated by 16 dB for the lens antenna and by only 11 dB for the column antenna. Note that spurious radiation

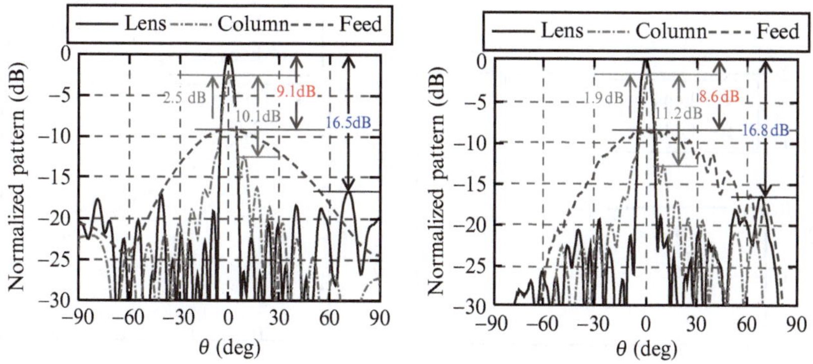

(a) Simulation (b) Measurement

Fig. 6.4 Farfield elevation patterns

from the microstrip feeding structure in the measurement setup caused a certain ripple in the feed antenna pattern, but this had no effect on the lens antenna pattern since it was out of focus.

One major drawback of the column antenna in conjunction with only a single RF substrate layer is the feeding at the endpoint of the column, which leads to an asymmetrical amplitude distribution and therefore increases sidelobe levels. Note that there is also a slight tilt in the column antenna pattern that could be eliminated by a corrected spacing d_E.

Generally, the main beam direction of the column antenna is dependent on the electrical length between the successive patches, as it is with the feeding antenna. But the latter mostly affects the illumination taper, whereas the lens itself is responsible for the beam direction. Therefore, the column antenna exhibits a so-called frequency scanning of approximately $1.5°/\text{GHz}$, as depicted in Fig. 6.5a this effect is almost negligible in case of the lens antenna. Furthermore the lens has also a smaller sensitivity to substrate losses, as shown in Fig. 6.5b For comparison, the gain sensitivity to the loss tangent of the lens material $\tan\delta_{\text{lens}}$ is also included in the figure, which is almost identical to the dependency on $\tan\delta_{\text{subs}}$.

6.5 Azimuth Plane Analysis

So far the lens antenna pattern has only been examined in the elevation plane, where the beam focusing of the lens occurs. Array processing algorithms usually assume identical element patterns, at least after the application of calibration methods (e.g. [8]). Figure 6.6 depicts the influence of the lens edge on the azimuth pattern for different feeding antenna offsets x_s from the symmetry axis (see Fig. 6.2), where λ_0 denotes the free space wavelength. All patterns are normalized to their mean

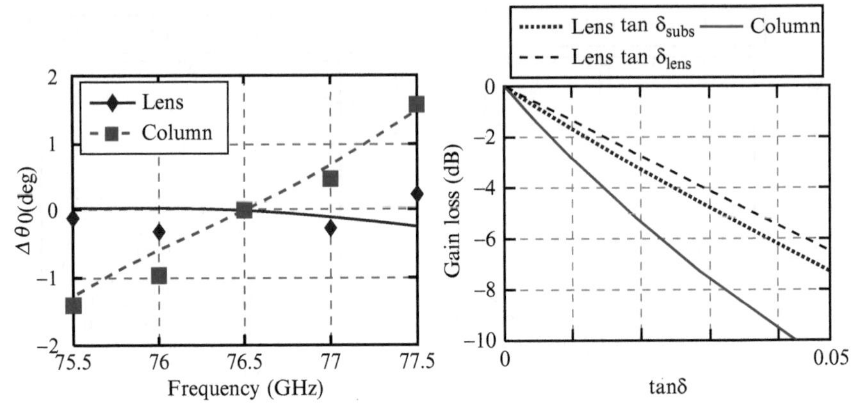

(a) Frequency scanning

(b) Simulated gain sensitivity to substrate and lens material loss tangent

Fig. 6.5 Comparison with column antenna

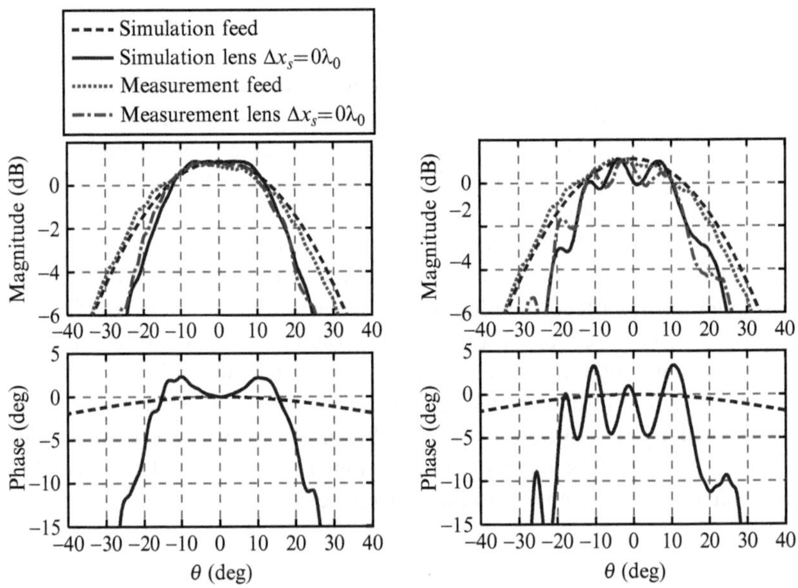

(a) Center position

(b) Shift $x_s = 7\lambda_0$

Fig. 6.6 Azimuth pattern perturbations due to lens edge effects

value within $\pm 7°$. The lens antenna patterns reveal some significantly increased perturbations at a feed antenna shift of $x_s = 7\lambda_0$ in comparison to the center position. Lens parameters were $L = 96\,\text{mm} \approx 24.5\lambda_0$, $F = 17.5\,\text{mm}$ and $D = 45\,\text{mm}$. No phase pattern measurements can be shown because of the high sensitivity of the measurement equipment to tolerances in this frequency band.

Let the complex pattern be defined as

$$g(\theta) = v(\theta)e^{i\alpha(\theta)}. \tag{6.1}$$

In order to capture the amplitude perturbations in numerical quantities, the root mean square error

$$\epsilon_{\text{rms},v} = \sqrt{\frac{1}{\theta_{\min} - \theta_{\max}} \int_{\theta_{\min}}^{\theta_{\max}} \left| \frac{v_{x_s}}{v_{\text{ref}}} - 1 \right|^2 d\theta'} \tag{6.2}$$

is evaluated within an integration area of $\theta_{\min/\max} = \pm 20°$ and the result is shown in Fig. 6.7 for both simulation and measurement. As can be seen, the error decreases with increasing distance d_{edge}.

Reasons for these perturbations are the coherent addition of the radiation directly transmitted through the lens with contributions from inner total reflections at the edge faces and with spillover radiation. Naturally, these contributions are dependent on the azimuth feed pattern, the lateral position of the feed antenna with respect to the lens edge and on the focal distance F.

6.6 Beam Pattern Analysis

The above perturbations impose important guidelines on the design of the Rx ULA. For a single target case the so-called "beam pattern"

$$p(\theta) = \left| \mathbf{w}^H(\theta_b)\, \mathbf{a}(\theta) \right|^2 \tag{6.3}$$

can be computed, where $\mathbf{w}(\theta_b) = \mathbf{t} \odot \boldsymbol{\zeta}(\theta_b)$ is the coefficient vector with the desired amplitude taper vector $\mathbf{t}$ and a progressive phase vector $\boldsymbol{\zeta}(\theta_b)$ leading to a beam direction of θ_b. The vector $\mathbf{a}(\theta) = [1, e^{k_0 d}, \ldots, e^{(M-1)k_0 d}]$ is the steering vector

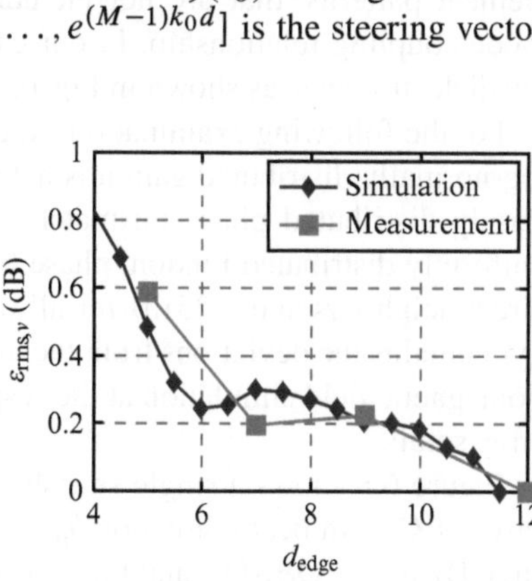

Fig. 6.7 Evaluation of azimuth pattern perturbations due to lens edge effects vs. distance of the feeding antenna from the lens edge $d_{\text{edge}} = (L/2 - \Delta x_s)/\lambda_0$

containing the geometry of the M-element ULA with spacing $d = d_{s,H}$ and the direction of arrival θ.

Next, let us extend this ideal array model to

$$\tilde{p}(\theta) = \left| \mathbf{w}^H(\theta_b)\, \tilde{\mathbf{a}}(\theta) \right|^2 \tag{6.4}$$

with the modified steering vector $\tilde{\mathbf{a}}(\theta) = \mathbf{G}(\theta)\, \mathbf{a}(\theta)$.

In a real array, $\mathbf{G} = \mathbf{G}_{\text{phys}}$ with

$$\mathbf{G}_{\text{phys}}(\theta) = g_{\text{Tx}}(\theta) \begin{pmatrix} g_{\text{Rx},1}(\theta) & & 0 \\ & \ddots & \\ 0 & & g_{\text{Rx},M}(\theta) \end{pmatrix} \tag{6.5}$$

containing the two-way pattern response of each array element, which is composed of the transmit antenna pattern $g_{\text{Tx}}(\theta)$ and the ith element pattern $g_{\text{Rx},i}(\theta)$ of the receiving array. Note that in the digital baseband signal only the modified beam pattern $\tilde{p}$ can be observed.

This diagonal matrix containing the physical two-way element patterns as well as mutual coupling effects can be modeled as follows:

$$\mathbf{G}_{\text{phys}}(\theta) \approx \bar{g}(\theta)\, \mathbf{Q}_{GP}\, \mathbf{Q}_C\, \mathbf{Q}_A(\theta). \tag{6.6}$$

Here, $\bar{g}(\theta)$ is the averaged two-way pattern response, $\mathbf{Q}_{GP}$ is a diagonal matrix attributing the gain and phase mismatches between different sensor channels, e.g. due to manufacturing tolerances, and $\mathbf{Q}_C$ is a full, but usually diagonally dominant matrix with ones on the main diagonal, describing the array coupling mechanisms assuming a single mode relationship [4]. Note that both $\mathbf{Q}_{GP}$ and $\mathbf{Q}_C$ are independent of the DOA. Now $\mathbf{Q}_A(\theta)$ is a DOA-dependent complex diagonal matrix that accounts for all remaining angle dependent differences between the observed element patterns, that are not included in $\mathbf{Q}_C$, i.e. that do not obey the single-mode coupling relationship. In our case, such additional differences occur due to the dielectric lens, as shown in Fig. 6.6.

For the following examinations, the matrix $\mathbf{Q}_{GP}$ is simulated using a zero-mean log-normally distributed gain mismatch with standard deviation of 1 dB and a uniformly distributed phase mismatch within $[-90^\circ\ 90^\circ]$. For the coupling matrix, uniformly distributed random phase between 0 and 2π, -20 dB coupling between direct neighbors and -25 dB for all other combinations was assumed. The matrix $\mathbf{Q}_A$ contains the deviations from the averaged two-way response computed by electromagnetic field simulation at the respective feed antenna position x_s given by the array setup.

Figure 6.8 shows a single snapshot of the beam pattern magnitude for different cases of $\mathbf{G}$ with beam direction $\theta_b = 10^\circ$ and a Chebychev taper for -80 dB SLL. Here $\mathbf{Q}_{GPC} = \mathbf{Q}_{GP}\mathbf{Q}_C$ and $\mathbf{I}_M$ denotes the identity matrix of size M. Lens parameters were $L = 23\lambda_0$, $F = 15$ mm and $D = 40$ mm. Receiving array parameters

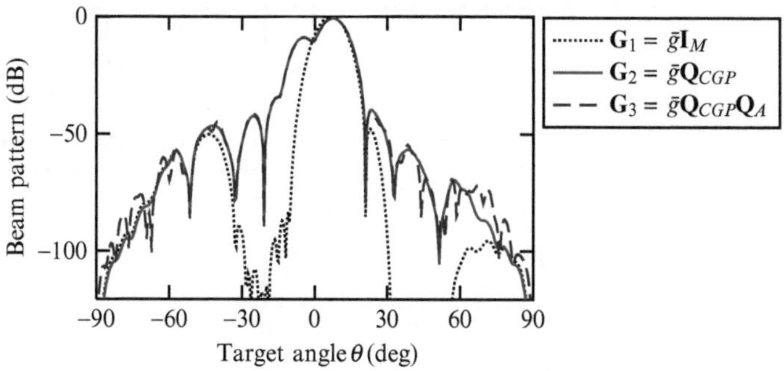

Fig. 6.8 Beam patterns without calibration

were $M = 8$, $d = \lambda_0$ and in the shown example the array center was positioned at $x_{s,a} = 4.5\lambda_0$. Parameters for the transmit array were $M_{\text{Tx}} = 6$, $d_{\text{Tx}} = 0.64\lambda_0$ and a Chebychev taper for -25 dB SLL. All patterns are normalized to the maximum value for the ideal array $\mathbf{G}_1$.

There exist several methods for array calibration [7]. In this paper a so called *global calibration* procedure is investigated that, tries to estimate the matrix $\hat{\mathbf{Q}}^{-1} = \mathbf{G}^{-1}$ from calibration measurements at angles θ^{cal}. Note that the angular region that needs to be covered is given by

$$\theta^{\text{cal}}_{\min/max} = \pm \arcsin\left(\frac{\pi}{k_0 d}\right). \tag{6.7}$$

For the simulations an angular step of $\Delta\theta^{\text{cal}} = 1°$ was used. Algorithm details can be found in [3]. In this context *global* means that the correction is applied independent of the a priori unknown DOA and therefore neglects angular dependencies of $\mathbf{G}$.

Figure 6.9 shows the effect of the calibration on the beam pattern with lens perturbations excluding ($\mathbf{G}_5$) and including ($\mathbf{G}_6$) random gain and phase deviations and mutual coupling. The ideal case ($\mathbf{G}_1$) as well as the uncalibrated beam pattern with lens perturbations ($\mathbf{G}_4$) are given as reference. The results show that the global calibration is very well able to compensate the extreme beam pattern deviations resulting due to $\mathbf{Q}_{GPC}$ from Fig. 6.8. But there always remains a residual error compared to the $\mathbf{G}_1$ case, because of the DOA-independent calibration model despite the DOA-dependent error $\mathbf{Q}_A$. Furthermore, there is almost no difference between the $\mathbf{G}_5$ and the $\mathbf{G}_6$ case, meaning that $\mathbf{Q}_A$ rather than $\mathbf{Q}_{GPC}$ is the dominant reason for the residual error. However, there is still a slight improvement in SLL attenuation compared to the uncalibrated $\mathbf{G}_4$ case.

In order to further examine the limits in SLL attenuation imposed by the lens effects $\mathbf{Q}_A$, a parameter study was conducted. The beam angle θ_b was varied between $[-10° \, 10°]$ in $1°$ steps and the digital Chebychev taper of the receiving array was set to values $[-30, -35, \ldots, -80]$. Moreover, the array position was laterally

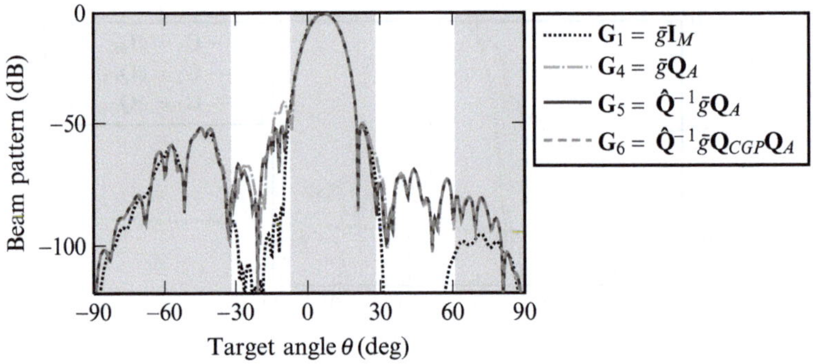

Fig. 6.9 Beam patterns with lens effects with and without calibration

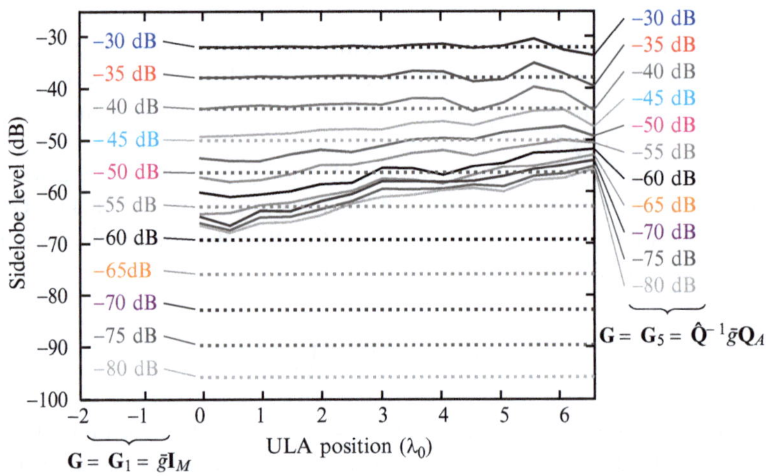

Fig. 6.10 Sidelobe levels in evaluation region of beam pattern

shifted from the center position $x_{s,a} = 0$ to $X_{s,a} = 6.5\lambda_0$. Figure 6.10 shows the sidelobel level evaluated outside the grating-lobe and main-lobe regions (see gray regions in Fig. 6.9), and averaged over the beam angles. Results for $\mathbf{G}_1$ are naturally independent of $x_{s,a}$ and contain both the Chebychev taper, the element patterns and the transmit pattern. However, there is a limit of approx. −65 dB for the center position and of approx. −55 dB at $x_{s,a} = 6.5\lambda_0$, due to the lens perturbations.

6.7 Conclusions

We presented a novel antenna concept for a 77 GHz LRR sensor relying on a cylindrical dielectric lens for elevation beam shaping. The lens antenna configuration has higher gain and a considerable lower SLL in the elevation plane compared to conventional planar antenna with longer patch columns. Furthermore, the concept has the potential for lower sensor costs due to reduced space consumption on the RF substrate. Almost no frequency scanning was observed and the reduced sensitivity to substrate losses allows for the utilization of inexpensive substrate materials while limiting performance degradations. However, the lens causes perturbations in the azimuth pattern, limiting the sidelobe attenuation in the DBF beam pattern. Since this limit cannot be overcome by calibration a good DBF performance relies on careful design of the overall lens antenna system, especially on the length of the lens and the position of the receiving array.

Acknowledgements This project is supported by the Ministry of Education and Research (BMBF) under contract nr. 16SV2181 (KRAFAS). The authors would like to thank all partners of the KRAFAS consortium, especially Robert Bosch GmbH and the University of Bremen for the support during the measurements.

References

1. J. Hilsebecher, G. Kühnle, H. Olbrich, Long-Range-Radar-Sensor für Fahrerassistenz-Systeme. www.elektroniknet.de (2004)
2. H. Iizuka, K. Sakakibara, T. Watanabe, K. Sato, K. Nishikawa, Millimeter-wave microstrip array antenna with high efficiency for automotive radar systems. R&D Rev. Toyota CRDL **37**(2), 7–12 (2002)
3. A. Kortke, Analyse und Kalibration von linearen Microstrip-Patch-Antennenarrays. Dissertation, Technische Universität Berlin (2006)
4. L. Kühnke, Realisierung und Kalibrierung aktiver Antennensysteme mit digitaler Strahlformung. Dissertation, Universität Hannover (2001)
5. M. Schneider, Automotive radar – status and trends. in *GeMiC*, (2005), pp. 144–147
6. M. Schneider,KRAFAS – Innovationen in der Mikrosystemtechnik und der Hochfrequenz-Mikroelektronik für kostenoptimierte Radarsensoren im Automotive-Bereich. in *VDE Kongress*, vol. 1. Aachen, 2006, pp. 275–282
7. M. Schoor, B. Yang, Local and global calibration for high-resolution DOA estimation in automotive radars. in *Proceedings of the 5th IEEE Sensor Array and Multichannel Signal Processing Workshop SAM 2008*, (2008), pp. 68–72
8. C. See, Sensor array calibration in the presence of mutual coupling and unknown sensor gains and phases. Electron. Lett. **30**(5), 373–374 (1994)
9. S. Tokoro, K. Kuroda, A. Kawakubo, K. Fujita, H. Fujinami, Electronically scanned millimeter-wave radar for pre-crash safety and adaptive cruise control system. in *Intelligent Vehicles Symposium*, (2003), pp. 304–309
10. P. Wenig, M. Schoor, O. Günther, B. Yang, R. Weigel, System design of a 77 GHz automotive radar sensor with superresolution DOA estimation. in *International Symposium on Signals, Systems and Electronics ISSSE '07*, (2007), pp. 537–540
11. P. Wenig, R. Weigel, M. Schneider, A dielectric lens antenna for digital beamforming and superresolution DOA estimation in 77 GHz automotive radars. in *Proceedings International ITG Workshop on Smart Antennas WSA 2008*, (2008) pp. 184–189

Chapter 7
High Precision Distance Measurement for Pedestrian Protection Using Cooperative Sensors

C. Morhart and E. Biebl

7.1 Introduction

For pedestrian protection in urban traffic scenarios precise localization in combination with reliable identification is needed. Both requirements are ideally met with cooperative sensor technology. Each road user gets equipped with an active microwave sensor combining communication and localization services. The following article describes a high precision distance measurement system, enabling car drivers to detect visually hidden pedestrians by exchanging signal data. A Round Trip Time of Flight measurement principle was implemented using bi-phase coded pulse compression. Signal compression is realized by correlation of pseudo random codes assuring secure time of arrival detection and clear burst identification. The SNR improvement of this method is utilized by spatial interpolation to get a highly precise distance measurement. The system is intended to address a large number of communication partners within each measurement cycle. This multi-user ability is achieved by ordering sensor transmit times in a Time Division Multiple Access scheme. The system performance was evaluated by use of a prototype system at 2.4 GHz that was able to achieve an accuracy of centimeters at a range of 450 m.

7.2 System Model

For distance measurement between two sensors, several measurement principles are common [13]. A very simple approach is to detect the received signal strength and calculate the distance out of the free space loss. However, a precise distance

C. Morhart (✉)
Technische Universität München, Fachgebiet Höchstfrequenztechnik, 80290 Munich, Germany
e-mail: morhart@tum.de

E. Biebl
Technische Universität München, Fachgebiet Höchstfrequenztechnik, 80290 Munich, Germany
e-mail: biebl@tum.de

S. Lindenmeier and R. Weigel (eds.), *Electromagnetics and Network Theory and their Microwave Technology Applications*, DOI 10.1007/978-3-642-18375-1_7,

measurement in a real traffic scenario using this method is not possible due to unpredictable attenuation effects. Another possibility is the Time of Arrival principle. Here, distance is measured by the signal propagation time between two sensors. This method depends on the accuracy of the system clocks and needs therefore appropriate synchronization algorithms. This is very difficult to apply in a multi-user environment with many car and pedestrian systems. Hence, a Round Trip Time of Flight system was implemented.

7.2.1 Round Trip Time of Flight

The Round Trip Time of Flight principle resembles the classical radar principle. A signal transmitted by the car sensor is received at the pedestrian side and sent back after a finite waiting time T_w. By knowing this waiting time the car can compute the distance Δs out of the time T_p passed since the start of transmission.

$$\Delta s = \frac{\Delta T}{2} c_0 = \frac{T_p - T_w}{2} c_0 \tag{7.1}$$

The advantage of this method is that no absolute time synchronization is needed but only a relative one to determine the exact delay time. By using short transmit signals and waiting times, the stability of a standard crystal oscillator is sufficient [8]. The variation of the waiting time T_w allows a differentiation between pedestrian sensors and therefore a Time Division Multiple Access.

7.2.2 Time Division Multiple Access

One requirement for a cooperative localization system in an automotive environment is the multi-user ability. It should be possible to address a huge number of pedestrians within each measurement cycle. One solution to achieve this is to order car and pedestrian signals in time slots (Fig. 7.1).

The car system initiates a measurement cycle by sending a coded data burst with length T_s in time slot T_1. At the same time all pedestrian systems are in receiving mode and listen in slot R_1. Following, a waiting time T_w is added to exclude reflections on scattering objects. In the next time slots T_{11} to T_{1n} consecutively all pedestrian systems answer to the car systems in a fixed multiple of the waiting time T_w. In that way the car system can subtract each individual waiting time as a multiple of $n\mathrm{T}_w$. To avoid the influence of clock errors the car system has to refresh the data burst after a time $\mathrm{T}_{\text{clock}}$. This time has to be chosen in accordance with the crystal precision and the maximum needed accuracy. By refreshing the data burst one can avoid that a longer waiting time causes a loss of noticeable distance precision.

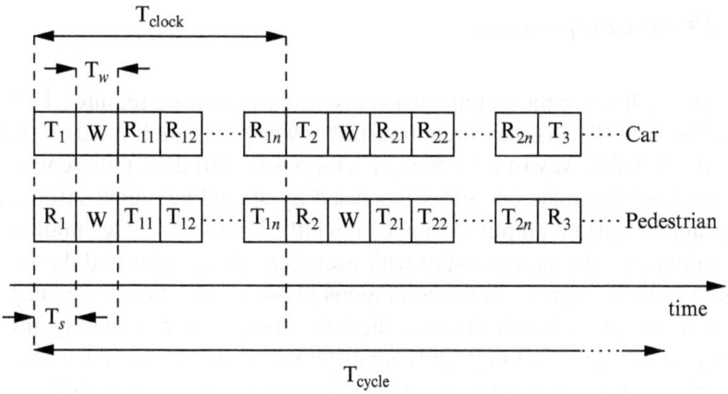

Fig. 7.1 Time slot order per transmission cycle

7.2.3 Channel Model

A linear time-variant multipath channel model is used. In dependency of the surroundings there are K multipaths having different time delay $T_k \in \mathbb{R}_0^+$ and different path attenuation $h_k(t) \in \mathbb{C}$. Therefore, each transmission produces K replicas of the same signal. Accordingly, the channel input response is defined by

$$h(t) = \sum_{k=0}^{K} h_k(T_k)\delta(t - T_k). \tag{7.2}$$

The complex channel input response factor $h_k(t)$ consists of a combination of attenuation loss $\alpha_k(t) \in \mathbb{R}$ and phase variation due to Doppler shift $f_{dk}(t)t$ and path delay $f_c t$:

$$h_k(t) = \alpha_k(t)\exp(j2\pi f_c t + j2\pi f_{dk}(t)t). \tag{7.3}$$

Dependent on the Doppler shift through moving objects in the transmission channel there is a Doppler frequency spread of the transmission signal. A coherence time of the channel can be defined in relation to this Doppler spread [10]. Regarding the speed of road users in traffic applications, the measurement time is much smaller than the coherence time. So, Doppler shift can be neglected.

Another major concern for distance measurement is the channel resolution Δs. This resolution is independent on the measurement principle directly connected with the signal bandwidth B [4].

$$\Delta s = \frac{c_0}{2B} \tag{7.4}$$

c_0 is the speed of light. Hence, multipaths can only be separated at a distance difference greater than Δs.

7.2.4 Pulse Compression

High precision distance measurement and insensitivity to noise require a high Signal to Noise Ratio (SNR). As maximum transmit power levels are limited by frequency regulation, a feasible way of increasing this ratio is to introduce pulse compression. By this method the pulse energy is increased by spreading signal information in time domain at constant amplitude and bandwidth. The two basic techniques are the phase modulated pulse compression with pseudo random codes and the frequency modulated pulse compression by continuous phase modulation for example with chirp waveforms [1]. The advantage of the pulse compression is an easy implementation in a multi-user system by applying different identification codes. In that way it is possible to distinguish between different car sensors at a higher SNR. Figure 7.2 shows an example for this method. The pulse $p(t)$ with time duration T_p is enlarged to T_c by a pseudo random code $c(t)$. The energy and average power of the pulse increase with the same ratio, called the compression factor L_c:

$$L_c = \frac{T_c}{T_p}. \tag{7.5}$$

The pulse shape can be regained by applying a compression filter. In the case of phase modulated codes this filter corresponds to the matched filter or correlation receiver of $c(t)$. Its output SNR is directly proportional to the compression factor L_c.

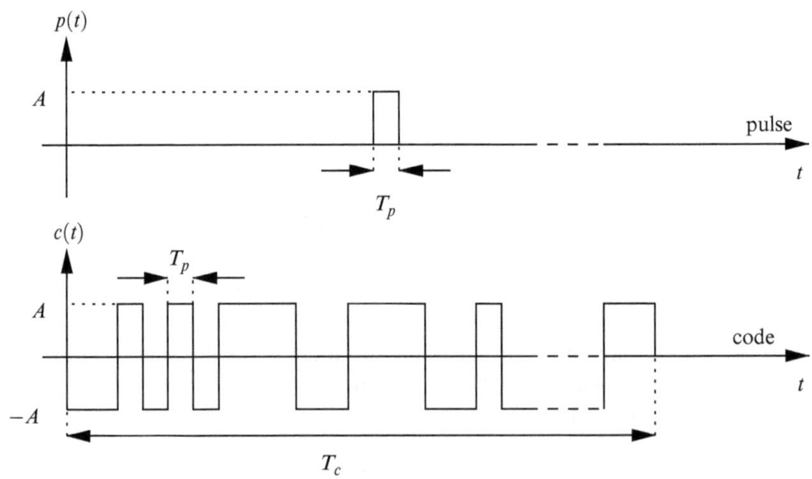

Fig. 7.2 Pulse spreading in time domain with a pseudo random code

7.2.5 Correlation Receiver

A distance measurement is comparable to channel estimation with limited bandwidth. The goal is to find out the fastest path as it is most likely the shortest distance. By implementing pulse compression, the channel impulse response and therefore the path information can be obtained by sending a coded impulse and by correlating the received channel response. In the case of the assumed channel model (7.2) a pulse compression with a correlation receiver can be described by Fig. 7.3.
At first, the following analysis is restrained to the assumptions of:

- No cross-correlation
- Side lobe free autocorrelation
- No channel fading

These points will be generalized in the Sects. 7.2.6 and 7.2.7. In this case the resulting signal $y(t)$ is described by

$$y(t) = c(t) * h(t) * \frac{1}{\varepsilon} c^*(-t) + n(t) * \frac{1}{\varepsilon} c^*(-t). \tag{7.6}$$

$\varepsilon = AT_p\sqrt{L_c}$ is a scaling factor necessary to maintain consistent units. It will cancel down in the final result.

$$\begin{aligned} y_c(t) &= \frac{1}{\varepsilon} \int_{-\infty}^{\infty} C(f) C^*(f) H(f)\, \mathrm{e}^{j2\pi f t}\, \mathrm{d}f \\ &= \frac{1}{\varepsilon} \int_{-\infty}^{\infty} |C(f)|^2 \sum_{k=1}^{K} h_k\, \mathrm{e}^{j2\pi f(t-T_k)}\, \mathrm{d}f. \end{aligned} \tag{7.7}$$

By the exclusion of fading the maximum result signals are obtained at time points T_k.

$$y_{ck}(T_k) = \frac{h_k}{\varepsilon} \int_{-\infty}^{\infty} |C(f)|^2 \mathrm{d}f = \frac{1}{\varepsilon} h_k \cdot E_c \tag{7.8}$$

E_c is the energy of the code, h_k the short form of $h_k(T_k)$. These points are equivalent to the correlation maxima of the K different transmission paths. In dependence of the channel attenuation $|h_k|$ the SNR of the kth path can be expressed by:

$$P_{sk} = |y_{ck}(T_k)|^2 = \frac{1}{\varepsilon^2} |h_k|^2 \cdot E_c^2 \tag{7.9}$$

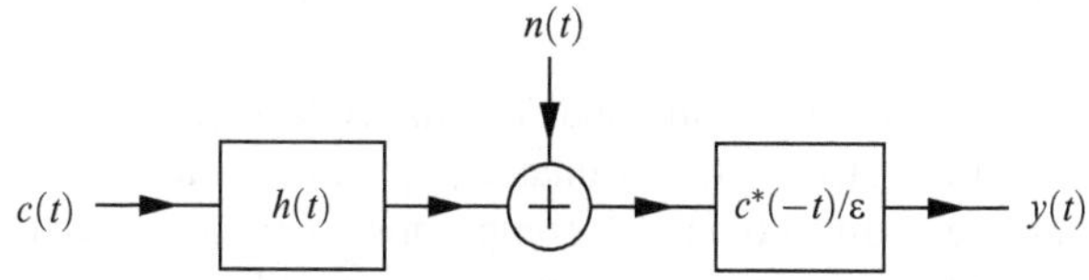

Fig. 7.3 Transmission model for a correlation receiver

$\frac{N_0 E_c}{(2\varepsilon)}$ is the receiver noise power, E_p the pulse energy. This result is equivalent to the optimum SNR of a matched filter. It is dependent on path attenuation, compression factor, pulse energy and noise power density. Following, the influence of non-idealities is included into this model.

7.2.6 Correlation Codes

It is theoretically shown [12] that there is no group of correlation codes offering perfect auto- and cross-correlation characteristics. Therefore, it is only possible to implement codes offering either optimum cross-correlation or optimum autocorrelation behavior.

Cross-correlation properties are important for the discrimination of appropriate measurement signals and co-channel interference. This interference can be caused by off-system communication services or non-synchronized distance measurements of other car systems having different codes. Depending on the received power their correlation result can overlay wanted information and disable distance measurement. This is a known issue in CDMA systems and called "near-far problem" [11]. Solution approaches exist by dynamic transmit power adjustment and by implementing multi-user receivers incorporating all possible transmit codes, but this would greatly increase digital sensor complexity. Also, the proposed system uses very short transmit signals occupying the channel only for a short period of time. By realizing an intelligent transmit protocol like "listen before talk" time overlap can be minimized. For that reason, it is more important to optimize the autocorrelation characteristics of transmit codes.

The disadvantage with correlation side lobes is their misinterpretation as ghost targets. This property limits the sensor sensitivity for detecting multipath transmission by the ratio between main and side lobe power called side lobe suppression ratio. Depending on the correlation technique there are different code families offering minimum side lobes. For standard aperiodic correlation only Barker Codes allow side lobe suppression [6]. Unfortunately, the number and the length of those codes are tightly limited to $L_c \leq 13$. This would restrain the performance of the pulse compression.

A better alternative is to use a periodic or cyclic correlation technique in combination with m-sequences. The length of those codes is unlimited and available as a multiple of 2. Their correlation output is two-valued with a maximum of N and constant side lobes -1.

$$\tilde{\varphi}[n] \in [-1, N] \qquad \text{with} \quad N = 2^r - 1, \quad r \in \mathbb{N}^+ \tag{7.10}$$

$\tilde{\varphi}[n]$ is the periodic correlation result and N is the code length. The respective side lobe suppression ratio is also N. Unfortunately, a pure periodic correlation is hard to realize as it is infinite in time domain. A compromise is to use a three times repeated overlapping standard correlation φ_{3xx} like it is shown in Fig. 7.4.

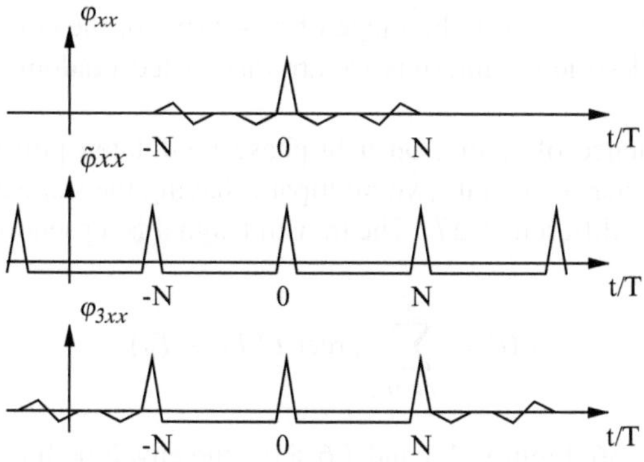

Fig. 7.4 Comparison between an aperiodic, periodic und three times repeated autocorrelation

This technique requires three times the time or processing complexity of an aperiodic correlation φ_{xx}, but offers an area of perfect side lobe suppression like a periodic correlation.

7.2.7 Channel Fading

Traffic scenarios exhibit relatively slow changing transmission channels. This is due to the fact that car speed is low compared to the measurement time of the proposed system. Nevertheless, there exist a lot of multipaths because of reflection and diffraction on buildings, road surface, other cars, etc. Therefore, location-dependent multipath fading occurs.

Fading in pulsed transmission appears if the multipath time difference is smaller than the pulse-width ($\varDelta T \leq T_p$). Depending on the phase difference between the signals, they combine additively or destructively. Regarding a channel with all multipaths having time difference much smaller than the pulse-width, the multipath channel input response converts to:

$$h(t) = h_{\text{eff}}(T_{\text{eff}})\delta(t - T_{\text{eff}}). \tag{7.11}$$

Multipaths become inseparable and channel parameters combine to an effective channel factor $h_{\text{eff}}(T_{\text{eff}})$.

$$h_{\text{eff}}(T_{\text{eff}}) \approx \sum_{k=0}^{K} \alpha_k(T_k)\, \mathrm{e}^{j2\pi f_c T_k + j2\pi f_{dk}(T_k)T_k} \tag{7.12}$$

In general, the attenuation factor $\alpha_k(t)$ is dependent on the physical surroundings of the channel and therefore changing slowly. In contrast, the phase is rotating by 360°

at a time factor of $\frac{1}{f_c}$, lying in the range of $\approx$ 400 ps. In the conclusion, points of constructive and destructive interference are distributed randomly in transmission space.

Next, the influence of fading on a bi-phase modulated pulse compression is examined. The scenario contains two multipaths having the same attenuation factor and a variable time difference ΔT. The transmit signal is a pseudo random code:

$$c(t) = \sum_{n=0}^{N} c_n \operatorname{rect}(t/T_p - T_n). \tag{7.13}$$

with length $N = 256$. Figures 7.5 and 7.6 show the envelope function of constructive and destructive interference. Apparently, the region of fading is limited by the pulse-width T_p. Furthermore, dependent on the signal shape and the time difference the influence of fading is changing. Thus, there are two possibilities to reduce the effect of fading. On the one hand, by minimizing pulse-width and hence maximizing bandwidth, the active region of fading is shortened. On the other hand, by applying antenna diversity in a selection combiner or maximum ratio combiner the remaining influence can be diminished [10].

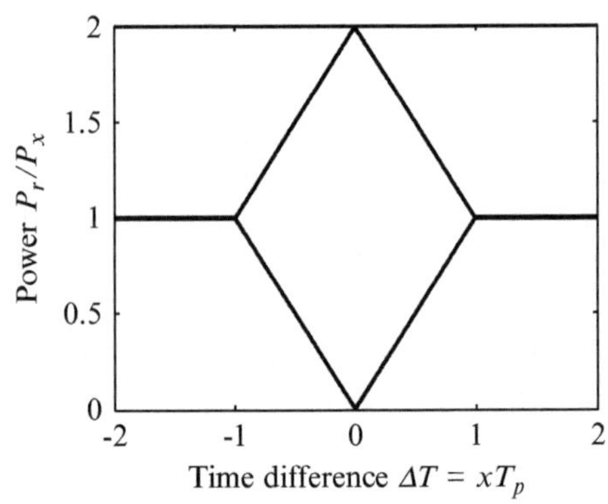

Fig. 7.5 Fading of two-path interference with a 256 Bit long m-sequence

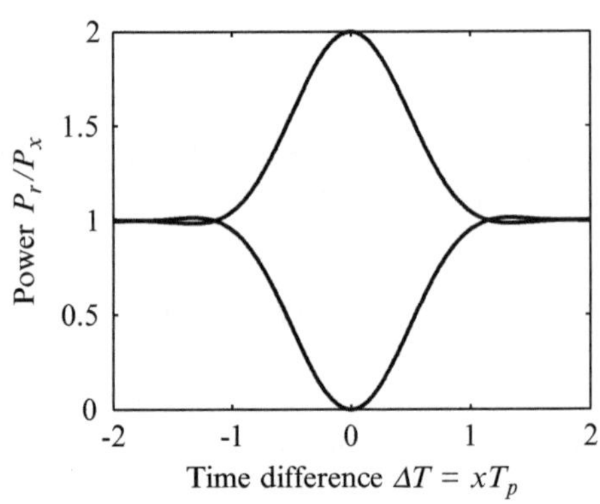

Fig. 7.6 Fading of two-path interference with a 256 Bit long, lowpass filtered m-sequence

7.2.8 Distance Accuracy

For the evaluation of the measurement accuracy an ideal distance estimator is assumed. With pulse compression the time of arrival is defined by the peak of the correlation result. Assuming a continuous signal this maximum can be found out by the zero crossing of the first derivative. Therefore, an optimum estimator consists of a correlation receiver, a differentiator and a zero crossing detector. For such a system the minimum received mean square (rms) error for estimating the time of arrival is defined by [2]:

$$\sigma_t = \frac{1}{\beta\sqrt{S/N}}. \tag{7.14}$$

S/N is the signal to noise ratio of the correlator output and β the rms bandwidth of the signal, which can be expressed as the normalized second moment of signal energy spectrum:

$$\beta^2 = \frac{\int_{-\infty}^{\infty} |\frac{\partial}{\partial t} c(t)|^2 \mathrm{d}t}{\int_{-\infty}^{\infty} |c(t)|^2 \mathrm{d}t} = \frac{\int_{-\infty}^{\infty} (2\pi f)^2 |C(f)|^2 \mathrm{d}f}{\int_{-\infty}^{\infty} |C(f)|^2 \mathrm{d}f}. \tag{7.15}$$

This result is reasonable in two ways. On the one side the accuracy is dependent on the second derivative of the maximum. The higher the gradient of the first derivative the easier the zero crossing is obtained. This property corresponds to the pulse shape and therefore to the pulse bandwidth. On the other side noise is leading to zero point deviations and accordingly to distance errors. Important is also that bandwidth is directly related to the precision whereas only the square root of the SNR is involved. Thus, it is easier to improve measurement accuracy by signal bandwidth than by SNR.

Regarding a numerical example for an optimum 2.4 GHz system, a uniform signal spectrum $C(f) = 1$ is assumed over the bandwidth of 80 MHz. Following, a minimum SNR of 46 dB is required to achieve a distance rms error of $\sigma_s = 1$ cm. Under more realistic conditions concerning modulation, bandwidth efficiency, non-ideal detection, even higher SNR values are needed. With respect to this result, a SNR improvement technique like pulse compression is very valuable to achieve a high accuracy measurement result with limited bandwidth.

7.3 System Implementation

The preceding section covered the basic properties of an unidirectional transmission. Following, these considerations are extended to a bidirectional distance measurement and implemented in a digital and analog prototype system.

7.3.1 Transmission Scheme

In Sect. 7.2.2 we characterized the Round Trip Time of Flight measurement. The distance is computed by the round trip time minus a fixed waiting time. To define this time as precisely as possible two transmission schemes concerning the properties of the pedestrian transponder were examined. The first one resembles the classical radar principle where the mobile transponder acts as an active back-scatterer, reflecting the car signal after a finite waiting time T_w. In the second model the transponder is realized identical to the car sensor. It analyzes the car signal, estimates the time of arrival by the fastest path, defines the waiting time and sends back an own code signal. Either implementation has its advantages. The first model is the easier one to realize and also needs less hardware resources. Figure 7.7 shows a diagram of this bidirectional communication scheme. A signal $x(t)$ is transmitted by the car sensor and detected at the pedestrian after traveling through the channel $h(t)$. If a valid car signal is detected, it is delayed by the waiting time T_w and amplified by $-G$ to compensate channel loss. The inversion of the signal is easy to implement and allows discrimination between up- and downlink. The resulting signal is sent back to the car sensor enabling the round trip time measurement. The benefit of this approach is relative simple and cheap transponder hardware. On the other hand this scheme has disadvantages in strong multipath scenarios. Combinations of different transmission paths in up- and downlink can cause distance errors. Furthermore, transponder amplification $|G|$ is adjusted to the strongest path and leading to SNR degradation for other multipaths. This is especially bad if the strongest path is not the fastest one.

The realization of the second model results in two separated transmissions for up- and downlink. Each sensor defines the time of arrival itself by the fastest path. Hence, the SNR of the bidirectional system is comparable to the unidirectional case (7.9). There's also no problem with multipath combinations as each transmission signal is generated newly on the sensor. Unfortunately, the overall system accuracy is limited by the accuracy of the worst sensor. Therefore both sensors have to be designed identically. This results in a higher hardware and software complexity and consequently a more expensive transponder design than in the first model.

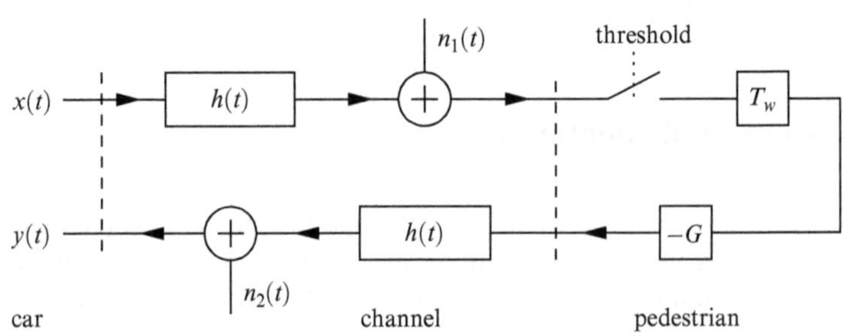

Fig. 7.7 Transmission scheme with reflector model

7.3.2 Correlation Filter

The core element of the proposed distance measurement system is the correlation algorithm for the pulse compression. To get an immediate result for triggering other communication services the correlation has to be carried out in real time and needs therefore a complex digital hardware like a FPGA or DSP.

In detail, the correlation is computed in digital form as

$$\varphi[n] = \sum_{N} c[n]s[n+N]. \tag{7.16}$$

$\varphi[n]$ is the correlation result, $c[n]$ the pseudo random code and $s[n]$ the received input signal. This equation can be carried out as a matrix vector product or implemented as a FIR-Filter in time or frequency domain. To fulfill the real time requirement a transposed FIR structure was chosen (Fig. 7.8). With the parallelization of the required additions and multiplications one gets a correlation result after each clock cycle and therefore the correlation peak is known immediately after the reception of the code signal. Another advantage of this method is the insensitivity to analog digital converter (ADC) clock jitter. Measurements of few ADC samples have a high standard deviation because of this jitter. By correlating over a long code sequence the influence of this error is minimized.

7.3.3 Interpolation

The correlation result offers a high quality sensor response which is almost free of jitter and has a SNR improvement in the length of the code sequence. Unfortunately, the pure correlation has relatively coarse distance accuracy as it is limited by the ADC sampling rate. As stated in (7.14) the system accuracy is not dependent on the ADC rate, but on the bandwidth and signal to noise ratio. Therefore, to increase the distance precision two signal interpolation techniques are used. Polynomial interpolation can be easily processed but has an interpolation error depending on the polynomial order and the flatness of the interpolated function [5]. A Shannon interpolation by upsampling the correlation data needs bandlimited signals but

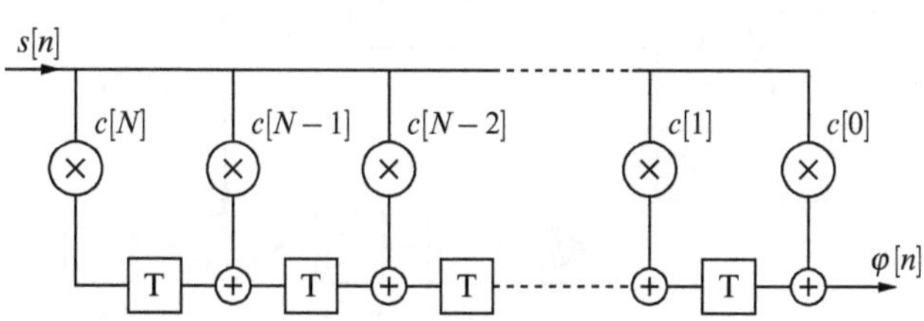

Fig. 7.8 Block diagram of correlation filter FIR structure

offers an ideal reconstruction at high computational effort. To combine the advantages, both methods are used. The first stage consists of a factor 10 upsampling filter flattening the correlation curve at reasonable computation cost. The second stage is a polynomial filter carrying out a second order Newton interpolation. The peak can then be found by the zero crossing of the first derivative of the polynomial function. This scheme is pictured in Fig. 7.9.
To process only signal data close to the correlation maximum, a threshold detector controls the input to the interpolation chain.

By applying these interpolation techniques the error of the polynomial interpolation can be neglected. Consequently, the result is an analytical expression of the time of arrival whose accuracy is only dependent on the signal parameters – SNR and bandwidth.

7.3.4 Prototype System

To demonstrate the functionality of the proposed measurement system a prototype at 2.4 GHz was built. This frequency is suboptimal in terms of detecting hidden persons [3] as lower frequencies have better diffraction characteristics. On the other side the ISM band at 2.4 GHz is the lowest frequency band available with sufficient bandwidth.

The combination of high bandwidth and maximum linearity could not be achieved with conventional transceiver modules. Therefore a proprietary analog sensor architecture was designed (Fig. 7.10). Both sensors, on the car and on the pedestrian side, are implemented symmetrically to check out different transmission

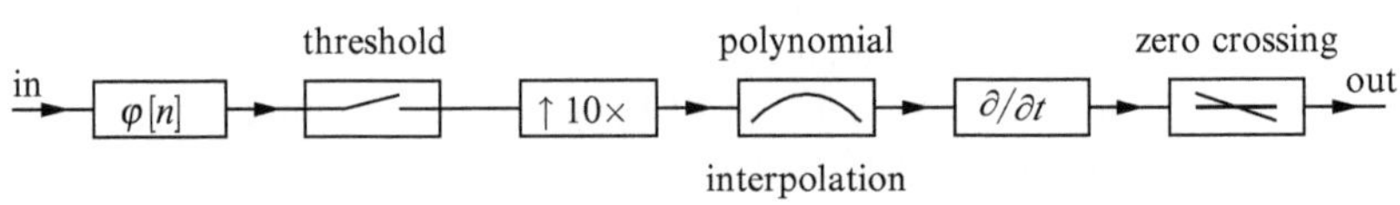

Fig. 7.9 Interpolation chain for correlation maximum estimation

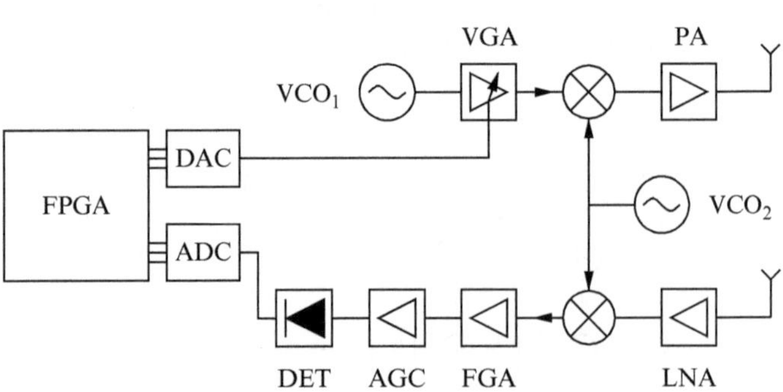

Fig. 7.10 Schematic capture of sensor architecture

schemes. Conventional amplitude modulation was used to get the simplest transmitter and receiver design. The analog modulation on the transmitter side is generated with a Variable Gain Amplifier (VGA). The amplification factor is controlled by the output of the Digital to Analog Converter (DAC). In that way it was possible to generate a very broadband linear modulation. The output signal is then mixed by a local oscillator into the 2.4 GHz ISM Band. The frequencies of the two oscillators (VCO_1, VCO_2) must be chosen in such a way that $f_{RF} = f_{VCO_1} + f_{VCO_2}$. To avoid crosstalk between transmitter and receiver both elements can be switched off individually.

An asynchronous AM demodulation was implemented using a full-wave envelope detector. This has two requirements on the IF stage, namely a tight filtering and an Automatic Gain Control (AGC). The filtering is necessary as the envelope detector is not frequency selective and therefore interference of out-of-channel communication services like digital television or GSM has to be suppressed. Furthermore, for the optimal performance of the demodulation the envelope demodulator has to be driven in its linear range. Thus, it is essential to use adaptive pre-amplification in the AGC amplifier. Fixed gain amplification was chosen in accordance to the desired distance range so that system dynamic from −80 dBm to −30 dBm at a maximum control time of 1 μs was achieved.

Digital signal processing was implemented on a FPGA. By maximum parallelization of correlation and interpolation algorithm, it was possible to carry out the correlation in real time at a clock rate of 125 MHz. For the interface to the analog front end an Analog to Digital Converter (ADC) and a DAC likewise with a clock rate of 125 MHz were used. The DAC and ADC rate has to be chosen in accordance to the signal bandwidth to fulfill the sampling theorem. This requirement was met by choosing bit duration of 16 ns. The overall code length of one correlation burst was 256 bits resulting in a SNR gain of ≈24 dB. In combination with the AGC control time minimum signal duration of 5 μs was obtained.

7.4 Verification

The accuracy of an optimum measurement system according to (7.14) is only limited by the SNR and signal bandwidth. In contrast, in a real system non-idealities like clock deviations or digital jitter have also to be regarded. Thus, the accuracy of the prototype implementation was checked in several measurement scenarios.

7.4.1 Influence of Clock Error

The properties of the system clock are a very important issue in a measurement system. In this context, especially phase noise and frequency drift have to be considered [7]. Phase noise or equivalently clock jitter indicate the rapid random fluctuations of the signal phase. This short term stability of the oscillator influences the quality of sampled ADC data for example. Frequency drift or wander is the arbitrarily oscillator offset from its nominal frequency. This attribute is dependent on the ambient temperature, supply voltage variations, manufacturing and aging of crystals for example. Accordingly, it is a long term characteristic that is changing slowly with the measurement process.

In the following the influence of these clock errors on the prototype is checked. On both sensors standard crystal oscillators were used which are non-synchronized to each other. For maximum SNR both systems were connected directly by cables and the waiting time of the pedestrian transponder was varied by steps of 1 μs from 7 μs to 1.1 ms. For each waiting time the appropriate error was calculated by the difference of the measurement result and the correct distance. This experiment was repeated 10,000 times to get a sufficient database for a statistical interpretation of the mean error μ_s and the standard deviation σ_s in dependence of the waiting time T_w. Figure 7.11 shows the results for the mean error and Fig. 7.12 for the error standard deviation. It can be seen that both parameters are increasing with growing waiting time. Furthermore, the absolute error caused by the oscillator frequency offset is much larger than the measurement uncertainty. In comparison, at a waiting time of 1 ms one gets an absolute distance error of 0.5 m and a standard deviation of 0.025 m. Nevertheless, there are several possibilities to overcome this error. First of all, as short transmit times of 5 μs are used, waiting times in the region up to 100 μs lead to a neglectable distance offset smaller 5 cm. In case of needing optimal accuracy or greater waiting times, signal refreshment (Sect. 7.2.1) or the employment of

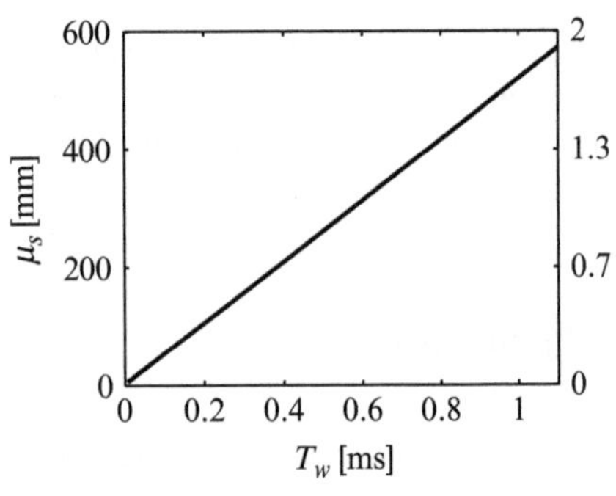

Fig. 7.11 Distance mean error μ_s dependent on the waiting time T_w

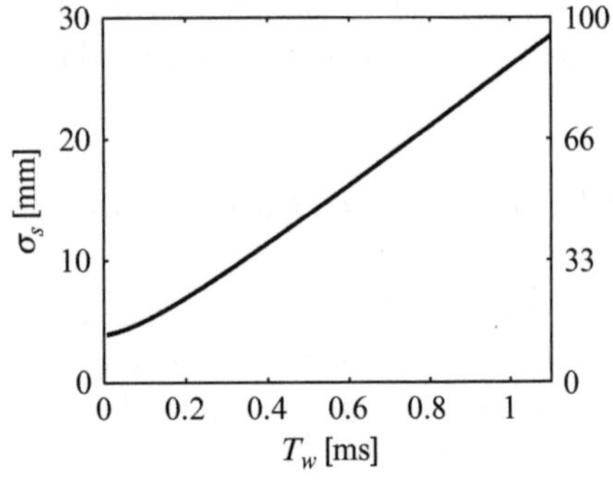

Fig. 7.12 Distance error standard deviation σ_s dependent on the waiting time T_w

calibration techniques are necessary. For example, calibrating on a fixed pedestrian position or transmitting frequency offset information are possible.

7.4.2 *Test-Scenarios*

For the evaluation of the reflector type transmission scheme extensive measurements were carried out with the prototype system [9]. To suppress the influence of clock errors a minimal waiting time T_w in combination with a position calibration was used. The prototype was specifically tested for its behavior in Line of Sight (LOS) and non-LOS scenarios. According to the considerations from Sect. 7.3.1, results showed that LOS measurements are comparable to one path transmissions. Thus, the measurement accuracy is solely dependent on the SNR. In a chosen distance range from 2 to 75 m, an error standard deviation of $\sigma_s = 4.7$ cm was achieved.

For non-LOS scenarios, especially those with pedestrian hidden between parked cars were selected. The results showed a degrading performance because of the influence of multipath transmission. On the one side the characteristic behavior of the reflector mode tends to deterministic errors by favoring the strongest transmission path. On the other side multipaths lead also to fading and SNR decrease causing a stronger measurement uncertainty. Combining these distance results with intelligent tracking technology, it was still possible to achieve a standard deviation of 30 cm. As stated before, the implementation of the alternate transmission scheme can improve this behavior. That model favors not the strongest but the fastest transmission path, but is not tested in practical application yet.

7.5 Conclusion

Cooperative sensor technology is a suitable way to protect visually hidden pedestrians. It combines sensor requirements of classification and localization of vulnerable road users. The proposed system is a highly accurate distance measurement sensor using a Round Trip Time of Flight principle. The developed algorithm allows fast target acquisition with high precision even in multi-user environment. By applying pulse compression in combination with interpolation the system sensitivity in LOS scenarios is only limited by the signal to noise ratio which can be enhanced by the compression factor. Furthermore, this system concept is also adaptable to the non-LOS case which can not be addressed by state of the art lidar or radar systems. First prototype implementation showed a high accuracy in the dimension of centimeters by a range up to 450 m in the LOS case. Even in non-LOS, the accuracy of the system was still sufficient to introduce autonomous braking in the focused car application. Ongoing research is concentrated on the implementation of hardware and software with higher performance detecting hidden pedestrians at a decreased requirement of space and power consumption.

References

1. D.K. Barton, *Modern Radar System Analysis* (Artech House, Inc, Norwood, 1988)
2. D.K. Barton, H.R. Ward, *Handbook of Radar Measurement* (Artech House, Inc, Dedham, MA, 1984)
3. A. Fackelmeier, C. Morhart, E. Biebl, Evaluation of diffraction effects for identifying hidden targets. in *GeMIC 2008 ITG-Fachbericht Band 206* (2008)
4. H. Klausing, W. Holpp, *Radar mir realer und synthetischer Apertur* (Oldenburg Verlag, Germany, 2000)
5. H. Kronmüller, *Digitale Signalverarbeitung* (Springer, Berlin, 1991)
6. H.D. Lüke, *Korrelationssignale* (Springer, Berlin, 1992)
7. H.H. Meinke, F.W. Gundlach, Taschenbuch der Hochfrequenztechnik, vol. 3, 5th edn. (Springer, Berlin, 1992)
8. C. Morhart, E. Biebl, Ein kooperatives, code-basiertes Abstandsmesssystem für eine große Anzahl simultaner Nutzer. Frequenz – J. RF-Eng. Telecomm. **62**(7–8), 175–179 (2008)
9. C. Morhart, E. Biebl, D. Schwarz, R. Rasshofer, Cooperative multi-user detection and localization for pedestrian protection. in *GeMIC 2009 ITG-Fachbericht Band 213* (2009)
10. J.G. Proakis, M. Salehi, *Communication Systems Engineering*, 2nd edn. (Pearson Education Limited, New York, 2002)
11. J.G. Proakis, M. Salehi, *Digital Communications*, 5th edn. (McGraw-Hill Education, Hightstown, NJ, 2008)
12. R. Scholtz, L. Welch, Group characters: Sequences with good correlation properties. Inf. Theory IEEE Trans. **24**(5), 537–545 (1978)
13. M. Vossiek, L. Wiebking, P. Gulden, J. Wieghardt, C. Hoffmann, P. Heide, Wireless local positioning. Microw. Mag. IEEE **4**(4), 77–86 (2003)

Chapter 8
A High-Precision Wideband Local Positioning System at 24 GHz

Stefan Lindenmeier, Christian Meier, Anestis Terzis, and Joachim Brose

8.1 Introduction

In an increasing range of indoor applications a precise localization of objects is required. The growing number of applications – for example object tracking or automatic gauging in industrial environment as well as safety facilities needs robust and highly accurate positioning systems. During the past years, several research efforts have been reported, achieving accuracies in the range of inches [1, 2]. These were mostly FMCW-based systems, detecting the transponder mounted at the object by solving the FFT and calculating the hyperbolisation of at least three receivers and one reference station. These state of the art systems were optimized for tracking objects over long distances with a high measurement repetition rate. For industrial applications in a highly reflective scenario demanding an accuracy of a few millimetres however a new concept is required. Thus a radio based positioning concept was introduced in [3] which can offer the required precision. In [4] a first 3D-demonstrator of this DSSS positioning system has been described. The achieved uncertainty of only around 0.1 mm in a resting position within a coverage volume of more than 2 m × 2 m × 2 m fulfilled the required accuracy. This was reached by a wideband spread spectrum concept in combination with a sophisticated high speed digital signal and data processing. The parameters of the radio system are listed in Table 8.1.

The transponders are mounted onto the object to be tracked. As in industrial use these transponders often get close to metal parts and the signals are highly affected by multipath propagation effects. Therefore, a high accuracy positioning is hard to get and a new improved signal processing method is necessary in order to suppress such multipath propagation effects. Several methods were investigated and verified by a software simulation with a specially designed system simulator [4]. Furthermore the recent algorithm [4] is combined with a new optimum method to suppress

S. Lindenmeier (✉), C. Meier, and J. Brose
Universität der Bundeswehr München, München, Germany

A. Terzis
Daimler AG, Konzernforschung/Group Research, Ulm, Germany

S. Lindenmeier and R. Weigel (eds.), *Electromagnetics and Network Theory and their Microwave Technology Applications*, DOI 10.1007/978-3-642-18375-1_8,

Table 8.1 System parameters of the local positioning system

Parameter	Value
Carrier frequency	24 GHz
Bandwidth	3.2 GHz
Modulation technique	BPSK
Chip rate $1/T_C$	1.6 Gchip/s
Output power	0 dBm
Coverage	10 m
Transmission technique	DSSS
Update rate	10 Hz
Accuracy	0.1...1 mm[a]
Resolution	≈10 mm[a]

[a]Values change in dependence to reflectivity scenario

multipath propagation effects. So the 3D-transponder can be detected with recent reproducibility, accuracy and improved robustness against multipath effects.

This contribution is organized as follows: Initially, we present the scenario the 3D-positioning system should be applied to. Then, we describe the system architecture and the hardware implementation, especially the PN code generator and the antennas. Finally, we introduce the post processor making use of the signal's PDF and show how the enhanced accuracy of the new algorithm can be successfully applied in a distance measurement scenario.

8.2 State of the Art

Actually, delay time measurements for positioning are typically carried out using short RF pulses, frequency modulated continuous waves or pseudo noise coded continues waves. For the described application a number of additional conditions have to be fulfilled besides basic accurate delay measurement, as:

- There are several transmitters the signals of which have to be separated
- The system has to cope with an outstanding multipath situation
- The LOS path length might be in between centimetres and several 10 m

Since the first days of Radar, monostatic and bistatic delay measurements of RF pulses (or "bursts") are successfully used for search and tracking of (moving) reflectors. Differing from these Radar applications now the transmitter and not some reflector has to be detected. So, a trigger link between the moving transmitter and the fixed receiver is necessary. For accurate positioning the pulse has to show mainly a very short rise time compared to the delay time. The rise time at the receiver is however depending on the carrier frequency and on the very complex transient behaviour of the whole system, including the propagation path. For acceptable resolution the pulse further on should be short and from the viewpoint of unambiguity the time interval between the pulses should be big enough to make sure, that all multipath signals have faded away. This means, that the RF energy, available for the measurement is inevitable small. For multi transmitter detection frequency division has to be used where the frequency slots have to show sufficient offsets in order to avoid crosstalk with the broadband spectra of the necessarily sharp RF pulses [5].

With a frequency modulated continues wave (FMCW) system the transmitted signal is (usually linearly) frequency modulated yielding in a frequency difference which depends on the signal delay between transmitter and receiver. Again a triggering link between transmitter and receiver is necessary. Accuracy and resolution depend mainly on the frequency deviation while the unambiguity depends on the repetition time. Counters and FFT are standard techniques for the determination of the frequency difference [6]. The signal energy for the measurement is available nearly continuously. The separate detection of different transmitters however needs either alternate use (time division) or far different frequencies (frequency division).

The pseudo-noise coded continuous waves (PNCCW) use a fixed carrier with usually extremely broadband modulation. Such systems are well known from earth navigation (GPS and Galileo, e.g.). The PN code acts as key for the delay detection. This is usually achieved evaluating the cross correlation peak of transmitted and received signal [7]. Further on, an additional carrier phase measurement allows ever higher accuracy. The signal energy is available during the whole measurement. The resolution depends on the length of a code chip and the unambiguity on the code length. Again, a trigger link is necessary. Multi transmitter detection is easily achievable by orthogonal codes. On the other hand, high mathematical effort is required for correlation, PN-code generation and for the filtering.

8.3 Selection of the Most Suitable Technique

The objective is to evaluate the exact position of the transponders mounted on a marker in a highly reflective area. This combination of transponder and marker is further called radio pen (pointer). The realization of this device is described in [8]. The reflective area might be for example the factory floor like in the scenario of Fig 8.1 which also represents the conditions for the radio positioning system.

In this indoor positioning scenario and considering [3, 4] the radio pen transmits the PN-coded signal. The receivers on the wall and ceiling get this signal with a time delay depending on the distance from the radio pen. So, with the time of arrival (TOA) measurement the distance between the receiver and the transmitter can be calculated. With at least four undisturbed receiver signals, one of them being the time reference the 3D position can be calculated. In Fig 8.1 there are also sketched some NLOS signals (dotted bright lines) which disturb the received LOS signal. Our goal now is to find ways within the limitations given by the system architecture to avoid the deterioration of the localization measurement caused by this disturbed reception.

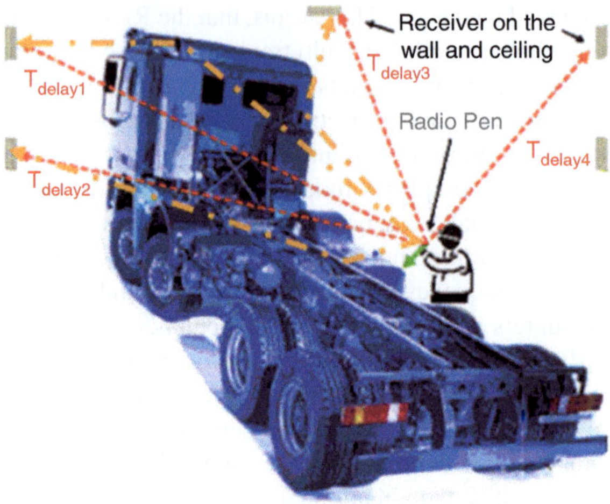

Fig. 8.1 Location and Multipath

8.4 High Precision Radio Location System

8.4.1 System Architecture

Figure 8.2 shows a simplified block diagram of the finally built-up 3D localization system. It consists of the transmit chain with the antenna to be localized and the receive chain with at least 3 antennas.

The 24 GHz transmitter output is BPSK modulated with the PN-signal from the code generator, amplified and connected to the active microwave pen pointer via a thin and highly flexible cable. The cable attenuation implies a further amplification at the antenna which is mounted on the pen pointer. The extreme wideband spread spectrum pseudo noise fast m-sequence code $c(t)$ with a code length of $N_C = 2047$ chips occupies a bandwidth of 1.6 GHz.

The localization of the transmit antenna is achieved with three ore more coherent receivers connected to optimum arranged antennas. All receivers are completely identical from input to their digital output. For the signal despreaded the delayed PN modulated local oscillator (LO) signal is multiplied with the received signals in the RF, IQ-demodulated, low pass filtered and then digitized for the following computational estimation of the path delays in the digital signal processor (DSP).

The LO frequency f_{c2} shows a frequency offset f_{IF} against the transmit frequency f_{c1}. This offset frequency can be easily obtained by mixing and low pass filtering f_{c1} and f_{c2}. f_{IF} is then multiplied with the received, despreaded and down converted signals in the IQ demodulators. The IF phase is identical to the RF phase and can also be used for distance measurements but with remarkably higher accuracy. The PN series at the receiver side is the same fast m-sequence code like with the transmitter but now with variable (RF) time delay τ So, the LO PN code can be

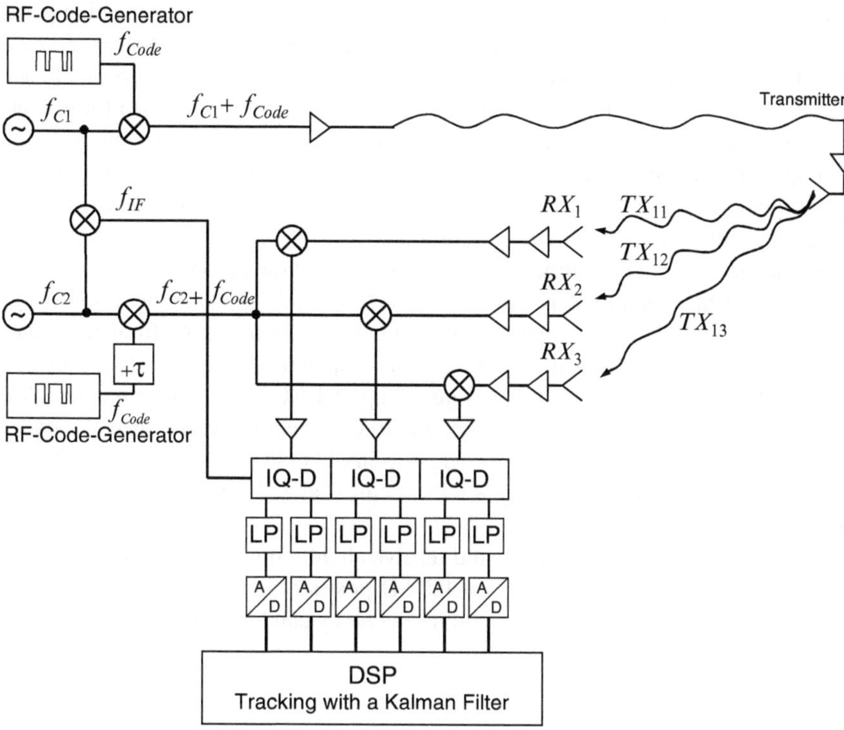

Fig. 8.2 Basic architecture of the 3D positioning system

written as $c(t - \tau)$. A further (digital) delay is possible with the post processing in the DSP. The combination of digital delay and RF-delay enables the shifting of the code $c(t - \tau)$ in steps of a third of the chip duration. The path delay Δt between transmitter and each receiver is best found with PN modulated signals with the help of the correlation between the receive signal and the electronically delayed transmit signal. In other words: the maximum of the correlation function between the electronically delayed code $c(t - \tau)$ and the path delayed received code $c(t - \Delta t)$ has to be found for each receiver. The corresponding distance, including the RF connection lines is $d = \tau \cdot c_0$. As long as the derived distance d is shorter than the unambiguity range $d_{\max} = T_{\text{code}} \cdot c_0 = N_C \cdot T_C \cdot c_0$ (where T_{code} is the code duration, N_C again the code length in chips and T_C the chip duration), then there exists only a unique solution for the distance measurement. With our system, using a code duration of $T_{\text{code}} = 1.28\,\mu\text{s}$ we obtain an unambiguity range of $d_{\max} = 384\,\text{m}$.

Especially for indoor application multipath effects are a limiting factor for acceptable measurement accuracy. For long-path deviations of more than $d_{\max}$ the multi reflected signal usually can be neglected due to the considerable free space attenuation and the limited reflection properties of the obstacles. The remaining short-path deviation waves show smaller peaks than the direct wave and occur at a later time in the correlation function and therefore can be suppressed taking the sidelobe-to-peak-ratio into account. As long as the deviation of the reflection paths towards the line-of-sight (LOS) path exceeds the chip resolution $d_{\mathrm{MP}} = T_{\mathrm{C}} \cdot c_0$, then the LOS waves are clearly distinguishable from the reflected waves and hence the latter can be ignored. So, not only for accurate localization but also for acceptable multipath suppression there exists the demand of a superior chip resolution d_{MP} leading to the high code bandwidth of $B = 1/T_{\mathrm{C}} = 1.6\,\mathrm{GHz}$. With the progressive signal processing used in our system the chip resolution is further improved by a factor of 2. So, our system's chip resolution of $d_{\mathrm{MP}} = T_{\mathrm{C}} \cdot c_0/2 = 9.3\,\mathrm{cm}$ allows for the separation of all multipath signals above 9.3 cm deviation.

The represented architecture's great advantage is the despread of the received signal with the help of the PN modulated LO signal directly in the RF front end rather than in the IF. So, the code rate can be chosen independently from IF frequency and bandwidth. With actual FPGAs code rates of up to 1.6 Gchip/s can be handled. We realized the code generator using a high end FPGA together with a special parallel serial converter. Further work is in progress to implement also the whole DSP- part of the receiver into this FPGA. Fig. 8.3 shows the realized concept as introduced in Fig. 8.2.

The synchronization of transmit and receive chain with a 10 MHz clock reduces the number of necessary receivers for the 3D-positioning to three. With a greater number of receivers, however, the system's accuracy can be further increased.

8.4.2 PN-Code Generator

As interesting parts of the system the 1.6 Gchip/s PN code generator and the various antennas are discussed in more detail.

The PN codes in the DSSS positioning system have to be different for each transmitter to allow channel separation. The codes must further follow specific

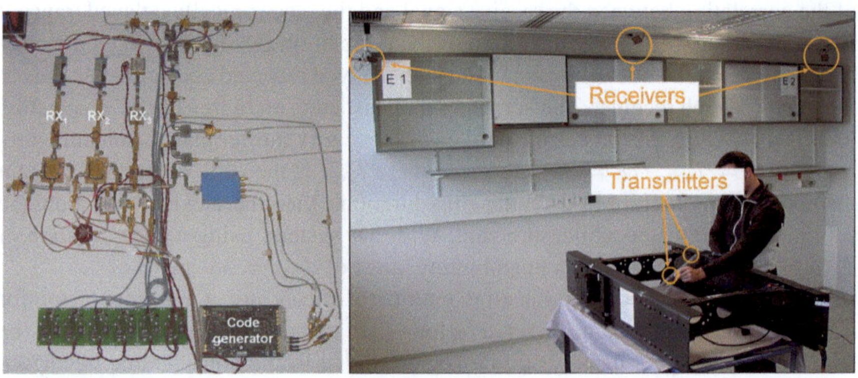

Fig. 8.3 Hardware prototype of the 3D measurement system (*left*: RF-circuit, *right*: System installation)

constraints: The autocorrelation peak must be much greater than the autocorrelation side lobes and cross correlation peaks in order to lock on the peak of the autocorrelation function.

There exist many different classes of PN-codes that are well suited for DSSS based radio location, among others the "maximal length sequences" (m-sequences). The m-sequences are cyclical with a period of $N_c = 2^L - 1$, where L denotes the order of the m-sequence code. This code has one more logical "one" than logical "zeroes" in a full sequence period. The DC component of the frequency spectrum is determined by the zero to one balance of the sequence. The logical zero is represented by a -1 in our implemented version (see Fig. 8.4). The autocorrelation peak of the discrete m-sequence codes is N_c and the autocorrelation function is unique. Figure 8.4 shows a part of the high speed m-sequence code.

The accompanying base band frequency spectrum of the m-sequence PN-Code is shown in Fig. 8.5.

The high speed PN-code generator is realized using a high end FPGA (field programmable gate array) including in-chip special parallel serial converter. The generator is implemented based on a linear feedback shift register. The basic block diagram is shown in Fig. 8.6.

The binary weighted modulo-2 addition of the taps is fed back to the input and the fed back weight coefficients g_i for any tap is either one, meaning fed back, or zero, meaning that it is not connected. The hardware implementation of the taps is performed with Flip-Flops and the modulo-2 addition with exclusive OR gates. In state of the art FPGAs elements can be clocked with some hundreds of MHz. For

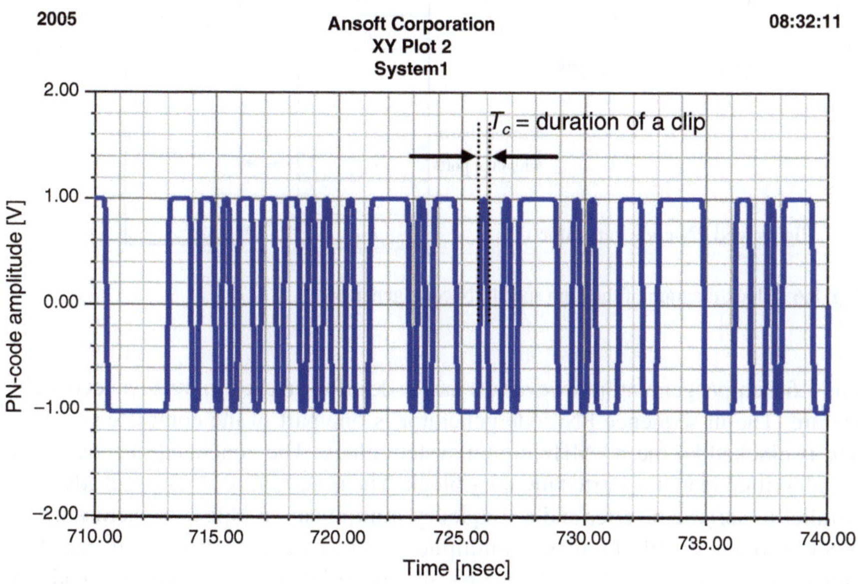

Fig. 8.4 Part of the high speed m-sequence PN-Code

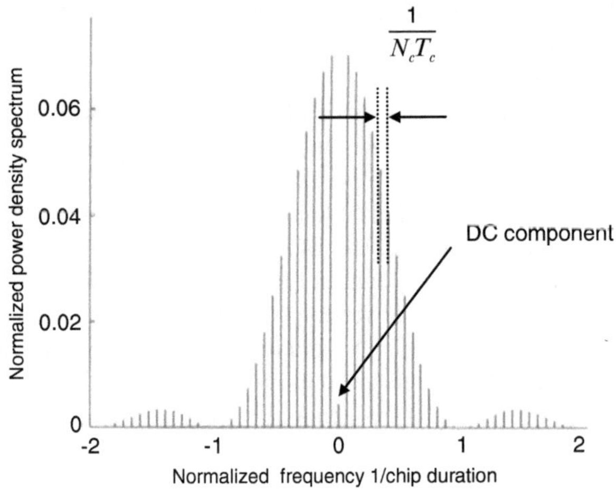

Fig. 8.5 Frequency spectrum of the high speed m-sequence PN-Code

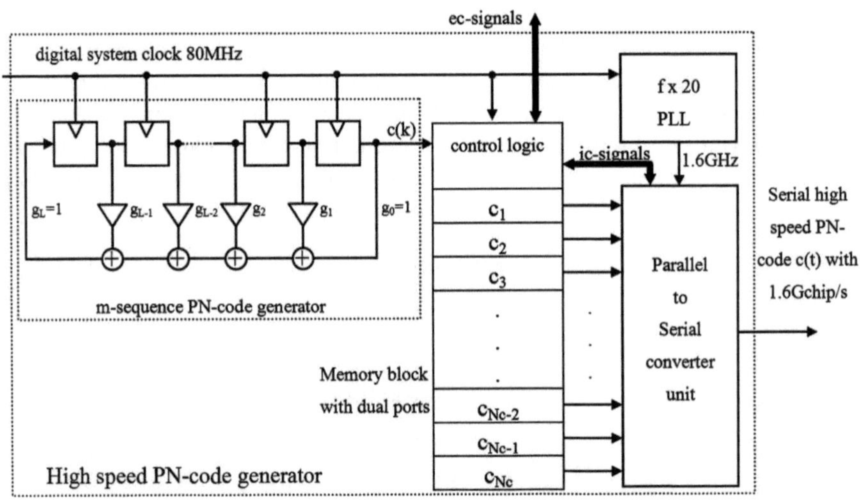

Fig. 8.6 Block diagram of the high speed PN-code generator

this 1.6 Gchip/s generator however we had to develop an even more advanced architecture. The m-sequence PN-code generator is one part of the complete generator and is clocked by the 80 MHz digital system clock. The generated m-sequence is serial written in a dual ported memory block. The used FPGA includes an advanced parallel to serial converter and a frequency multiplication block for increasing the clock frequency. This frequency multiplier produces a new PLL stabilized clock signal of 1.6 GHz, which clocks the parallel to serial converter unit that reads the PN-code out of the memory block in a parallel method. The resulting output is consequently a serial high speed PN-code with a chip rate of the desired 1.6 Gchip/s

$N_C = 2047$ chips. There is an internal control interface between the control logic of the memory block and the parallel to serial converter unit, which is used to synchronize the read process. Its configuration can be modified, using the external control signals (ec-signals). The control signals include the configuration of the PN-code length and the timing for the read process. The developed architecture is completely implemented in a single FPGA and enables a chip rate of up to 3.2 Gchip/s for future extensions of the system.

8.4.3 Transmit and Receive Antennas

The antenna patterns for reception and transmitting should be adapted to their application: hemispherical for the flexible pen and suitable directional for the fixed receive antenna. We used ruggedized and lightweight square micro strip patch antenna with hybrid feed for the pen pointer and micro strip arrays with single feed CP patches as directional antennas for reception.

Figure 8.7 shows the construction of the test implementation with 90°-Hybrid at the rear side and the measured pattern with a gain of about 3.5 dB above dipole. These antennas additionally contain power amplifiers.

Figure 8.8 shows two of the developed antennas for different half-power beam widths (HPBW) with gains of 20 and 24 dB, respectively.

8.4.4 Post Processing

To compensate for signal delay caused e.g. by the RF transmission lines, the system needs to be calibrated with a fixed transmitter-receiver distance r_{cal}. With the

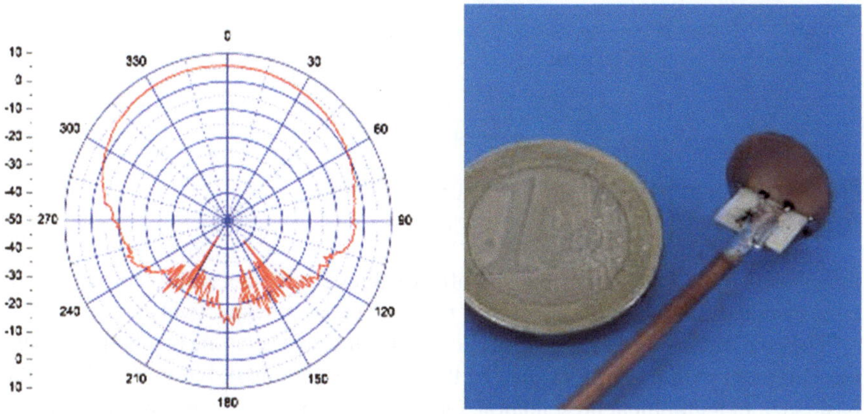

Fig. 8.7 Antenna construction and measured pattern

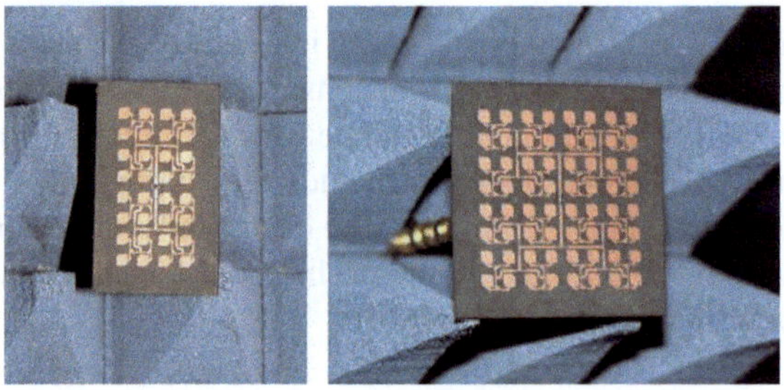

Fig. 8.8 Directional antennas for 10° (*right*) and 23° (*left*) HPBW

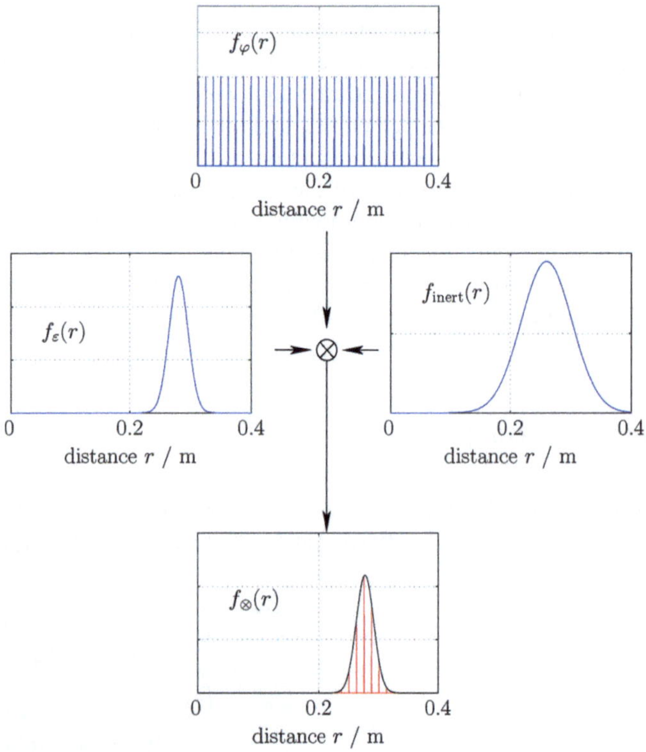

Fig. 8.9 Graphical presentation of the post processing pattern

indicated distance r and phase φ the offset values for the necessary corrections of edge distance and phase are $r_{\mathrm{off}} = r - r_{\mathrm{cal}}$ and $\varphi_{\mathrm{off}} = \varphi$, respectively.

The goal of the post processing is to combine the most accurate result of the phase measurement with the unambiguity of the edge measurement. Figure 8.9 shows the basic idea in a graphical presentation. On top is the ambiguous periodic probability density function (PDF) of the distance measurement using the phase.

The plot on the left shows the PDF from the edge detection. If both are multiplied to the joined PDF, then the most plausible peak is scored higher. We found, that in case of two rather similar values of two related spikes the maximum detection works definitely better if a third PDF is introduced based on the mean values from the previous distance estimate (right curve).

8.5 Simulation, Evaluation and Measurements

In order to test and to optimize the microwave PN-coded positioning system it was completely simulated in MatLab. The simulators consist of the functional blocks: transmitter, propagation channel and receiver. The transmitter contains the code generator, the oscillator, the amplifiers and the filters. Each of these components has its own adaptable subset of parameters such as noise value, jitter and frequency response. Each propagation channel is modeled assuming a single LOS signal, several unwanted NLOS (no line of sight) signals from obstacles and additional white Gaussian noise (AWGN). Suitable modifications of the channel parameters allow handling highly reflective scenarios as well as scenarios without or with only negligible reflections. For the receiver simulation not only superheterodyn arrangements were investigated but also other architectures, like direct down conversion.

The task of the digital filter is to select an appropriate estimate for position and orientation of the transmitter-equipped pen pointer. This kind of (result-) filtering is done in two stages: At first the distances from each receiver to the transmit antennas are estimated from (ambiguous) carrier phase measurements and from delay measurements of corresponding chip edges (ambiguous up to d_{max}). From the distance estimates then in a second stage the transmitter's 3D position is found by geometrical trilateration.

The DSP offers two different measurement signals from the processing of each receiver output: the correlation at the different delay steps and the phase measurement output from the down converted RF. Both are additionally disturbed by noise, multipath and interference. While the correlation peak is wide (variance of 10 mm) in terms of distance resolution, mainly because of the finite chip length, the phase measurement offers an accuracy of close to only 0.1 mm as shown in the probability density plots of Fig. 8.10.

If the LOS path between transmitter and receiver is obstructed or if the signal is interfered, no (useable) measurement data are available and during the next steps even no previous estimates. So, the PDF's variance is increased. We took this in consideration using an adaptive variance with the Kalman filter. The result of tracking after a drop-out is shown in Fig. 8.11. While the edge distance gives only a rough image of the real trace, the overall distance the estimate including phase information yields to precise tracking. After a signal loss introduced between the 300th and the 600th measurement, the estimated distance is tracked precisely again after a gap of only about 50 measurements.

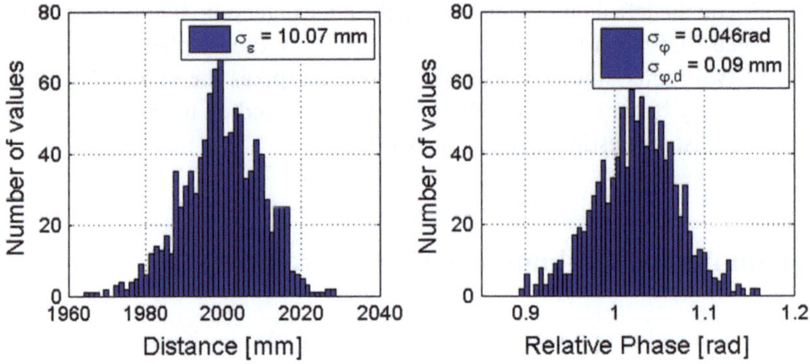

Fig. 8.10 Histograms of edge distance and phase measurements

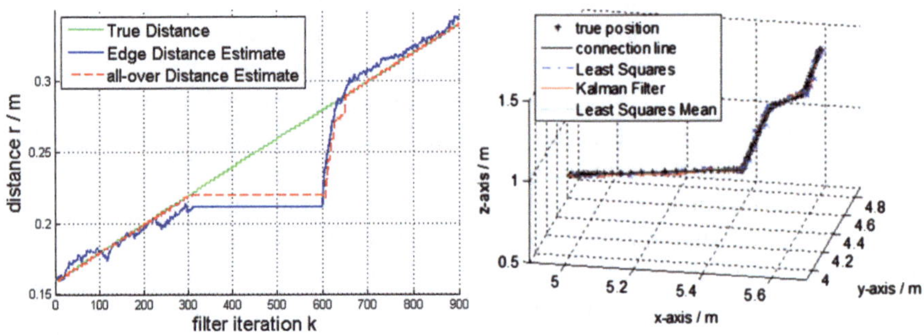

Fig. 8.11 Distance estimation when signal gets lost (*left*) and 3-dimensional tracking (*right*)

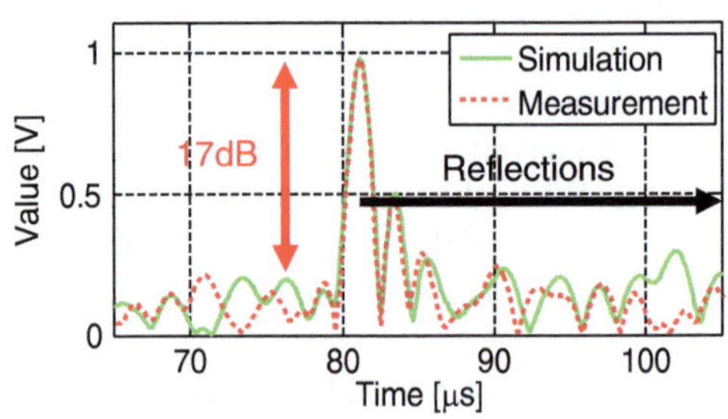

Fig. 8.12 Correlation peak: Simulation vs. Measurement

Another result is shown in Fig. 8.12 where the magnitude of the correlation function over the code phase of the simulator output is plotted. It can be seen that simulation and measurement in a reflective environment show good agreement even though noise will always cause differences in the results.

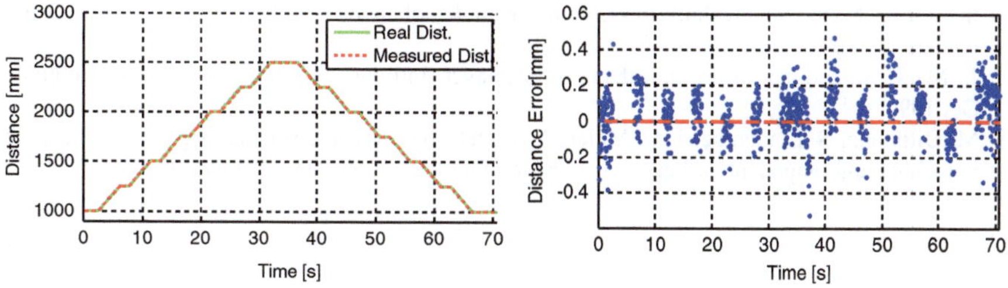

Fig. 8.13 Achieved accuracy for the distance estimation

8.6 Results

The plots at Fig. 8.13 show the results of distance measurements with moving transmitter position and the algorithm presented in this contribution. As a result, the position can be determinate with a deviation of 0.1 mm.

8.7 Conclusion

A high repetition code generator with a chip rate of up to several Gchip/s has been investigated for its feasibility in a DSSS positioning system as well as for radar applications. Measurements show a high stability in the code generation and an achievable accuracy in the millimetre range. The high resolution which is based on a fast chip rate enables good object separation while the fast long spreading codes yield to a high insensitiveness against interferers as well as to a wide unambiguousness range.

References

1. A. Stelzer, K. Pourvoyeur, A. Fischer, Concept and application of LPM – a novel 3-D local position measurement system. IEEE Trans. Microw. Theory Techn. IEEE-MTT **52**(12), 2664–2669 (2004)
2. L. Wiebking, *Entwicklung eines zentimetergenauen mehrdimensionalen Nahbereichs-Navigationssystems* (VDI, München, 2003)
3. C. Meier, A. Terzis, S. Lindenmeier, A high precision wideband local positioning system at 24GHz. in *Microwave Symposium Digest, 2006. IEEE MTT-S International*, June 2006, pp. 1580–1583
4. C. Meier, A. Terzis, S. Lindenmeier, A robust 3D high precision radio location system. in *Microwave Symposium Digest, 2007, IEEE MTT-S International*, 3–8 June 2007, pp, 397–400
5. J. Sachs, P. Peyerl, R. Zetik, Stimulation of UWB sensors: pulse or maximum sequence? International Workshop on UWB Systems, Oulu, Finland, June 2003
6. A. Stelzer, Aufbau eines Mikrowellenabstandsmeßsystems mit Submillimeter Genauigkeit unter Verwendung direkter Frequenzmessung und Six-Port Phasenbestimmung. Dissertationsschrift,

Institut für Nachrichtentechnik/Informationstechnik, Johannes Keppler Universität Linz, Österreich, Feb 2000
7. J. Meel, Spread spectrum introduction. DeNayer Intituut, Sint-Katelijne-Waver, Belgium, Oct 1999
8. M. Dittmann, S. Lindenmeier, C. Meier, A. Terzis, System zur Lokalisierung und Vermessung von Fahrzeugkomponenten. Patent Application DE102006059 804 A1, July 2007

Chapter 9
Monitoring of Electrochemical Processes in Catalysts by Microwave Methods

Gerhard Fischerauer, Andreas Gollwitzer, Alexander Nerowski, Matthias Spörl, Sebastian Reiß, and Ralf Moos

9.1 Introduction

Modern vehicles driven by a gasoline engine make use of the three-way catalyst (TWC) to eliminate noxious gas components like CO, HC or NO_x from the exhaust gas. Such a catalyst consists of a dielectric honeycomb-like matrix coated with an oxygen-storing component such as ceria (an n-type semiconductor) and possibly a catalytically active noble metal such as platinum. The efficient removal of noxious gases calls for a normalized air-to-fuel ratio λ of one. This cannot be achieved in practice, but by running the engine alternately in lean mode (more air than needed for complete combustion; $\lambda > 1$) and in rich mode (more fuel than needed for complete combustion; $\lambda < 1$) at least the time average $\overline{\lambda(t)}$ of λ can be made to take on the desired value of one. Under proper working conditions, the amplitude of the oscillations in $\lambda(t)$ around $\overline{\lambda(t)} = 1$ is much smaller downstream of the catalyst than upstream.

The oxygen storage capacity of the catalyst an important state variable as it is a measure of its conversion efficiency. It is commonly inferred indirectly from the λ oscillations upstream and downstream, which are monitored by lambda probes. This indirect approach is disadvantageous and inaccurate because the engine can only be switched from lean to rich operation or vice versa after the breakthrough of noxious gas components has been detected. The desire to avoid such breakthroughs explains the interest in direct measurement methods for the TWC state.

G. Fischerauer (✉)
Bayreuth Engine Research Center (BERC), Faculty of Engineering Science,
University of Bayreuth, Bayreuth, Germany
e-mail: mrt@uni-bayreuth.de

A. Gollwitzer
University of Applied Sciences Furtwangen

A. Nerowski
Technical University of Dresden

M. Spörl
Bayreuth

S. Lindenmeier and R. Weigel (eds.), *Electromagnetics and Network Theory and their Microwave Technology Applications*, DOI 10.1007/978-3-642-18375-1_9,

A straightforward direct measurement approach is to insert sensors into the catalyst. Suitable sensors are interdigital capacitors with coatings identical to that of the catalyst. Their impedance mirrors the oxygen loading of the coating, hence of the catalyst, as demonstrated for lean NO_x traps in [1]. However, this method incurs the drawback that the sensors must be mounted and contacted inside the catalyst.

Our work therefore aims at a cable-less solution to the in-situ catalyst state observation problem. It is based on the perturbation of a microwave cavity resonator as a function of TWC oxygen loading and resembles the technique widely used for determining the dielectric properties of materials at microwave frequencies [2,3]. We do, however, not consider the perturbation of a previously empty cavity by a small specimen of the material under test but rather the perturbation of a catalyst-filled cavity by material parameter changes in the catalyst [4–6].

9.2 Theory and System Setup

9.2.1 Catalyst Structure and Material Parameters

Figure 9.1 shows a cross-sectional view of a typical catalyst with its ceramic (dielectric) matrix, the thin-film coating, or "washcoat", the conductivity of which depends on its oxygen content, and the channels filled with exhaust gas. The noble-metal content in the thin-film coating is so small that the percolation limit is not reached, i.e., no continuous current paths exist.

An electromagnetic RF field probing such an inhomogeneous structure will not "see" the microscopic details, but rather their spatial averages (as if the structure were a continuous medium). There exist numerous models for the effective relative permittivity ε_{rc} and conductivity σ_c of composite dielectric materials depending on the dielectric properties and the shape and distribution of high-permittivity or conductive inclusions [7, pp. 133–177] . Most of these effective-media theories are empirical and approximate. Unfortunately, data on the frequency characteristics of the effective parameters and on their dependence on the shape of the inclusions are very limited.

In a simplified treatment, we compute the effective material parameters as the volume-weighted averages of the corresponding parameters of the various phases. The data listed in Table 9.1, which describe the typical TWCs investigated in our work, then result in an effective relative permittivity of $\varepsilon_{rc} \approx 2.6$. This value, although only an estimate, seems reasonable when compared to the values of

Fig. 9.1 Cross-sectional view of the typical honeycomb-like structure of a catalyst

Table 9.1 Model parameters of the various materials present in a TWC

Property	Material				
	Air	Ceramic substrate	Al_2O_3	Oxidized ceria (CeO_2)	Reduced ceria (Ce_2O_3)
Volume fraction ν	1	0.35	0.044	0.054	0.054
Relative permittivity ε_r	1	5	9	10	10
Conductivity σ in S/m	0	0	0	$2 \cdot 10^{-3}$[9]	10[9]

$\varepsilon_{rc} \approx 3\ldots9$ for various powder catalysts, which contain fewer air inclusions than our catalysts, at temperatures between 20 and 500°C [8]. By analogy to [8], the frequency dependence of the effective permittivity in the lower GHz range can be assumed weak enough to be negligible.

In like manner, the data in Table 9.1 lead to effective conductivities of $\sigma_c \approx 7.4 \cdot 10^{-5}$ S/m and 0.4 S/m, respectively, for the oxidized and the reduced catalyst. This calculation neglects the discrete Pt inclusions in the washcoat, which is justified by the fact that the complex permittivity of a dielectric with discrete conducting inclusions is dominated by permittivity (rather than conductivity) effects at high frequencies [7, p. 169].

It is obvious that the effective conductivity of the catalyst is determined by electrochemical processes in the ceria film. In lean atmospheres, ceria is oxidized whereas it is reduced in the presence of reducing gases such as CO or H_2. The relevant chemical reactions are

$$2Ce_2O_3 + O_2 \rightarrow 4CeO_2 \quad \text{and} \tag{9.1}$$

$$2CeO_2 + CO \rightarrow Ce_2O_3 + CO_2 \quad \text{or} \quad 2CeO_2 + H_2 \rightarrow Ce_2O_3 + H_2O \tag{9.2}$$

This leads to a strong dependence of the ceria conductivity σ_{ceria} on the oxygen partial pressure pO_2 in the ambient atmosphere. This dependence can be written as [4,9,10]

$$\sigma_{\text{ceria}} \sim e^{-E_A/(k_B T)} \cdot (pO_2)^{-m} \tag{9.3}$$

with Boltzmann's constant k_B, a thermal activation energy E_A, and $m \approx 1/4$. As pO_2 changes by up to 20 orders of magnitude between rich and lean gases, the ceria conductivity may easily vary by several orders as listed in Table 9.1.

Based on relation (9.3), one may measure the oxygen partial pressure of interest by way of the ceria conductivity, which in turn has to be extracted from the effective catalyst conductivity. In the following, we will discuss how this can be done by RF measurements.

9.2.2 *Canned TWC as Cavity Resonator*

Figure 9.2 shows a three-way catalyst in its stainless steel housing. Typical dimensions are $2a = 125$ mm and $\ell_T = 110$ mm The empty housing represents a cylindrical electromagnetic waveguide with a cutoff frequency of the dominant (TE_{11}) mode of

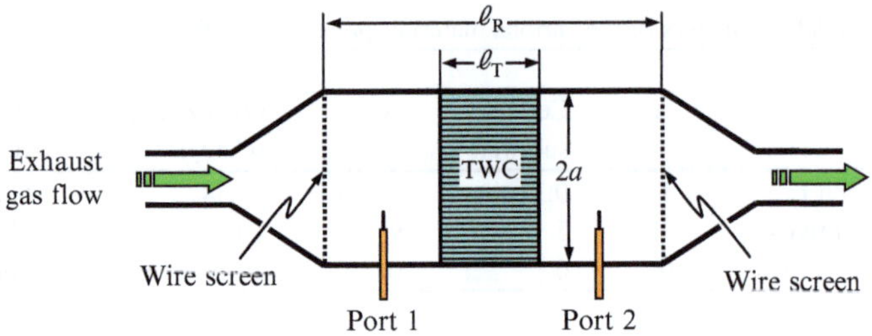

Fig. 9.2 Canned three-way catalyst with coaxial feeds

$$f_{c,\mathrm{TE11}} = \frac{x'_{11} c_0}{2\pi a} \approx 1.4\,\mathrm{GHz}. \tag{9.4}$$

Here, $x'_{11} \approx 1.841$ denotes the first zero of the first-order Bessel function derivative $J'_1(x)$, and c_0 is the vacuum speed of light [11, p. 206]. The insertion of any dielectric material into the housing decreases the cutoff frequency. One therefore concludes that the catalyst-filled housing constitutes a highly overmoded cylindrical waveguide in the easily accessible frequency band around 2.5 GHz.

Although the catalyst possesses a certain conductivity σ_c, the maximum values to be expected after Sect. 9.2.1 are only on the order of 1 S/m. This is seven orders below the conductivity of metals and does not cause excessive attenuation to the electromagnetic waves propagating in the waveguide.

As the diameter of the exhaust pipe surrounding the catalyst housing is much smaller than $2a$, its cutoff frequency considerably exceeds the value of (9.4). Between 1.5 GHz and several GHz, any electromagnetic wave propagating in the housing is evanescent in the exhaust pipe. In this frequency range, the housing acts as a filled cavity resonator. However, because of the cones at either end of the cavity used to flange-mount the housing to the exhaust pipe, the resonator geometry is not very well defined and may be subject to mounting tolerances. For this reason, we inserted additional steel meshes permeable to the exhaust gas flow but effectively short-circuiting the electric field. This resulted in a cylindrical cavity resonator of length $\ell_R = 375$mm.

Let us assume for the time being that the catalyst fills the entire housing and that the housing is a perfect electric conductor (PEC). As discussed in Sect. 9.2.1, the TWC is treated as a homogeneous dielectric with effective relative permittivity ε_{rc} and effective conductivity σ_c. The latter is so small as to be negligible in a first discussion. Under these assumptions, the cavity supports transverse magnetic (TM) and transverse electric (TE) resonances at the frequencies

$$f_{nmp} = \frac{c_0}{2\pi\sqrt{\varepsilon_{rc}}}\sqrt{k^2_{\rho,nm} + (\frac{p\pi}{\ell_R})^2} \tag{9.5}$$

with the radial wave numbers

$$k_\rho = \frac{x'_{nm}}{a}\ (\mathrm{TE}_{nm}\ \text{modes}) \text{ or } k_\rho = \frac{x_{nm}}{a}\ (\mathrm{TM}_{nm}\ \text{modes}). \tag{9.6}$$

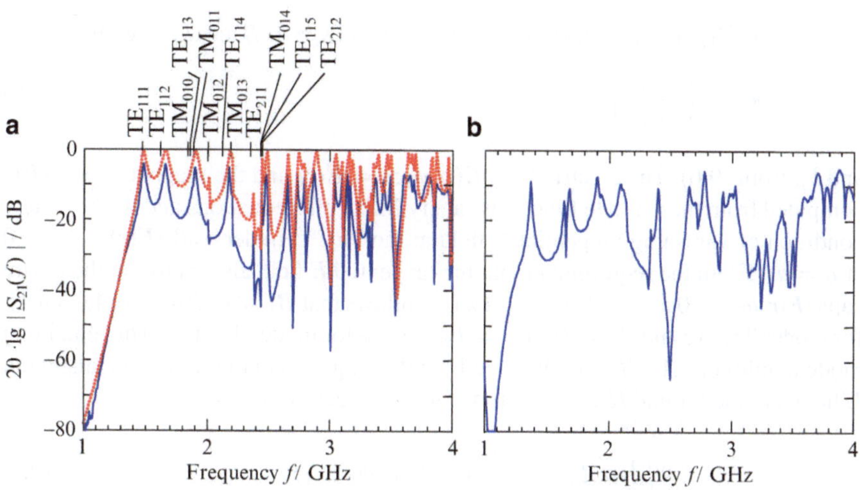

Fig. 9.3 Transmission coefficient magnitude $|\underline{S}_{21}(f)|$ and mode resonance frequencies for catalyst housing acting as two-port cavity resonator. (**a**) Empty housing. *Solid line*: measurement; *dotted line*: finite-element simulation with Ansoft HFSS (from [6], with permission; © 2009 IEEE). (**b**) Measurement for a housing filled with catalyst canned by third party (corrected from [5])

Here, x_{nm} and x'_{nm} respectively denote the m-th zeros of the Bessel functions $J_n(x)$ and the Bessel function derivatives $J'_n(x)$, and p is the longitudinal mode index (zero or positive integer for TM modes, positive integer for TE modes) [11, p. 214].

When the cavity is coupled to a source and a load, one obtains a microwave two-port network. Its S-parameters will become locally extreme at the resonance frequencies (9.5), at least at lower frequencies at which only a few waveguide modes are non-evanescent. For instance, the transmission coefficient magnitude $|\underline{S}_{21}|$ will be large whenever the probe feed couples well to the resonating mode and small when it couples weakly. In either case, one may extract the value of $\varepsilon_{\mathrm{rc}}$ from the measured S-parameter spectrum. From a measurement point of view, the many details of spectra such as the one shown in Fig. 9.3 are quite advantageous as one can expect them to be strongly affected by the catalyst state.

9.2.3 Computation of Cavity Resonance Frequencies

Since the catalyst does not fill the entire housing and usually is only partially loaded with oxygen, one no longer deals with a homogeneously filled cavity. We assume that the problem stays cylindrically symmetric and lossless, but allow a longitudinal inhomogeneity. The cavity can then be thought of as made up of, say, N homogenous sections of length ℓ_i. Let the cavity axis be aligned with a cylindrical coordinate system (coordinates ρ, φ, z), and let section i extend from $z = z_{i-1}$ to $z = z_i = z_{i-1} + \ell_i$ with $z_0 = 0$ and $z_N = \ell_{\mathrm{R}}$. Then, the field in section i can be derived from the scalar mode function [11, pp. 202 ff.]

$$\Psi_{nm}^{(i)}(\rho,\phi,z) = J_n(k_\rho \rho)\cdot \sin n\phi \cdot (A_i \sin k_{zi}z + B_i \cos k_{zi}z) \text{ with} \tag{9.7}$$

$$k_{zi}^2 = \left(\frac{2\pi f}{c_0}\right)^2 \varepsilon_{\mathrm{rc},i} - k_\rho^2 \tag{9.8}$$

and k_ρ from (9.6). The electric field $\boldsymbol{E}(\rho,\varphi,z) = \boldsymbol{E}(\boldsymbol{r})$ and the magnetic field $\boldsymbol{H}(\boldsymbol{r})$ computed from these mode functions satisfy Maxwell's equations and the boundary condition of vanishing tangential $\boldsymbol{E}$ field at the PEC cylinder wall ($\boldsymbol{E}(\boldsymbol{r}) \circ \boldsymbol{e}_\rho = 0$ at $\rho = a$). From the requirement that the tangential $\boldsymbol{E}$ field also vanish at the cavity caps ($\boldsymbol{E}(\boldsymbol{r}) o \boldsymbol{e}_z = 0$ at $z = 0$ and $z = \ell_R$), it follows that $B_1 = 0$, $B_N = -A_N \cdot \tan \ell_{\mathrm{R}}$ for modes TE to z and $A_1 = 0$, $A_N = B_N \cdot \tan \ell_{\mathrm{R}}$ for modes TM to z. The remaining mode amplitudes A_i, B_i are obtained by enforcing the continuity of the tangential $\boldsymbol{E}$ field and the normal $\boldsymbol{H}$ field at the section interfaces through

$$\Psi_{nm}^{(i)}(\rho,\phi,z_{i-1}) \overset{!}{=} \Psi_{nm}^{(i-1)}(\rho,\phi,z_{i-1}), \tag{9.9}$$

$$F_i \frac{\partial \Psi_{nm}^{(i)}(\rho,\varphi,z_{i-1})}{\partial z} \overset{!}{=} F_{i-1} \frac{\partial \Psi_{nm}^{(i-1)}(\rho,\varphi,z_{i-1})}{\partial z} \tag{9.10}$$

with $F_i = 1$ for TE modes and $F_i = 1/\varepsilon_{\mathrm{rc},i}$ for TM modes ($i = 2, 3, \ldots, N$). The above corresponds to a system of homogeneous linear equations for the mode amplitudes A_i, B_i. With $\boldsymbol{x} = (A_1\ A_2\ B_2\ A_3\ B_3 \ldots A_{N-1}\ B_{N-1}\ A_N)^{\mathrm{T}}$ for TE modes and $\boldsymbol{x} = (B_1\ A_2\ B_2\ A_3\ B_3 \ldots A_{N-1}\ B_{N-1}\ B_N)^{\mathrm{T}}$ for TM modes, one obtains:

$$\underbrace{\begin{pmatrix} \mathbf{A}_{11}\mathbf{a} & -\mathbf{A}_{21} & & & & \mathbf{0} \\ & \mathbf{A}_{22} & -\mathbf{A}_{32} & & & \\ & & \mathbf{A}_{33} & -\mathbf{A}_{43} & & \\ & & & \ddots & \ddots & \\ \mathbf{0} & & & & \mathbf{A}_{N-1,N-1} & -\mathbf{A}_{N,N-1}\boldsymbol{b} \end{pmatrix}}_{=:\boldsymbol{M}(k_{zi},z_i)} \cdot \mathbf{x} = \begin{pmatrix} 0 \\ 0 \\ 0 \\ \vdots \\ 0 \end{pmatrix} \tag{9.11}$$

where

$$\mathbf{A}_{ij} = \begin{pmatrix} 1 & 0 \\ 0 & F_i k_{zi} \end{pmatrix} \cdot \mathbf{B}_{ij},\ \mathbf{B}_{ij} = \begin{pmatrix} \sin k_{zi}z_j & \cos k_{zi}z_j \\ \cos k_{zi}z_j & -\sin k_{zi}z_j \end{pmatrix},$$

$$\mathbf{a}^{(\mathrm{TE})} = \begin{pmatrix} 1 \\ 0 \end{pmatrix},\quad \mathbf{a}^{(\mathrm{TM})} = \begin{pmatrix} 0 \\ 1 \end{pmatrix},\quad \text{and} \quad \mathbf{b} = \mathbf{B}_{N,N} \begin{pmatrix} 0 & 1 \\ 1 & 0 \end{pmatrix} \mathbf{a}. \tag{9.12}$$

A non-trivial solution exists iff the determinant of the system matrix vanishes:

$$\det \mathbf{M}(k_{zi}, z_i) = \det \mathbf{M}(f; k_\rho, \varepsilon_{\mathrm{rc},i}, z_i) = 0. \tag{9.13}$$

This is a transcendental equation for the frequency f. Its infinitely many solutions f_{nmp} are the wanted resonance frequencies, which must be computed numerically. The result of such a computation is shown in Fig. 9.4 for a catalyst divided into two

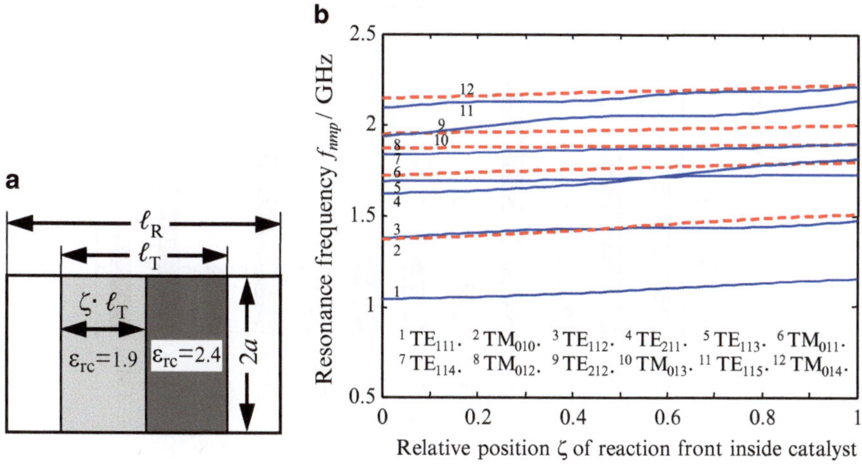

Fig. 9.4 Cavity resonator filled with a two-section catalyst. (**a**) Geometry. The lower-permittivity section occupies a fraction ζ of the catalyst length. (**b**) Resonance frequencies f_{nmp} of the lowest-order modes (from [6], with permission; © 2009 IEEE)

sections with slightly different relative permittivities, the section interface presumably coinciding with a reaction front. As expected, the resonance frequencies are the higher, the more the lower-permittivity section extends into the catalyst.

Two interesting features of Fig. 9.4b strike the eye. First, all curves increase monotonically such that the reaction front position can be uniquely determined from the resonance frequencies. Second, the mean sensitivities of the resonance frequencies to the relative reaction front position,

$$\bar{S}_{\zeta}^{f_{nmp}} := \frac{f_{nmp}(\zeta = 100\%) - f_{nmp}(\zeta = 0)}{100\%}, \tag{9.14}$$

vary by almost a factor of ten between the various modes (Fig. 9.5). The modes with the highest $\boldsymbol{E}$ field amplitudes in the catalyst region will be particularly sensitive to permittivity changes in the catalyst. In any case, these numerical experiments prove that even modest material parameter changes within the catalyst can be easily measured by RF methods.

In the presence of losses, the eigenvalue problem (9.11–9.13) becomes complex-valued. The real part of the resulting complex eigenfrequency is interpreted as resonance frequency, the imaginary part is inversely proportional to the resonance quality factor Q. When, in addition, local inhomogeneities destroy the cylindrical geometry, analytical solutions become so involved as to be of no advantage compared to purely numerical approaches. However, as the frequency shifts brought about by catalyst state changes do not exceed a few percent, one may have recourse to perturbation formulas. When a homogeneously reduced catalyst with effective relative permittivity ε_{rc} and conductivity σ_c is associated with the electric field $\boldsymbol{E}_0(\boldsymbol{r})$ and the resonance frequency f_r, the parameter changes $\Delta\varepsilon_{rc}(\boldsymbol{r})$ and $\Delta\sigma(\boldsymbol{r})$ due to (possibly localized) oxidization will cause a fractional frequency shift of [11, pp. 322 ff.],

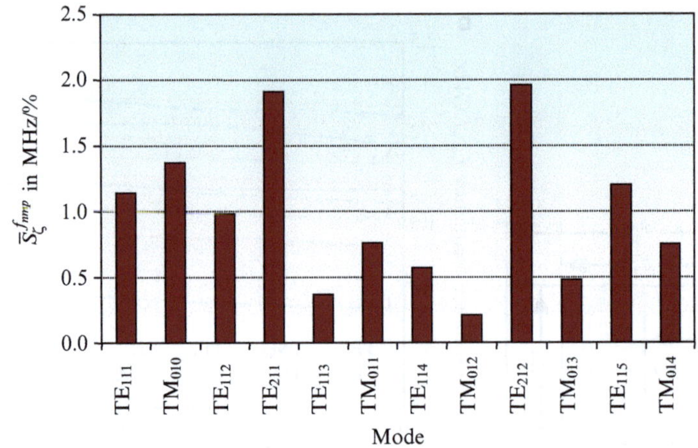

Fig. 9.5 Mean sensitivities of the resonance frequencies f_{nmp} from Fig. 9.4 to the normalized reaction front position ζ (from [6], with permission; © 2009 IEEE)

$$\frac{\Delta f_\mathrm{r}}{f_\mathrm{r}} = -\frac{1}{W}\iiint_V \varepsilon_0 \Delta\varepsilon_\mathrm{rc}(\boldsymbol{r})|\boldsymbol{E}_0(\boldsymbol{r})|^2 \mathrm{d}V \tag{9.15}$$

where W denotes the total energy stored in the original cavity:

$$W = 2\iiint_V \varepsilon_0 \varepsilon_\mathrm{rc}(\mathbf{r})|\mathbf{E}_0(\mathbf{r})|^2 \mathrm{d}V. \tag{9.16}$$

This is the first-order approximation. A term involving $\Delta\sigma$ will have to be included on the right-hand side of Equation (9.15) when $\Delta\sigma$ becomes appreciably large. Likewise, the quality factors Q_0 and Q before and after the perturbation, equal to the full resonance curve bandwidths at half maximum, are related by

$$\frac{1}{Q} - \frac{1}{Q_0} = \frac{1}{\pi f_\mathrm{r} W}\iiint_V \Delta\sigma(\mathbf{r})|\mathbf{E}_0(\mathbf{r})|^2 \mathrm{d}V. \tag{9.17}$$

As the catalyst oxidization decreases its effective conductivity ($\Delta\sigma(\boldsymbol{r}) < 0$), oxidization will increase the resonance Q. A partial oxygen loading of the catalyst will only result in partial frequency and Q factor shifts because spatial restrictions of the parameter changes $\Delta\varepsilon_\mathrm{rc}(\boldsymbol{r})$ and $\Delta\sigma(\boldsymbol{r})$ reduce the effective integration volume. This proves that the specific analytical results from figures 9.4 and 9.5 can be generalized: the reaction front inside the catalyst must be observable via the measurable quantities Δf_r and Q.

The complete RF problem not only involves the mode structure of the cavity, but also the coupling of the resonator to the signal source and sink. Therefore, the probe feeds (design, position) are just as important to the observable S-parameters as the cavity geometry and filling. The solution to this full-fledged problem requires numerical techniques such as the finite-element method (FEM) [6].

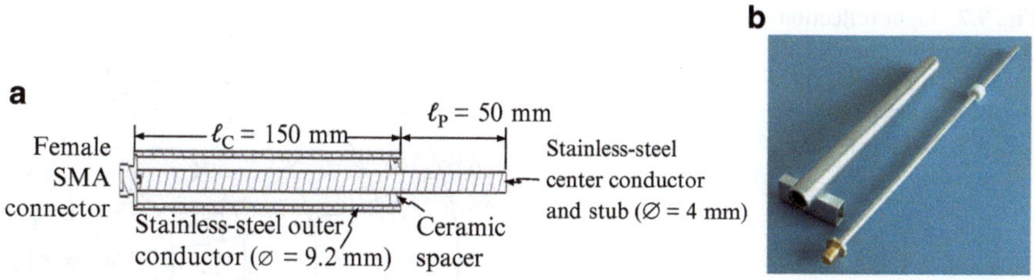

Fig. 9.6 High-temperature probe feed with adjacent coaxial line. (**a**) Schematic (not to scale). (**b**) Photograph of the assembly before mounting (from [6], with permission; © 2009 IEEE)

9.2.4 *High-Temperature Cavity Feeds*

To excite and observe the cavity resonances, we used two thin probe feeds (short stubs) connected to an automatic vector network analyzer (VNA) via coaxial lines to obtain a two-port resonator (see Fig. 9.2).

The coupling efficiency is restricted for several reasons. First, when observing many modes, one cannot simultaneously place the probe feeds at electric field maxima of all modes. Second, there apply mounting restrictions in the catalyst monitoring application. And third, the exhaust gas and catalyst operating temperatures are as high as 600°C. No commercial probes are available for this temperature range.

Figure 9.6 shows a second-generation custom-made high-temperature probe feed. They consist of a short stub and an air-filled coaxial line with stainless steel conductors. The coaxial line length was chosen such that the temperature at its far end as seen from the catalyst was low enough to enable connection to a commercial flexible coaxial cable. The center conductor was fixed by a ceramic spacer and by soldering to an SMA connector at the far end. The chosen center and outer conductor radii resulted in a wave impedance of almost exactly 50 Ω.

The measured input reflection coefficients of two lines short-circuited at the stub side (with an appropriate electrical delay set after the VNA calibration) are shown in Fig. 9.7. Ideally, one should see a point at unit distance from the center of the Smith chart at all frequencies. The measured data are close enough to this ideal, the more so as the small observable imperfections do not interfere with the primary purpose of our work, viz., to investigate the influence of the catalyst state on the cavity resonances.

9.2.5 *Complete Measurement Setup*

The function of the complete measurement system is visualized in Fig. 9.8. The indirect effect of the normalized air-to-fuel ratio λ on the catalyst constitutive parameters leads to changes in the resonant cavity which are observed via the scattering matrix $\boldsymbol{S}$ of the cavity with reference ports placed at the SMA connectors of

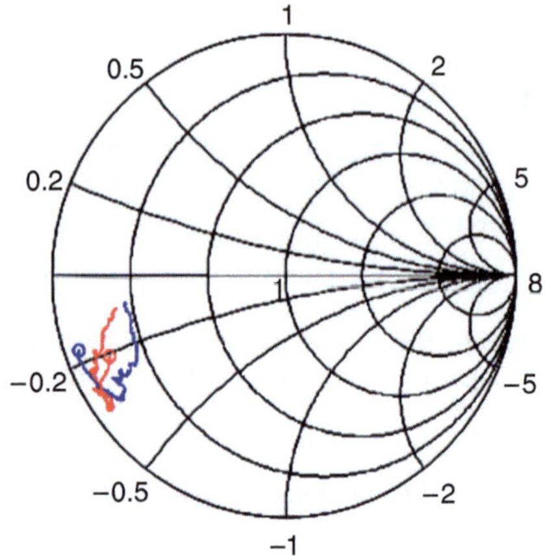

Fig. 9.7 Input reflection coefficients of the two coax-to-cavity feeds for short-circuited stubs ($f = 1 \ldots 4$ GHz)

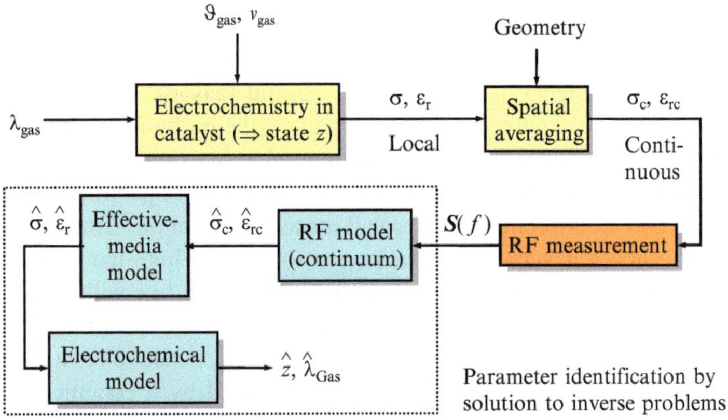

Fig. 9.8 Functional diagram of the measurement system (from [6], with permission; © 2009 IEEE)

the high-temperature cavity feeds. Note that the identification of the electrochemical state of the catalyst from the measured S-parameters requires the inference of:

- Spatially averaged material parameters from RF data ($\ell_{scale} \sim 10$ cm)
- Local material parameters from spatial averages ($\ell_{scale} \sim \mu m \ldots mm$)
- The O_2 content in the catalyst from local material parameters ($\ell_{scale} \sim$ sub$-\mu m$)

As indicated, these three inverse problems involve quite different scale lengths – the entire problem is of a multiscale nature.

Figure 9.9 shows a photograph of the laboratory setup of the housed catalyst with the surrounding infrastructure. The catalyst in this setup was a standard product by Umicore (400 cells per square inch; cell density times cell wall thickness = 6 mil). To make sure that the reaction enthalpy did not affect the catalyst temperature, the latter was monitored by two thermocouples inserted into the catalyst housing. The housing was heated from the outside by a heating mat not shown in Fig. 9.9.

Fig. 9.9 Photograph of the test setup at the gas mixing unit [5]. 1 = TWC housing; 2 = threads for cavity feeds; 3 = threads for thermocouples; 4 = isolated gas feed line

Different gas mixtures representing synthetic lean and rich exhaust gases ($\lambda \approx$ 1.12 and $\lambda \approx$ 0.90...94, respectively) were made to flow through the catalyst at a rate of 20 l/min; at this low rate, thermal effects caused by reaction enthalpy could be neglected. The gas mixture contained water (10 vol.-%), CO_2 (10 vol.-%) as well as reducing agents like H_2, CO, propane, and 0.2 vol.-% O_2 in the rich gas.

In the lean gas, 2 vol.-% O_2 and NO were added. The gas was preheated to some hundred °C and λ was monitored upstream and downstream of the catalyst by broadband lambda probes (Bosch LSU 4.2) connected to an ETAS LA4 Lambda Meter. A conventional exhaust gas analysis served to control the experiments.

9.3 Experimental Results

In first experiments, the catalyst was exposed to either lean or rich synthetic exhaust gases for time intervals long enough to make sure that the catalyst was either fully oxidized or fully reduced. The measured S-parameters clearly depend on the oxidation state (Fig. 9.10). Especially near the maxima and minima, the changes in the magnitude of S_{21} can reach values of 10–20 dB.

A typical result of a dynamic measurement is shown in Fig. 9.11. While the lambda probe upstream of the TWC reacts instantaneously to switches from lean to rich gas or vice versa (left edges of the shaded time intervals A to D), the lambda probe downstream exhibits the expected delayed response (right edges of the time intervals). This corresponds to the wanted buffering action of the catalyst: when it is void of oxygen and suddenly subjected to lean (oxygen-rich) gas, it gradually incorporates oxygen, and only when the reaction front reaches the downstream end of the catalyst, the downstream lambda probe indicates lean gas. Conversely, when the catalyst is fully loaded with oxygen and suddenly subjected to rich (oxygen-poor) gas, it gives off oxygen (used to oxidize and thus neutralize the rich gas components),

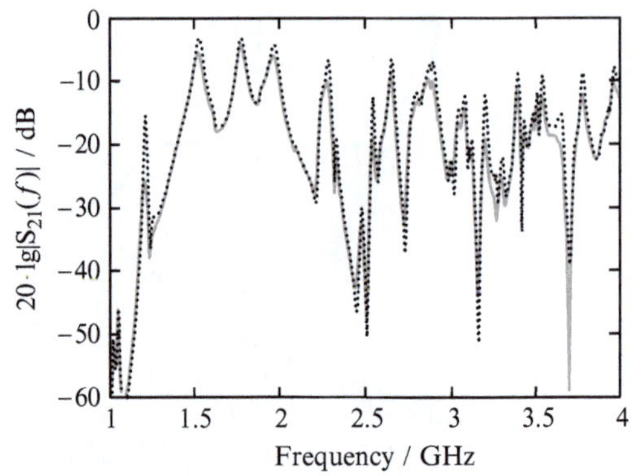

Fig. 9.10 Transmission coefficient amplitude $|S_{21}|$ of the catalyst-filled cavity resonator when exposed to lean (*dotted line*) and rich (*solid line*) synthetic exhaust gases. Catalyst canned in-house, temperature $\vartheta = 430°\mathrm{C}$ (modified from [5])

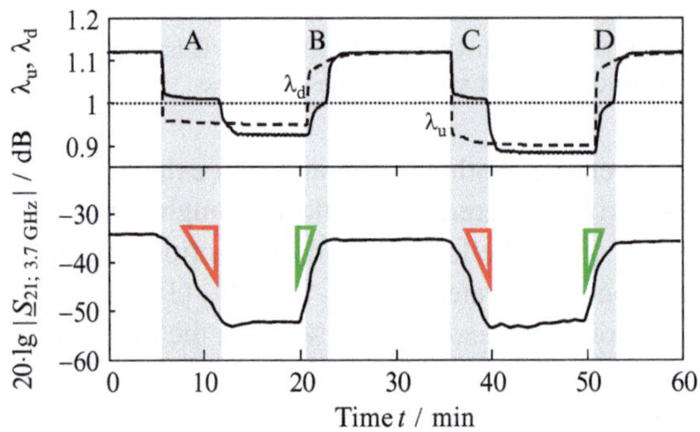

Comp.	Conc. / % Lean	Conc. / % Rich
H_2O	10	
CO_2	10	
CO	0	x
H_2	0	x
NO	x	0
Prop.	0	x
O_2	2	0.2

Fig. 9.11 Transient experiments. *Top*: λ after and before the TWC as a function of time. *Bottom*: Transmission coefficient magnitude $|S_{21}|$ at $f = 3.70\,\mathrm{GHz}$, also as a function of time (from [6], with permission; © 2009 IEEE)

and only when the reaction front reaches the downstream end of the catalyst, the downstream lambda probe indicates rich gas.

The buffering action of the TWC in response to λ switches is mirrored by various signal characteristics of the cavity resonator S-parameters. Figure 9.11 shows but one example of this: the magnitude of $|S_{21}|$ at the fixed frequency $f = 3.70\,\mathrm{GHz}$. It is obvious that the RF signal is correlated with the progression of the reaction front in the catalyst during the shaded time intervals. As it turns out, the time rate of change in $\lg |S_{21}|$ after a λ switch in the upstream gas is an almost linear function of λ_u (Fig. 9.12). Other signal characteristics such as the phase or amplitude of any of the four S-parameters or a combination of them such as the "loss function" $1 - |S_{11}|^2 - |S_{21}|^2$ might also be used.

To check these results under real-life conditions, similar experiments were carried out on an engine test dynamometer. In these experiments, an Audi V6-3.2-l-FSI engine running at 1,000 rpm and at a torque of 230 Nm was used. In this operating point, the engine produced rich exhaust gas. These tests, too, confirmed the feasibility of the approach [4]. The same holds true for dynamometer experiments involving automotive electrochemical systems other than three-way catalysts.

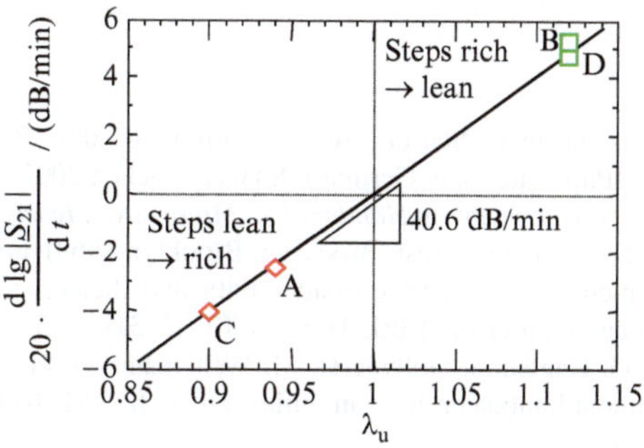

Fig. 9.12 Time rate of change in lg $|S_{21}(f = 3.70\,\text{GHz})|$ after a λ switch in the upstream gas, when the catalyst was either fully oxidized or fully reduced before the switching. The capital letters A through D refer to the shaded time intervals in Fig. 9.11 (from [6], with permission; © 2009 IEEE)

9.4 Conclusion

We have shown by way of an example (automotive three-way catalyst) that important state variables of electrochemical processes such as the oxygen loading of a catalyst can be observed in situ by microwave measurements. In particular, the progression of a reaction front through a catalyst shows up in the S-parameters of the cavity resonator formed by the catalyst and its metal housing. The frequencies involved in this problem (a few GHz) coincide with the frequencies used in modern wireless communications systems. Hence, cost-effective chipsets, with which one could design stand-alone measurement systems, are readily available. Such measurement systems would, for instance, pave the way for lambda control and on-board diagnosis without lambda probes.

The identification of the electrochemical state of the catalyst from the measured S-parameters involves three inverse problems associated with different scale lengths, viz., the inference of spatially averaged material parameters from the RF data, the inference of the local material parameters from their spatial averages, and the inference of the oxygen content in the catalyst from the local material parameters. We have discussed various analytic as well as numeric approaches to the solution of these inverse problems. While the forward problem leading from spatially averaged material parameters to RF parameters may be considered solved, the specifics of the electrochemical and effective-media models call for further research.

Acknowledgements This work was supported by the German Research Foundation (DFG), grants number Fi 956/3–1 and Mo 1060/6–1. The authors are indebted to Drs. Ulrich Göbel, Jürgen Gieshoff, and Martin Rösch from Umicore, Hanau, Germany who provided TWC samples.

References

1. C. Zimmermann, Neuartiger Sensor zur Bestimmung des Zustandes eines NOx-Speicherkatalysators (Ph.D. thesis, in German). Shaker, Aachen 2007
2. HM. Altschuler Dielectric constant. in *Handbook of Microwave Measurements*, vol. II, 3rd edn ed. by M. Sucher, J. Fox (Polytech. Inst. Brooklyn, Brooklyn 1963), pp. 495–548
3. SH. Chao, Measurements of microwave conductivity and dielectric constant by the cavity perturbation method and their errors IEEE Trans. MTT **33**, 519–526 (1985)
4. R. Moos, M. Spörl, G. Hagen, A. Gollwitzer, M. Wedemann, G. Fischerauer, TWC: lambda control and OBD without lambda probe – an initial approach. SAE Technical Paper Series No. 2008–01–0916 (2008)
5. G. Fischerauer, M. Spörl, A. Gollwitzer, M. Wedemann, R. Moos, Catalyst state observation via the perturbation of a microwave cavity resonator. Frequenz **62**, 180–184 (2008)
6. G. Fischerauer, A. Gollwitzer, A. Nerowski, M. Spörl, R. Moos, On the inverse problem associated with the observation of electrochemical processes by the RF cavity perturbation method. in *Proceedings of SSD'09*, Djerba (2009)
7. P.S. Neelakanta, *Handbook of Electromagetic Materials* (CRC Press, Boca Raton, 1995)
8. G. Roussy, J.M. Thiebaut, F. Ename-Obiang, E. Marchal, Microwave broadband permittivity measurement with a multimode helical resonator for studying catalysts. Meas. Sci. Technol. **12**, 542–547 (2001)
9. H.L. Tuller , A.S. Nowick, Defect structure and electrical properties of nonstoichiometric CeO_2 single crystals. J. Electrochem. Soc. **126**, 209–217 (1979)
10. P. Jasinski, T. Suzuki, H.U. Anderson, Nanocrystalline undoped ceria oxygen sensor. Sens Actuators B bf 95, 73–77 (2003)
11. R.F. Harrington, *Time-Harmonic Electromagetic Fields* (McGraw-Hill, New York, 1961)

Part III
Communication Technology

Chapter 10
Mobile Phones: The Driving Force Towards the Integration of Analog and Digital Blocks for Baseband and RF Circuitry

Josef Hausner and Christian Drewes

10.1 Introduction

The introduction of the Global System for Mobile Communications (GSM) almost two decades ago has evolved to be the most successful second generation cellular system, followed by its successor Universal Mobile Telecommunications System (UMTS)/Wideband Code Division Multiple Access (WCDMA). Since the introduction of GSM, until recently, most of the effort was spent on integration and improvement of circuit switched voice services on mobile handsets. In the meantime, data services have been gaining importance, marked by data extensions to GSM and UMTS, such as General Packet Radio Service (GPRS), Enhanced Data Rates for GSM Evolution (EDGE), and High Speed Packet Access (HSPA). Next generation systems, such as Long Term Evolution (LTE) that is being standardized under the auspices of the 3rd generation partnership project (3GPP), are based solely on the Internet Protocol (IP). Due to its tremendous success and worldwide roaming capability, GSM is still the choice for voice and low data rate applications. Currently, operators have either launched or are planning to launch GSM/EDGE in 165 countries and HSPA in 96 countries. The global subscriber base of these technologies is nearly 90% worldwide.[1] Considering these facts, all future mobile terminals will have to support the legacy standards.

Traditionally, terminal implementations have reused previously developed blocks up to a complete cellular system. For instance, an LTE implementation may reuse complete 2G and 3G modems. The advantage is that legacy modems are verified, certified, and field-operational. However, since not all of these protocols have to be active simultaneously, they could be implemented primarily in software on a

[1] Global Mobile Suppliers Association, www.gsacom.com

J. Hausner (✉)
Institute for Integrated Systems, Ruhr-Universität Bochum, 44780 Bochum, Germany
e-mail: josef.hausner@rub.de

C. Drewes
Intel Mobile Communications GmbH, 85579 Neubiberg, Germany
e-mail: christian.drewes@intel.com

S. Lindenmeier and R. Weigel (eds.), *Electromagnetics and Network Theory and their Microwave Technology Applications*, DOI 10.1007/978-3-642-18375-1_10,

programmable or reconfigurable platform, in particular those protocols that are computationally less demanding. Such an approach is commonly referred to as a software defined radio [1].

Mobile phones can be classified as either low-cost or high-end devices. This paper focuses on the latter: development of cellular modem platforms for the feature-rich and smart phone market segment. Low-cost phones leverage modem developments done earlier, and focus on integrating as many components as possible into a single silicon die [2]. Thus they allow cost optimized cellular phones with some basic functionality. High end modems, implementing the latest and greatest cellular system, typically use the most advanced silicon technology available, and are less integrated as ultra-low-cost devices, although the trends towards higher integration are visible [3,4].

This paper is structured as follows: In Sect. 10.2 we provide a brief overview on the evolution of the digital cellular radio systems. Implementation aspects of cellular radio terminals, including their respective computational complexity and integrated circuits (IC) area are highlighted in Sect. 10.3. In Sect. 10.4 we provide an outlook on the implementations of future cellular terminal modems. Finally, some concluding remarks in Sect. 10.5 close this paper.

10.2 Cellular Radio Systems

Figure 10.1 shows the evolution of major digital cellular radio systems. An obvious trend is that of all those different radio systems only a few became mainstream and dominated cellular technology. All of these belong to the 3GPP family. More precisely, only a single system per generation is truly mainstream. For 2G systems, the clear winner is GSM and its enhancements GPRS and EDGE. Japanese

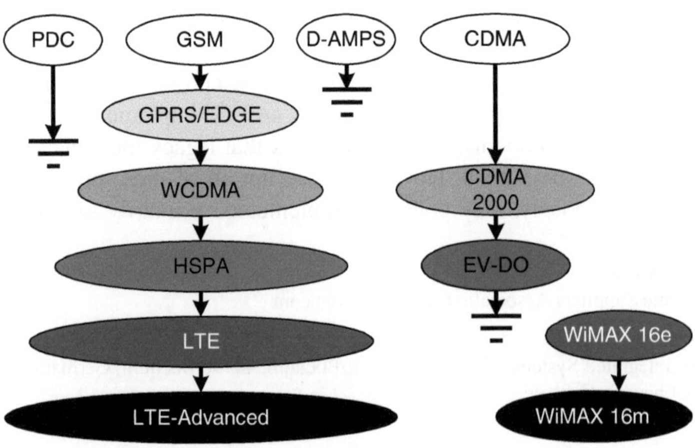

Fig. 10.1 Evolution of digital radio systems

Personal Digital Cellular (PDC) system was phased out a couple of years ago and replaced by WCDMA and CDMA2000. Similarly, the American Digital Advanced Mobile Phone System (D-AMPS) was replaced by 3GPP systems. For 3G systems, 3GPP-based WCDMA including HSPA has the dominant market share, in part due to its strong integration into the overall cellular GSM network. The main competition to WCDMA was CDMA2000, which had evolved from cdmaOne, and was further refined as Evolution-Data Optimized (EV-DO) to accommodate high speed data. Work on a successor to CDMA2000 as a pre-4G system similar to LTE, called Ultra Mobile Broadband (UMB), has not yet been successful in getting any network design wins. The real competition to LTE will come from two standards belonging to the IEEE Worldwide Interoperability for Microwave Access (WiMAX) family: IEEE 802.16e and IEEE 802.16m. LTE itself comes in two flavors: frequency and time division duplexing (FDD, TDD) for operation in paired and unpaired spectrums respectively. Currently, LTE has achieved significantly more market shares and design wins than its main competitor WiMAX. Main focus of LTE is on the FDD version, with TDD getting some attention in China.

Figure 10.2 highlights those 3GPP systems together with some further technology evolutions. Regarding data rates it can be observed that these double roughly every 16–18 months. Thus, having today's 7.2 Mb/s HSPA devices in mind, we can expect 50 Mb/s LTE devices in 2012 and 150 Mb/s LTE devices in 2014. That means that 1 Gb/s would be available around 2018. Those data rate enhancements can be

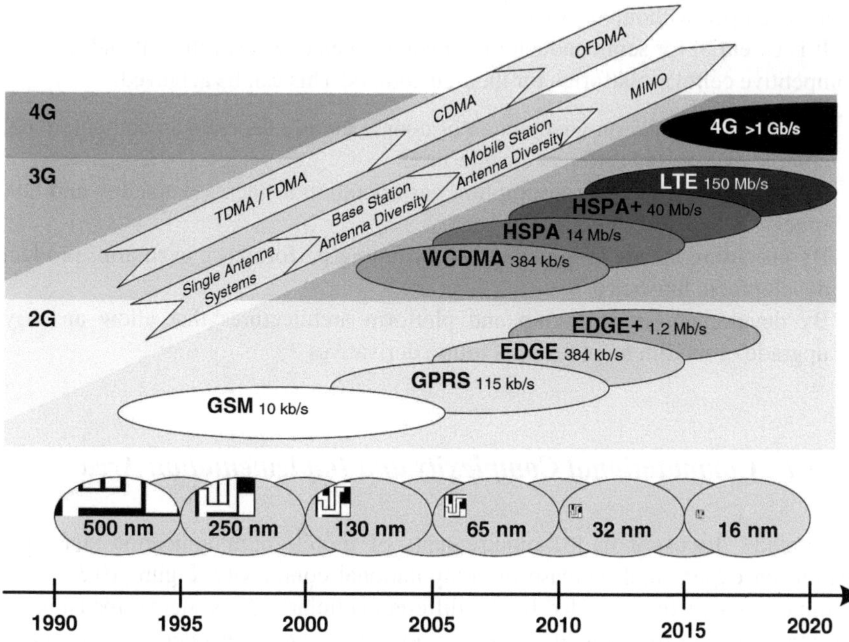

Fig. 10.2 Evolution of 3GPP radio systems, semiconductor technology nodes, antenna techniques, and implementation technologies

attributed to a few effects, all increasing some kind of parallelism in the frequency, time, or spatial domains:

- System bandwidth increases from 200 kHz (GSM) up to 100 MHz for 4G systems
- Scheduling allows continuous allocation of radio resources in contrast to GSM, where multiple users where multiplexed
- Multiple-input multiple-output (MIMO) antenna systems, allows multiple simultaneous transmissions towards a single terminal

10.3 Implementation Aspects

When developing a new mobile phone platform, it needs to be considered that it takes roughly 2 years from start of project to availability of first commercial products, such as phones or data cards. For new air-interfaces being implemented for the first time, product development needs to allow an initial period to develop the basic modem concepts and architectures, which adds to the overall project time. Since the life-time of a mobile platform generation within different products is approximately three years, this implies that at the start of project features that will get implemented have to anticipate market requirements by at least 5 years. The implementation of unnecessary features can lead to a significant waste of resources. Even worse, if essential features are not implemented, complete platforms might have been developed without any revenues.

It is essential for semiconductor companies to enable cost-efficient and feature-competitive cellular platforms for their customers. This can be achieved:

- By minimizing silicon area, which of course greatly depends on computational complexities of the different air interfaces
- By minimizing energy consumption per operation to increase standby and talk times
- By providing means of updating and of tuning performance even after product development finished
- By developing scalable chip and platform architectures that allow an easy upgrade or modification towards future derivatives

10.3.1 Computational Complexity and Implementation Area

As already discussed in [5] requirements of mobile communication standards exhibit an exponential increase in computational complexity. Figure 10.3 shows arithmetic instructions needed by the different cellular radio systems, focusing on the inner modem functionalities, i.e., without the channel encoders and decoders. Although the total complexity increases from generation to generation, the complexity per received bit decreases. An exception was UMTS due to its WCDMA

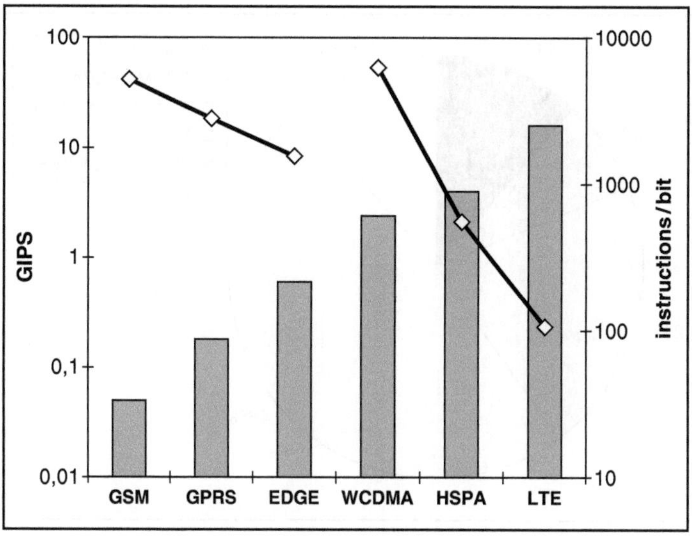

Fig. 10.3 Computational complexity of the inner receiver per cellular standard: total instructions (*columns*) and instructions per received bit (*lines*)

technology: a relatively narrowband data transmission with up to only 384 kb/s occupies 5 MHz bandwidth on up to six radio links simultaneously. Considering a time span of roughly 20 years between GSM and 150 Mb/s LTE systems, computational complexity of the inner receiver doubled roughly every 2.5 years, considerably slower than data rates, which doubled every 1.5 years.

When looking at the silicon area of a state-of-the-art HSPA/EDGE baseband modem (Fig. 10.4), it can be observed that the silicon area required by WCDMA and HSPA physical layers is approximately four times that of GSM and EDGE, which is also roughly inline with Fig. 10.3 when comparing the computational complexity of different systems. Modem area includes dedicated modules for 2G and 3G with their respective required logic and memory blocks ("GSM&EDGE" and "WCDMA&HSPA"). Other blocks are audio processing, modem controller subsystem, and general modem peripherals such as external interfaces or trace modules. The controller subsystem includes its bus system and memory required for protocol stack processing. Overall complexity corresponds to roughly 50 million transistors.

By extrapolating the factor between 3G and 2G implementation area to include also future cellular generations, we can approximate the total area of a baseband implementation by a geometric series:

$$a = \sum_{k=0}^{K} n^k \cdot a_0 = \frac{1 - n^{K+1}}{1 - n} \cdot a_0 \qquad (10.1)$$

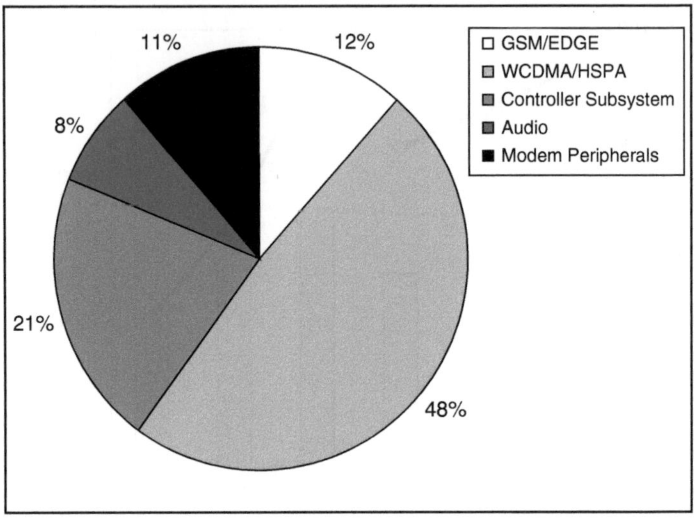

Fig. 10.4 Area breakdown of a state-of-the-art dual mode 2G (GSM/EDGE) and 3G (WCDMA/HSPA) digital baseband

with a_0 being the silicon area required by the most advanced cellular system, n is the scaling factor (e.g. $\frac{1}{4}$), and K the total number of cellular systems being implemented in a multi-mode terminal. The total area a is upper-bounded by $a_0/(1\text{–}n)$ i.e. $\frac{4}{3}a_0$ for $n = \frac{1}{4}$. That means, all older cellular systems' physical layers jointly contribute only 33% to the total area.

10.3.2 Energy Consumption

Besides algorithmic computational complexity the largest problem of mobile computing is based on the fact that the mobile phones are battery driven. Mobiles that feature a large number of functionalities and capabilities, such as camera and display(s), or short-range communication such as Bluetooth and WLAN, and higher data rates over the cellular air interface, are leading to a serious problem: a dramatic and critical increase of energy consumption. This has two major impacts: first, the absolute power consumption could make active cooling of the mobile device necessary, and secondly, the standby time decreases, as the developments in battery capacity and efficiency are very slow compared to the increase in energy consumption. When purchasing, the standby time of mobile devices is one of the top criteria. Thus, solutions have to be found that decouple the problem of more complex mobile devices and the need for higher data rates.

Initially one might think that advances in semiconductor technology might enable such features. This is partly true, since the active power in the digital domain is described by the following equation:

$$P_{active} = V_{DD}^2 \cdot f \cdot C_L \cdot k; \quad \text{k} = \text{const.} \tag{10.2}$$

Thanks to technology scaling, the supply voltage V_{DD} has dropped down to 1 V or below. Furthermore, values for the load capacitance C_L have reduced as well. Applying different values for the supply voltage within one technology node also affects the switching speed of the digital circuitry. Depending on the performance requirements of the microcontroller, the digital signal processor (DSP), and active logic, an appropriate clock frequency has to be selected. This scheme is known as dynamic frequency scaling. And, in case of on chip feedback loops, this scheme can be extended to adaptive voltage (V_{DD}) scaling.

Examining the power consumption figures for 2G and 3G talk modes, it can be seen that despite the dramatically higher signal processing complexity of WCDMA compared to GSM, the power consumption in the digital baseband is still acceptable. Interestingly, the culprit in 3G power dissipation is the radio frequency (RF) transceiver. This, to a large degree, can be attributed to a wider system bandwidth (5 MHz for WCDMA versus 200 kHz for GSM) and continuous data reception. In future systems, the trend towards wider bandwidths will continue. However, with the ongoing HSPA standardization activity, discontinuous reception and transmission has been introduced recently to arrest the high energy consumption of 3G systems.

However, in standby mode the increasing leakage currents, caused by extremely short channel lengths of the metal–oxide–semiconductor (MOS) field-effect transistor combined with very thin gate oxide, are becoming the dominating factor of standby power dissipation, in particular on chips containing millions of such transistors:

$$P_{leakage} = V_{DD} \cdot \sum I_{Leakage} \tag{10.3}$$

During sleep phases in mobile idle mode only very little switching activity occurs in the system resulting in extremely low switching currents while leakage currents are not affected. In fact, already at room temperature leakage dominates the current budget. Furthermore, leakage increases by about a factor of one hundred at maximum case temperature compared to room temperature contributing significantly also in other power use cases with higher switching activity. Therefore, designs incorporate a number of power down features to minimize leakage currents. By using a strategy of optimized mixed threshold voltage (V_t) synthesis, the leaky regular V_t-devices are only used where required due to performance requirements. For all other logic gates, high V_t-devices are used with significantly lower leakage. These necessary countermeasures lead to sophisticated on-chip power management and power down schemes in modern System-on-Chip (SoC) designs, such as voltage scaling, numerous power islands, and sophisticated data retention mechanisms during sleep mode.

10.3.3 Scalability and Flexibility

Especially for new air interfaces, like LTE today, it is essential to develop initial platforms in a way that allows late changes to the implementation, since

(a) standardization might not be finalized, (b) operator and terminal manufacturer requirements might not be stable, (c) a learning curve with a new air interface might be required. Flexibility is a key value for modem platforms since it enables faster time-to-market allowing late changes and ability to customize. Furthermore, baseband modem algorithms and protocol stack implementation can be tuned in a late phase.

When developing modems for new cellular systems, simple receiver algorithms are used for initial implementations, which then get replaced by more sophisticated receivers in later products. Simple 2G equalizers were replaced by interference cancelling receivers; simple 3G rake receivers were replaced by more powerful chip-rate equalizers, which are currently being enhanced to perform interference cancellation. A similar development approach is expected for LTE MIMO detectors [6]. Therefore, platforms need to include chipsets that have an architecture which, in principle, is scalable to allow future receiver upgrades.

Initially, LTE MIMO detectors will be based on linear equalizers, optimized under the minimum mean squared error (MMSE) criterion. Such linear equalizers, while relatively easy to implement, however do not fully exploit the potential gains of MIMO technology, which is only achievable by maximum likelihood (ML) detectors. To approach ML performance efficiently, tree-search schemes known from sequential decoding have been proposed [7], e.g., sphere decoding or the M-algorithm. Choosing an algorithm with deterministic complexity, the algorithmic effort increases approximately by a factor of four compared to MMSE equalizers. Although OFDM equalization is highly parallelizable, it constitutes a challenge for a power-efficient and area optimized implementation. Note that even MMSE-based equalization already contributes roughly 25% to the overall LTE efforts indicated in Fig. 10.3.

Moreover, quantities of early platforms supporting new cellular standards might be very low. This would not justify typical development costs of a new cellular platform only for those early years. Efforts spent have to be reused for subsequent platforms, also ending in requiring scalability of developed platforms.

10.4 Implementation Outlook

10.4.1 Economic Facets, System Verification

Considering implementation of future cellular phones, also economic aspects need to be taken into account. Figure 10.5 highlights cost trends in semiconductor industry. Recent 45 nm fabs require an investment in the range of 3 billion US Dollars. Technology development accounts to another 1 billion US Dollars. Besides the fact that those investments need to generate revenues significantly higher than 10 billion US Dollars, also product R&D costs grow significantly. They will reach 100 million US Dollars (including mask sets), also for next generation cellular terminal chip sets. Therefore, high volumes are essential, raising the need for further

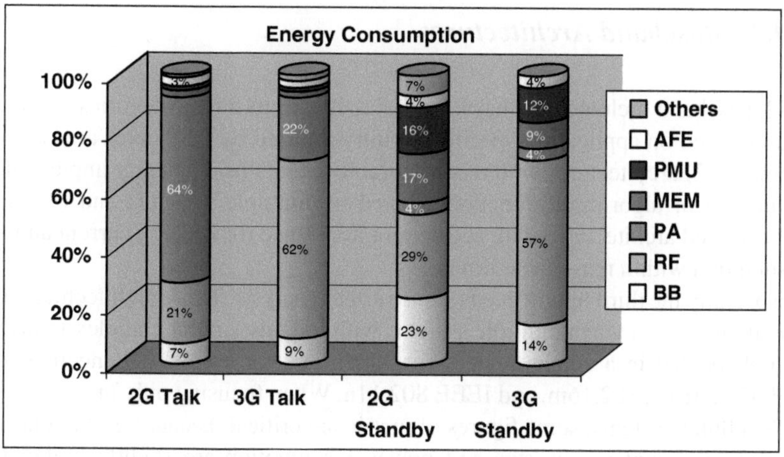

Fig. 10.5 Cost trends in semiconductor industry [8]. Investments in wafer fabs, R&D, and mask sets follow an exponential growth

standardization and consolidation of cellular systems as well as for multimode capable designs.

Since a big part of the R&D efforts are not directly related to silicon implementation, but to system integration and verification, multimode design helps only partly. When taking a look at recent 3G developments, approximately only one third of the overall efforts were directly consumed by chip development. The rest was spent on software driver development, system bring-up, system verification, and conformance and interoperability testing.

Testing includes a couple of thousand conformance tests distributed roughly equally between 2G and 3G and a similar number for interoperability tests at network vendors and operators. Legacy systems even require more tests than new systems – new tests are permanently added without necessarily deleting old tests. Accordingly, even if the hardware platform supports multiple cellular systems and efforts spent for silicon implementation can be reused, additional verification efforts have to be spent for every individual cellular system. By reusing developments already done for legacy systems, it is sufficient to run a test regression, which usually is much less costly than a complete new test run, which would be required for new system architectures covering also legacy systems.

Even if complete reuse of legacy systems is done without exploiting any possible synergies with new cellular systems, the overhead area is upper bounded by roughly one third of the area needed for a new cellular system potentially also including modules for connectivity or for some media processing, such as audio, video, or graphics (see above). More importantly, by reusing legacy modules, test and verification efforts are mainly reduced to stabilizing the new cellular system. Any modification of legacy implementations should be restricted to low level optimizations, like reuse of on-chip memories for different standards or tweaking performance for some special use cases.

10.4.2 Baseband Architectures

In [9], baseband architectures have been classified into three categories: (1) architectures based on application-specific circuitry assisted by DSP cores are currently dominant; (2) architectures with reconfigurable data-paths efficiently implementing a multitude of algorithmic functions shared by multiple systems; and (3) multi-DSP-centered architectures with accelerator assistance for keeping current and area consumption within reasonable limits.

Especially the third option needs some further analysis. Those architectures show their strengths best when multiple systems with roughly similar complexity need to be implemented in a single device, but which are not used simultaneously, e.g., 3GPP LTE, IEEE 802.16m, and IEEE 802.11n. When focusing only on mainstream 3GPP cellular systems, some figures of metric are critical: because of the relatively high software content, die size and power consumption are slightly higher than with implementations corresponding to the first category [9]. Pure die size might not be the most important criterion anymore, since silicon technology scales a bit faster than processing requirements. For current consumption, most important criteria are standby and talk times, both not maxing out processing capabilities. By extensively applying sophisticated low-power design principles, current consumption could likely be tamed for those use cases. Since baseband power consumption is only one of many current consumers (see Fig. 10.6), a moderate increase might be accepted.

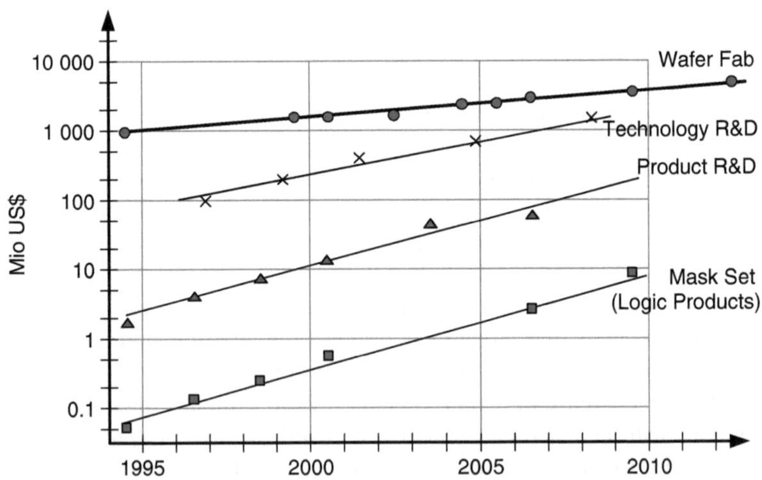

Fig. 10.6 Energy consumption of a state-of-the-art dual mode 2G/3G (GSM/EDGE/UMTS) modem: 3G standby current is ca. 20% higher than 2G standby current; 3G talk current is ca. 25% higher than 2G talk current (all for GSM-A conditions, GSM-PA at PCL 7 (+29 dBm) 3G-PA at +10 dBm, low band, no display). *PA* Power amplifier, *RF* RF-transceiver, *BB* digital baseband, *AFE* analog front end, *PMU* power management unit, *MEM* memory

Due to their programmability, DSP-centered architectures are inherently flexible. Not to constraint itself, this flexibility needs also to be transferred to the assisting accelerators. Flexibility is mostly needed for new systems; legacy systems are mature: any change of legacy systems is rather incremental and evolutionary, not to compare with a complete new air interface like WCDMA and LTE introduced. Especially during the early days of a new system, the path to take might be ambiguous. For instance, future roles of LTE-TDD and IEEE 802.16m might be not so clear, yet. Thus, an implementation potentially leveraging synergies between those different systems, like those based on DSP-centered architectures would do, might initially be expedient. In contrast, especially for those new systems some hardware acceleration support is required to control area and current consumption.

To fully exploit signal processing capabilities of those architectures, legacy systems might need to be ported to software. This requires much higher efforts in verifying system functionality than a simple reuse approach, where a key design criterion is minimization of modifications.

Overall, only if DSP-centered architectures manage to cope more efficiently and more economically with growing complexity of future multi-system implementations, they might substitute the current reuse-based architectures.

10.4.3 SoC Integration

The trend towards packing functionality onto a single die will continue and move from low-cost phones into the higher phone categories. This is enabled by digitalizing big parts of RF processing, so that complementary MOS (CMOS) technology scaling enables an overall shrink of the RF processing. Since bandwidths and number of antennas increase, the interface between RF and baseband signal processing gets increasingly broadband: 2G requires a single 26 MHz signal for both, transmit and receive path, 3G requires two 312 MHz signals, one for transmit path, one for receive path, and LTE already requires three 1.248 GHz signals, one for transmit path and two for receive path. Future LTE-Advanced systems might require interface data rates considerably beyond 10 Gbit/s. Clearly, that interface does not scale equally nicely as the rest of the modem implementation. Apart from the high amount of energy required for transferring high-speed data across chip boundaries, this gives a further push for single-chip integration.

Also regarding RF signal processing support for more and more frequency bands will be common. Current mobile phones support up to four 2G bands and three 3G bands. Table 10.1 lists currently specified bands for 3G services. 2G bands are fully covered by respective 3G bands. It is expected that the number of frequency bands will be growing in future too, e.g., also including bands in the areas of 3.5 GHz, 2.3 GHz, and 450 MHz. Today's low-cost implementations support only a single band (e.g., band I), typical triple-band implementations support bands I, II, and V, and typical future penta-band implementations might additionally support bands IV and VIII.

Table 10.1 3G frequency bands

3GPP band	TX band (MHz)	RX band (MHz)	Region
Band I	1920–1980	2110–2170	Europe/Asia/Japan
Band II	1850–1910	1930–1990	Americas
Band III	1710–1785	1805–1880	Europe
Band IV	1710–1755	2110–2155	Americas
Band V	824–849	869–894	Americas, Australia
Band VI	830–840	875–885	Japan
Band VII	2500–2570	2620–2690	(Worldwide, LTE)
Band VIII	880–915	925–960	Europe
Band IX	1749.9–1784.9	1844.9–1879.9	Japan
Band X	1710–1770	2110–2170	Americas
Band XI	1427.9–1452.9	1475.9–1500.9	Japan
Band XII	698–716	728–746	Americas
Band XIII	777–787	746–756	Americas
Band XIV	788–798	758–768	Americas

3GPP TS 25.101 v8.2.0 (2008–03), User Equipment (UE) radio transmission and reception (FDD).

10.5 Connecting the Unconnected

As can be seen in Fig. 10.7, the importance of mobile radio is even higher for emerging markets than for industrialized markets. Different regions of the world are sorted according to their per capita gross domestic product (GDP). Even if the GDP is not the only or the most appropriate economic wealth indicator, it is used here because it is well measurable. The lower the GDP, the higher the growth rates of cellular subscriptions. The subscriptions per inhabitant are still lower as in wealthier regions of the world, but the markets are not yet saturated. The importance of mobile phones is pronounced by the ratio of cellular to fixed-line subscribers. This ratio is about 2 in North America and in Western Europe, whereas there are 6 (India) or 9 (Africa) cellular subscriptions per fixed line in regions with the smallest per capita GDP. Hence, the added value of a mobile phone must be much higher there. By enabling affordable cellular phones via SoC integration, semiconductor technology contributes to increasing the level of living and creates value not just economically, but also perceptible in every day's life.

Roughly 3 billion "unconnected" people are currently living in markets with little or no wired communications infrastructure. Likewise, trouble regions, which got into a crisis by either a natural cause (droughts, floods) or a human cause (wars) usually do not have a reliable wired infrastructure. The basic benefits of cellular handsets for those people include access to information, trading, bargaining, financial transactions, and even a kind of identity by a unique phone number. In a nutshell, access to basic communications helps people to survive.[2]

All this requires intuitive user interfaces without the need for literacy in the right language, independence of electrical power supplies, and affordable handsets

[2] Chipchase J. Future Perfect, www.janchipchase.com.

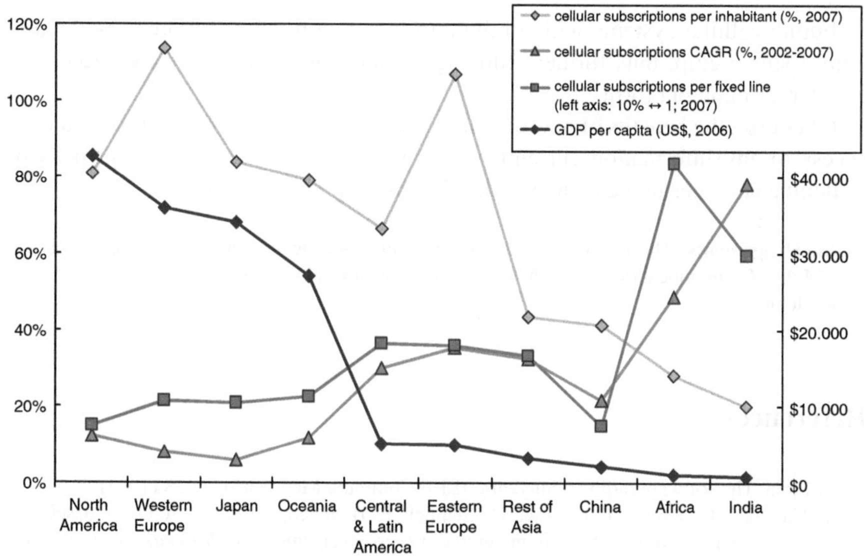

Fig. 10.7 Relation between some basic telecommunications indicators in different world regions (World Telecommunication Indicators, ITU, www.itu.int/ITU-D/ict/index.html.)

supporting that basic functionality. The above presented SoC integration is key in enabling those technologies for emerging markets. On the other hand, only the extremely high volumes of phones for emerging markets justify the development of the corresponding semiconductor technology for SoC integration.

In a next step, this high integration of functionality onto a single die will continue and move from basic phones into the higher phone categories.

10.6 Summary

To develop platforms for future cellular systems that are cost, area, and power efficient will be one of the major challenges for semiconductor companies. One of the hurdles to cope with is cost increase due to exponentially growing modem complexity and fabrication cost.

Whereas silicon implementation will likely always be dominated by the latest cellular standards, verification complexity is more equally distributed between different systems. Since verification of today's systems is already equally expensive as chip manufacturing, it is essential to minimize those efforts by reusing proven legacy implementations to the fullest possible extent. Software-centric architectures will only be used, once they allow implementing cellular systems more efficiently than current reuse-based architectures including all system verification tasks.

Future cellular systems will continue to integrate more and more functionality onto a single chip, thus further reducing component count and board space of a cellular terminal platform.

Of course, the benefits of new cellular systems with ubiquitous and instantaneous access to any information currently required should be worth to accept perhaps difficult challenges in the course of developing competitive solutions.

Acknowledgements The authors wish to gratefully acknowledge the support of their colleagues at Intel Mobile Communications and Ruhr-Universität Bochum. This work was supported by Infineon Technologies.

References

1. J. Mitola, The software radio architecture. IEEE Commun. Mag. **33**(5) 26–38 (1995)
2. M. Hammes, C. Kranz, J. Kissing, D. Seippel, P. Bonnaud, E. Pelos, A GSM baseband-radio in 0.13 μm CMOS with fully integrated power-management. in *Proceedings of the IEEE International Solid State Circuits Conference (ISSCC) 2007*, San Francisco, 2007, pp. 18–20
3. H. Eul, ICs for mobile multimedia communications. in *Proceedings of the IEEE International Solid State Circuits Conference (ISSCC) 2006*, San Francisco, 2006, pp. 21–39
4. Y. Neuvo, Cellular phones as embedded systems. in *Proceedings of the IEEE International Solid State Circuits Conference (ISSCC) 2004*, San Francisco, 2004, pp. 32–37
5. J. Hausner, Integrated circuits for next generation wireless systems. in *Proceedings of the European Solid-State Circuits Conference (ESSCIRC) 2001*, Villach, 2001, pp. 26–28
6. J. Berkmann, C. Carbonelli, F. Dietrich, C. Drewes, W. Xu, On 3G LTE terminal implementation – standard, algorithms, complexities and challenges. in *Proceedings of the International Wireless Communications and Mobile Computing Conference (IWCMC) 2008*, Crete, 2008, pp. 970–975
7. A. Murugan, H. El Gamal, M.O. Damen, G. Caire, A unified frame-work for tree search decoding: Rediscovering the sequential decoder. IEEE Trans. Inform. Theory **52**(3), 933–953 (2006)
8. W. Ziebart, Technical and economical trends in microelectronics. in *Proceedings of the European Solid-State Circuits Conference (ESSCIRC) 2007*, Munich, 2007, pp. 1–10
9. U. Ramacher, Software-defined radio prospects for multistandard mobile phones. IEEE Comput. **40**(10), 62–69 (2007)

Chapter 11
Wireless for Industrial Automation: Significant Trend or Overrated?

F. Krug and L. Wiebking

11.1 Introduction

Wireless technology is currently experiencing a boom in wireless personal communications. Nevertheless, applications for industrial sensor systems have to overcome some different challenges.

Major automation vendors are increasingly integrating wireless applications into their products. New opportunities for plant improvements are being seen in efficiency, safety, security, and productivity. But in order to work in the difficult and changing industrial environment, the wireless technologies must deliver reliable performance, cost effectiveness, and ease of use.

The idea to use wireless technology to reduce costs and improve efficiency is not new. Manufacturers for example have been using wireless in the warehouse for asset tracking, materials handling, and supply chain management for a while now.

The new wireless network technologies address the specific challenges of using wireless in large manufacturing facilities. By offering an extended range and lower costs of plant and process network communications, significant improvements in the overall efficiency of the plant can be realized.

With the growing list of wireless applications, the numbers of wireless devices and systems that support these applications grow as well. Although most of the systems are using unlicensed frequencies, which are shared across the different technologies and applications, complexity arises from using multiple wireless technologies. The emergence of robust standards simplifies the sharing of these frequencies. However, no single wireless technology or standard is perfectly suited for being the single best solution for every application.

F. Krug (✉) and L. Wiebking
Siemens AG, Munich, Germany
e-mail: florian.krug@siemens.com, leif.wiebking@siemens.com

S. Lindenmeier and R. Weigel (eds.), *Electromagnetics and Network Theory and their Microwave Technology Applications*, DOI 10.1007/978-3-642-18375-1_11,

11.1.1 Requirements for Industrial Wireless

Industrial environments differ heavily from office or home environments. It is important to recognize that although industrial networks carry far less data than business networks, they carry the data through much harsher conditions. High temperatures, excessive airborne particulates, long distances between the equipment and systems, and other challenges make it difficult to place and reach data communication devices.

Wireless sensor systems can revolutionize industrial processing and help industry meet the demands of increased competitiveness. Intelligent wireless sensors built for ubiquitous use in industrial environments will enable real-time data sharing throughout a facility to increase industrial efficiency and productivity. Wireless sensor technology offers reliable, autonomous process control to improve product quality, increase yield, and reduce costs. By using electromagnetic waves as their transmission medium, wireless systems avoid the limitations of wired networks and offer competitive advantages in terms of cost, flexibility, and ease of use [1].

The costs associated with installing, maintaining, troubleshooting, and upgrading wiring have escalated while costs for wireless technology have continued to drop, particularly in the areas of installation and maintenance [2].

Some industrial applications require absolute reliability in systems control to avoid serious consequences such as injury, explosions, and material losses. Emerging wireless sensor systems can offer built-in redundancy and capabilities for anticipatory system maintenance and failure recovery. Demonstration of reliability will pave the way for deployment in these applications.

Integrated wireless sensor systems with distributed intelligence can enable operator-independent control of industrial processes. Sensor nodes can dynamically adapt to and compensate for device failure or degradation, manage movement of sensor nodes, and react to changes in task and network requirements. They can locate themselves in 3-D space and correlate their positions with on-line plant maps to assure correct placement. Continuous, high-resolution, ubiquitous sensing systems have the potential to autonomously monitor and control industrial processes. Based on the application, such systems will be capable of maximizing product quality and yield while minimizing waste, emissions, and cost.

Manufacturers and industrial companies have become increasingly concerned about threats of industrial espionage and cyber terrorism. New strategies for encrypting and even hiding wireless data transmissions promise a level of security that equals or surpasses that of wired systems. Upgradeability is essential to maintain security as technologies evolve and new threats emerge.

Recent advances in materials technology should enable integrated wireless sensor systems to meet durability and reliability requirements in harsh industrial environments. Integrated sensor nodes encased in advanced materials should be able to endure repeated exposure to caustic gases and high temperatures. Some applications may require components designed to withstand highly specific environmental challenges.

With the wide range of potential applications and broad diversity of physical devices, the software components will need to be highly modular and efficient. A generic development architecture should allow specialized applications from a wide spectrum of devices without requiring cumbersome interfaces. This will also enable connection to existing sensors and easy upgrades to incorporate more advanced modules in the future [1].

11.2 Wireless Standards for Industrial Applications

Wireless technologies can be separated into two general groups: high data rate and low data rate. The most known wireless standards for industrial applications are WLAN, Bluetooth, and ZigBee. In the following, the pros and cons of these state-of-the-art wireless network technologies are being discussed.

Currently, the most prominent specification for 802.11 WLAN standards is Wi-Fi alliance. Wi-Fi operates in the license-free 2.4 GHz industrial, scientific, and medical (ISM) band [3]. However, 802.11 WLAN is only the standard for high data rate applications.

Generally accepted as the most advanced and widely used wireless technology, the 802.11 product family is applicable to numerous IT and process related applications as a wireless extension of Ethernet. The high degree of standardisation and low costs of the technology, coupled with broad availability of enabled products and an increasingly sophisticated level of security make it a primary wireless technology for higher bandwidth devices such as mobile operator terminals, video surveillance cameras and handheld data loggers [4].

Nevertheless, compared to wired networks, Wi-Fi requires excessive overhead in terms of power consumption, software, processor resources, short ranges (160 m max) and size of physical components, making it less than effective in most industrial situations. Additionally, the channels to support high density sensor and condition monitoring networks are limited.

The Bluetooth technology is originally designed as a short-range wireless connectivity solution for personal, portable, and handheld electronic devices. The Bluetooth radio also operates on the 2.4 GHz ISM band. Notably, Bluetooth employs a fast, frequency-hopping spread spectrum (FHSS) technology to avoid the interference in the ISM band and ensure the reliability of data communication. For industrial applications, the employment of Bluetooth is rather limited.

With extensive applications of Bluetooth for wireless data communication in hand-held devices and wireless computing, researchers also have drawn on Bluetooth for local positioning. Similar to Wi-Fi, Bluetooth can provide several meters of localization accuracy based on the popular received signal strength indicator (RSSI) methodology. Strong multipath interference is identified as one of the key factors that affect positioning accuracy [3].

In field trials, it was found that the communication range of Bluetooth modules may reduce from the nominal 100–20 m due to complex site conditions. Moreover,

Bluetooth has a relatively high duty cycle and a minimal data throughput; currently a maximum of 3 Mbit/s is possible.

Sensors for industrial applications mainly rely on low data rate solutions. In the low data rate area, network structures are either arranged as Star or Mesh networks.

Star networks are usually realized with 802.15.4 standard in the field of factory automation. Advantages of this standard are its real-time data character and low power consumption.

ZigBee is a global standard for wireless mesh network technology that addresses remote monitoring and control applications. The technology defines the physical and medium access control (MAC) layers for low cost and low rate WPAN. Important features of ZigBee include a low data rate, extremely low power consumption, low complexity and high reliability and security [3].

A disadvantage is Zigbee's low data rate of up to 720 kbit/s and its poor interoperability. However, because it is relatively new, hardware developers are still refining and defining their systems.

Best practice for mesh networks for industrial process sensors is Wireless HART which employs 802.15.4 and offers a multihop mesh network layer.

The goal is to provide a standard, yet extensible, protocol stack for use with 802.15.4 radios with enough flexibility for use in limited power environments for low latency, single hop networks as well as longer distance, multihop mesh network configurations [4].

11.3 Power Technologies

Even as wireless sensor technology continues to benefit from advances in other commercial wireless products, system developers will need to overcome significant hurdles unique to industrial applications. In industry, uninterrupted production has always been of paramount importance. Plant managers will not adopt a new technology until they are certain it can deliver real value to their operations. Many manufacturing industries operate on narrow profit margins, so any system downtime can have major consequences for profitability. Industrial facilities require systems that perform quickly, reliably, and cost-effectively [1]. As a matter of fact, innovations in power technologies are critical to wireless' further spread.

One obstacle for wireless technology is the fact that the available battery life time for devices is not yet sufficient for many applications. Of course, the ideal solution would be to not require a battery change throughout the whole device life cycle. Therefore, two approaches exist: the first one is to reduce power usage of the devices; the second one is to develop innovative technologies for power distribution.

Further problems to be solved are the reliability of power generation, the maintenance need of power storage, and the emission and range of power transmission. One solution for these problems can be found in ABB's WISA (Wireless Interface for Sensors and Actuators). It offers a wireless real-time capable sensor/actuator interface for industrial applications, magnetic fields based power supply through alternating fields, and one magnetic coil supplied by power source and smaller coils as receivers that enforce the magnetic flow.

11.4 Architecture

Wireless monitoring sensors allow better, real-time data for the control system, predictive maintenance or asset management application. Operators in the field are now able to see the control system and review standard operating conditions, procedures, and corrective actions in real-time as they make field adjustments [5] (Fig. 11.1).

11.5 Self Energized Sensors

Through design improvements, wireless sensor systems of the future will require less power and therefore less maintenance (e.g., battery replacement) than today's systems. By 2010, costs associated with operating and maintaining these systems (sensing and transmission) will decrease by 90%. In the long term, systems will be self-powering, capturing energy (e.g., thermal, solar, or vibrational energy) from the

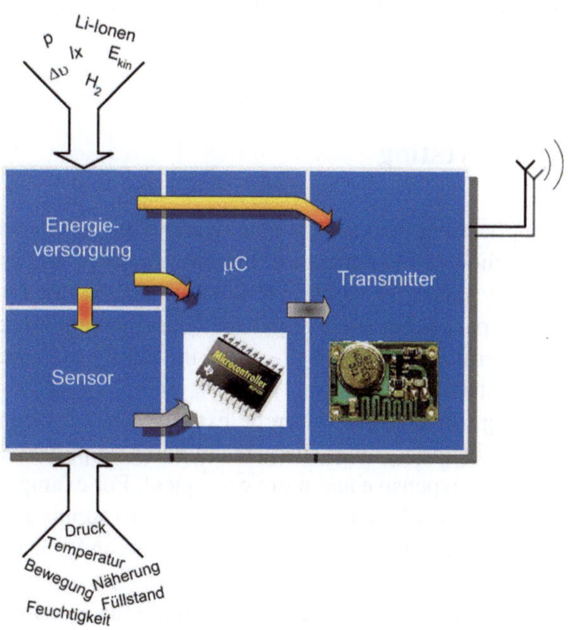

Fig. 11.1 Wireless sensors, architecture

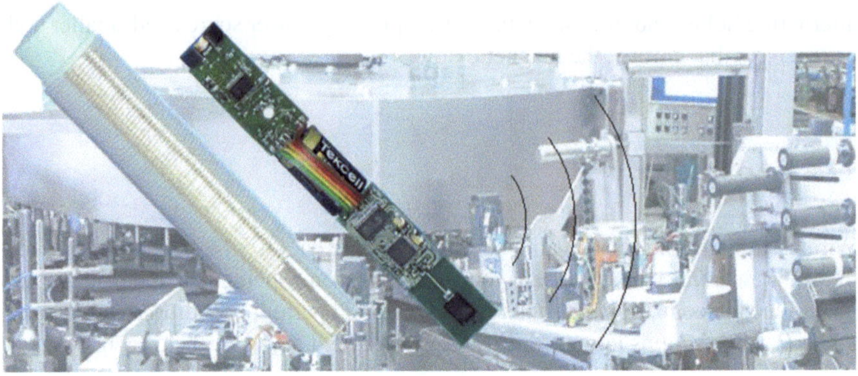

Fig. 11.2 Self energized sensors: proximity switch

industrial environment and virtually eliminating power maintenance activities and related costs [1] (Fig. 11.2).

The lifetime of innovative self energized sensors exceeds 5 years thanks to an ingenious energy management. An integrated sleep-mode guarantees power reduction and with the help of power saving electronics, the sensors draw less than 20 μA. Other features of the sensors, which use standard M18 housing, include event triggered communication by radio link (modified ZigBee), online programmable sample rate from 5 to 100 ms, and indication of switch status by an intelligent LED driver.

11.6 Energy Harvesting

Since industrial applications increasingly employ miniaturization and require longer intervals between scheduled maintenance, the power source and power conservation strategies are key issues for wireless sensor systems. Some of today's wireless systems rely on solar panels, but many require batteries that require periodic replacement. For the long term, developers will extend the ability to scavenge or harvest power from the industrial environment [1].

For the powering of sensors, different external energy sources exist (see Fig. 11.3). Instead of batteries, other chargers are being employed; chargers which are expected to be less expensive and more ecological. For example, some devices can convert vibration and solar energy into electricity to supply power to the sensors. In general, five different energy harvesting devices can be distinguished: solar, mechanical, thermal, radio frequency (RF), and wind.

Research on energy harvesting is valuable although the amount of energy to be harvested from the environment is typically very low. Nevertheless, several prejudices concerning energy harvesting exist. A first concern is that it is only applicable to low power devices. Secondly, energy harvesting is not very reliable since it

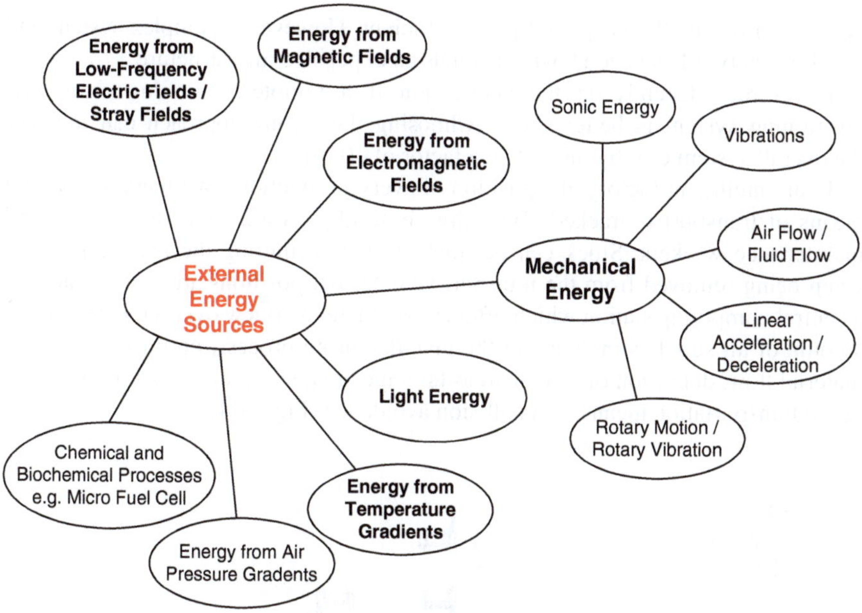

Fig. 11.3 Energy harvesting: powering of sensors with ambient energy

depends on the availability of environmental energy and no generic solution is possible. Finally, the integration of energy converters will cause additional costs.

11.7 Wireless Local Positioning and Self Organizing Wireless Sensor Networks

Local positioning will be one of the most exciting features of the next generation of wireless systems. Completely new concepts and features for wireless data transmission and transponder systems will emerge. Self-organizing sensor networks, ubiquitous computing, location sensitive billing, context dependent information services, tracking and guiding are only some of the numerous possible application areas [6].

Local positioning of a mobile device works in both ways: it can either gather information about its position or it can be localized from elsewhere. In a self-positioning system, the measuring unit is mobile. It receives the signals of several transmitters in known locations and can calculate its actual position based on measured signals. Remote-positioning systems work exactly the other way round: the signal transmitter is mobile and several fixed measurement units receive the transmitter's signals. The position of the transmitter is then calculated by a master station which collects all measurement units. In a remote-positioning system, the mobile device can be small, cheap, and power efficient. However, a complex system and backbone network is needed, which equals an expensive infrastructure.

It severely depends on the application if a remote-positioning or a self-positioning system is better suited. Choosing the wrong approach can increase the overall system cost by more than a factor of 10 [6].

In an intelligent factory, the position of every production machinery, stock, and means of transport is tracked. Typically, these objects are fork-lifts, cranes, and maintenance workers. Stock can be tracked by transmitting the precise position when being removed from the transport vehicle. All positions are consolidated in a central computing station which offers a complete overview over the location and amount of all supplies included in the manufacturing process, optimization of the material flow, definition of virtual areas (storage area, etc.), restriction of operation of the transportation means, and collision avoidance (Fig. 11.4).

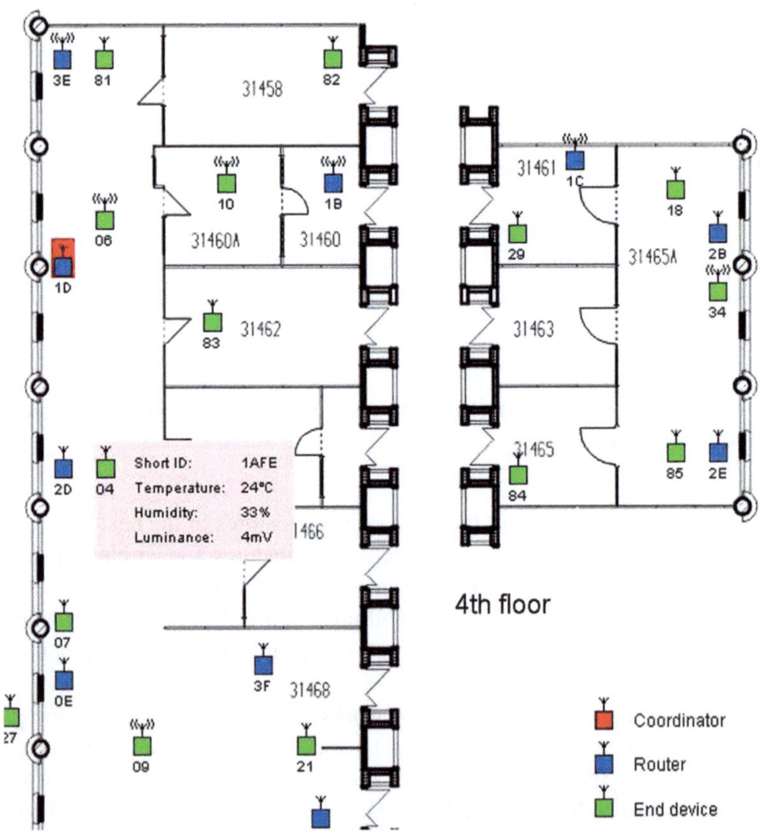

Fig. 11.4 Self organizing wireless sensor network with highly integrated low-power sensor boards

Naturally, this application requires a faster update rate adjusted to the speed of the transportation means. Furthermore, the precision of the position measurements must be similar or better than the physical dimension of the transported objects [6].

Self organizing wireless sensor networks allow monitoring of large areas with lowest infrastructure effort. The wireless networks connect embedded sensors, actuators, and processors. A wireless sensor network refers to a group of sensors, or nodes, linked by a wireless medium to perform distributed sensing tasks.

A platform for industrial sensor applications, which uses a small generic and standardized hardware, optimizes power management. The network structures are thereby either arranged as Star or Mesh networks and the highly integrated low-power sensor boards include signal processing of sound, images, radar signals, and others signals. Furthermore, the robust localization (802.15.4a) of the sensor nodes support the maintenance and installation processes.

By integrating sensing, signal processing, and communications functions, a sensor network provides a natural platform for hierarchical information processing [7].

11.8 Industrial Applications for Wireless Sensors

There is not a "one-size-fits-all" wireless networking technology that adequately supports the diverse and demanding requirements of industrial applications and environments [5]. Nevertheless, employing the best wireless application for any given plant ensures a cost-effective industrial application implementation.

One application of wireless sensors in an industrial environment is the continuous monitoring of perishable food, beverages, and pharmaceuticals. Each pallet of perishable goods is equipped with one intelligent, self-energizing wireless sensor module. These wireless modules measure sensor values like temperature, humidity, CO2, and more. The sensor information is then transferred on from one pallet to another and provided to a central database. Moreover, the wireless modules can be located globally -via GPS or locally- by means of transmission delay. Condition information is generated and forwarded via self-organizing wireless network to a (distant) access-point (Fig. 11.5).

Another industrial application for wireless sensors can be found in the temporary environmental monitoring in oil, gas, and mining installations. The monitored area is equipped with intelligent, self-energizing wireless modules that measure gas concentration (CO, NOx, CxHy), temperature, and humidity. In addition to that, the wireless module can locate itself by means of transmission delay measurement. Condition information will be generated in the wireless network and forwarded wirelessly to an access-point (Fig. 11.6).

In a similar way work the energy autonomous sensor nodes which are deployed along pipelines. Here, sensors are monitoring noise and vibration of the pipe or the environment and can thus identify unusual sound and vibration patterns (locally and in collaboration with neighboring sensor nodes) due to cracks or digging activities.

Fig. 11.5 Food and beverage: monitoring of perishable groceries

Fig. 11.6 Sensor information is transferred from one node to another and passed on to a central controller or database

Finally, wireless sensors can also be quite useful in logistics. They offer a general solution for container and wagon tracking as well as their surveillance. Again, each container or wagon is equipped with one intelligent, self-energized wireless module. The module controls defined local functions and parameters such as temperature, vibration, and localization. Additionally, the wireless module can localize itself either globally via GPS or locally via transmission delay. The gained information

Fig. 11.7 Information is passed on from one container to the next

will be forwarded via self-organizing wireless network to a distant access-point (Fig. 11.7).

11.9 Swot Analysis

Wireless technology offers large opportunities for many applications but also some threats in the field of industrial automation. In the following, a brief SWOT (strengths, weaknesses, opportunities, threats) analysis will be performed to identify major drivers and threats.

The potential reduction of cost is a major strength of wireless technology. Wiring, maintenance, planning, and installation costs can be minimized and thus increase a company's profitability. Additionally, wireless enhances the mobility and flexibility of both- devices and users. Finally, with the help of wireless technology a higher degree of personalization is possible.

Despite these remarkable strengths, a number of weaknesses persist. Firstly, wireless is vulnerable against interference, noise (EMI), and other environmental effects. Also, the problems of high power consumption and engineering complexity remain unsolved. Furthermore, the real-time responsiveness under all conditions needs further improvements.

The employment of wireless technology in industrial automation offers tremendous opportunities. Through the enhanced technologies safety and information availability increase and location-based-services are facilitated. Moreover, a new organization of work using the given mobility and a customization become possible. Finally, the new business opportunities for established supplier companies in the worldwide market should not be underestimated.

Threats concerning wireless technologies can be found in the public perception of radiation exposure including the fear of a possible health risk. Skepticism against the returns on investments persists and needs to be further investigated. The time to market and the time to standardization remain a big question mark for analysts and specialists.

11.10 Conclusion

This paper presents the potentials of wireless sensor systems in industrial applications. In these applications, sensors mainly rely on low data rate solutions. Wireless technologies for industrial automation create new opportunities to make plants more efficient, productive, and secure. Nevertheless, it is important to recognize that although industrial networks carry far less data than business networks, they carry the data through much harsher conditions. Different standards for industrial applications are being presented and evaluated. The paper concludes by analyzing industrial applications and new innovations on wireless sensors.

Acknowledgements The authors would like to thank Alexander Franck for supporting the publication.

References

1. "Industrial Wireless Technology for the 21st Century", U.S. Department of Energy, Workshop notes, USA, 2002
2. W.W. Manges, G.O. Allgood, S.F. Smith, It's time for sensors to go wireless, Part 1: technological underpinnings, sensors. J. Appl. Sensing Technol. **16**(4), 10–20 (1999)
3. X. Shen, W. Chen, M. Lu, Wireless sensor networks for resources tracking at building construction sites. Tsinghua Sci. Technol. **13**(S1), 78–83 (2008)
4. I. McPherson, Industrial wireless: hope, help or hype? *The Industrial Ethernet Book* (2006)
5. Apprion^{TM}, Open industrial wireless solutions: realizing the full potential of wireless (2008)
6. M. Vossiek, L. Wiebking, P. Gulden, J. Weighardt, C. Hoffmann, P. Heide, Wireless local positioning. IEEE Microw. Mag. **4**, 77–86 (2003)
7. G.J. Pottie, *Hierarchical Information Processing in Distributed Sensor Networks* (ISIT, Cambridge, MA, 1998)

Chapter 12
Sub-Microsecond Contactless Ultra-Wideband Data Transmission in Rotating Systems Using a Slotted Waveguide Ring

Christoph Seifarth and Gerd Scholl

12.1 Introduction

Real-time communications and interference free, robust wireless operation are required in many wireless sensor networks (WSN), especially in factory automation [1].

In factory automation real-time capability is often compared with the performance of the wired AS-Interface (AS-i), a field bus with a maximum response time of 5 ms, or an AS-i compliant wireless fieldbus gateway [2]. However, in some wireless sensor networks, especially in wireless control loop applications, a sub-real-time communication with sub-microsecond data transmission is essential.

Although the amount of control data is low, usually a few bits or bytes, the requirement of a sub-microsecond data transmission leads to data rates of several tens or even hundred Mbit/s. These data rates could be easily achieved by modern wireless standards such as Wireless LAN (IEEE 802.11) or Wireless USB with data rates of up to 480 Mbit/s, but they require large protocol overhead and signal processing effort. Hence, they lack short transmission and latency times and cannot realize sub-microsecond communications [3].

Ultra-wideband (UWB) systems are a promising alternative to common narrow band transceivers operating in the unlicensed ISM-bands for interference free and robust wireless data transmission, since UWB signals show a high robustness against multipath and frequency selective fading, even in dense multipath environments [4].

In impulse based UWB systems the information to be transmitted is carried by short pulses or pulse trains, either without the use of additional carrier modulation, i.e. a carrierless (baseband) transmission [5, 6], or using a single-tone carrier [7, 8]. UWB devices, as defined by the European Commission for the member countries of the European Union in 2007, spread their radio-frequency energy over a

C. Seifarth (✉) and G. Scholl
Institute of Electrical Measurement Engineering, Helmut Schmidt University
(University of the Federal Armed Forces Hamburg), 22039 Hamburg, Germany
e-mail: c.seifarth@ieee.org, gerd.scholl@hsu-hh.de

S. Lindenmeier and R. Weigel (eds.), *Electromagnetics and Network Theory and their Microwave Technology Applications*, DOI 10.1007/978-3-642-18375-1_12,

frequency range wider than 50 MHz with a low power spectral density of, e.g., −41.3 dBm/MHz in the upper UWB frequency band of 6.0–8.5 GHz [9].

Data transmission to and from moving guided vehicles [10, 11] or moving parts in machines [12] is a well known application for slotted waveguides, which provide a robust operation even in harsh and interference-prone industrial environments [13]. Slotted rectangular waveguides offer wave guidance and data transmission for large bandwidths and exhibit approximately the same transmission properties as conventional closed rectangular waveguides [14].

If used in machines with rotating parts slotted waveguides can be a promising alternative to maintenance intensive slip-rings, which produce mechanical abrasion and dust [15], or capacitive data links [16] which are very sensitive to mounting tolerances. Rotating couplers [17] or conventional rotary joints [18, 19] as used in radar devices for decades mostly demand mounting on the rotary axis which is not suitable for applications where the rotary axis must not be filled by any machine parts. Optical slip-rings and fiber optic rotary joints have also been present for some decades and provide very wide signal bandwidths [20] but are sensible to dust and grime.

Especially when several independent data channels have to be implemented and no time division multiple access is possible, a slotted waveguide ring benefits from its frequency division multiple access potential, whereas in other RF transmission systems additional data channels have to be added mechanically.

In this chapter we describe a data transmission system with ultra-low latency times using UWB technology for time-critical control loop applications in rotating systems. The implemented transceiver is fabricated using commercial off-the-shelf components offering the designer flexibility and lower costs for industrial non-mass-market products and applications.

12.2 System Description and Principle of Operation

12.2.1 Transmitter and Receiver

In the fast UWB transmitter design as shown in Fig. 12.1 a digital pulse generator using D-type flip-flops was implemented. Fast D-type flip-flops offer a low-cost generation of ultra-wideband pulses and a convenient LVTTL or LVCMOS logic-state input for binary data signals [7].

A bit stream of up to 100 Mbit/s feeds the first flip-flop which generates a Gaussian-shaped pulse and triggers the second flip-flop generating a second pulse. Depending on the data rate, those two pulses provide a Return-to-Zero (RZ) or Non-Return-to-Zero (NRZ) On-Off-Keying (OOK) modulated signal. The generated pulses are differentiated to get Gaussian monocycle shaped pulses and amplified by a gain block. Using a voltage controlled oscillator (VCO) and a broadband double-balanced mixer the signal is upconverted to the desired frequency band between 6.0 and 8.5 GHz and subsequently bandpass filtered.

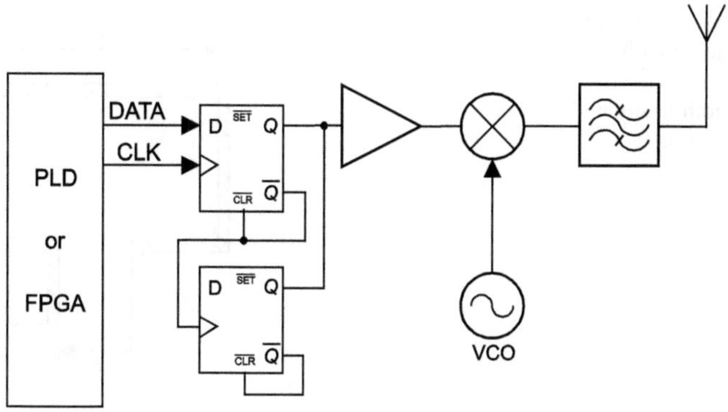

Fig. 12.1 Design of ultra-wideband transmitter for data transmission in rotating systems

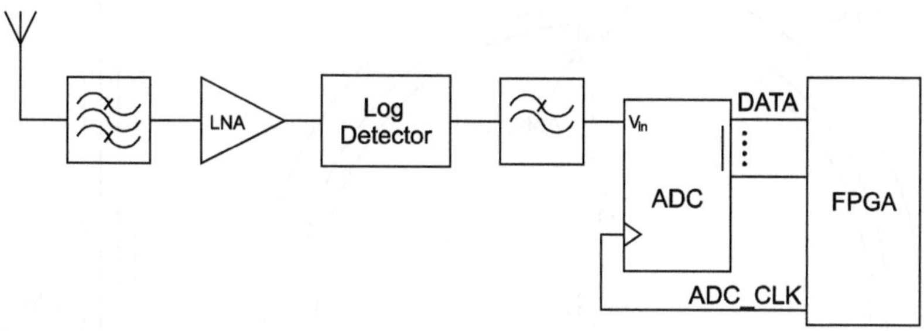

Fig. 12.2 Design of ultra-wideband receiver for sub-microsecond data transmission in rotating systems

The design of the corresponding fast ultra-wideband receiver is shown in Fig. 12.2. The received signal is bandpass filtered and amplified using a low-noise amplifier (LNA). A logarithmic detector is employed for signal detection, where the received signal energy corresponds directly with the envelope of the transmitted OOK-signal, which is sampled using an analog-digital-converter at a sample rate of 300 MS/s.

12.2.2 Slotted Waveguide Ring

Figure 12.3 shows a cross section of a slotted waveguide ring for which the inner waveguide width a and height b were chosen to provide guidance only for the fundamental TE_{10} waveguide mode across the desired frequency range of 6.0–8.5 GHz. The slot of the waveguide faces in radial direction towards the rotating axis and its width s is chosen to $0.3a$ to ease mounting and fabrication tolerances as far as possible. The height h_s of the slot is designed following the design rules in [11] to achieve the best wave guidance inside the waveguide and a minimum outside electric field.

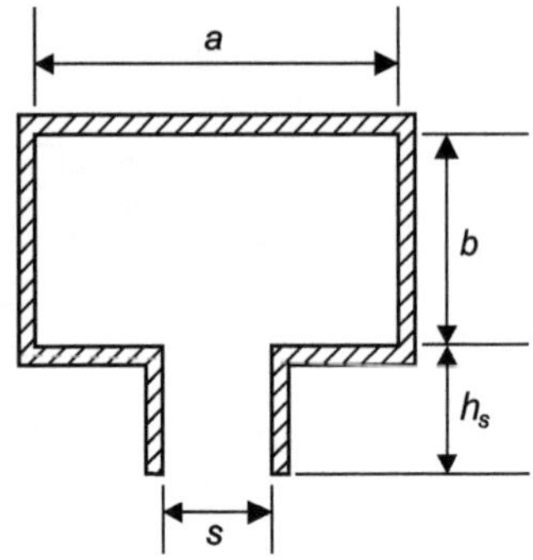

Fig. 12.3 Cross section of slotted waveguide with waveguide height a, waveguide width b, slot width s and slot height h_s

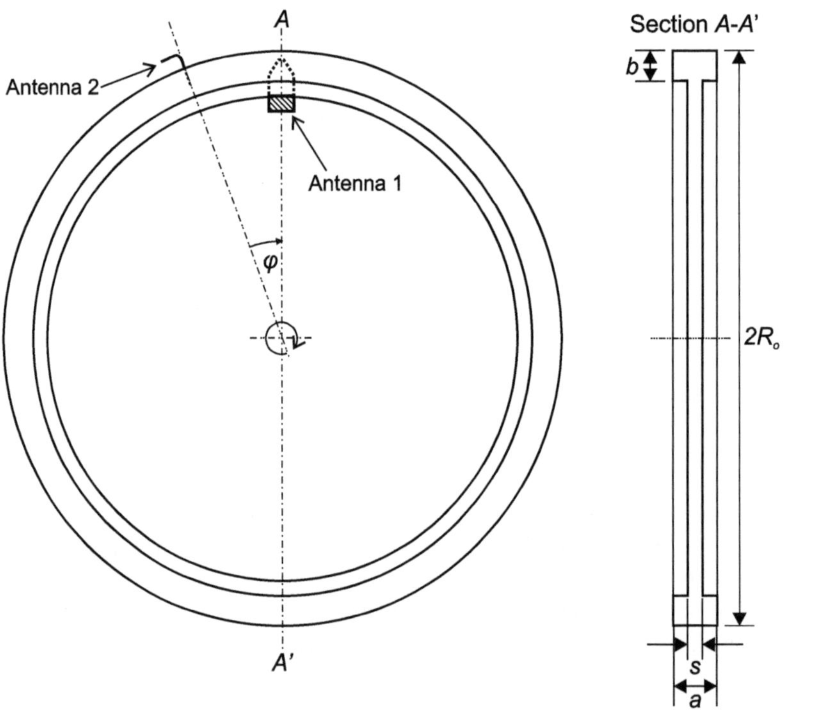

Fig. 12.4 Sketch of slotted waveguide ring of radius R developed for UWB data transmission in rotating systems with rotation angle φ between stationary antenna 2 and moving antenna 1

Since the slotted waveguide is designed as a closed ring structure as shown in Fig. 12.4 and the UWB signal inside the ring is excited using antenna structures with symmetrical radiation pattern in the E-plane, a multipath, i.e. a multiple clockwise and a multiple counter-clockwise, propagation of the signal can be observed if no countermeasures are taken.

The frequency-selective and dispersive channel impulse response $H(\omega, \varphi)$ of the slotted waveguide with multipath propagation can be expressed as [21]:

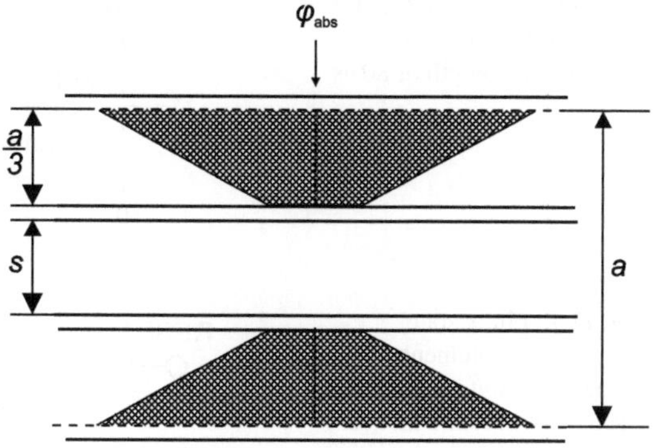

Fig. 12.5 Sketch of microwave absorber foam placed inside the slotted waveguide ring of waveguide width a and slot width s with indication of placement angle φ_{abs}

$$H(\omega,\varphi) = \sum_{\nu=0}^{\ell-1} \rho_\nu(\varphi)\mathrm{e}^{-\mathrm{i}\omega_0\cdot\tau_\nu(\omega,\varphi)} \tag{12.1}$$

with ℓ as the number of paths, ρ_ν as the amplitude and τ_ν as the time delay of the ν-th multipath component and ω_0 as the carrier frequency of the UWB signal.

To reduce multipath propagation and standing wave effects on the one hand but to still ensure a continuous data transmission on the other hand, four pieces of common microwave pyramid absorber foam material were fixed at a certain rotation angle φ_{abs} to the sidewalls of the waveguide ring as shown in Fig. 12.5.

12.2.3 Data Encoding, Decoding and Synchronization

Data is encoded in packets of one byte including three start and two stop bits as shown in Fig. 12.6 to minimize transmission and latency times as well as detection and decoding errors [22].

The pattern of start and stop bits is chosen to be unique compared to any occurable data pattern to minimize decoding errors.

An automatic calculation of the threshold value for data bit recovery as well as an error detection are performed in a field programmable gate array (FPGA). The finite state diagram of the implemented finite state machine (FSM) for data bit recovery is shown in Fig. 12.7.

State 1 is the idle state in which the state machine waits for the sample no. 0 to cross the threshold value to signalize the detection of the first start bit “1”. Since the received signal with a data rate of 100 Mbit/s is digitized with 300 MS/s, every bit sent is sampled trifold.

With the next sample no. 1 the state machine changes to state 2, in which the samples no. 0 to no. 8 are compared to given templates of the start bits “101” and are stored in a shift register. If the samples do not match to one of the start bit templates, the state machine returns to state 1.

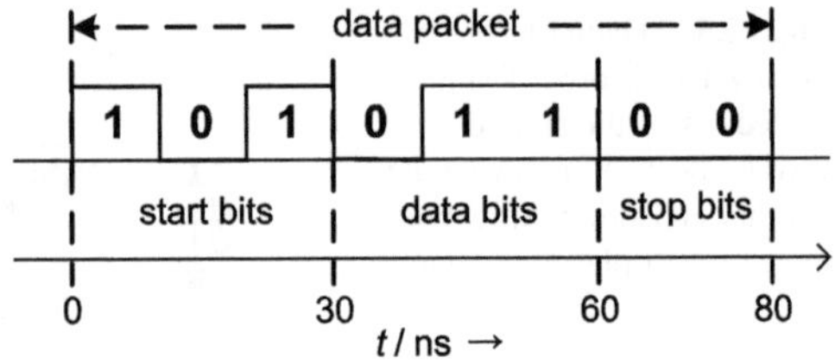

Fig. 12.6 Data encoding for ultra-wideband transceiver with a packet length of 80 ns

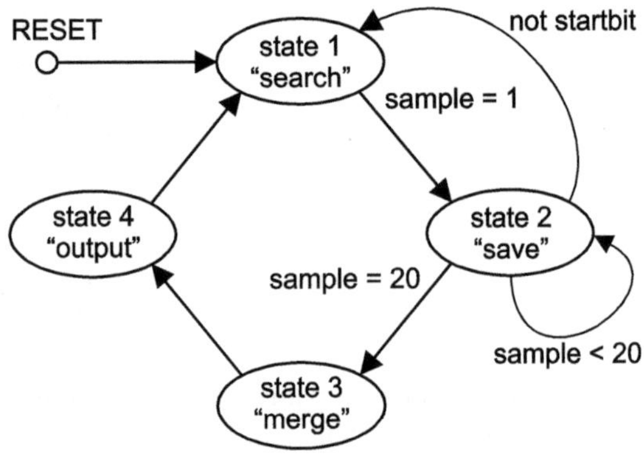

Fig. 12.7 Finite state diagram of implemented Moore type finite state machine with four states for data bit recovery in the UWB receiver

During state 2 a total of 21 samples are stored in the shift register. After sample no. 20, which represents the last sample of the first stop bit, samples no. 9 to no. 17 are converted to three parallel data bits. State 4 outputs the recovered bits to three I/O pins of the FPGA if the stop bits are fully recognized.

A complete cycle of the finite state machine requires only 24 clock cycles of 3.33 ns leading to a time duration of 80 ns, which is exactly the length of one data packet.

For immediate synchronization purposes and a continuous data transmission three finite state machines are implemented, the first starting on the first rising edge of a data bit, the second on the second rising edge and the third on a third rising edge. With this implementation and the unique start and stop bit pattern only one FSM gives a valid output and the receiver synchronizes itself continuously even when the continuous data transmission is interrupted [22].

12.3 System Fabrication

12.3.1 Transmitter

A standard printed circuit board (PCB) process and the 20 mil thick high-frequency substrate Rogers 4350B with a relative dielectric constant of $\varepsilon_r = 3.46$ and a dielectric loss tangent of $\tan\delta = 0.004$ was chosen for fabrication of the RF-frontend of the UWB transmitter shown in Fig. 12.8.

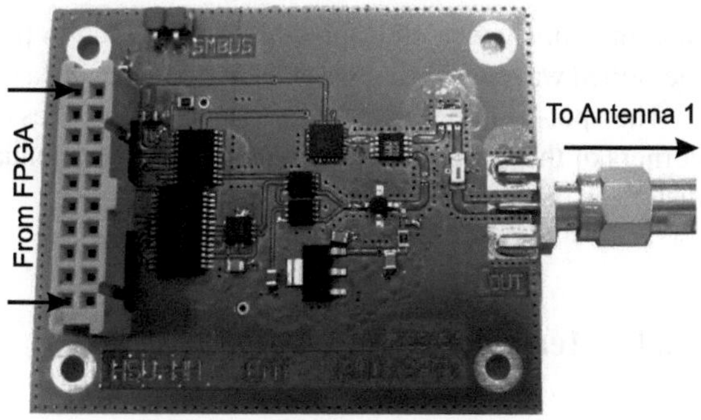

Fig. 12.8 Photograph of fabricated RF-frontend of ultra-wideband transmitter for sub-microsecond data transmission in rotating systems

For digital pulse generation the SN74AUC1G74 D-type flip-flops from Texas Instruments with a propagation delay of typically 1.1 ns were used.

The SiGe gain block HMC479MP86 distributed by Hittite Microwave Corporation amplifies the generated pulses by 15 dB. The implemented frequency mixer HMC220MS8 from Hittite offers a low conversion loss of typically 7 dB and RF and IF bandwidths of 5–12 GHz and DC-4 GHz, respectively. Its LO input is fed by a 12.5 dBm oscillator signal from Hittite's HMC532LP4 VCO.

The ceramic highpass and lowpass filters, HFCN 5500+ and LFCN 7200+, respectively, from Mini-Circuits form a bandpass filter and feature a insertion loss of 1 dB in the passband region. If steeper band edges are needed and a higher insertion loss is acceptable, a customized substrate integrated waveguide (SIW) filter like the one presented in [23] can be implemented into the same PCB.

12.3.2 Receiver

The bandpass filter implemented in the receiver consists of the same ceramic lowpass and highpass filters as in the transmitter. The LNA HMC565LC5 is from Hittite Microwave Corporation and offers a gain of 21 dB and a noise figure of 2.5 dB.

For signal energy detection the logarithmic detector AD8317 from Analog Devices was implemented and for anti-aliasing filtering Mini-Circuits' LFCN 80+ lowpass filter. Texas Instruments' 14 bit analog-digital-converter ADS5474 featuring a low conversion time of only 3.5 clock cycles and a sample rate of up to 400 MS/s was used on the evaluation board ADS5474EVM.

Digital signal processing was realized in a VHDL program on Xilinx's Virtex-II Pro XC2VP30 FPGA on the XUPV2P development system board.

12.3.3 Slotted Waveguide Ring

Top, bottom and side walls as well as the slot walls of the waveguide ring were manufactured separately using stainless steel and then spot-welded together instead of continuously welded to prevent skewing due to thermal expansions. Afterwards the slotted waveguide ring was nickel-coated using a standard galvanization process.

The spot-welding process left small slots of less than 0.1 mm in width in the corners of the waveguide leading to additional attenuation of the waveguide due to distortion of the surface currents.

12.4 Measurement Results

12.4.1 Transmitter and Receiver

Figure 12.9 shows the transmitted UWB signal with a random data bit sequence of "011" consisting of two double-pulses per logical "1" bit.

The received signal at a rotation angle of $\varphi = \pi/2$ after the logarithmic detector and anti-aliasing filter is shown in Fig. 12.10. Since a falling output voltage level of the logarithmic detector corresponds with a rising input power level, a logical "1" is represented by a low voltage level and a logical "0" by a high voltage level.

The slope of the logarithmic detector corresponds to 22 mV/dBm. Thus, the worst signal-to-noise ratio (SNR) shown in Fig. 12.10 leads to 3 dB, where for proper and secure decoding an SNR of 2 dB is required in the current design of the receiver.

12.4.2 Slotted Waveguide Ring

In Fig. 12.11 the measured magnitude of the channel impulse response at a rotation angle of $\varphi = \pi/2$ is shown with and without absorber material placed inside the waveguide ring at the rotation angle $\varphi_{abs} = \pi/4$.

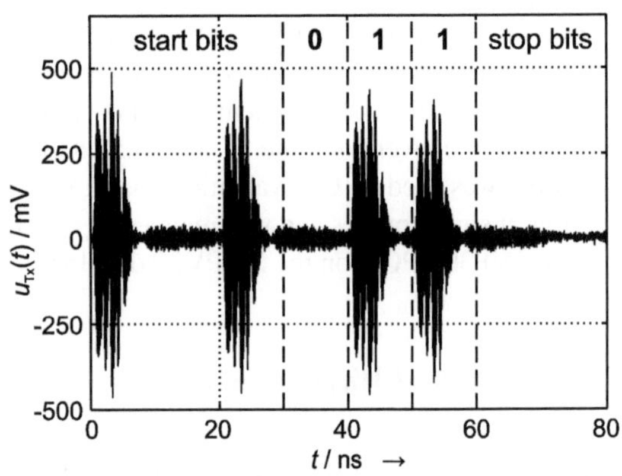

Fig. 12.9 Measured transmitted ultra-wideband signal for data transmission in rotating systems for the bit sequence "011"

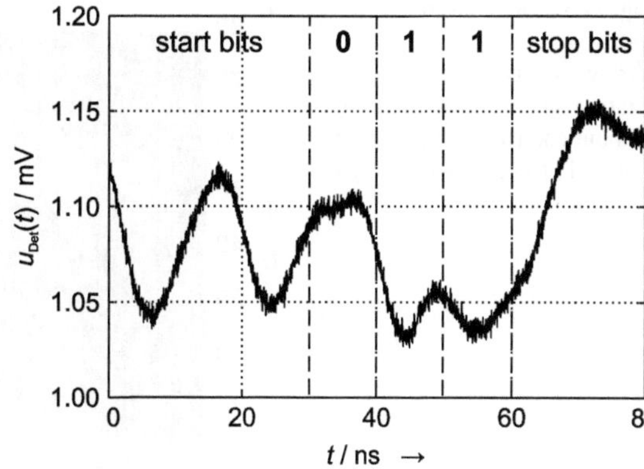

Fig. 12.10 Measured detected ultra-wideband signal before analog-digital conversion at a rotation angle of $\varphi = \pi/2$ for the bit sequence "011"

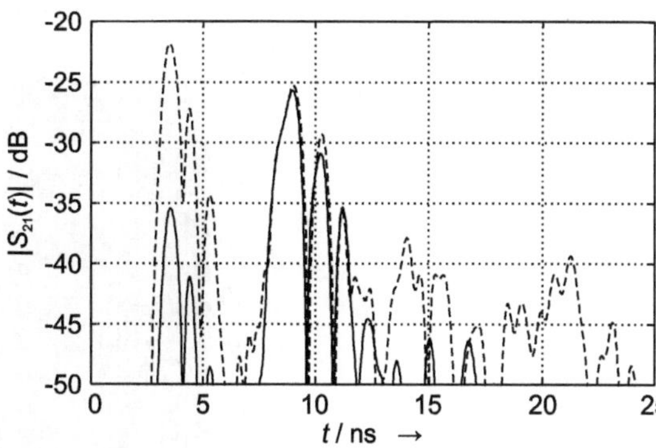

Fig. 12.11 Measured magnitude of channel impulse response $|S_{21}(t)|$ at a rotation angle of $\varphi = \pi/2$ versus time with (solid line) and without (dashed line) absorber material placed at $\varphi_{abs} = \pi/4$

When using the absorber material the peaks due to multipath propagation at $t = 3.4$ ns, $t = 14$ ns and $t = 21$ ns are attenuated by 13.6 dB and the peak at $t = 9$ ns is the direct and unobstructed, strongest path, leading nearly to a single-path channel transfer function.

Figure 12.12a and b show the measured spectrograms of two received 2.5 ns long UWB pulses with a Gaussian monocycle shape as described in [7] at a rotation angle of $\varphi = \pi/2$ without and with microwave absorber material placed at $\varphi_{abs} = \pi/4$, respectively.

In Fig. 12.12a, where no absorber material is used, the multipath propagation of the UWB pulse and the corresponding time delays can be observed very well. The direct and strongest path is shown as the dispersed pulses between $t = 1$ ns and $t = 5$ ns, $t = 50$ ns and $t = 55$ ns and between $t = 60$ ns and $t = 65$ ns followed by the pulses propagated in counter-clockwise direction between $t = 6$ ns and $t = 10$ ns, $t = 56$ ns and $t = 60$ ns and $t = 66$ ns and $t = 68$ ns, respectively.

The signal energies shown at $t \approx 16$ ns and $t \approx 22$ ns correspond to multiple circulations along the ring, which also lead to partly-destructive interference during the third pulse between $t = 60$ ns and $t = 70$ ns. Any multipath propagation inside the slotted waveguide ring is suppressed effectively when absorber material is used

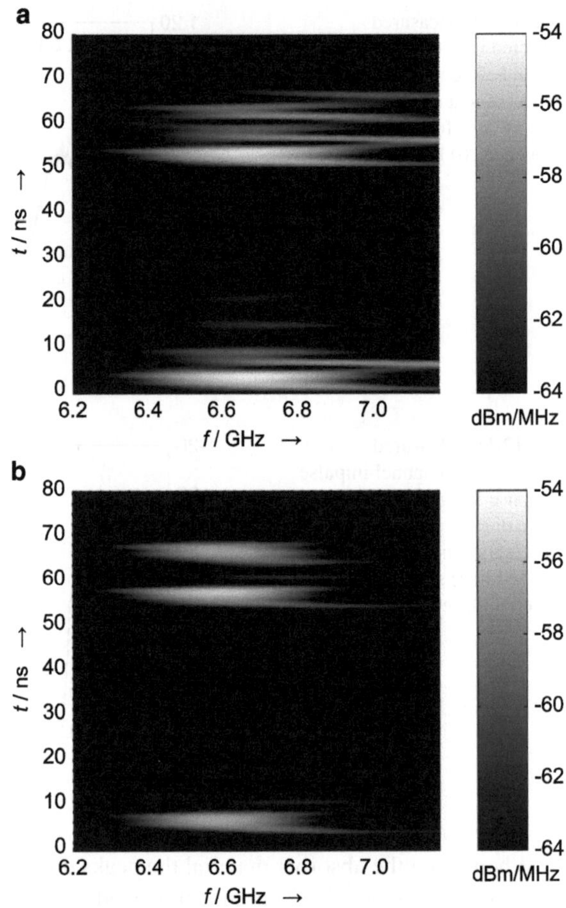

Fig. 12.12 Measured spectrogram of received ultra-wideband pulses at a rotation angle of $\varphi = \pi/2$ (**a**) without and (**b**) with absorber material placed at $\varphi_{abs} = \pi/4$

as can be seen in Fig. 12.12b, where only the slightly dispersed originally sent pulses are visible.

12.4.3 Timing Analysis

As can be seen in Table 12.1, the overall delay between providing the first bit of the data packet at the input of the transmitter and the output of the 3 parallel bits at the receiver sums up to 144 ns at the longest distance between both antennas.

The digital signal processing in the FPGA was designed and optimized to take only as long as the length of a data packet, i.e. 80 ns. In addition to those 80 ns required for signal processing another 6 ns are needed for signal routing inside the FPGA. The chosen fast analog-digital-converter provides a latency of only 3.5 clock cycles and, hence, needs only 12 ns for conversion of the detected signal.

Table 12.1 Timing analysis of ultra-fast ultra-wideband transceiver for

Process	Time consumption (ns)
Modulation in transmitter	17
Waveguide propagation	2–15
Detection in receiver	14
Analog to digital conversion	12
Digital signal processing in FPGA	86
Overall delay	131–144

Table 12.2 Performance overview of sub-microsecond UWB system for rotating systems

Parameter	Value
Frequency range	6.0–8.5 GHz
Bandwidth	1.0 GHz
Data rate	100 Mbit/s
Control data update rate	12.5 MHz
Control data time delay	$\leq$ 150 ns
Modulation	RZ-OOK
Output power	−8.6 dBm

Those two receiver components, i.e. the analog-digital-converter and the FPGA, keep the demodulation and decoding time as short as possible to realize a 3 bit data transmission in less than 150 ns. Due to a continuous data transmission an updating rate of 12.5 MHz for the control data and a net data rate of 37.5 Mbit/s is achieved.

12.4.4 Performance Summary

Table 12.2 summarizes the performance of the ultra-wideband system for ultra-fast data transmission in rotating systems. Due to a variable center frequency of the UWB system and the wideband performance of the slotted waveguide ring, a frequency division multiple access can be implemented as well as a full-duplex operation.

12.5 Conclusion

An ultra-fast ultra-wideband system for sub-microsecond communication between rotating parts of a machine was presented. The system design and principle of operation was explained in detail and its functionality proven by measurements in the analog and digital time domain.

Microwave absorber material suppresses multipath propagation inside the slotted waveguide ring but still ensures an uninterrupted data transmission. A continuous packet-based 3 bit control data transmission in less than 150 ns with a data rate of 100 Mbit/s was achieved using impulse based ultra-wideband technology and high-speed digital signal processing.

References

1. H.J. Körber, H. Wattar, G. Scholl, Modular wireless real-time sensor/actuator network for factory automation applications. IEEE Trans. Ind. Informat. **3**, 111–119 (2007)
2. R. Heynicke, D. Krüger, H. Wattar, G. Scholl, Modular wireless fieldbus gateway for fast and reliable sensor/actuator communication. in *Proceedings of the IEEE International Conference on Emerging Technologies and Factory Automation*, Hamburg, 2008, pp. 1173–1176
3. C. Seifarth, T. Jurenz, G. Scholl, Sub-microsecond ultra-wideband transceiver for time-critical wireless sensor networks. Frequenz **62**(7–8), 191–194 (2008)
4. M.Z. Win, R.A. Scholtz, On the robustness of ultra-wide bandwidth signals in dense multipath environments. IEEE Commun. Lett. **2**, 51–53 (1998)
5. J. Han, C. Nquyen, On the development of a compact sub-nanosecond tunable monocycle pulse transmitter for UWB applications. IEEE Trans. Microw. Theory Tech. **54**(1), 285–293 (2006)
6. M.Z. Win, R.A. Scholtz, Impulse radio: How it works. IEEE Commun. Lett. **2**(2), 36–38 (1998)
7. C. Seifarth, R.G. Heynicke, G. Scholl, Electronically tunable pulse generator with programmable pulse repetition rate for 6.0-8.5 GHz ultra-wideband communications. Microw. Opt. Technol. Lett. **50**(6), 1649–1651 (2008)
8. R. Xu, Y. Jalin, C. Nguyen, Power-efficient switching-based CMOS UWB transmitters for UWB communications and radar systems. IEEE Trans. Microw. Theory Tech. **54**, 3271–3277 (2006)
9. European Commission: Commission decision 2007/131/EC of 21 February 2007 on allowing the use of the radio spectrum for equipment using ultra-wideband technology in a harmonised manner in the community. Off. J. Eur. Union **L55**, 33–36 (2007)
10. H. Dalichau, Offene Wellenleiter für die Nachrichtenübertragung zu spurgeführten Fahrzeugen. Fortschritt-Berichte der VDI-Zeitschriften. Reihe 9, Nr. 28. (VDI-Verlag, Düsseldorf, 1981)
11. K.P. Lange, H. Dalichau, Ein Schlitzhohlleiter für breitbandige Nachrichtenübertragung zu Schienenfahrzeugen. Nachrichtentechnische Zeitschrift **30**(1), 92–94 (1977)
12. M. Liess, R. Hau, K. Elsenaar, Microwave communication to moving parts inside machines. Int. J. Adv. Manuf. Technol. **20**, 58–62 (2002)
13. H. Dalichau, Übergänge und Fahrzeugkoppler für Schlitzhohlleiterstrecken. Frequenz **36**(6), 169–175 (1982)
14. J. Bretting, H. Dalichau, H. Groll, K. Petermann, J. Siegl, Hochfrequenz-Wellenleiter – Transmission lines and waveguides. in *Taschenbuch der Hochfrequenztechnik*, vol. 2:Komponenten, 5th edn. ed. by K. Lange, K.H. Löcherer. (Springer, Berlin, 1992), chap. K, pp. K 1–K 49
15. R. Holm, *Electric contacts: Theory and Application*, 4th, repr. ed. edn. (Springer, Berlin, 1979)
16. G. Roberts, P. Hadfield, M.E. Humphries, F. Bauder, J.M.G. Izquierdo, Design and evaluation of the power and data contactless transfer device. in *Proceedings of the IEEE Aerospace Conference*, vol. 3. Aspen, CO, 1997, pp. 523–533
17. C.W. Allen, H.L. Krauss, A wide-band rotating coupler for VHF use. IEEE Trans. Microw. Theory Tech. **MTT-24**(5), 267–269 (1976)
18. E.D. Evans, An analysis of a coupled-ring rotary joint design. IEEE Trans. Microw. Theory Tech. **40**(3), 577–581 (1992)
19. H.J. Riblet, R.L. Williston, X-band rotary joint. IEEE Trans. Microw. Theory Tech. **MTT-1**(1), 23–24 (1953)

20. G.F. Dorsey, Fiber optic rotary joints – A review. IEEE Trans. Compon. Hybrids Manuf. Technol. **CHMT-5**(1), 37–41 (1982)
21. C. Seifarth, G. Scholl, Wideband microwave rotary joint using a slotted waveguide ring. IEEE Trans. Microw. Theory Tech. **57** (2009). (submitted for publication)
22. C. Seifarth, G. Scholl, Sub-microsecond UWB data transmission in time-critical wireless control loops. in *Proceedings of the 2008 IEEE International Mini-Symposium on Electromagnetic and Network Theory and their Microwave Technology Application*, Munich (2008)
23. C. Seifarth, R. Draheim, G. Scholl, C-band transceiver testbed with substrate integrated waveguide (SIW) filter for ultra-wideband (UWB) communications. in *Proceedings of the 5th European Radar Conference*. Amsterdam, 2008, pp. 328–331

Chapter 13
"Green" Inkjet-Printed Wireless Sensor Nodes on Low-Cost Paper, Liquid and Flexible Organic Substrates

M.M. Tentzeris, L. Yang, A. Traille, and A. Rida

13.1 Ionic Antennas: Biosensors – RFID

The explosive growth of the biosensors and health-related wearable monitoring devices has accentuated the need for miniaturized, high-efficiency conformal materials that can operate over a wide range of frequencies, while they can be integrated in wearable and lightweight configurations. One of the major issue for the implementation of Wireless Body Area Networks (WBAN) is the very limited range of commonly used metal antennas. Due to the high dielectric constant between the metal antenna material (as well as the metal-based circuitry) and the mostly "ionized-water" human body parts, the near-field gets significantly disturbed, while local reflections due to the dielectric mismatch further shorten the operation range. Even wearable bracelet-like sensing devices have a very low range due to this reason.

Ida has demonstrated the dependence of the efficiency bandwidth on the permittivity of the dielectric surrounding a metal conductor. This dielectric slows the velocity of the electromagnetic energy in the dielectric and leads to physically smaller than the thin wire counterparts for the same frequency. An antenna with a salt solution radiator was published by Hatch, who coined the term "Ionic Liquid Antenna" in 2000, but only indirectly demonstrated its operability in HF frequencies. Encapsulating the proposed liquid antennas in flexible plastic containers makes them quite easily wearable. In addition, corrosion resistance is another advantage of the ionic-liquid, glass/plastic-enclosed antennas, while the easy elimination of air gaps permits shape manipulation and an improved electromagnetic coupling between the probe and the probed dielectric. Since the dielectric property and the

M.M. Tentzeris (✉), A. Traille, and A. Rida
GEDC/ECE, Georgia Institute of Technology, Atlanta, GA 30332-250, USA

L. Yang
Texas Instruments, Dallas, TX, USA

S. Lindenmeier and R. Weigel (eds.), *Electromagnetics and Network Theory and their Microwave Technology Applications*, DOI 10.1007/978-3-642-18375-1_13,

conductivity of these solutions are a function of the salinity level, these antennas can be easily reconfigured for different areas of mounting/observation.

RFID is an emerging compact wireless technology for the identification of objects, and is considered as an eminent candidate for the realization of a completely ubiquitous "ad-hoc" wireless networks. RFID utilizes electromagnetic waves for transmitting and receiving information stored in a tag or transponder to/from a reader. This technology has several benefits over the conventional ways of identification, such as higher read range, faster data transfer, the ability of RFID tags to be embedded within objects, no requirement of line of sight, and the ability to read a massive amount of tags simultaneously [1]. A listing of applications that currently use RFID are: retail supply chain, military supply chain, pharmaceutical tracking and management, access control, sensing and metering application, parcel and document tracking, automatic payment solutions, asset tracking, real time location systems (RTLS), automatic vehicle identification, and livestock or pet tracking.

The demand for flexible RFID tags has recently increased tremendously due to the requirements of automatic identification/tracking/monitoring in the various areas listed above. Compared with the lower frequency tags (LF and HF bands) already suffering from limited read range (1–2 feet), RFID tags in UHF band see the widest use due to their higher read range (over 10 feet) and higher data transfer rate [2]. The major challenges that could potentially hinder RFID practical implementation are: (1) Cost; in order for RFID technology to realize a completely ubiquitous network, the cost of the RFID tags have to be extremely inexpensive in order to be realized in mass production amounts (2) Reliability; and that extends to primarily the efficiency of the RFID tag antennas, readers, and the middleware deployed, (3) Regulatory Situation; meaning tags have to abide to a certain global regulatory set of requirements, such as the bandwidth allocations of the Gen2 Protocols defined by the EPC Global regulatory unit [3] and [4]) Environmentally-friendly materials, in order to allow for the easy disposal of a massive number (in the billions) of RFID's.

This article demonstrates for the first time how inkjet-printing of antennas/matching networks on low-cost paper-based materials can tackle all four challenges enabling the easy implementation of ubiquitous RFID and wireless biosensing networks. It starts by discussing how we can use conductive inkjet-printing technology for the fast fabrication of RF/wireless circuits, introduces a flexible wearable magnetic material, and eventually shows the capability of integrating sensors with RFID tags and discusses how added this functionality could revolutionize data fusion and real-time environmental cognition.

13.2 Flexible Magnetic Material

The technology for RFID systems continuously improves and extends to structures of non-planar shapes and to conformal sensors for wireless body-area networks (WBAN). Also, there is an increased demand for miniaturization, potentially addressed by the choice of substrate material with high dielectric and/or high magnetic constant, particularly magnetic materials. Three-dimensional transponder antennas that utilize wound coil inductors do make use of magnetic cores, but they are quite bulky and impractical. On the other side, flexible magnetic materials for two-dimensional embedded conformal planar antennas have not yet been successfully realized for standard use. This paper introduces a novel, mechanically flexible magnetic composite for printed circuits and two-dimensional antennas, which can reap the same miniaturization and tuning benefits as the non-flexible magnetic cores used for three-dimensional antennas.

One of the most significant challenges for applying new magnetic materials is understanding the interrelationships of the properties of the new materials with the design and performance of the specific topology (e.g. radiation pattern, scattering parameters). In previous studies, it has often been cited that the objectives of miniaturization and improved performance are tempered by the limited availability of materials that possess the required magnetic properties, while maintaining an acceptable mechanical and conformality performance. Recently, formulation of nano-size ferrite particles has been reported and formulation of magnetic composites comprised of ferrite filler and organic matrix has been demonstrated. The first step for this work was to develop a magnetic composite that provides the advantage of low temperature processing for compatibility with organic substrate.

13.3 Multi-Hop Algorithms

Recent advances in wireless communications and digital electronics have rendered the construction of relatively low-cost, low-power, multifunctional sensor nodes feasible. However, the deployment of large scale WSN infrastructures based on the collaboration of a large number of nodes has only become a reality through the embedded software implementing the different layers of the protocol stack developed the last years.

Specifically, since a generally large number of sensor nodes are densely deployed in WSN fields, multi-hop communication is exploited in the interconnection between nodes. First, since the transmission power of a wireless radio is proportional to the square of the distance or an even higher-order due to the close proximity of the antennas of the sensor nodes to the ground, multi-hopping both leads to less power consumption and lower cost than the traditional single hop communication and allows high spatial frequency reuse. Furthermore, multi-hop routing can effectively overcome shadowing and path loss effects offering coverage over large geographical regions. Improved sensing accuracy by distributed processing of large quantities of sensing information is also feasible through multi-hop communication. Finally, the ability to sustain sensor network functionalities without any interruption due to sensor node failures can also be achieved because of the multiple paths available for the data to flow offered by multi-hop routing.

13.4 Inkjet-Printing

A fast process, like inkjet printing, can be used efficiently to print electronics on/in organic substrates. This also enables components such as: antennas, IC, memory, batteries and/or sensors to be easily embedded in/on organic modules. Modern inkjet printers operate by propelling tiny droplets of liquid down to several pL. This new technology of inkjet printing utilizing conductive paste may rapidly fabricate prototype circuits without iterations in photolithographic mask design or traditional etching techniques, that have been widely used in industry. Printing is completely controlled from the designer's computer and does not require a clean room environment. A droplet's volume determines the resolution of the printer, for e.g. a droplet of 10 pL gives $\sim$25 μm minimum thickness or gap size of printed traces/lines. The cartridge consists of a Piezo-driven jetting device with integrated reservoir and heater.

Inkjet Printing; unlike etching which is a subtractive method by removing unwanted metal from the substrate surface, jets the single ink droplet from the nozzle to the desired position, therefore, no waste is created, resulting in an economical fabrication solution. Silver nano-particle inks are usually selected in the inkjet-printing process to ensure a good metal conductivity. After the silver nano-particle droplet is driven through the nozzle, sintering process is found to be necessary to remove excess solvent and to remove material impurities from the depositions. Sintering process also provides the secondary benefit of increasing the bond of the deposition with the paper substrate [5]. The conductivity of the conductive ink varies from 0.4 to 2.5×10^7 Siemens/m depending on the curing temperature and duration time. At lower curing temperature, larger gaps exist between the particles, resulting in a poor connection. When the temperature is increased, the particles begin to expand and gaps start to diminish. That guarantees a virtually continuous metal conductor, providing a good percolation channel for the conduction electrons to flow. To ensure the conductivity performance of microwave circuits, such as RFID modules, curing temperatures around 120°C and duration time of 2 h were chosen in the following fabrication to sufficiently cure the nano-particle ink. Alternatively, much shorter UV heating approaches can achieve similar results.

13.5 Conformal Performance

In order to verify the performance of the conformal RFID antenna, measurements were performed by conforming the same RFID tag onto a foam cylinder. The radius of the cylinder was chosen to be very small at 27 mm, in order to explore the limits of the design. The return loss of the fabricated antenna is shifted down by 22 MHz with a center frequency at 458 MHz. Previous results showed a shift of 6 MHz for a lower curvature of 54 mm radius, which proves that the shift is increasing with the curvature level. Overall the antenna still has good performance if the shift in frequency is considered at the beginning of the design process, even for such a

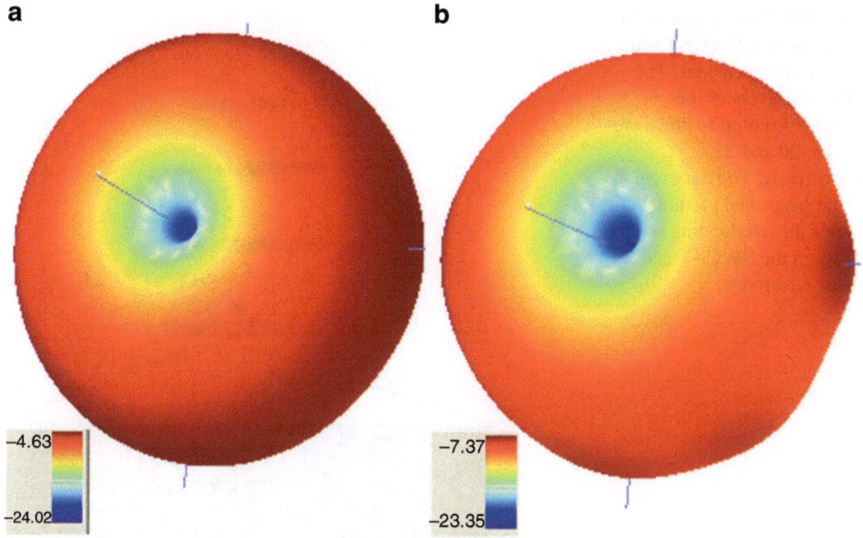

Fig. 13.1 Measured radiation pattern of (**a**) the flat RFID tag and (**b**) the conformal RFID tag. Max gain drops from −4.63 to −7.37 dBi

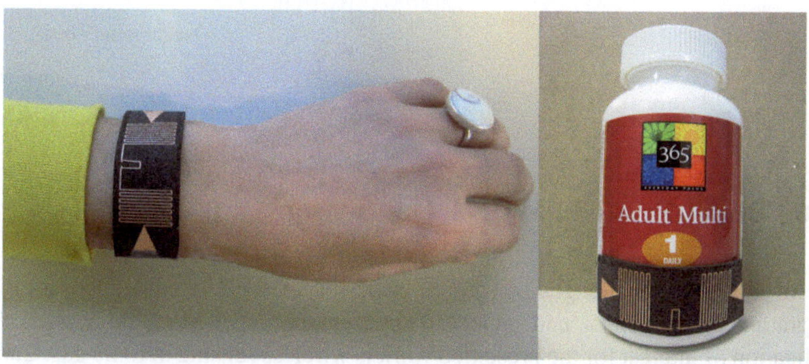

Fig. 13.2 Embodiments of the conformal RFID tag prototype in the applications of wireless health monitoring and pharmaceutical drug bottle tracking

large bend Figure 13.1 shows the radiation patterns for the straight and conformal antennas. The doughnut shape is slightly degraded for the conformal antenna and the maximum gain drops from −4.63 to −7.37 dBi.

The flexible nature of the substrate enables the RFID tag module's application in diverse areas. Figure 13.2 demonstrates the conformal RFID tag prototype in the applications of wireless health monitoring and pharmaceutical drug bottle tracking [6].

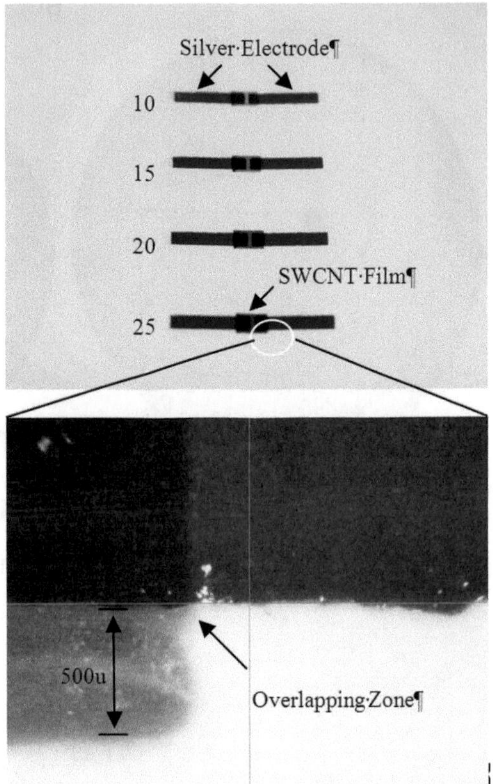

Fig. 13.3 Photograph of the inkjet-printed SWCNT films with silver electrodes. The SWCNT layers of the samples from up to down are 10, 15, 20 and 25, respectively. The dark region in thc magnified picture shows the overlapping zone between the SWCNT and the silver electrodes

13.6 Inkjet-Printed SWCNT Gas Sensor

One of the major challenges of "green" paper-based RFID-enabled sensors is the integration of the sensor and nanostructures on the paper substrate as well. The application of interest for the presented work is wireless sensing of toxic gas. Carbon Nanotubes (CNT) composites were found to have electrical conductance highly sensitive to extremely small quantities of gases, such as ammonia (NH_3) and nitrogen oxide (NO_x), etc. at room temperatures with a very fast response time [7]. The conductance change can be explained by the charge transfer of reactive gas molecules with semiconducting CNTs [8]. Previous efforts have shown the successful utilization of CNT-based sensors employing the change in resistance [9]. However, due to the insufficient molecular network formation among the inkjet-printed CNT particles at micro-scale, instabilities were observed in both the resistance and, especially, the reactance dependence on frequency above several MHz, which limits the CNT application in only DC or LF band [10]. To enable the CNT-enabled sensor to be integrated with RFID antenna at UHF band, a special recipe needs to be developed.

This section presents a conformal CNT-based RFID-enable sensor node for gas sensing applications, fully printed directly on paper substrate [11]. Specifically, in this study one benchmarking RFID tag was designed for the European UHF RFID band centering at 868 MHz. The printed CNT particles were Single-Walled Carbon Nanotubes (SWCNT) from Carbon Solutions, which were dispersed in dimethylformamide (DMF) solution and sonicated to meet the viscosity requirement for the inkjet printer. The SWCNT composite is printed directly on the same paper as the antenna, for a low cost, flexible, highly integrated module. The impedance of the SWCNT film forms the sensor part. The antenna was printed first, followed by the 25 layers of the dispersed SWCNT as a load with "gas-controlled" value. When 4% consistency ammonia was imported into the gas chamber, the SWCNT impedance changed from 51.6-j6.1 Ω to 97.1-j18.8 Ω at 868 MHz, resulting in a 10.8 dBi variation in the backscattered power from the RFID antenna, that can be easily detected by the RFID reader to realize the "real-time" gas detection. As a direct-write technology, inkjet printing transfers the pattern directly to the substrate. Due to its capability of jetting one single ink droplet in the amount as low as 1 pl, it has widely drawn attention from the industrial world as a more accurate and economic fabrication method than the traditional lithography method.

CNT composites have been found to have a very unique resistance performance that can enable the realization of the next generation of sensors with a very high sensitivity up to 1 ppb (part per billion), an improvement of 2–3 orders to traditional sensors. The electrical resistance of the fabricated device was measured by probing the end tips of the two electrodes. The DC results in air are shown in Fig. 13.4. The resistance goes down from when the number of SWCNT layers increases. Since a high number of SWCNT overwritten layers will also help the nano particle network formation, 25-layer film is expected to have the most stable impedance-frequency response and selected for the gas measurement. In the experiment, 4% consistency ammonia was guided into the gas flowing chamber, which includes gas inlet, outlet and exhaust hood. The SWCNT film was kept in the chamber for 30 min.

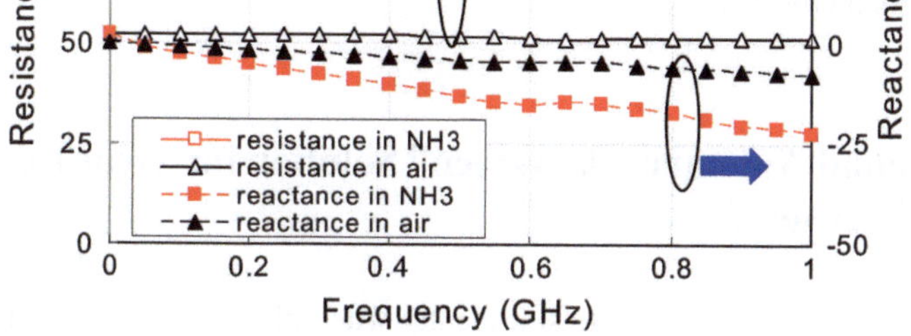

Fig. 13.4 Measured electrical resistance of SWCNT gas sensors

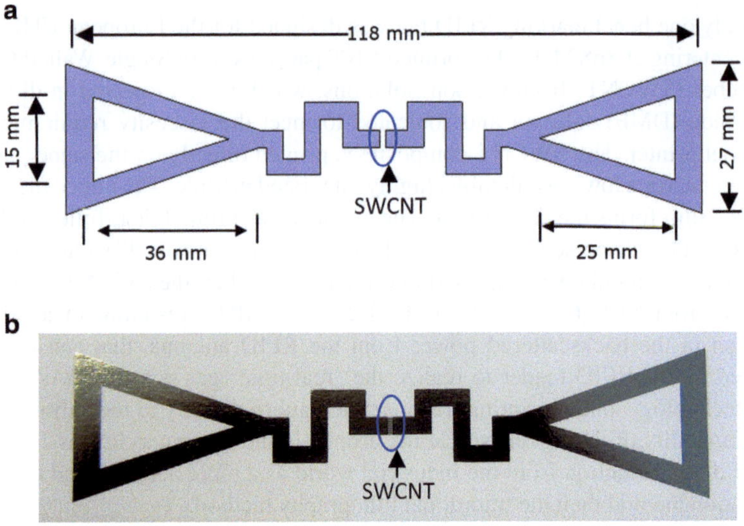

Fig. 13.5 The RFID tag module design on flexible substrate: (**a**) configuration (**b**) photograph of the tag with inkjet-printed SWCNT film as a load in the middle

A network vector analyzer (Rohde & Schwarz ZVA8) was used to characterize the SWCNT film electrical performance at UHF band before and after the gas flowing. In Fig. 13.4, the gas sensor of SWCNT composite shows a very stable impedance response up to 1 GHz, which verifies the effectiveness of the developed SWCNT solvent recipe. At 868 MHz, the sensor exhibits a resistance of 51.6 Ω and a reactance of −6.1Ω in air. After meeting ammonia, the resistance was increased to 97.1Ω and reactance was shifted to −18.8 Ω.

The CNT-film was inkjet-printed a gas-sensitive load for a bow-time antenna designed to operate for RFID tags around 868 MHz. (Figs 13.5 and 13.6) [11]. In the air, the SWCNT film exhibited an impedance of 51.6-j6.1 Ω, which results in a low power reflection at −18.4 dB. When NH_3 is present, SWCNT film's impedance was shifted to 97.1-j18.8 Ω. The mismatch at the antenna port increased the power reflection to −7.6 dB, a 10.8 dBi increase at the received backscattered power level. By detecting this backscattered signal difference on the reader's side, the sensing function can be fulfilled.

13.7 Liquid Antennas: A "Green" Solution for Wearable Biosensors?

Metallic antennas do not operate sufficiently when planted extremely close to the human body due to the dielectric discontinuity against human tissue (Metal : $\varepsilon_r = 1$, Blood: $\varepsilon_r = 58$, Skin: $\varepsilon_r = 37$), that causes the disruption of their near field.

Fig. 13.6 Photograph of the paper-based conformal tag

The problem of matching "human tissue" to "air" is commonly encountered in ultrasound techniques, which led to the research and development of tissue mimicking dielectric phantom models. In addition to matching, metallic antennas are heavy, vulnerable to corrosion, toxic to the human body and bending them introduces unwanted resonances. Liquid Antennas on the other hand, enclosed in glass would possess the biocompatible properties that would be useful for health monitoring devices, especially when they are implanted into human tissue. Plus, liquid (e.g. aquatic) solutions can be enclosed in flexible plastic, and bent in various configurations without introducing holes or air gaps, thus allowing them to operate sufficiently while worn as clothes. Liquid antennas would also be smaller as well as lighter allowing them to be easily integrated into everyday mobile human activities.

Material Characterization: Dielectric Liquids: The first step for the realization of practical liquid antennas is the accurate determination of how, fluid composition (ion species), electrolyte concentration, electrode polarization, geometry (e.g. toroid curvature), transport resistance, frequency and other parameters will affect the current distribution, radiation pattern, and efficiency of the liquid antenna. It is essential to characterize the electric properties of various electrolytes, however complicated is the process both theoretically and experimentally. As of now, approximations of electrical properties can be derived using a combination of Force Field equations, MD (Molecular Dynamics) Simulators, Debye or Cole–Cole theories of molecular relaxation as well as some experimental data to create empirical models that characterize the trend. Measurement uncertainties are discussed later in the paper.

Principle of Operation: In electrolyte solutions, current is created by ions which migrate under the influence of an electrical field. In the case of an electrolytic liquid-filled loop antenna, the voltage gradient is due to the Lorentz force, that can be generated by two electrodes connected on opposite sides of the loop, similar to that of a toroid shaped battery. The antenna must be designed and tuned so that the charge-discharge-charge cycle occurs at a specific resonant frequency, that determines the antenna frequency.

As a proof of concept, free-space, as well as in vitro (in the proximity of a "SEP" human head phantom) simulations benchmarking one liquid-loop antenna performance were performed using FEKO [12]. The antenna geometry (Fig. 13.7) consists of a dielectric toroid operating at 915 MHz with an aquatic solution 5 mol/L NaCl, that is connected with an edge port between two hollow metal plates. The simulation os performed for the toroid placed on top of a human head "phantom", as shown in Fig. 13.8.

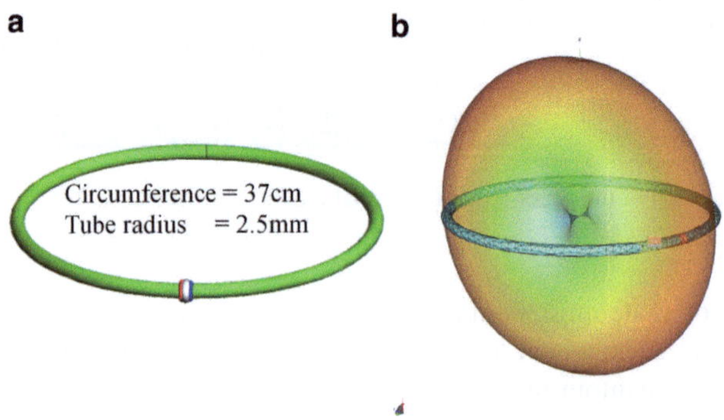

Fig. 13.7 (**a**) Ionic Loop Antenna (**b**) E-Pattern (Free space)

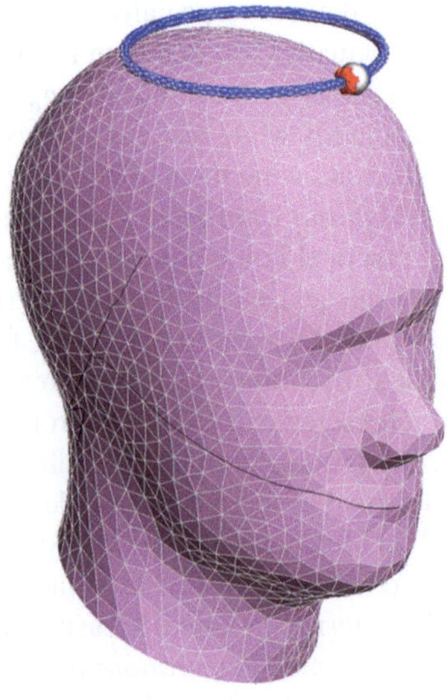

Fig. 13.8 Human Head in FEKO (Skin $\varepsilon_r = 41$, $\tan\delta = 0.414$ @ 915 MHz)

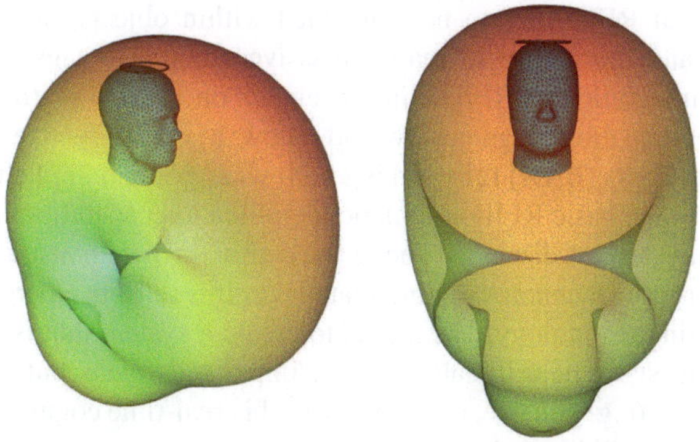

Fig. 13.9 Liquid Antenna on "SEP" Human Phantom (NaCl $\varepsilon_r = 40$, $\tan\delta = 0.175$ @ 915 MHz)

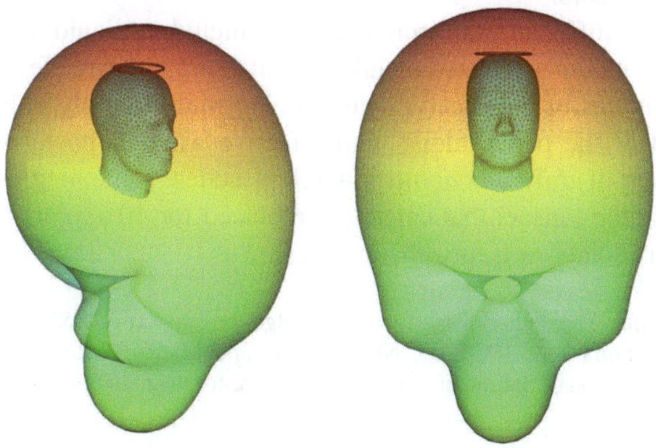

Fig. 13.10 Metallic antenna on "SEP" Human Phantom (PEC $\varepsilon_r = 1$)

The radiation pattern (Fig. 13.9) was compared with that of a metallic loop antenna with the same dimensions (Fig. 13.10). It is clear that the liquid antenna pattern is much more concentrated in the area of the phantom eliminating stray radiation and enhancing "high-focus" bio-diagnostic applications. In addition, the material "near-field" matching capability of the liquid antenna can be readjusted "on the fly" for positioning close to different tissues by modifying the molarity of salt, thus changing the dielectric constant, something impossible for the conventional metal antenna.

13.8 Conclusions

RFID is an emerging compact wireless technology for the identification of objects, and is considered as an eminent candidate for the realization of a completely ubiquitous "ad-hoc" wireless networks. This technology has several benefits over the conventional ways of identification, such as higher read range, faster data transfer, the ability of RFID tags to be embedded within objects, no requirement of line of sight, and the ability to read a massive amount of tags simultaneously. The effective integration of RFID's in biosensors on flexible platforms (e.g. LCP and other biocompatible organics) would allow a very effective realization of Body-Area-Networks (BAN) fully linking both wearable and implantable devices.

In addition to the basic RFID automatic Identification capabilities along with the technologies and designs discussed above, the authors will demonstrate the capabilities of inkjet-printing technology in integrated wireless sensors on organics bridging RFID and sensing technology. The aim is to create a system that is capable of not only tracking the status of artificial/prosthetic/implanted organs, but also monitoring critical biosignals (e.g. glucoze, oxygen). With this real-time cognition of the status of a certain object will be made possible by a simple function of a sensor integrated in the RFID tag. The ultimate goal is to create a secured "intelligent network of RFID-enabled sensors."

There will be different platforms presented including liquid antennas and circuits, as well as modules realized on inkjet-printed organic flexible substrates. Last, but not least, issues of enhanced-range utilizing multi-hop algorithms will be an important part of the paper, along with effective ways of power scavenging for the development of truly autonomous wireless nodes, something especially critical for implantable sensors that cannot be replaced for 10+ years (e.g. placed on pacemakers, artificial hearts, prosthetic knees).

Acknowledgements The authors wish to acknowledge the support of NSF ECS-0801798, NSF ECS-0313951, Georgia Tech IFC/SRC, NEDO Japan and Microsoft Research Center. Special thanks to Kim Rutkowski of Satimo in Kennesaw, Georgia, for the radiation pattern measurements.

References

1. K. Finkenzeller, *RFID Handbook*, 2nd edn. (Wiley, Chichester, 2004)
2. S. Basat, S. Bhattacharya, A. Rida, S. Johnston, L. Yang, M.M. Tentzeris, J. Laskar, Fabrication and assembly of a novel high-efficiency UHF RFID tag on flexible LCP substrate. in *Proceedings of the 56th IEEE-ECTC Symposium*, 2006. pp. 1352–1355
3. UHF Gen-2 System Overview. Texas Instruments, Sept 2005, available http://rfidusa.com/superstore/pdf/UHF_System_Overview.pdf
4. Y. Kurokawa, T. Ikeda, M. Endo, H. Dembo, D. Kawae, T. Inoue, M. Kozuma, D. Ohgarane, S. Saito, K. Dairiki, H. Takahashi, Y. Shionoiri, T. Atsumi, T. Osada, K. Takahashi, T. Matsuzaki, H. Takashina, Y. Yamashita, S. Yamazaki, UHF RF CPU's on flexible and glass substrates for secure RFID systems. IEEE J. Solid-State Circuits **43**(1), 292–299 (2008)
5. L. Yang, A. Rida, R. Vyas, M.M. Tentzeris, RFID tag and RF structures on a paper substrate using inkjet-printing technology. IEEE Trans. Microw. Theory Tech. **55**(12), Part 2, 2894–2901 (2007)
6. L. Yang, L.J. Martin, D. Staiculescu, C.P. Wong, M.M. Tentzeris, Conformal magnetic composite RFID for wearable RF and bio-monitoring applications. IEEE Trans. Microw. Theory Tech. **56**(12-2), 3223–3230 (2008)
7. K.G. Ong, K. Zeng, C.A. Grimes, A wireless, passive carbon nanotube-based gas sensor. IEEE Sens. J. **2**, 82–88 (2002)

8. C. Cantalinia, L. Valentini, L. Lozzic, I. Armentano, J.M. Kenny, L. Lozzi, S. Santucci, Carbon nanotubes as new materials for gas sensing applications. J. Eur. Ceram. Soc. **24**, 1405–1408 (2004)
9. J.-H. Yun, H. Chang-Soo, J. Kim, J.-W. Song, D.-H. Shin, Y.-G. Park, Fabrication of carbon nanotube sensor device by inkjet printing. in *2008 Proceedings of IEEE Nano/Micro Engineered and Molecular Systems*, Jan. 2008, pp. 506–509
10. J. Song, J. Kim, Y. Yoon, B. Choi, J. Kim, C. Han, Inkjet printing of singe-walled carbon nanotubes and electrical characterization of the line pattern. Nanotechnology **19** (2008)
11. L. Yang, R. Zhang, D. Staiculescu, C.P. Wong, M.M. Tentzeris, A novel conformal RFID-enabled module utilizing inkjet-printed antennas and carbon nanotubes for gas detection applications. IEEE Antennas Wireless Propag. Lett. **8**, 653–656 (2009)
12. A. Traille, L. Yang, A. Rida, M.M. Tentzeris, A novel liquid antenna for wearable biomonitoring applications. in *Proceedings of the 2008 IEEE-IMS Symposium*, Atlanta, GA, June 2008. pp. 923–926

Chapter 14
A Joint Matlab/FPGA Design of AM Receiver for Teaching Purposes

Hikmat N. Abdullah and Alejandro A. Valenzuela

14.1 Introduction

Implementing an AM Receiver using analog electronics has always been the norm. However, with the improvement of digital systems, it has become easier to emulate analog with digital circuitry. The digital AM Receiver is a digital system that attempts to achieve the same analog AM Receiver functionality by just an FPGA and a small of amount of analog electronics.

The motivation for this work came from the work done on software radio by companies like Vanu [1]. Software radio allows a single device to receive many different wireless transmissions. By using digital signal processing techniques in FPGAs, the software radio could possibly be achieved in digital systems. However, since building an AM Receiver is quite easy to learn, it was sensible to focus on AM transmission instead of FM and other more intricate wireless transmission. This work is worthwhile also because it develops digital design techniques that can be applicable to more advanced communication systems. For example, this work could be expanded to receive FM and other wireless transmissions if the necessary modifications are made on the Embedded Matlab™ code that describes fundamental blocks.

Matlab programming language is one of the well known design tools in Engineering projects [2]. This design tool is normally used to obtain simulation waveforms to verify the functionality of system under consideration. Due to its high efficiency, it is widely used for teaching purposes of Engineering students. For hardware design consideration, for instance using FPGA, a special hardware descriptive languages like VHDL, Verilog, ... etc. are used to realize designs. One of the problems

A.A. Valenzuela (✉)
University of Applied Sciences, Bonn-Rhein-Sieg, Germany
e-mail: alejandro.valenzuela@h-bonn-rhein-sieg.de

H.N. Abdullah
University of Al-Mustansiryah, Baghdad, Iraq
e-mail: hikmat_04@yahoo.com

S. Lindenmeier and R. Weigel (eds.), *Electromagnetics and Network Theory and their Microwave Technology Applications*, DOI 10.1007/978-3-642-18375-1_14,

that students or beginner designers face is the long cycle design flow and many languages to learn with the necessity of careful manual translation [3, 4].

In this chapter, an efficient short cycle design flow is used. With this design flow, the designer could implement his design models originally written as Matlab codes using FPGA board without the need to learn VHDL or even other FPGA design entries. As well as this approach reduces the time required to complete the hardware implementation, it will give the beginner designer, for instance the student, a better and easy understanding of how different design parts functions using his written Matlab codes. However, the automatic translation of Matlab code to VHDL 1 requires extra precaution. The written Matlab code should take into account what is so called fixed point arithmetic notations, i.e. each design parameter used in the code should be declared initially. This defines a special kind of Matlab kown as *Embedded Matlab™*[5]. This requires that fixed point toolbox and simulink fixed point products be available in Matlab™-Simulink environment. However, there are other recommended products like signal processing blockset, signal processing toolbox, filter design toolbox, stateflow and EDA simulator link that may also be used to carry out more advanced designs.

14.2 Traditional FPGA and Joint MATLAB/FPGA Waveform Design Methods

SDR waveform design has typically been extremely inefficient. In the past, system-level specifications and simulations were "thrown over the wall" to the hardware designers who then started coding in their favorite Hardware Definition Language (HDL). There were, of course, some challenges with this approach. First, the system designer had no insight into the implementation details of the FPGA and, therefore, could not best optimize the system design without lengthy communications with the engineers implementing the design. Secondly, the designer needed to be an expert in HDL–not the sort of expertise an engineer was likely to pick up overnight. Third, this approach involves manual code generation, which is time-consuming and tedious, as well as likely to require extensive debugging–all of which increases development time and cost. This approach also contains some inherent tendencies towards inefficiency, since the system must be created twice, first on the system-level tool and then on the implementation tool once again increasing the time and cost of system development. Figure 14.1 shows an example flowchart for this traditional waveform development flow.

As FPGAs increase in complexity, it is necessary to have system-level tools that can aid the designer in simplifying the design methodology. Tools such as Simulink HDL Coder have been developed to address the issues found when performing complex system development such as waveform design. With this tool, the new design flow would consists of 3 segments: implementing/designing modules in Matlab-Simulink™ Environment where each module function is described by Embedded Matlab code with the aid of fixed point toolbox, translating

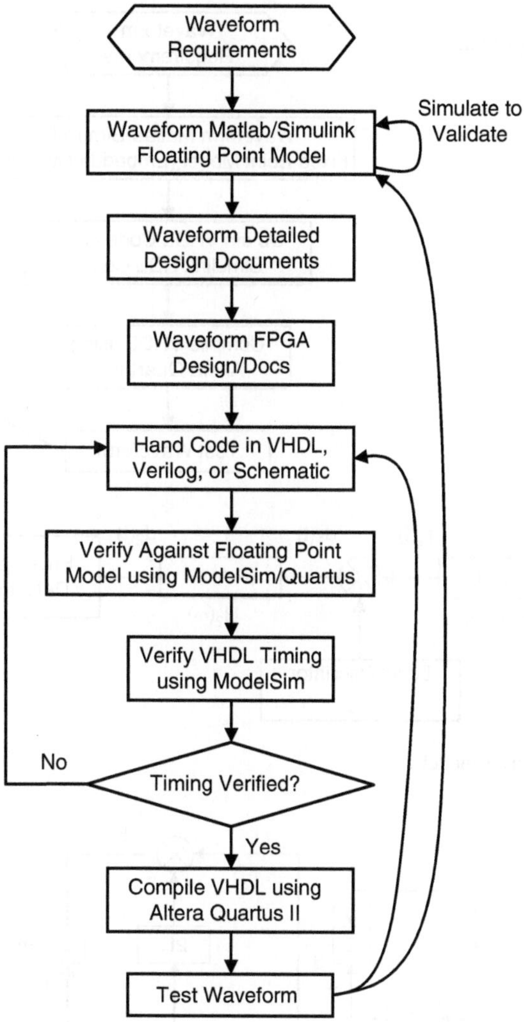

Fig. 14.1 Traditional Waveform Design Flow

Matlab-Simulink™ design using Simulink HDL Coder to get VHDL code, and finally performing pin assignment, compilation and programming of the Altera FPGA board using Quartus II® environment. Figure 14.2 shows the design flow of joint Matlab/FPGA approach that uses Simulink HDL coder tool.

14.3 System Model

Figure 14.3 shows the block diagram of the digital AM Receiver under consideration. The first segment in the receiver is the programmable bandpass filter. It is used to receive the test transmissions within normal AM frequency band 530–1,600 kHz.

Fig. 14.2 Joint Matlab/FPGA Design Flow

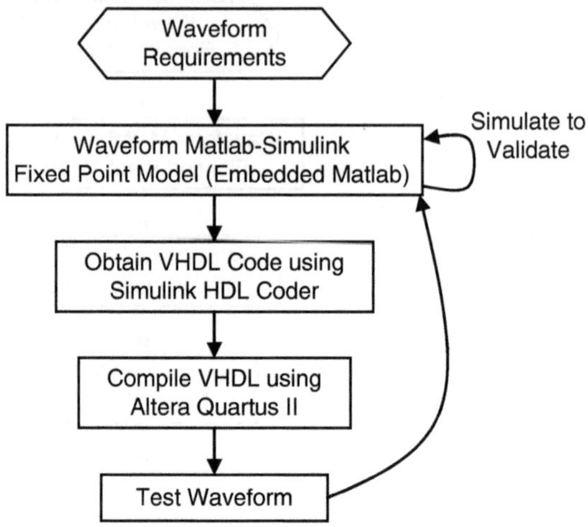

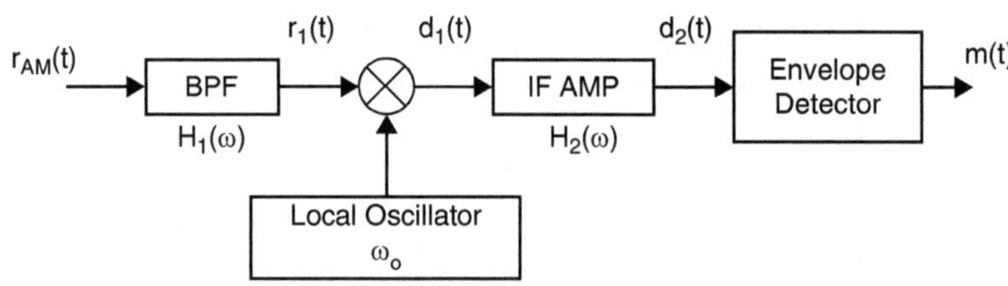

Fig. 14.3 AM Receiver model

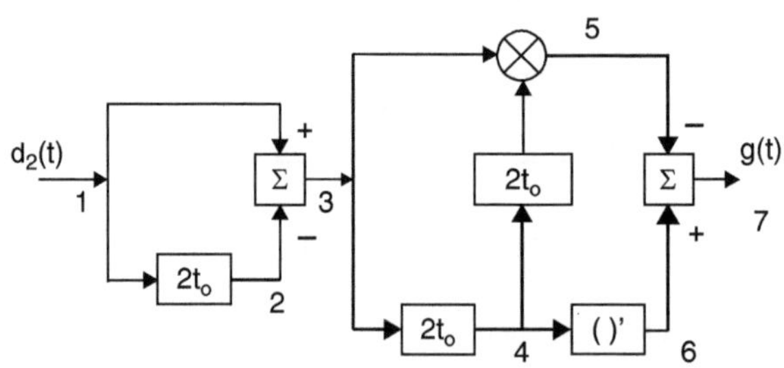

Fig. 14.4 Digital envelope detector

A digital down converter is used then to translate the received spectrum to 455 kHz intermediate frequency. The demodulation of the digital signal is performed using an efficient digital envelope detector shown in Fig. 14.4 [6]. The digital version of the envelope detector is a nonlinear filter based upon a discrete version of the recently introduced Teager–Kaiser energy operator, but also closely resembles a complex digital sampling demodulator [7].

The first step in joint MATLAB/FPGA design approach is to write embedded Matlab code for each design parts. This requires firstly writing mathematical models for each part in the design. This is an important step since the main objective of this study is teaching purposes. Referring to Fig. 14.3, the received AM signal $r_{AM}(t)$ is described as:

$$r_{AM}(t) = A_c(1 + f(t))\cos(\omega_c t + \theta_1) + n(t) \tag{14.1}$$

where f(t) is the message signal, A_c is the carrier amplitude, ω_c is the carrier angular frequency, n(t) is the additive white Gaussian noise and θ_1 is the phase offset. The spectrum of signal at the output of the tunable BPF with the frequency transfer function $H_1(\omega)$ would be:

$$R_f(\omega) = R_{AM}(\omega) \cdot H_1(\omega) \tag{14.2}$$

when mixed with the local oscillator signal, $R_f(\omega)$ would be shifted by $\pm\omega_o$ where ω_o is the local oscillator frequency resulting the signal $d_1(t)$ whose spectrum is given by:

$$D_1(\omega) = (\pi R_f(\omega + \omega_o) + \pi R_f(\omega - \omega_o)) \tag{14.3}$$

the IF amplifier is a BPF that is used to pass the intermediate frequency. Normally, the term that contains the difference between incoming and locally generated frequencies is filtered out and other frequencies are rejected. Mathematically speaking, the spectrum of the signal at the IF amplifier output with the frequency response $H_2(\omega)$ would be:

$$D_2(\omega) = (\pi R_f(\omega + \omega_o) + \pi R_f(\omega - \omega_o)) \cdot H_2(\omega) \tag{14.4}$$

the input signal to the digital envelope detector would take the following form in time domain

$$d_2(t) = a(t) + b(t)\cos(\omega_{IF} t + \theta_2) \tag{14.5}$$

where a(t) is the (slowly varying) signal offset, b(t) is the envelope, ω_{IF} is the intermediate carrier frequency and θ_2 is the phase offset. According to Fig. 14.4 the output signal is given by:

$$m(t) = [d_2(t - t_o) - d_2(t - 3t_o)]^2 - [d_2(t) - d_2(t - 2t_o)] \cdot [d_2(t - 2t_o) - d_2(t - 4t_o)] \tag{14.6}$$

where t_o is a one sample delay.

The Matlab codes that simulate the behavior of different receiver parts according to the previous equations are to be simulated correctly in Matlab environment before proceeding to the FPGA implementation phase.

14.4 Design and Implementation of Digital AM Receiver

As stated in the design flow of Fig. 14.2, the design starts with Matlab-Simulink™. Different receiver blocks in Fig. 14.3 have been implemented using Embedded Matlab™ codes. Embedded Matlab function blocks can be found in Matlab-Simulink™ *user defined functions*. Furthermore, the following toolboxes should be

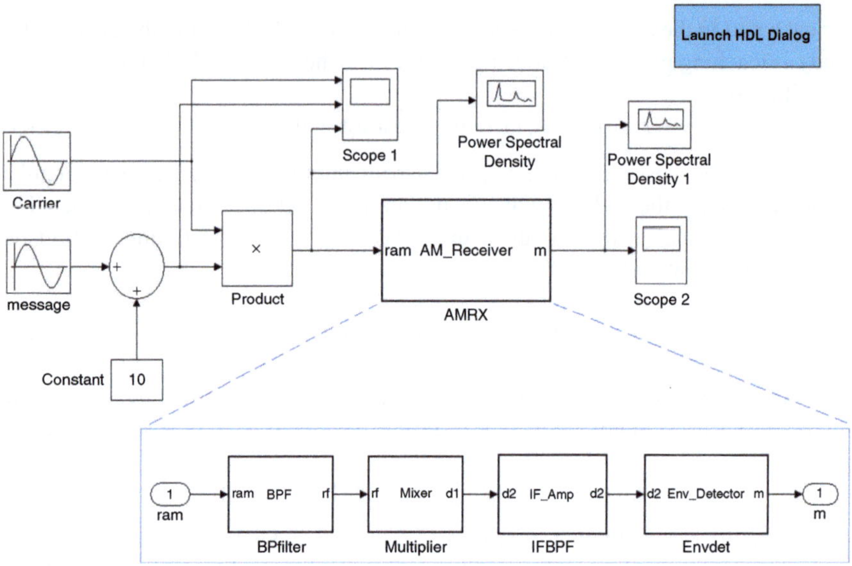

Fig. 14.5 Implementation of the AM Receiver in Matlab-Simulink™

available to successfully complete the design: fixed point toolbox, Simulink fixed point and of course Simulnik HDL coder. Figure 14.5 shows the Matlab-Simulink™ window of the implemented receiver.

As it is clear in this figure, the design has a top-level system "AM_Receiver" and all receiver parts "BPF", "Mixer", "IF_Amp" and "Env_Detector" are represented as sub-systems. Each subsystem may include further subsystems. The function of each subsystem is described by Embedded Matlab code. This organization allows us to get a set of VHDL files controlled or called by a top-level VHDL file which is the normal case in FPGA environment. The functionality of different receiver parts has been tested using Matlab-Simulink simulator. VHDL files have then been generated using Simulink HDL coder which could be launched easily from the special icon *"launch HDL dialog"* shown at the top right corner of Fig. 14.5.

The generated VHDL files would take a form of many entities. Each entity represent a hardware construction unit of the corresponding Matlab code in hierarchal form. At the top-level of the design, there is the entity AM-RX which implicitly uses other receiver entities (normally called components). The VHDL files are fed to Altera Quartus II® version 7.2 environment for the purpose of FPGA implementation. The top-level entity which defines the input and output of the overall receiver is used to perform the pin assignments as shown in Fig. 14.6. A compilation and synthesis are then carried out. A double crosscheck of the waveform is carried out using Quartus II® simulator. The bit steam files generated from compilation and synthesis processes are then downloaded to an Altera Cyclone III EP3C120 DSP development board with 50 MHz clock frequency.

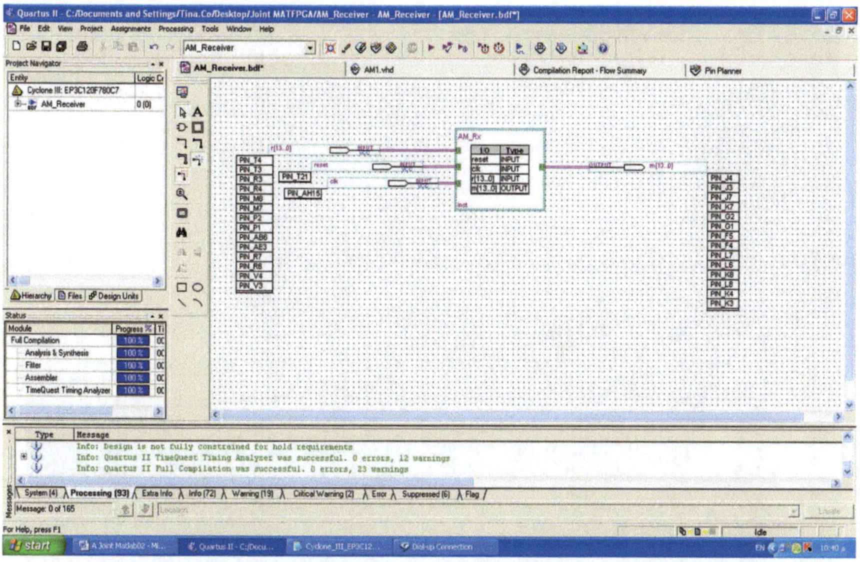

Fig. 14.6 Top-level entity of digital AM Receiver (Quartus II environment)

To test the board with an analog input signal, a High Speed Mezzanine Card (HMSC) data converter (Altera product) is interfaced to Cyclone III FPGA board. This card has a pair of 14 bit, 150-MSPS ADC/DAC converters. The 14 bits of the channel A the ADC converter have been used to demodulate the received AM signal. Similarly, 14 bits of the channel A DAC converter have been used to produce the analogue detected message signal. The pin assignments of ADC and DAC converters as well as the clock and reset signals are shown in Fig. 14.6. Figure 14.7 shows the overall implemented system.

14.5 Implementation Results

This section presents the implementation results of the digital AM receiver using a joint Matlab/FPGA design flow approach. A 1,200 kHz carrier frequency and 60 kHz single tone message were used to test the functionality of the implemented receiver. Figure 14.8 shows the received AM modulated signal in both time and frequency domains while Fig. 14.9 shows the demodulated waveform with its corresponding spectrum. Figure 14.10 depicts the waveforms during the envelope detection process. The different numbers on the left side of this figure refer to the waveforms obtained on the corresponding node numbers in Fig. 14.4. All above results were obtained using time and power spectral density scopes available in Matlab-Simulink™ simulator. When VHDL code was generated for different

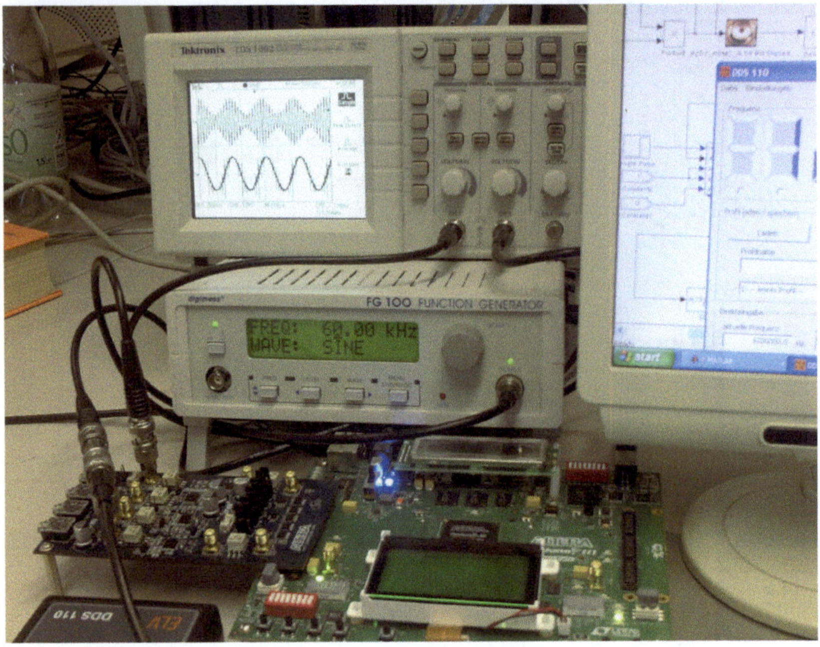

Fig. 14.7 Test Hardware of the implemented AM Receiver

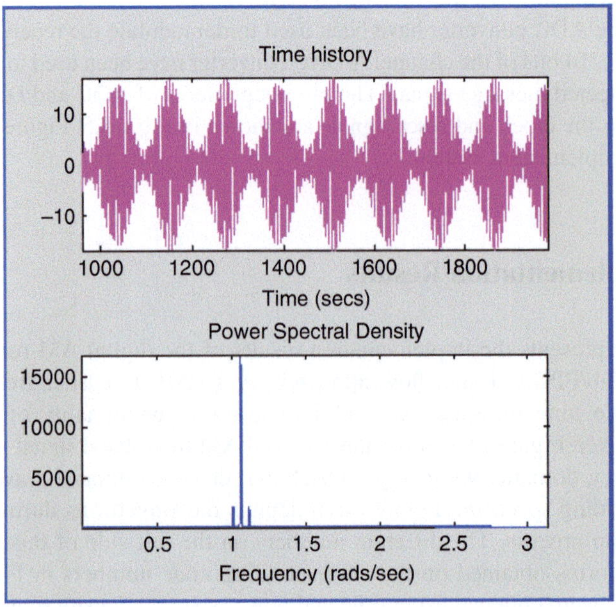

Fig. 14.8 Received AM signal

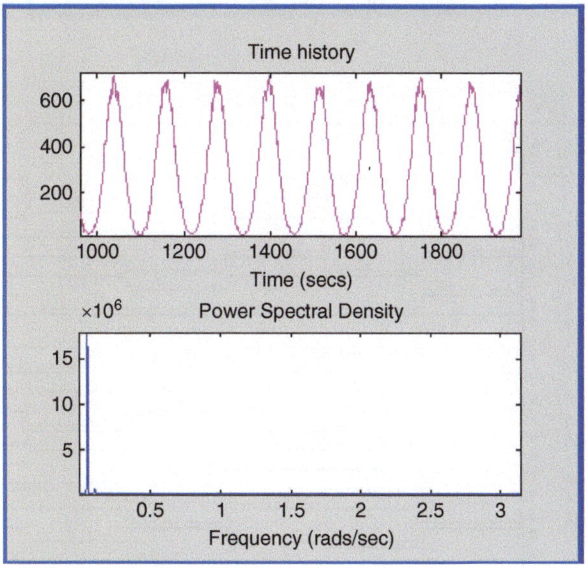

Fig. 14.9 Digitally demodulated message signal

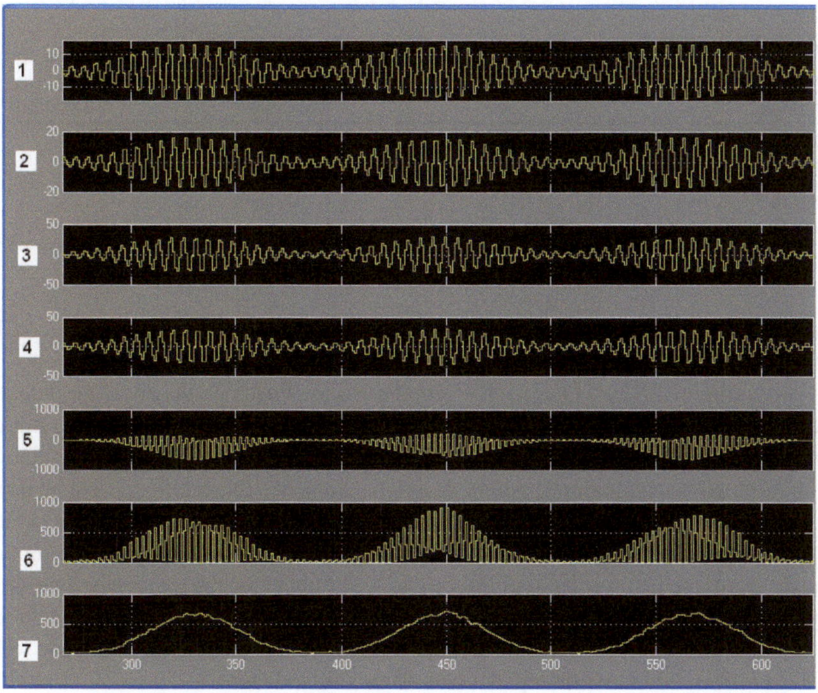

Fig. 14.10 Waveforms at different points of envelope detector in Fig. 14.4

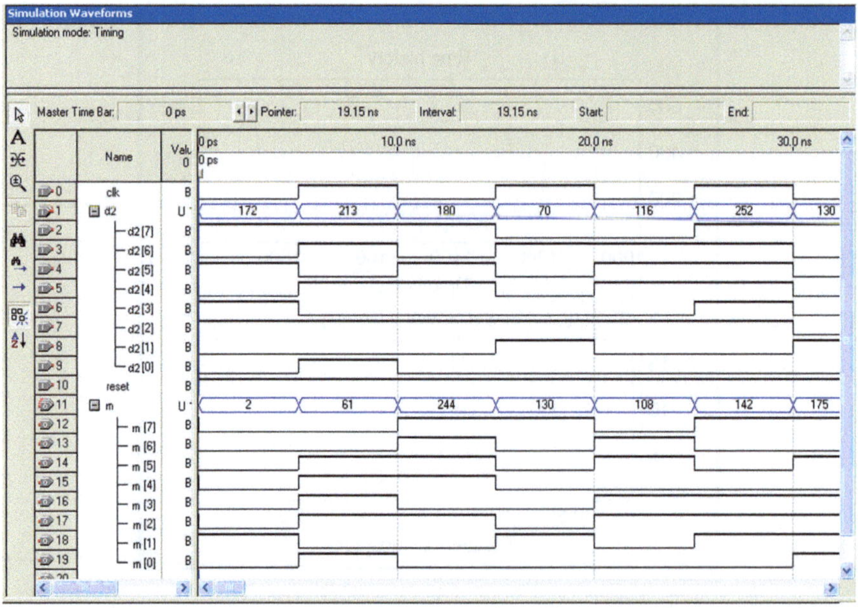

Fig. 14.11 Quartus II simulation of envelope detector

Table 14.1 Summary of synthesis reports of implementing the AM receiver using the Altera CycloneIII kit

Resource	Used	Available	Utilization (%)
IOs	31	532	6.1
Total logic element	1,874	119,088	1.8
Total memory bits	64,321	3,981,312	<1
Total PLLs	1	4	25

receiver components using HDL coder tool, other crosscheck waveforms can be obtained using Altera Quartus II® simulator.

An example of Altera Quartus II® simulator waveforms are the envelope detector waveforms shown in Fig. 14.11. The waveforms of this simulator are purely digital since FPGA environment deals only with digital signals. To get faster simulation with more advanced features, Altera design package ModelSim SE-EE can be used.

To know the amount of the FPGA resources used in the design, the maximum possible operating frequency and other hardware characteristics, synthesis reports can be generated using the synthesis tool. A summary of the synthesis reports of CycloneIII EP3C120F780N device utilization for the AM receiver implementation is given in Table 14.1. The maximum possible operating frequency was found to be 191.2 *MSPS* while maximum path delay from any node was 5.23 *ns*.

As it is obvious from Table 14.1, the programmable digital AM receiver uses less than 2% of the total logic gates and 6.1% of the total I/O pins available in the Altera board. This confirms that FPGA is a promising technology to implement one chip software defined radio for both practical and teaching purposes.

14.6 Conclusions

The example design of the digital AM receiver using Matlab coding and then its conversion into VHDL proved that such an idea be a very attractive way of teaching beginner designers very quickly. The main objective of this chapter was to concentrate on that the beginner designers could easily see the action of their written Matlab codes on hardware. So, they don't need to spend a lot of time to learn DSP or VHDL design skills and sophisticated hardware design issues. This teaching strategy requires the designer be familiar with Matlab-Simulink and Altera Quartus II® software packages. The designer should also know a special sort of Matlab programming "Embedded Matlab" which is a mixture of normal Matlab instructions and fixed-point definitions. An efficient method of digital envelope detection is also presented and its successful implementation was described. This method shows the possibility of implementing the digital versions of well known analogue modulation and demodulation schemes so that both digital and analogue applications could be implemented in a unified development kit reducing the hardware count required for teaching purposes.

Acknowledgements This work was financed by the DAAD (German Academic Exchange Service).

References

1. www.vanu.com
2. www.mathworks.com
3. S.W. Cox, FPGA based waveform design techniques for software defined radios. in *SDR Forum Technical Conference, HW-1-005* 2003
4. A. Krukowski, I. Kale, Simulink/Matlab-to-VHDL route for Full-Custom/FPGA rapid prototyping of DSP algorithms. in *Matlab DSP Conference (DSP'99)*, Tampere, Finland, 1999
5. Embedded MATLAB™ User's Guide (2008) The MathWorks, Inc.
6. K.G. Larkin, Efficient demodulator for bandpass sampled AM signals. IEEE Electron. Lett. **32**(2), 101–102 (1996)
7. P. Maragos, J.F. Kaiser, T.F. Quatieri, On amplitude and frequency demodulation using energy operators. IEEE Trans. Signal Process. **42** (4), 1532–1550 (1993)

Chapter 15
MoM Based EMI Analysis on Large Wind Turbine GSM Communication System

F. Krug and B. Lewke

15.1 Introduction

Wind turbines (WT) cause electromagnetic interference (EMI) via three principal mechanisms namely, near field effects, diffraction and reflection/scattering [1–4]. Near field effects refers to the potential of a wind turbine to cause interference to radio signals due to electromagnetic fields emitted by the generator and switching components in the turbine nacelle or hub. Diffraction occurs when an object modifies an advancing wavefront by obstructing the wave path of travel. Diffraction effects can occur when the object not only reflects part of the signal but also absorbs the signal. Reflection/scattering interference occurs when turbines either reflect or obstruct signals between a transmitter and a receiver. This occurs because when the rotating blades of a turbine receive a primary transmitted signal they act to produce and transmit a scattered signal. In this situation the receiver may pick up two signals simultaneously, with the scattered signal causing EMI because it is delayed in time (out of phase) or distorted compared to the primary signal.

Other important events for the electromagnetic field distribution of a wind turbine are lightning impacts [5]. These lightning events have strong impact on the electronic systems in a wind turbine. Because of the increasing availability requirements for wind turbines there is a trend of more complex electronic monitoring equipment for large wind turbines [6,7]. State of the art WT control communication is realized via low-bandwidth slip-rings and main-shaft between hub and nacelle. The trend to increase electronic equipment leads to the requirement of higher communication bandwidth. Wireless communication links present one solution to this task. Generator failures and lightning strikes may lead to a loss of communication,

F. Krug (✉)
Siemens AG, Munich, Germany
e-mail: florian.krug@siemens.com

B. Lewke
Laboratory for High-Voltage Technology and Power Transmission, Technische Universität München, 80290 Munich, Germany
e-mail: bastian.lewke@ieee.org

S. Lindenmeier and R. Weigel (eds.), *Electromagnetics and Network Theory and their Microwave Technology Applications*, DOI 10.1007/978-3-642-18375-1_15,

resulting in an uncontrollable WT. A GSM transceiver backup system installed to the hub would allow operators to access the control systems even in case of communication loss via nacelle. To optimize such electronic systems in wind turbines an EMI analysis is necessary. On the other hand to proof such complex models efficient measurement methods like the time-domain measurement principle give a deeper understanding of the EMI effects on the electrical energy systems [8].

This paper presents an analytical method and a tool suitable for the analysis of EMI effects on sensitive electronic components in a wind turbine generator. The method of moments (MoM) is used to analytically describe the electro-magnetic fields caused by a GSM 900 MHz transmitter mounted on a hub of the wind turbine that can be used as communication backup system for the control systems of the hub. Using a commercial simulation tool based on MoM, the electromagnetic field distribution will be analysed to determine an optimized wireless communication link to a base station.

15.2 Method-of-Moments Model

15.2.1 State-of-the-Art Wind Turbine Communication System

Modern multi-megawatt wind turbines are equipped with a pitch control system for adjusting the blades' pitch angle. Rotation speed of the turbine is controlled by this system. Communication between the pitch control system in the rotating hub and wind turbine operator is realized over slip-rings at the turbine's main shaft.

Due to lightning strikes or generator failures, fire might break out in the nacelle of the wind turbine. In this case, communication possibilities between turbine operators and the pitch control system might be interrupted. An emergency stop of the facility is no longer possible. This might lead to severe damage of the wind turbine.

By a GSM transmitter with an operating frequency of 900 MHz that is installed on the wind turbine hub, a novel emergency communication is established.

15.2.2 FEKO Model

Using the commercial method-of-moments simulation tool FEKO, a general simulation model of a multi megawatt wind turbine hub was generated [5]. A Hertzian dipole according to the approximation

$$\Pi(\underline{x}) = \frac{e^{-jkr}}{4\pi\varepsilon_0 r}\int_V \underline{P_0}(\underline{x}')dV' \tag{15.1}$$

was used to excite the hub model with a GSM based frequency of 900 MHz, with polarization P_0 and the resulting radiated power P

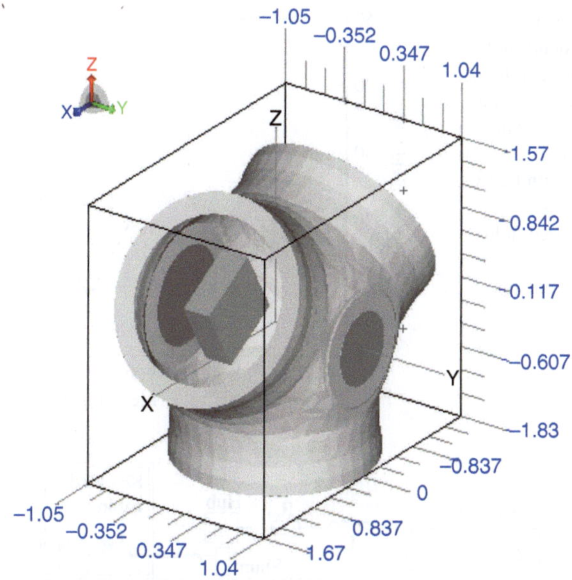

Fig. 15.1 Electromagnetic model of wind turbine hub. Figures on the grid show the dimensions of the hub in meters

$$P = \Re\{2\pi \int_0^{\pi} T_r(r,\vartheta) r^2 \sin\vartheta d\vartheta\} \tag{15.2}$$

and the Poynting vector T_r [9]

$$T_r = \frac{1}{2}\left(E_\vartheta H_\varphi^* - E_\varphi H_\vartheta^*\right) \tag{15.3}$$

with the electric and magnetic field components E_φ, E_ϑ, H_φ and H_ϑ.

The electromagnetic model is depicted in Fig. 15.1. According to the wavelength of the GSM signal, the model has 67,744 elements. Model dimensions are $2.09 \times 2.60 \times 2.50$ m. Material is cast iron with a relative permeability of $\mu_r = 1{,}500$ and a conductivity of $\sigma_i = 1.03 \times 10^7$ S/m. Control boxes are simulated as stainless steel with a conductivity of $\sigma_s = 1.1 \times 10^6$ S/m. The man entrance to the hub is sealed by an aluminium plate with conductivity $\sigma_m = 3.816 \times 10^7$ S/m.

For the calculation, the fast multipole method (FMM) was used in combination with the incomplete LU-matrix decomposition. The maximum number of iterations was set to be 10,000.

15.2.3 Cast Iron Material Under Electro-Magnetic Load

For each of the three materials, the skin effect has been taken into account for the simulation according to [10]

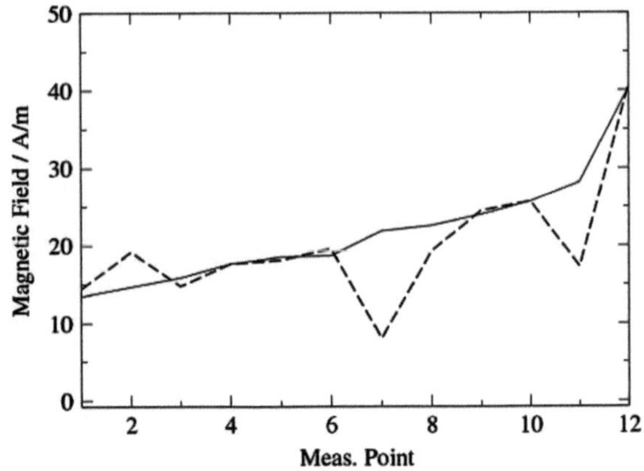

Fig. 15.2 Comparison between measurements and simulations of magnetic fields inside a wind turbine hub due to injected current [5]. The *solid line* represents simulation results with FEKO while the *dashed line* connects the measured field values at discrete measurement points inside the hub

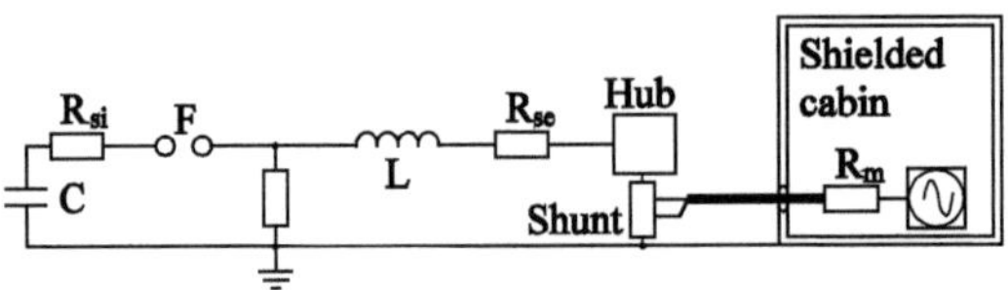

Fig. 15.3 Laboratory setup for field measurements inside the wind turbine hub. A $1.2/50\,\mu\text{s}$ impulse current with 1.3 kA amplitude was injected into the hub by an impulse generator

$$Z_{s,k} = \frac{1-j}{2\sigma_k \delta_k} \frac{1}{\tan((1-j)d_k/2\delta_k)} \tag{15.4}$$

with thickness d_k, σ_k as conductivity and skin depth δ_k for k representing either cast iron, stainless steel or aluminum.

Magnetic field measurements inside the hub for injected currents of up to 1.3 kA were used to verify the simulation model, see Fig. 15.2. Nonlinear material parameters have to be taken into account only for injected currents higher than 40 kA [5]. Therefore, nonlinear effects may be neglected for the GSM 900 MHz analysis of the cast iron hub.

Field measurements were performed according to Fig. 15.3. An impulse current was injected into the cast iron hub in order to derive the field distribution inside. Field measurements inside the hub were made with a field probe based on the principle of induction. The measurement points 1–12 (see Fig. 15.2) covered the maximum space possible due to probe requirements [5]. Injected currents were generated using a 1 MV impulse generator and were measured using a 4.2 mΩ shunt. The higher deviation between measurement and simulation at positions 7 and 11 in Fig. 15.2 is due to EMI from the impulse generator and the connection lines which could only partly be implemented into the simulation model.

15.3 Simulation Results

15.3.1 Antenna Inside Hub

For the first analysis, the electric Hertzian dipole used as GSM 900 MHz transmitter was placed in the center of the hub. This position would allow for the highest protection of the transmitter against all kinds of EMI, especially against lightning. Excitation is a sinusoidal wave with a magnitude of $I \cdot dl = 1$. The resulting Poynting vector inside the hub is depicted in Fig. 15.4. According to (15.2), the Poynting vector holds responsible for the radiated power of the electric Hertzian dipole.

As can be seen in Fig. 15.4, the signal damping of the metallic structure is in the range of 20–40 dB. A good communication link between the transmitter inside the hub and an external receiver and vice versa can not be installed. Therefore

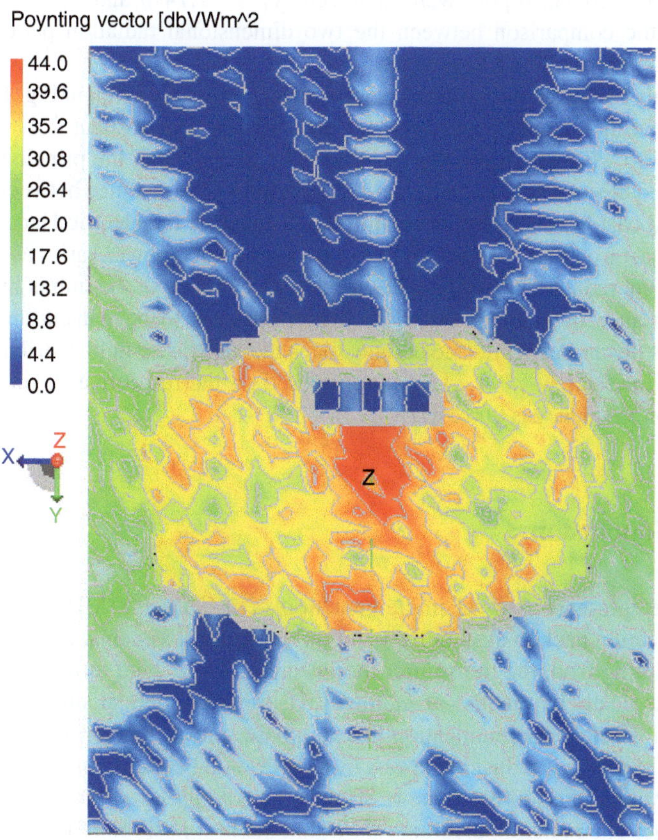

Fig. 15.4 Poynting vector radiation pattern due to sinusoidal excitation of the Hertzian dipole with $f = 900$ MHz and $I \cdot dl = 1$. View is in negative z-direction according to Fig. 15.1

the placement of communication transmitters and receivers inside the hub is not practicable.

15.3.2 Antenna Outside Hub

In contrast to the former model, the transmitting Hertzian dipole will now be installed outside the hub.

Because of the described scattering and diffraction effects that occur at wind turbines due to their rotation, the best possible installation point for the electric Hertzian dipole antenna is at the man entrance. An installation between the blades of the wind turbine is therefore not recommended.

For the simulation model this means that the communication unit is placed at positions with $y > 1$ m, for $z = 0$ m and $x = 0$ m. Two different positions of the electric Hertzian dipole were analyzed: $y_1 = 1.14$ m and $y_2 = 1.24$ m. In Fig. 15.5 the comparison between the two dimensional radiation patterns of the Hertzian dipole at y_1 and y_2 is given.

For both y-coordinate positions, the total electric field is given in Fig. 15.5. It can be seen that the angular dependency of both signals is equivalent for $\varphi = 0°$ and $\vartheta = 0°–360°$. For variation of $\varphi = 0°–180°$ and $\vartheta = 90°$, the principal shape of the radiated electric field is no longer independent of the distance between Hertzian dipole and cast iron hub. By increasing the distance of the electric Hertzian dipole from the cast iron body of the hub by 0.1 m, the radiated power varies as well.

While the radiated power parallel to the man entrance of the hub is higher for a greater distance, the radiated electric field perpendicular to the man entrance is lower for greater distance. Figure 15.6 demonstrates the electric far field distribution of the Hertzian dipole at $y_2 = 1.24$ m superimposed on the respective Poynting vector. Simulation frequency is 900 MHz.

15.4 Conclusion

This paper presents a novel implementation of a GSM 900 MHz transceiver as communication backup-system for wind turbine control systems. Electro-magnetic fields resulting from the transmitter mounted on a large wind turbine hub are analyzed analytically by method-of-moments. Using a commercial simulation tool, an optimized wireless communication link to a base station is determined. Placement of the GSM transmitter inside the cast iron hub would be preferable in order to minimize lightning interference. Due to strong signal damping of 20–40 dB this is not practicable. The radiation diagrams show that the best position of the transmitter is at the man entrance. Simulations of the Hertzian dipole positioned at different locations show a strong directed radiation pattern, allowing for a good communication link between wind turbine hub and base station.

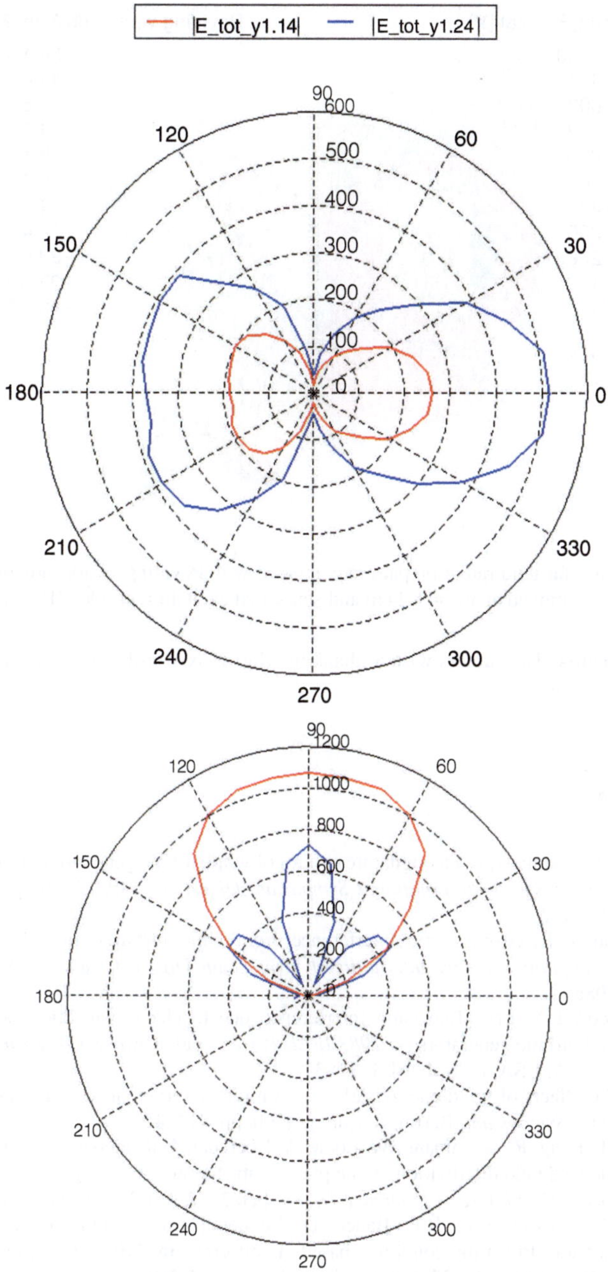

Fig. 15.5 Comparison of the 2D radiation pattern of the electric Hertzian dipole at the positions $y_1 = 1.14\,\mathrm{m}$ and $y_2 = 1.24\,\mathrm{m}$. Radiation frequency is 900 MHz with an amplitude of $I \cdot dl = 1$. *Top*: Radiation pattern for $\varphi = 0°$ and $\vartheta = 0°–360°$. *Bottom*: Radiation pattern for $\varphi = 0°–180°$ and $\vartheta = 90°$

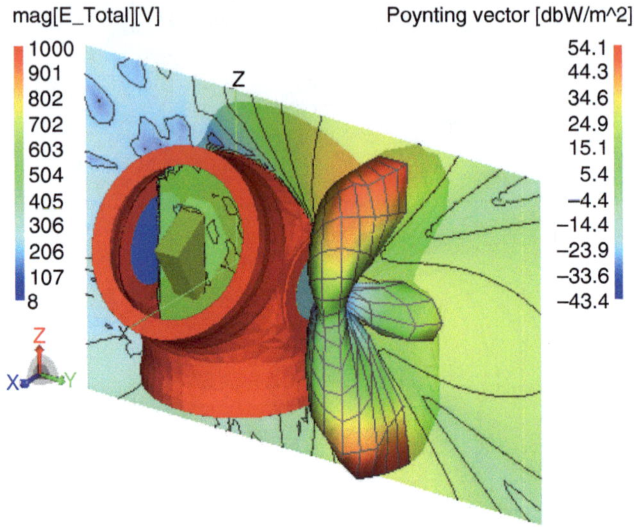

Fig. 15.6 Electric far field radiation pattern compared with Poynting vector radiation for electric Hertzian dipole positioned at $y_2 = 1.24$ m and sinusoidal excitation of 900 MHz and $I \cdot dl = 1$

Acknowledgements The authors wish to thank the Karl-Max-von-Bauernfeind Verein for financial support of their work.

References

1. A. Tennat, B. Chambers, Radar signature control of wind turbine generators. in *IEEE Antennas and Propagation Society International Symposium Digest*, vol. 4A, Washington, USA, July 2005, pp. 489–492
2. D.L. Sengupta, Electromagnetic interference from wind turbines. in *1999 IEEE Antennas and Propagation Society International Symposium Digest*. Orlando, USA, July 1999, pp. 1984–1986
3. K.H. Cavecey, L.Y. Lee, Television interference due to electromagnetic scattering by the MOD-2 wind turbine generators. in *1983 IEEE Power Engineering Society Summer Meeting*. Los Angeles, CA, USA, 83 SM 461-1, 1983
4. A. Frye, The effects of wind energy turbines on military surveillance radar systems. in *2000 German Radar Symposium*, Berlin, Germnay, 2000, pp. 415–422
5. B. Lewke, F. Krug, R. Teichmann, W. Loew, A. Oberauer, J. Kindersberger, The influence of lightning-induced field distribution on the pitch-control-system of a large wind-turbine hub. in *2006 European Wind Energy Conference Digest*, Feb 27–2 March, Athens, Greece, 2006
6. F. Krug, J.R. Rasmussen, R.F. Bauer, D. Lemieux, Ch. Schram, U. Ahmann, Wind turbine/generator drivetrain condition based monitoring. in *2004 European Wind Energy Conference Digest*, Nov 22–25, London, United Kingdom, 2004
7. R. Matsuzaki, A. Todoroki, Wireless detection of internal delamination cracks in CFRP laminates using oscillating frequency changes. Compos. Sci. Technol. **66**, 407–416 (2005)
8. F. Krug, P. Russer, Quasi-peak detector model for a time-domain measurement system. IEEE Trans. Electromagn. Compatibility, **47**(2), 320–326 (2005)
9. O. Zinke, H. Brunswig, *Hochfrequenztechnik 1 – Hochfrequenzfilter, Leitungen, Antennen,* 6th edn. (Springer, Berlin, 2000)
10. EM Software and Systems, *FEKO User Manual*, Suite 5.2, Stellenbosch, South Afrika, 2006

Part IV
Numerical Methods for Electromagnetic Field Modeling

Chapter 16
Novel Frequency-Domain and Time-Domain Techniques for the Combined Maxwell–Dirac Problem in the Characterization of Nanodevices

Tullio Rozzi, Davide Mencarelli, and Luca Pierantoni

16.1 Electromagnetic Field and Mass-Less Dirac Equation: Overview

In the framework of the emerging interest in nanoelectromagnetics and nanoelectronics, scientists are considering the equations of charge transport holding at the nanoscale, namely the Schrödinger and Dirac equations. The discovery and experimental characterization of new materials, such as carbon nanotubes, nanoribbons, and graphene, that feature as low-dimensional systems, seem to open new possibilities [1–8]. In fact, the above nanostructures, owing to their typical absence of crystal defects, behave as ideal transport channels. These in turn require understanding and accurate simulation of their behavior under RF and optical signals. Beside the technological aspects, intriguing theoretical challenges assume now novel significance. Over the last years, in particular, renewed efforts have been made in order to establish a clear correspondence between Dirac and Maxwell equations [9–16]. Classically, Maxwell equations describe the evolution of the electromagnetic fields generated by charge and current sources, and are relativistically covariant. On the other hand, the Dirac equation governs the coherent transport of quantum particles by the space-time evolution of a four component vector, usually referred to as "spinor". The state of a system is thus expressed by a state vector, the spinor, in a linear space: usually, the two pairs of components of the spinor describe the two possible spin states of a quantum particle. The Dirac formulation constitutes the relativistic and vector counterpart of the celebrated scalar Schrödinger equation and can be formally derived by the Klein–Gordon equation, which stands, in quantum mechanics, for the familiar D'Alembert wave equation. For an exhaustive discussion see, for example, [7]. The aim of unifying the above two systems of equations in a simple and elegant formulation, able to explain phenomena of such apparently different nature, promoted investigation of their possible analogies and connections.

T. Rozzi (✉), D. Mencarelli, and L. Pierantoni
Dipartimento di Ingegneria Biomedica, Elettronica e Telecomunicazioni, Università Politecnica delle Marche, Ancona, Italy
e-mail: t.rozzi@univpm.it, d.mencarelli@univpm.it, l.pierantoni@univpm.it

S. Lindenmeier and R. Weigel (eds.), *Electromagnetics and Network Theory and their Microwave Technology Applications*, DOI 10.1007/978-3-642-18375-1_16,

For example, in [9] and [10] a careful study of the mathematical form of Maxwell and Dirac equations pointed out that electromagnetic fields and spinors are joint by many symmetry properties, invariant forms and conservation laws. Some attempts to find other analogies have been made using the Clifford numbers, which can be seen just as a particular way to express the Dirac matrices (reported in Sect. 16.2), to which the Clifford algebra is isomorphic [12, 13]. In [13], for instance, the far field radiated from a simple dipole antenna is calculated from an integral equation based on Clifford algebras. Other authors [17] point out that solutions of Maxwell system can be associated to solutions of Dirac equation through some nonlinear relation. It is also worth mentioning the Majorana representation of the electromagnetic fields, reported for example in [18]. In this formulation, however, a significant lack of generality arises from the assumption that current-source terms have the form of gradients. The above references suggest a close correlation to hold between the zero mass Dirac and Maxwell equations. In this context, a major role could be played by the electromagnetic (*EM*) simulators, which are usually employed to solve very complex and challenging EM field problems. In the literature, a few examples of application of EM solvers to quantum problems have already been reported [4], but these have never been actually applied in connection with the Dirac equation. Although, in the present work, we are not concerned with the use of EM solvers for solving the spinor space-time dependence, it will be shown how the spinor solutions of the Dirac equation can be directly used to express the solutions of the Maxwell equations. An equally important open question is to be able to proceed the opposite way, i.e. to pass from Maxwell to Dirac solutions. Eventually, our analysis may be a first step in the direction of a profitable use of EM solvers for a rigorous description of quantum problems, as well as the development of EM simulators based on the Dirac equation.

16.2 Tensor Form of Maxwell's Equations

In order to expedite further analysis we shall briefly recall the ordinary tensor form of the Maxwell equations.

It is convenient to introduce indexed coordinates x_μ, $\mu = (0, 1, 2, 3)$, defined by [19]

$$x^0 = ct, \quad x^1 = x, \quad x^2 = y, \quad x^3 = z \tag{16.1}$$

which are the so called contravariant components of the position vector, whereas the covariant components x_μ read, by definition,

$$x_0 = ct, \quad x_1 = -x, \quad x_2 = -y, \quad x_3 = -z \tag{16.2}$$

In a similar fashion, it is possible to define the components A^μ and A_μ of the contravariant and covariant 4-vector potentials respectively, in terms of the usual

3-vector potential **A** and the scalar potential ϕ. The contravariant potential is

$$(\phi/c, \mathbf{A}) = (\phi/c, A^x, A^y, A^z) \tag{16.3}$$

whereas the covariant form is

$$(\phi/c, -\mathbf{A}) = (\phi/c, -A^x, -A^y, -A^z) \tag{16.4}$$

Now, it is quite usual to express the partial derivatives with respect to the space-time components defined in [1] and [2] respectively

$$\partial_\mu \equiv \frac{\partial}{\partial x^\mu} \quad \text{and} \quad \partial^\mu \equiv \frac{\partial}{\partial x_\mu} \tag{16.5}$$

In the following, in order to recover a compact and elegant analytical derivation of the electromagnetic fields, we will make use of the electromagnetic tensor, defined as

$$F_{\mu\nu} \equiv A_{\mu,\nu} - A_{\nu,\mu} \tag{16.6}$$

where the subscripts (superscript) after commas indicate partial derivatives in terms of the contravariant (covariant) components. The six components of the electromagnetic field are easily related to the six nonzero elements of the antisymmetric electromagnetic tensor, which reads

$$[F_{\mu\nu}] \equiv \begin{bmatrix} 0 & -E^x/c & -E^y/c & -E^z/c \\ E^x/c & 0 & B^z & -B^y \\ E^y/c & -B^z & 0 & B^x \\ E^z/c & B^y & -B^x & 0 \end{bmatrix} \tag{16.7}$$

In absence of sources, it can be shown that the four Maxwell equations are equivalent to just two equations for F, [19]

$$\begin{aligned} &F^{\nu}_{\mu\nu,} = 0 \quad \Leftrightarrow \quad \begin{cases} \nabla \times B = \mu_0 \varepsilon_0 \partial_t E \\ \nabla \cdot E = 0 \end{cases} \\ &F_{\mu\nu,\sigma} + F_{\nu\sigma,\mu} + F_{\sigma\mu,\nu} = 0 \quad \Leftrightarrow \quad \begin{cases} \nabla \times E = -\partial_t B \\ \nabla \cdot B = 0 \end{cases} \end{aligned} \tag{16.8}$$

In (16.8), and throughout the following, the standard Einstein repeated index summation rule is used. The components of the 4-potential are assumed to satisfy the Lorentz gauge condition, that can be written as

$$\frac{\partial A^x}{\partial x} + \frac{\partial A^y}{\partial y} + \frac{\partial A^z}{\partial z} + \frac{1}{c}\frac{\partial \phi}{\partial t} = A_{\mu,}{}^{\mu} = 0 \tag{16.9}$$

Under the Lorentz condition, the 4-potential satisfy the relation

$$\nabla^2 A = \left(\frac{\partial^2}{\partial x^2} + \frac{\partial^2}{\partial y^2} + \frac{\partial^2}{\partial z^2} - \frac{1}{c^2}\frac{\partial^2}{\partial t^2} \right) A = 0 \tag{16.10}$$

which is the usual wave equation.

16.2.1 Dirac Equations

The Dirac system of equations is briefly described in this subsection. The Dirac matrices are defined as

$$\gamma^0 = \begin{bmatrix} I & \cdot \\ \cdot & -I \end{bmatrix} \quad \gamma^\nu = \begin{bmatrix} \cdot & \sigma^\nu \\ -\sigma^\nu & \cdot \end{bmatrix} \tag{16.11}$$

where I is the 2-dimensional unit and zero matrices, and σ^ν are the Pauli matrices given by

$$\sigma^1 = \begin{bmatrix} 0 & 1 \\ 1 & 0 \end{bmatrix}, \quad \sigma^2 = \begin{bmatrix} 0 & i \\ -i & 0 \end{bmatrix}, \quad \sigma^3 = \begin{bmatrix} 1 & 0 \\ 0 & -1 \end{bmatrix} \tag{16.12}$$

where ($i = -j$). The fundamental algebraic structure of the Dirac matrices, that is the same as the Clifford algebras, is described by the relation

$$\gamma^{\mu\nu} + \gamma^{\nu\mu} = 2g^{\mu\nu} \tag{16.13}$$

where

$$g^{\mu\nu} = \begin{bmatrix} 1 & 0 & 0 & 0 \\ 0 & -1 & 0 & 0 \\ 0 & 0 & -1 & 0 \\ 0 & 0 & 0 & -1 \end{bmatrix} \tag{16.14}$$

is the metric tensor. In the zero-mass case [1], the 4-spinor ψ satisfies the equation

$$\gamma^\nu \partial_\nu \psi = 0 \qquad \psi = \begin{bmatrix} \psi_0 \\ \psi_1 \\ \psi_2 \\ \psi_3 \end{bmatrix} \tag{16.15}$$

which is the zero mass Dirac equation in the standard representation. The representation is not unique, for, the same relation is satisfied, together with (13), if γ^ν is replaced by $\underline{\underline{M}}\gamma^\nu \underline{\underline{M}}^{-1}$, where $\underline{\underline{M}}$ is an arbitrary non-singular matrix, as in [7].

Writing out (16.15) explicitly, the Dirac equation reads

$$\begin{pmatrix} 0 & 0 & \partial_z & \partial_x - i\partial_y \\ 0 & 0 & \partial_x + i\partial_y & -\partial_z \\ \partial_z & \partial_x - i\partial_y & 0 & 0 \\ \partial_x + i\partial_y & -\partial_z & 0 & 0 \end{pmatrix} \begin{pmatrix} \psi_0 \\ \psi_1 \\ \psi_2 \\ \psi_3 \end{pmatrix} = -\frac{\partial_t}{c} \begin{pmatrix} \psi_0 \\ \psi_1 \\ \psi_2 \\ \psi_3 \end{pmatrix} \tag{16.16}$$

By simple matrix product, one can see that the repeated application of the operator on the left side of the above (16.16) yields

$$\nabla^2 \psi - \frac{\partial^2}{c^2 \partial t^2} \psi = 0 \tag{16.17}$$

which is usually referred to as the Klein–Gordon equation. As a matter of fact, the Dirac equation has been historically derived trying to find an expression for the "square root" of the Klein–Gordon equation of [7]. Note that the latter equation can be factorized as it is done for a difference of perfect square, using the Dirac matrices: in "Reconstruction of the 4-potential from the Dirac spinor" of *Appendix* 1, we will show how the positive and negative spinor solutions can be used to express the four-potential.

16.3 Electromagnetic Fields from Dirac Spinors

For the sake of readability, we report in Appendix 1 the proof of the basic result, mentioned earlier, highlighting the close relation between the solutions of Maxwell and Dirac equations.

16.3.1 Solution of the Dirac Equation

In this section, we will establish the method for the derivation of the electromagnetic field solutions, in the source-free case, starting from (16.15). This, in fact, constitutes the main conceptual result of the present work, i.e. associating an electromagnetic field to a spinor, solution of the Dirac equation. To this aim we let, in (16.16), the 4-potential A play the role of the 4-spinor ψ, and subsequently we will derive the resulting electric and magnetic field components in the frequency domain:

$$\gamma^\nu \partial_\nu A = 0 \tag{16.18}$$

Since our task is to extract the Maxwell–Dirac connection, we will make use of ω instead of E, thus restoring the conventional electromagnetic symbol for the frequency. For any fixed frequency $\omega = ck_0$, being c the vacuum light velocity, (16.18) will be solved in its original form, explicitly shown in (16.16), as well as with the reversed sign of the time derivative: in fact, in connection with what discussed at the end of the previous section, the general electromagnetic solution is formed by a combination of positive and negative spinor solutions, that we indicate with **a** and **b** respectively.

We assume first a "plane wave" form for the space-time dependence of the electromagnetic potential, that may be conveniently rewritten as

$$A \equiv (\mathbf{a}+\mathbf{b})\, e^{i\left(k_x x + k_y y + k_z z - k_0 t\right)} = \left(\begin{bmatrix} a_0 \\ a_1 \\ a_2 \\ a_3 \end{bmatrix} + \begin{bmatrix} b_0 \\ b_1 \\ b_2 \\ b_3 \end{bmatrix} \right) e^{i\left(k_x x + k_y y + k_z z - k_0 t\right)} \tag{16.19}$$

where

$$k_0 = \omega / c \tag{16.20}$$

and **a**, **b** satisfy

$$\begin{aligned} \begin{pmatrix} 0 & 0 & k_z & k_x - i k_y \\ 0 & 0 & k_x + i k_y & -k_z \\ k_z & k_x - i k_y & 0 & 0 \\ k_x + i k_y & -k_z & 0 & 0 \end{pmatrix} \begin{pmatrix} a_0 \\ a_1 \\ a_2 \\ a_3 \end{pmatrix} &= k_0 \begin{pmatrix} a_0 \\ a_1 \\ a_2 \\ a_3 \end{pmatrix} \\ -\begin{pmatrix} 0 & 0 & k_z & k_x - i k_y \\ 0 & 0 & k_x + i k_y & -k_z \\ k_z & k_x - i k_y & 0 & 0 \\ k_x + i k_y & -k_z & 0 & 0 \end{pmatrix} \begin{pmatrix} b_0 \\ b_1 \\ b_2 \\ b_3 \end{pmatrix} &= k_0 \begin{pmatrix} b_0 \\ b_1 \\ b_2 \\ b_3 \end{pmatrix} \end{aligned} \tag{16.21}$$

In the above we made use of position (16.19). In order to obtain non-trivial solutions of (16.21), the following relation, expressing the wave number conservation, should hold

$$k_x^2 + k_y^2 + k_z^2 = k_0^2 \tag{16.22}$$

Restricting the analysis, for the moment, to a 2-dimensional domain, and thus setting, for instance, $k_y = 0$, a general solution of (16.21) is found

$$\mathbf{a} = \begin{pmatrix} a_0 \\ a_1 \\ \frac{1}{k_0}\left(k_z a_0 + k_x a_1\right) \\ \frac{1}{k_0}\left(k_x a_0 - k_z a_1\right) \end{pmatrix} \qquad \mathbf{b} = \begin{pmatrix} b_0 \\ b_1 \\ -\frac{1}{k_0}\left(k_z b_0 + k_x b_1\right) \\ -\frac{1}{k_0}\left(k_x b_0 - k_z b_1\right) \end{pmatrix} \tag{16.23}$$

Forming the sum $\mathbf{a} + \mathbf{b}$, one obtains for the 4-vector potential

$$\mathbf{a} + \mathbf{b} = \begin{pmatrix} a_0 + b_0 \\ a_1 + b_1 \\ \frac{1}{k_0}\left(k_z(a_0 - b_0) + k_x(a_1 - b_1)\right) \\ \frac{1}{k_0}\left(k_x(a_0 - b_0) - k_z(a_1 - b_1)\right) \end{pmatrix} \tag{16.24}$$

The components in (16.24) are not independent, since the Lorentz condition (16.9) still remains to be applied: the central point to be stressed is that, in order to retain the most general solution, the Lorentz gauge (16.9) has to be applied to the combination of positive and negative solutions, and not separately to each of them. This yields

$$k_0^2(a_0 + b_0) + k_x k_0(a_1 + b_1) + k_x k_z(a_0 - b_0) - k_z^2(a_1 - b_1) = 0 \tag{16.25}$$

Using (16.6), (16.7), and (16.24), and omitting the common exponential term of (16.19), the following expressions for the electric field components are derived

$$\begin{cases} E^x = -ic\left(k_x(a_0 + b_0) + k_0(a_1 + b_1)\right) \\ E^y = -ic\left(k_z(a_0 - b_0) + k_x(a_1 - b_1)\right) \\ E^z = -ic\left(k_z(a_0 + b_0) - k_z(a_1 - b_1) + k_x(a_0 - b_0)\right) \end{cases} \tag{16.26}$$

the associated magnetic field components are given by

$$\begin{cases} B^x = ik_z/k_0\left(k_z(a_0 - b_0) + k_x(a_1 - b_1)\right) \\ B^y = -ik_z(a_1 + b_1) + ik_x/k_0\left(k_x(a_0 - b_0) - (a_1 - b_1)\right) \\ B^z = -ik_x/k_0\left(k_z(a_0 - b_0) + k_x(a_1 - b_1)\right) \end{cases} \tag{16.27}$$

16.3.2 Example I: Uniform Plane Waves

As a first example, the derivation of the plane-wave solution is shown. By setting $k_z = 0$ in (16.26)–(16.27) and applying the conditions (16.25), we obtain the usual expression of electromagnetic plane waves propagating along the x-direction

$$\begin{cases} E^x = 0 \\ E^y = -ick_x(a_1 - b_1) \\ E^z = -ick_x(a_0 - b_0) \end{cases} \qquad \begin{cases} B^x = 0 \\ B^y = k_x^2/k_0(a_0 - b_0) \\ B^z = -k_x^2/k_0(a_1 - b_1) \end{cases} \tag{16.28}$$

Imposing in (16.28) $a_0 - b_0 = 0$ or $a_1 - b_1 = 0$ allows to distinguish between the $y-$ and $z-$polarized electric field solutions, respectively. The case of $k_x = 0$, with propagation along the z-direction, may be recovered in a similar way.

16.3.3 Example II: 2-Dimensional Homogeneous Waveguides

We now pass to illustrate the more significant case of waveguide modes. As an example, let us search for *TE* (Transverse Electric) modes propagating along the x-direction and confined by metallic parallel plates normal to the z-direction. Beside the Lorentz constraint (16.9), the following conditions are to be imposed on the field components (16.26), (16.27)

$$\begin{array}{lll} E^x = 0 & \text{yielding} & a_1 + b_1 = -k_x/k_0(a_0 + b_0) \\ E^z = 0 & \text{yielding} & a_1 - b_1 = a_0 + b_0 + k_x/k_z(a_0 - b_0) \\ B^y = 0 & \text{yielding} & a_1 + b_1 = k_x/k_0\left(k_x/k_z(a_0 - b_0) - (a_1 - b_1)\right) \end{array} \tag{16.29}$$

As a matter of fact, two of the four conditions (16.25)–(16.29) are independent. Thus, the electric and magnetic fields can be rewritten in terms of $a_0 + b_0 = 0$ or $a_0 - b_0 = 0$ only, and their expressions are given by

$$\begin{cases} E^x = E^z = H^y = 0 \\ E^y = -ic\left(k_x(a_0 + b_0) + \frac{k_0^2}{k_z}(a_0 - b_0)\right) \\ H^x = \frac{i}{\mu_0}\left(k_0(a_0 - b_0) + \frac{k_x k_z}{k_0}(a_0 + b_0)\right) \\ H^z = -\frac{i}{\mu_0}\left(\frac{k_x^2}{k_0}(a_0 + b_0) + \frac{k_x k_0}{k_z}(a_0 - b_0)\right) \end{cases} \tag{16.30}$$

As in the case of the plane wave, the above expressions can be simplified by an appropriate choice of the arbitrary constants. We may set, for example, $a_0^+ + a_0^- = 0$, so that (16.30) becomes

$$\begin{array}{l} E^y = -ic\frac{k_0^2}{k_z}(a_0 - b_0) \\ H^x = \frac{i}{\mu_0}k_0(a_0 - b_0), \quad H^z = -\frac{i}{\mu_0}\frac{k_x k_0}{k_z}(a_0 - b_0) \end{array} \tag{16.31}$$

Imposing $a_0 - b_0 = 0$ in (16.30) would produce the same final result, apart from a multiplication constant. In order to recover the correct standing-wave behavior of TE fields confined in the z-direction, we just need to subtract expression (16.31) from the same expression where k_z is replaced by $-k_z$. Then, imposing E_y to vanish on the parallel metallic plates yields the usual transverse wave-number quantization. For the sake of brevity, these simple final passages are omitted. Nonetheless, it is easy to show that the ratio between the transverse field components equals, as expected, the characteristic impedance of *TE* modes

$$Z_0 = \frac{E^y}{H^z} = \frac{\omega\mu_0}{k_x} \tag{16.32}$$

The analysis of *TM* (Transverse Magnetic) modes may be carried out by analogy. The general case of 3-dimensional waveguides can be developed through the same steps as above, and will be reported in the Appendix 2.

16.3.4 Example III: Dielectric Waveguides

The analysis of the discrete and continuous spectrum of inhomogeneous waveguides, such as, for example, dielectric slabs or multilayers, can be carried out starting from the results of Sect. 16.3.3. Let us consider, as an example, the even *TE* modes of a symmetric dielectric slab propagating in the x-direction, shown in Fig. 16.1, where a magnetic-wall boundary condition holds at $z = 0$.

Indicating by ω/k^I and ω/k^{II} (with $k^I < k^{II}$) the phase velocities in the cladding region I and in the underlying region II, respectively, (16.21) can be rewritten separately for each region

$$\begin{pmatrix} 0 & 0 & k_z^{I,II} & k_x \\ 0 & 0 & k_x & -k_z^{I,II} \\ k_z^{I,II} & k_x & 0 & 0 \\ k_x & -k_z^{I,II} & 0 & 0 \end{pmatrix} \mathbf{a}^{I,II} = k^{I,II}\mathbf{a}^{I,II} \tag{16.33}$$

The same equation with appropriate sign, as in (16.21), can be written for $\mathbf{b}$. In the above, $k_z^{I,II}$ is the transverse wavenumbers in region I, II and k_x is the longitudinal wavenumber: they are related by the wavenumber conservation holding in both regions. In order to recover *TE* modes from the 4-spinor solution of (16.21), (16.24) is rewritten separately for cladding and core

$$\mathbf{a}^I + \mathbf{b}^I = \begin{pmatrix} a_0^I + b_0^I \\ a_1^I + b_1^I \\ \frac{1}{k^I}\left(k_z^I(a_0^I - b_0^I) + k_x(a_1^I - b_1^I)\right) \\ \frac{1}{k^I}\left(k_x(a_0^I - b_0^I) - k_z^I(a_1^I - b_1^I)\right) \end{pmatrix}$$
$$\mathbf{a}^{II} + \mathbf{b}^{II} = \begin{pmatrix} a_0^{II} + b_0^{II} \\ a_1^{II} + b_1^{II} \\ \frac{1}{k^{II}}\left(k_z^{II}(a_0^{II} - b_0^{II}) + k_x(a_1^{II} - b_1^{II})\right) \\ \frac{1}{k^{II}}\left(k_x(a_0^{II} - b_0^{II}) - k_z^{II}(a_1^{II} - b_1^{II})\right) \end{pmatrix} \tag{16.34}$$

By following the same steps of the previous section, we derive the counterpart of (16.30), showing the components of the electromagnetic field in the region I

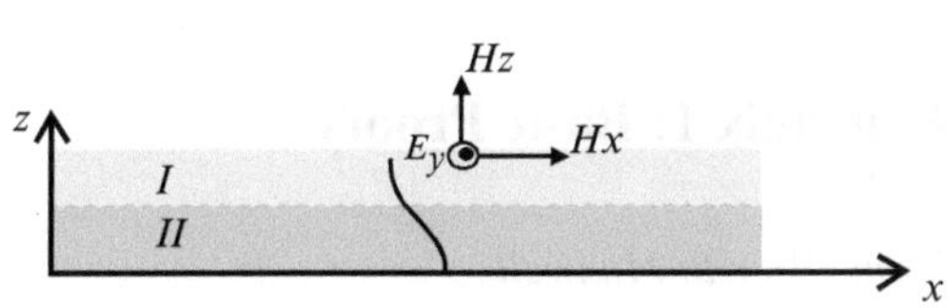

Fig. 16.1 Dielectric slab

$$\begin{bmatrix} E^{y^I} \\ H^{y^I} \\ H^{z^I} \end{bmatrix} = \begin{bmatrix} -ic\left(k_x(a_0^I + b_0^I) + \frac{k^{I^2}}{(-i\gamma)}(a_0^I - b_0^I)\right) \\ \frac{i}{\mu_0}\left(k^I(a_0^I - b_0^I) + \frac{k_x(-i\gamma)}{k^I}(a_0^I + b_0^I)\right) \\ -\frac{i}{\mu_0}\left(\frac{k_x^2}{k^I}(a_0^I + b_0^I) + \frac{k_x k^I}{(-i\gamma)}(a_0^I - b_0^I)\right) \end{bmatrix} e^{-\gamma z} \qquad (16.35)$$

where the z-dependencc is shown explicitly and an exponential decay is imposed: $k_z^I = -i\gamma$. Also, we derive the standing wave solution in region II as indicated at the end of the Sect. 16.3.3:

$$\begin{bmatrix} E^{yII} \\ H^{xII} \\ H^{zII} \end{bmatrix} = (a_0^{II} - b_0^{II}) \begin{bmatrix} -ick^{II^2}/k_z^{II} \\ (i/\mu_0)\,k^{II} \\ -(i/\mu_0)\,k_x k^{II}/k_z^{II} \end{bmatrix} \cos(k_z^{II} z) \qquad (16.36)$$

The above expressions can be simplified by an appropriate choice of the arbitrary constants. Setting, for instance, $a_0^I + b_0^I = 0$ in (16.34), and imposing the field continuity at the dielectric interface, one obtains the usual dispersion relation of *TE* modes and the expression of the field components as a function of the 4-potential, solution of (16.33).

16.4 Conclusions

In conclusion, we reconsider the relation between Maxwell and Dirac equations. First, we look at the connection between four-spinor and four-potential, instead of the electromagnetic field: this choice appears to be a more natural bridge between the above two equations. The central point is that the spinor has to be assumed as a combination of positive and negative solutions of the Dirac equation, satisfying the Lorentz condition. Second, we report some practical examples of how EM fields can be reconstructed, through the four-potential, from the knowledge of the spinor. In particular, we demonstrate the case of EM fields propagating in standard waveguides, such as dielectric slab and rectangular waveguide. In the light of present results, we may consider proceeding the other way round, that is, from Maxwell to Dirac and applying EM simulators to the Dirac equation. In fact, the spinor is constrained by the Lorentz condition, and this seems to be the actual limit of the above correspondence. This aspect and the practical application of the EM simulators warrant further investigation.

Appendix 1: Basic Proofs

From Dirac to Maxwell

In this subsection, the electromagnetic tensor $F_{\mu,\nu}$ is shown to satisfy Maxwell equations, provided that the 4-potential, from which $F_{\mu,\nu}$ is defined in (16.6), satisfies the Dirac equation and the Lorentz gauge condition. Hence, the hypotheses are:

$$\gamma^{\kappa}\partial_{\kappa}A = 0, \quad A^{\mu}_{\mu} = 0 \tag{16.37}$$

and the thesis is given by (16.8). Starting from the hypotheses (16.37), we can write:

$$\begin{aligned}&\gamma^{\kappa}\partial_{\kappa}\left(\gamma^{\nu}\partial_{\nu}A\right) = \tfrac{1}{2}\left(\gamma^{\kappa}\partial_{\kappa}\left(\gamma^{\nu}\partial_{\nu}A\right) + \gamma^{\nu}\partial_{\nu}\left(\gamma^{\kappa}\partial_{\kappa}A\right)\right) = \\ &= \tfrac{1}{2}\left(\gamma^{\kappa}\gamma^{\nu}A_{,\kappa\nu} + \gamma^{\nu}\gamma^{\kappa}A_{,\nu\kappa}\right) = g^{\nu\kappa}A_{,\nu\kappa} = A^{\nu}_{,\nu} = 0\end{aligned} \tag{16.38}$$

Then, each component of A satisfies the above equation:

$$A^{\nu}_{u,\nu} = 0 \tag{16.39}$$

By subtracting in the last equation the following quantity

$$\partial_{\mu}\left(A^{\nu}_{\nu,}\right) \tag{16.40}$$

which is vanishing due to the second of (16.37), we obtain the final result:

$$A^{\nu}_{\mu,\nu} - A^{\nu}_{\nu,\mu} = F^{\nu}_{\mu,\nu} = 0 \tag{16.41}$$

that is one of the two tensor Maxwell's equations (16.8). The second one follows directly from (16.6), in fact:

$$\begin{aligned}&F_{\mu\upsilon,\sigma} + F_{\sigma\mu,\upsilon} + F_{\upsilon\sigma,\mu} = \partial_{\sigma}(A_{\mu,\upsilon} - A_{\upsilon,\mu}) + \\ &+ \partial_{\upsilon}(A_{\sigma,\mu} - A_{\mu,\sigma}) + \partial_{\mu}(A_{\upsilon,\sigma} - A_{\sigma,\upsilon}) = \\ &= A_{\mu,\upsilon\sigma} - A_{\upsilon,\mu\sigma} + A_{\sigma,\mu\upsilon} - A_{\mu,\sigma\upsilon} + A_{\upsilon,\sigma\mu} - A_{\sigma,\upsilon\mu} = 0\end{aligned} \tag{16.42}$$

Reconstruction of the 4-potential from the Dirac spinor

Starting from Maxwell equation, we can follow back the same passages of the previous subsection, obtaining:

$$\begin{aligned}&F_{\mu\nu},^{\nu} = A^{\nu}_{\mu,\nu} - A^{\nu}_{\nu,\mu} = A^{\nu}_{\mu,\nu} = 0 \\ &\Rightarrow \quad \tfrac{1}{2}\left(\gamma^{\kappa}\gamma^{\nu} + \gamma^{\nu}\gamma^{\kappa}\right)A_{,\nu\kappa} = \gamma^{\kappa}\partial_{\kappa}\left(\gamma^{\nu}\partial_{\nu}A\right) =^{2} A = 0\end{aligned} \tag{16.43}$$

The four potential expressing the electromagnetic tensor is a solution of the wave equation, equivalent to a repeated application of the Dirac equation, and not to the Dirac equation only. However, for the case of zero mass, the latter can be based on three anticommuting matrices

$$\gamma^{\nu}, \quad \nu = 1, 2, 3 \tag{16.44}$$

In fact, defining the operator in terms of (16.44)

$$\bar{D}A = \gamma^{\nu}\partial_{\nu}A \tag{16.45}$$

which satisfies the equation

$$\bar{D}^2A = \left(\partial_x^2 + \partial_y^2 + \partial_z^2\right)A \tag{16.46}$$

we obtain the following factorization of the wave equation:

$$\begin{aligned}\gamma^{\kappa}\partial_{\kappa}\left(\gamma^{\nu}\partial_{\nu}A\right) &= \left(\bar{D} + \tfrac{\partial t}{c}\right)\left(\bar{D} - \tfrac{\partial t}{c}\right)A = \\ &= \left(\bar{D} - \tfrac{\partial t}{c}\right)\left(\bar{D} + \tfrac{\partial t}{c}\right)A = 0 \quad \kappa, \nu = 0, 1, 2, 3\end{aligned} \tag{16.47}$$

The general form for A is given by linear combination of: A^+, solution of:

$$\left(\bar{D} + \frac{\partial t}{c}\right)A^+ = 0 \tag{16.48}$$

and A^+, solution of:

$$\left(\bar{D} - \frac{\partial t}{c}\right)A^- = 0 \tag{16.49}$$

Appendix 2: Rectangular Waveguide

In order to carry out the analysis of a uniform rectangular waveguide, shown in Fig. 16.2, we follow the same line of reasoning of Sect. 16.3.3, having restored the y-dependence in (16.19). The conservation of the wavenumber (16.22) becomes then

$$k_x^2 + k_y^2 + k_z^2 = k_0^2 \tag{16.50}$$

The general solution of (16.1) is now given by

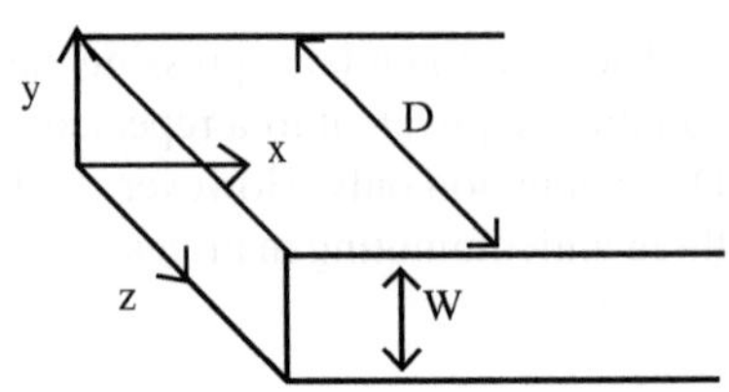

Fig. 16.2 The usual rectangular waveguide

$$\mathbf{a} = \begin{pmatrix} a_0 \\ a_1 \\ \frac{1}{k_0}\left(k_z a_0 + (k_x - ik_y)a_1\right) \\ \frac{1}{k_0}\left((k_x + ik_y)a_0 - k_z a_1\right) \end{pmatrix}, \quad \mathbf{b} = \begin{pmatrix} b_0 \\ b_1 \\ -\frac{1}{k_0}\left(k_z b_0 + (k_x - ik_y)b_1\right) \\ -\frac{1}{k_0}\left((k_x + ik_y)b_0 - k_z b_1\right) \end{pmatrix} \tag{16.51}$$

Forming the sum $\mathbf{a} + \mathbf{b}$ yields:

$$\mathbf{a} + \mathbf{b} = \begin{pmatrix} a_0 + b_0 \\ a_1 + b_1 \\ \frac{1}{k_0}\left(k_z(a_0 - b_0) + (k_x - ik_y)(a_1 - b_1)\right) \\ \frac{1}{k_0}\left((k_x + ik_y)(a_0 - b_0) - k_z(a_1 - b_1)\right) \end{pmatrix} \tag{16.52}$$

The Lorentz condition (16.9) leads to the following expression:

$$\begin{aligned} &k_0^2(a_0 + b_0) + k_x k_0(a_1 + b_1) + k_y k_z(a_0 - b_0) + \\ &+ k_y(k_x - ik_y)(a_1 - b_1) + (k_x + ik_y)k_z(a_0 - b_0) - \\ &- k_z^2(a_1 - b_1) = 0 \end{aligned} \tag{16.53}$$

Again, we make use of (16.6), (16.7), and (16.52), to derive the electric and magnetic field components:

$$\begin{cases} E^x = -ic\left(k_x(a_0 + b_0) + k_0(a_1 + b_1)\right) \\ E^y = -ic\left(k_y(a_0 + b_0) + k_z(a_0 - b_0) + (k_x - ik_y)(a_1 - b_1)\right) \\ E^z = -ic\left(k_z(a_0 + b_0) - k_z(a_1 - b_1) + (k_x + ik_y)(a_0 - b_0)\right) \end{cases} \tag{16.54}$$

$$\begin{cases} B^x = ik_z/k_0\left(k_z(a_0 - b_0) + k_x(a_1 - b_1)\right) - \\ \qquad - ik_y/k_0\left((k_x + ik_y)(a_0 - b_0) - k_z(a_1 - b_1)\right) \\ B^y = -ik_z(a_1 + b_1) + ik_x/k_0\left((k_x + ik_y)(a_0 - b_0) - k_z(a_1 - b_1)\right) \\ B^z = ik_y(a_1 + b_1) - ik_x/k_0\left(k_z(a_0 - b_0) + (k_x - ik_y)(a_1 - b_1)\right) \end{cases} \tag{16.55}$$

In order to focus on a particular example, let us search for a *TE* modes propagating along the x-direction, which requires that

$$E^x = 0 \quad \text{yielding} \quad a_1 + b_1 = -k_x/k_0(a_0 + b_0) \tag{16.56}$$

By using, in (16.54) and (16.55), the condition (16.56), and imposing the Lorentz constrain (16.53), that can be slightly simplified as

$$\begin{aligned} &(k_0^2 - k_x^2)(a_0 + b_0) = \\ &= (k_z^2 - k_y k_x + ik_y^2)(a_1 - b_1) - (k_x + k_y + ik_y)k_z(a_0 - b_0) \end{aligned} \tag{16.57}$$

one obtains the general expression for *TE* modes. In the following equation we report the electric field component:

$$\begin{cases} E^x = 0 \\ E^y = -ic\left(k_z(k_y k_x + k_y^2 - k_z^2)(a_0 - b_0) + k_z^2(k_y + k_x - ik_y)(a_1 - b_1)\right) \\ E^z = -ic\left((-k_z^2 k_y + k_y^2 k_x + ik_y^3)(a_0 - b_0) + k_z(-k_x k_y + k_y^2 - ik_y^2)(a_1 - b_1)\right) \end{cases} \tag{16.58}$$

The magnetic field are not shown because, as for the 2-dimensional case, the ratio between the transverse field components simply returns the expected characteristic impedance:

$$Z_0 = \frac{E^y}{H^z} = -\frac{E^z}{H^y} = \frac{\omega\mu_0}{k_x} \tag{16.59}$$

In order to recover standing wave dependence of the field in the transverse direction, we need to sum (16.58) to the expression itself, where the following substitutions have been made

$$k_y \to -k_y, \quad k_z \to -k_z \tag{16.60}$$

Finally, an appropriate choice of the arbitrary constants of (16.58) is needed to obtain the correct spatial dependence of the fields with respect to the x- and y-directions:

$$k_z k_x(a_1 - b_1) = (-k_z^2 + k_y k_x + k_y \gamma)(a_0 - b_0) \tag{16.61}$$

Hence, employing (16.61) in (16.58) yields the electric fields in terms of the negative and positive solutions of the Dirac equation:

$$\begin{cases} E^x = 0 \\ E^y = -ick_z^2(k_y - ik_y)(a_1 - b_1)\cos(k_z z)\sin(k_y y) \\ E^z = -ick_z(k_y^2 - ik_y^2)(a_1 - b_1)\sin(k_y y)\cos(k_z z) \end{cases} \tag{16.62}$$

Of course, imposing E_y and E_z to vanish respectively at $z = D$, and $y = W$, yields the usual transverse wavenumber quantization. The magnetic field may be recovered by using (16.59).

References

1. P.L.M. Euen, M.S. Fuhrer, H. Park, Single-walled carbon nanotube electronics. IEEE Trans. Nanotech. **1**(1), 78–85 (2002)
2. A.M. van der Zande, S.S. Verbridge, I.W. Frank, D.M. Tanenbaum, J.M. Parpia, H.G. Craighead, P.L. McEuen, Electromechanical resonators from graphene sheets. Science **315**, 490–492 (2007)

3. F. Miao, S. Wijeratne, Y. Zhang, U.C. Coskun, C.N.L.W. Bao, Phase-coherent transport in graphene quantum billiards. Science **317**, 1530–1533 (2007)
4. D. Mencarelli, T. Rozzi, L. Pierantoni, Coherent carrier transport and scattering by lattice defects in single- and multi-branch carbon nanoribbons. Phys. Rev. B 75, 085402 **77**, 1 954 351–11 (2008)
5. G. Pirio, P. Legagneux, D. Pribat, K.B.K. Teo, M. Chhowalla, G.A.J. Amaratunga, W.I. Milne, Fabrication and electrical characteristics of carbon nanotube field emission microcathodes with an integrated gate electrode. Nanotechnology **13**(1), 1–4 (2002)
6. P.J. Burke, An RF circuit model for CNTs. IEEE Trans. Nanotech. **2**, 55–58 (2002)
7. F. Schwable, *Advanced Quantum Mechanics*, 3rd edn. (Springer, Berlin, 2000)
8. L. Pierantoni, D. Mencarelli, T. Rozzi, Boundary immittance operators for the Schrödinger-Maxwell problem of carrier dynamics in nanodevices. IEEE Trans. Microw. Theory Tech. **57**(5), 1147–1155 (2009)
9. V.M. Simulik, I.Y. Krivsky, Relationship between maxwell and dirac equations: symmetries, quantization, models of atom. Rep. Math. Phys. **50**(3) (2002)
10. V.M. Simulik, Connection between the symmetry properties of dirac and maxwell equations. Theor. Math. Phys. **87**(1), 386–393 (1991)
11. V. Simulik, *What is electron?* V. Simulik (Apeiron, Montreal, 2005)
12. A. Chantaveerod, A.D. Seagar, T. Angkaew, Calculation of electromagnetic field from integral equation based on clifford algebra. in *Piers Proceedings*. Czech Republic, Prague, Aug 2007, pp. 71–71
13. A. Chantaveerod, T. Angkaew, Numerical computation of electromagnetic far-field from near-field using integral equation based on clifford algebra. in *Proceedings of Asia-Pacific Microwawve Conference 2007, APMC 2007, Asia-Pacific*. Prague, Czech Republic, Dec 2007, pp. 1–4
14. H. Torres-Silva, The close relation between the maxwell system and the dirac equation when the electric field is parallel to the magnetic field. Ingeniare, Revista chilena de ingenierìa **16**(1), 386–393 (2008)
15. A. Campollattoro, New spinor representation of maxwell equations. Int. J. Theor. Phys. **29**(2), 141–155 (1990)
16. A. Campollattoro, New spinor representation of maxwell equations. Int. J. Theor. Phys. **29**(2), 141–155 (1990)
17. J. Vaz, Jr., W.A. Rodrigues, Jr., Equivalence of the dirac and maxwell equations and quantum mechanics. Int. J. Theor. Phys. **32**(6), 945–959 (1993)
18. R. Mignami, E. Recami, M. Baldo, About a dirac-like equation for the photon according to ettore majorana. Lett. Nuov. Cim. **11**, 572–586 (1974)
19. J. Foster, J.D. Nightingale, *A short course in General Relativity*, 2nd edn. (Springer, New York, 1995)

Chapter 17
Electromagnetic Partitioning Methodology Towards Multi-Physics Chip-Package-Board Co-Design and Co-Simulation

Sidina Wane and Damienne Bajon

17.1 Introduction

In the context of semiconductor integration technology solutions [1], co-design and co-simulation between chip, package and board design levels, as well as between multiple physics are essential in today's electronic designs. Chip (IC) and package/board design and simulation have been considered as separate and disjoint activities in common semi-conductor product development. The key challenges of simultaneous co-design and co-simulation of chip, package and board are multiple. Among such challenges, are cultural barriers between chip designers and package/board designers. Chip designers are used to work with design tools which are intrinsically restricted to a dedicated flow environment (*Digital, Analog, Mixed-signal*) in reference to a specific technology (*CMOS, GaAs, BiCMOS,... etc*), while package/board designers, on the other hand, often use totally different tools by different computer platforms. In addition, there is a lack of standardized distributed database exchange formats (beyond conventional file-based extensions: gds2/Lef/Def for chip level design, mcm/sip for package design and gerber/ndd/hkp for board design) across the different existing design environments. Requirement of common design environment, where constraints from Chip level are propagated to package-level and even to board-level, results from the necessity to facilitate bridging various design domains (Analog, Digital and Mixed) that used to be driven by different tools/flows. Classically such design domains are tackled separately often without a single system-level view for global simulation. An *integrated methodology* with IC, package, and PCB in one physical single model is the most accurate way to simulate today's complex system-on-chip, system-in-package, package-on-package designs. Although single model EM

S. Wane (✉)
NXP-Semiconductors, Esplanade Anton Philips 14906, Colombelles Caen, France
e-mail: sidina.wane@ieee.org

D. Bajon
ISAE-Université de Toulouse, 10 avenue Edouard Belin, Toulouse, France
e-mail: damienne.bajon@isae.fr

S. Lindenmeier and R. Weigel (eds.), *Electromagnetics and Network Theory and their Microwave Technology Applications*, DOI 10.1007/978-3-642-18375-1_17,

simulation analysis of unified chip-package-board assembly design would be the most accurate approach, this is not possible using existing design tooling suites. The main reason is related to the very high complexity resulting from the merging of IC level, package level and board level specific database informations such as layout description, connectivity assignment, layers stack definition, and etc. The *integrated methodology* often requires huge computing resources. To render possible simultaneous analysis of selected/identified complete paths across the different integration levels (chip, package and even board), one needs to use *segmentation* approaches [2, 3] with *segregated methodology*. This means to combine different simulation techniques (frequency-domain, time-domain, [8] and mixed-signal) with extraction models (broadband equivalent circuit representations [3–8], digital activity model derivation, transistor IPs for analogue blocks, and etc) for system-level analysis. Such *segmentation approaches* are based on divide-and-conquer techniques. Therefore, efficient hybridization of various simulation tools with different assumptions (analytical/semi-analytical, quasi-static, and full-wave), thus different accuracy levels, is required. In order to accurately model the higher order effects resulting from discontinuities (bending, power-ground shapes, via-hole transition connections, bond-wiring, bumps) in the analysis, proper boundary conditions for each segmentation in association with the appropriate accuracy assumptions need to be defined. the *segregated methodology* treats chip, package and board designs independently and obtains overall solutions by cascading solutions from IC, package and PCB. While the *segregated* (divide-and-conquer) approach can make EM simulations more manageable in terms of computing resources, it also introduces some technical difficulties, such as techniques of partitioning and deembedding. Recently *global methodologies* [2, 3] extending classical cascading techniques with introduction of *residual S-parameters* to account for coupling between different sub-partitions have been proposed. In [9] topological and functional partitioning in EM analysis is discussed in the scope of system-on-chip applications.

To meet the requirement of active analog and digital co-simulations, scalable hierarchical approaches that allow coupled analysis between different abstract views (schematic/symbol/physical) are necessary. Transistor level description or behavioral-modeling for particular noisy block could be sufficient in capturing analog active parts intrinsic responses. However for digital dies – generally considered as aggressors (noise injectors)- additional details on their power consumption and dynamic switching activities are important to properly deal with global power and signal integrity analysis and time-budgeting considerations. In the published research work various approaches have been proposed for the estimation of time-domain switching activity profiles for digital active modules, with restriction to microprocessors and micro-controllers. Among such approaches are analytical waveform profile calculation, numerical macro-modeling and/or statistical techniques, and measurement methods [10–17]. Analytical calculations based on peak-value assumption referring to simple canonical waveforms (triangular/trapezoidal shapes) for rough model representation of digital dies internal current profile are unable to derive temporal and spatial distribution of power activity through chip partitions accordingly to multi-clock frequency domains.

Main limitation of macro-modeling techniques concerns difficulties to extract time-domain current activity phase information of derived waveform profiles. Extraction of current activity phase information in complement to magnitude responses requires the use of advanced numerical techniques such as wavelet transforms [16]. Measurement based methods, in order to be exploitable and easily correlated with simulation results, require efficient on-board setting protocols and efficient numerical de-embedding algorithms backed-up by signal processing analysis.

This paper discusses electromagnetic functional and topological (geometry) partitioning methodologies towards multi-physics Chip-Package-Board co-simulation and co-analysis, with application to real-world mixed-signal multi-chip module systems. Limits of the segregated methodology in comparison with the integrated global methodology are discussed in full-wave and quasi-static assumptions for the co-design and co-simulation of chip, package and board. A *global distributed iterative* co-simulation methodology for concurrent/simultaneous analysis of passive and active parts is proposed.

The organization of the paper is articulated around three sections. The first section discusses state of the art EM partitioning approaches and techniques for large-scale problems, with an illustration of marco-pixel partitioning concept [18, 47] introduced in Transverse Wave Formulation Approach (TWF) [42] to assess the importance of couplings between partitioned sub-domains. The couplings between macro-pixels where contributions of both higher and lower order modes are accurately taken into account can be efficiently computed using an original NUFFT (Non-Uniform Fast Fourier Transform) [19] offering the use of non-uniform multi-grid space stepping. The second section presents application of full-wave and quasi-static partitioning methodologies to component-level, function bloc-level and system-level applications. Different segmentation strategies are compared to draw limits of cascade assumptions in comparison with full-EM model approaches and experimental results. Guideline and design rule derivation towards standard segmentation-based scalable wideband model synthesis for electromagnetic Interference (EMI) aware design analysis, are discussed. In the third section a system-level global analog-digital co-simulation methodology is proposed based on power-signature concept to model high-speed digital modules temporal and spatial distribution of their power switching activity, including thermal-electrical co-analysis.

17.2 EM Partitioning Approaches for Large/Multi-Scale Structures: State of the Art

To bring full-wave EM simulation accuracy to applications including structures with increasing complexity demands innovative electromagnetic computational techniques. Numerical electromagnetic methods generally distinguish two procedural steps that drive their computational burden and associated complexity. The first step concerns the construction and filling of the characteristic matrix which encodes the spatial distribution or the description of the structure under simulation in the appropriate algebraic form, impedance or admittance matrix for the Method of Moment (MoM), stiffness matrix for the Finite element method (FEM), connectivity and scattering matrix for TLM-like methods [45] etc...) according to the selected method. The second step deals with solving the resulting matrix equation using the appropriate numerical technique, direct or block inversion, iterative resolutions being often difficult to circumvent in larger scale problems.

In the integral equation techniques associated to the method of moments [23], the computational effort required for the filling of impedance or admittance matrix results from the calculation of the interaction between each discretizing element. Classically the interaction between all the discretizing elements are computed by evaluating spatial Green's functions expressed in the form of improper Sommerfeld integrals [20]. The oscillatory behaviour together with the slowly decaying variations of integrands involved in Sommerfeld integrals renders their numerical evaluation time consuming. For accurate and fast calculation of such integrals different techniques have been proposed [24–26]. Among these techniques are discrete image method (DIM), fast Hankel transforms (FHT), interpolation and tabulation approaches, steepest descent path (SDP) techniques and Krylov subspace based-techniques in conjunction with other related subspace based-approaches. Using such techniques the filling of the MoM matrix can be achieved with reasonable accuracy on the matrix terms for acceptable time effort in moderately large problems. However the computational complexity to fill the MoM matrix remains $O(N^2)$, where N represents the number of unknowns. Once the MoM matrix is filled the direct resolution of the linear system using standard LU decomposition requires a $O(N^3)$ processing time. The Fast Multipole Method (FMM), introduced to electromagnetics by Rokhlin, leads to significant reduction of computational complexity [27]. In the FMM approach, the interactions between discretizing elements are not calculated or stored in an explicit manner. Discretizing elements, current sources, are divided into sub-groups and interactions between distant sub-groups are calculated using plane wave expansions. The multi-level formulation of the FMM, MLFMM, limits the storage requirement to $O(N \log N)$ and the computation time to $O(N \log N)$. However FMM is only suitable for free space problems and find difficulties in multilayered medium applications while another limitation with the traditional FMM, is related to low-frequency instabilities [28]. The Fast Inhomogeneous Plane Wave Algorithm (FIPWA) has been introduced by Hu et al. [29] as an alternative to FMM, to overcome difficulties related to multilayered structures. The FIPWA [29–31] scales as *O(NlogN)* in run time and $O(N)$ in memory. The basic algorithm of the FIPWA relies on the decomposition of the discritizing current elements into sub-scatters. The interactions between nearby groups are computed in an explicit way while, for long range interactions, the Green's functions are expanded over an integral summation of inhomogeneous plane waves following properly chosen SDP. To limit the homogeneous plane wave translation from source sub-scatter to observation sub-scatter, interpolation and extrapolation procedures are used. Issues related to error control of the FIPWA with lossy and active media are discussed in [32].

Characteristic Basis Function Method (CBFM) [33], designed for solving by MoM-type methods large scale electromagnetic problems involving both microwave circuits as open region and scattering problems may be regarded as applying an alternative implicit grouping strategy, based on physical considerations. After subdividing the original structure in sub-blocks of amenable complexity, Characteristic Basis Function are extended basis functions attached to sub-blocks and generated by Singular Value Decomposition (SVD) technique applied to the current solution of the sub-block MoM matrix. The rank of the SVD being by orders of magnitude less than the dimension of the sub-block MoM matrix, the dimension of the re-assembled overall MoM matrix is considerably reduced and renders possible, iteration-free inversion.

Domain decomposition methods (DDM), used for both multi-processor and single processor configurations, multi-grid approaches (MGA) or macro-modeling techniques (MMT) can be used for expediting EM analysis of multiscale structures in conjunction with finite methods (FEM, FDTD, FIT, etc…) [34–40]. The basic idea of the DDM referring to a technique of "divide-and-conquer", primarily developed for solving partial differential equations [41], is to decompose the computational domain into smaller sub-domains with manageable computation complexity. In order to enforce the matching of the local solutions after solving equations on each sub-domain, interface conditions referred as "transmission" conditions are written on the artificial boundaries introduced by the division process. Important advantages of the DDM concern on one hand, saving memory requirement and, re-use of previously computed results in case of modifications on local sub-domains, on the other.

In [49–52] Russer, Mongiardo and Felsen discuss systematic approach to compute electromagnetic field in complex structures through network theory. The field problem is translated into an equivalent network problem where the electromagnetic fields at boundaries are represented using the Tellegen's theorem for fields, based on generalized transformer network connection. The use of network theory is introduced as an unifying approach for combining different approaches and methods, each method being considered in the appropriate computational sub-domain. This leads to potential directions for complexity reduction of EM analysis through the hybridization of various methods. In this prospect, hybridization of Transmission Line Matrix method (TLM) with Transverse Wave Formulation (TWF) method [42], have been suggested in [43]. Table 17.1 shows formal similarities of TLM and TWF methods among FDTD and MoM formulations.

Challenges of power-waves formulation [21] for nonlinear systems mainly lies in three major aspects. The first challenge concerns the necessity to extend classical S-parameter definition (*restricted to mono-modal assumptions*) with the notion of large signal reflection coefficients, properly incorporating nonlinear effects. The second challenge is in link with proper transfer and conversion analysis between active, reactive reactive and apparent power-wave components. The third challenge states the need for unified representation formalism of systems electrical behavior in terms of power-waves energy transfer and conversion. Such unified formalism will allow for effective low power design optimization and power-energy oriented control not easy to achieve with conventional voltage-current based approaches. In addition power-waves formulation establishes the required bridging connections between different disciplinary fields (*mechanics, aerodynamics, acoustics,*

Table 17.1 Formal similarities of TLM and TWF among FDTD and MoM approaches

Conjugate Quantities	DIFFERENTIAL Methods with rectangular grid mesh in TIME-DOMAIN	Corresponding INTEGRAL METHODS in FREQUENCY DOMAIN
Fields E, H	FDTD 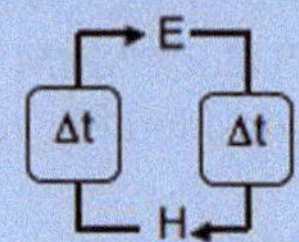	MoM 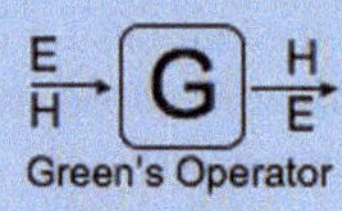
Waves A, B	TLM 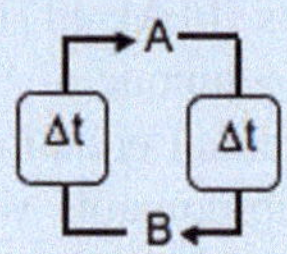	TWF 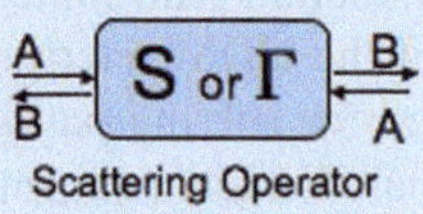

etc...) for multi-physics cosimulation and co-analysis. Physics-based equivalent circuit derivation [22] bridging geometry aspects with network representations is seen promising in order to combine different energy-domains in one unified model analysis. Important efforts have been devoted to attempts for deriving generalized power-waves definition that unifies linear and nonlinear representations. For large-signal measurement, recently X-parameters with PHD (Poly- Harmonic-Distorsion) have been introduced to extend S-parameters to nonlinear devices.

17.3 Challenges of Electromagnetic Partitioning Methodologies

17.3.1 Sub-Domains Selection and Interfacing Conditions

Challenges of electromagnetic partitioning methodologies include formulation and derivation of techniques for proper segmentation of overall simulation domain into sub-partitions to analyze separately with manageable memory and complexity. Specification of electromagnetic boundary conditions to impress on the interfacing junctions between the partitioned sub-domains are of paramount importance. When dealing with circuit analysis the nature of considered boundary conditions can give means for defining global ground references using electrical walls or perfect conducting metal interfaces. Use of periodic walls in the investigation of invariant periodic structures can lead to tremendous complexity reduction by restricting the analysis to elementary canonical cells. When dealing with scattering problems plane waves polarized excitations are generally considered based on diffraction analysis following modal decomposition formalism. Classical boundary interfaces

Fig. 17.1 Sub-domains selection and interfacing boundaries

commonly used in electromagnetic formulations are Dirichlet, Neumann, or Robin [44] type of conditions. Significant amount of research work with the DDM, MGA and MMT concern the development of numerical algorithms for ensuring low reflection losses, for power conservation purposes, at the frontiers between sub-domains or at the interfaces of macro-elements to incorporate in EM computational domains. Overlap between partitioning sub-domains are generally considered in the implementation of domain-decomposition formulation [33,44]. Once the segmentation of the overall simulation domain into sub-domains (as illustrated in Fig. 17.1) is performed and each sub-domain is analyzed using the right approach, determination of the global system responses requires accurately accounting for coupling between the different contributions.

17.3.2 Sub-Domains Coupling and Interaction Analysis

To illustrate the importance of couplings between sub-domains resulting from full-wave partitioning analysis, the concept of macro-pixel [47] is introduced and developed with Transverse Waves Formulation (TWF). Attributes of observed coupling range distribution and potential control of truncation orders at macro-pixel levels are investigated, and give potentialities for hybridisation with time-domain formulations.

17.3.2.1 Concept of Macro-Pixel Partitioning

The concept of macro-pixels (regrouping micro-pixels) described in Fig. 17.2, is introduced as extension, based on physical considerations, of the notion of pixel with possible spatial variations within its sub-domain to capture both lower and higher order variations. The electromagnetic interactions between macro-pixels are efficiently computed [47] using an original NUFFT (Non-Uniform Fast Fourier Transform). In [48], a compact-cell (*C-Cell*) methodology is proposed and implemented using the TWF method, aiming at accurate incorporation in EM analysis of elements that exhibit sharp contrast in dimensions. The continuity operator $\tilde{S}^{(uv)}$ [48] averaging effect on higher order modes is represented in Fig. 17.2e. The associated transfer function is characterized through a scattering approach as illustrated in Fig. 17.2d or by means of circuit approach (multi-port excitation) [48].

The representation of the reflection operator through its expansion on the two components transverse TE and TM mode basis function of the overall domain, is, using Dirac notation [46]

$$\hat{\Gamma} = \sum_{m=0}^{M} \sum_{n=0}^{N} |\mathbf{f}_{m,n}^{TM}\rangle \Gamma_{m,n}^{TM} \langle \mathbf{f}_{m,n}^{TM}| + \sum_{m=0}^{M} \sum_{n=0}^{N} |\mathbf{f}_{m,n}^{TE}\rangle \Gamma_{m,n}^{TE} \langle \mathbf{f}_{m,n}^{TE}| \tag{17.1}$$

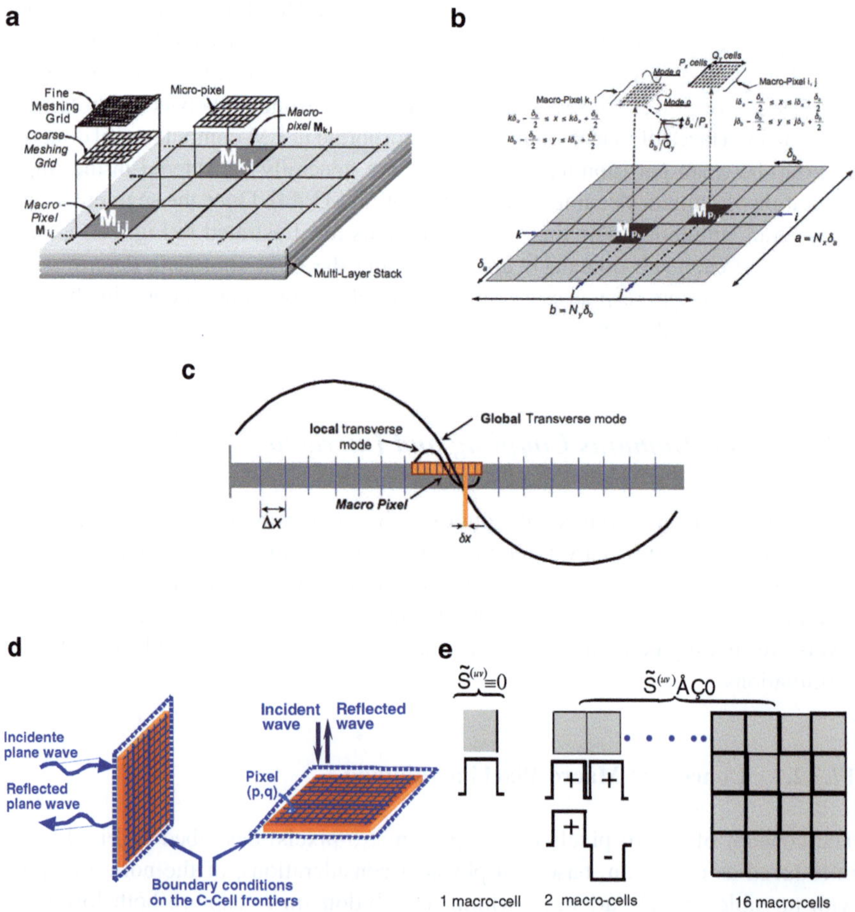

Fig. 17.2 Multi-grid partitioning of a complex circuit into macro-pixels (composed of micro-pixels) (**a**) coupling between Macro-pixel $Mp_{i,j}$ to macro-pixel $Mp_{k,l}$ (**b**), *local (to macro-pixels) and global modal basis functions* (**c**). Scattering approach (**d**) transfer function of the C-Cell relating, at the macro-pixel scale, the number of high-frequency component to the number of macro-cells (**e**)

To investigate the properties of this operator in the spatial domain it is convenient to express the components of the transverse TE and TM mode basis functions as the product of a complex amplitude and a real normalized structure function [46] as:

$$f_{m,n}^{M,u=x,y}(x,y) = K_{m,n}^{M,u=x,y} \times \Phi_{m,n}^{u=x,y}(x,y) \tag{17.2}$$

where $M = TE, TM$ and $K_{m,n}^{M,u=x,y}$ results from normalizing condition of $|\mathbf{f}_{\mathbf{m,n}}^{\mathbf{M}}\rangle$.

Expanded on the normalized structure functions $\Phi_{m,n}^{u=x,y}(x,y)$ the expression of the reflection operator in the spatial domain has the following dyadic form:

$$\hat{\Gamma} = \sum_{m=0}^{M}\sum_{n=0}^{N}\begin{bmatrix} \left|\Phi_{m,n}^{x}\right\rangle \Gamma_{m,n}^{xx}\left\langle\Phi_{m,n}^{x}\right| & \left|\Phi_{m,n}^{x}\right\rangle \Gamma_{m,n}^{xy}\left\langle\Phi_{m,n}^{y}\right| \\ \left|\Phi_{m,n}^{y}\right\rangle \Gamma_{m,n}^{yx}\left\langle\Phi_{m,n}^{x}\right| & \left|\Phi_{m,n}^{y}\right\rangle \Gamma_{m,n}^{yy}\left\langle\Phi_{m,n}^{y}\right| \end{bmatrix} \tag{17.3}$$

Considering now the multi-grid partitioning shown in Fig. 17.2a where the overall domain is subdivided into macro-pixels, each of ones being subdivided at the lower scale in micro-pixels. The restriction to a macro-pixel $Mp_{i,j}$ of order (i,j), Fig. 17.2b of a transverse incident or reflected wave component $w_{u=x,y}^{Mpi,j}$ expanded on the global transverse basis functions is given by:

$$\begin{aligned} w_{u=x,y}^{Mpi,j}(x,y) &= H^{Mpi,j}\Big(\sum_{m}^{M}\sum_{n}^{N} w_u^{TE,m,n}\Big|f_{m,n}^{TE,u=x,y}(x,y)\Big\rangle \\ &\quad + w_u^{TM,m,n}\Big|f_{m,n}^{TM,u=x,y}(x,y)\Big\rangle\Big) \\ &= H^{Mpi,j}\left(\sum_{m}^{M}\sum_{n}^{N} w_u^{m,n}\left|\Phi_{m,n}^{u=x,y}(x,y)\right\rangle\right) \end{aligned} \tag{17.4}$$

where $H^{Mpi,j}$ represents the windowing *Heaviside* characteristic function of the $Mp_{i,j}$ macro-pixel domain. On the other hand, the restriction of the component $w_{u=x,y}^{Mp_{i,j}}$ in (17.4) on the macro-pixel $Mp_{i,j}$ sub-domain, can be expanded on local basis functions $|g_{M_{i,j},p,q}^{u=x,y}(x,y)\rangle$ following:

$$w_{u=x,y}^{Mpi,j}(x,y) = \sum_{p,q}^{P_x,Q_y} w_{u,p,q}^{Mp_{i,j}}\left|g_{Mp_{i,j},p,q}^{u=x,y}(x,y)\right\rangle \tag{17.5}$$

the dimension of the local basis being $P_x Q_y$ Fig. 17.2b.

At the macro-pixel scale, reflected waves $b_{u=x,y}^{M_{i,j}}(x,y)$ on a macro-pixel $Mp_{i,j}$ and incident waves $a_{u=x,y}^{M_{k,l}}(x,y)$ on a macro-pixel $M_{k,l}$ are related, in the spatial domain, by the reflection operator $\hat{\Gamma}_{Mp_{i,j}}^{Mp_{k,l}}$:

$$\mathbf{a}^{Mp_{k,l}} = \hat{\Gamma}_{Mp_{i,j}}^{Mp_{k,l}}\mathbf{b}^{Mp_{i,j}} \tag{17.6}$$

Expanding the components of the incident and reflected waves on the local basis functions, the coupling between a local mode of order (p,q) on $Mp_{i,j}$ and a local mode of order (p_0,q_0) on $Mp_{k,l}$ as depicted in Fig. 17.2b, can be expressed through:

$$\begin{aligned} a^{Mpk,l}_{x,p_0,q_0}(x,y) &= \sum_{m,n}^{M,N}\sum_{p,q}^{P_x,Q_y}\left\langle g^x_{Mp_{k,l},p_0,q_0}(x,y)\middle|\Phi^x_{m,n}(x,y)\right\rangle \\ &\quad\times\Gamma^{xx}_{m,n}\left\langle \Phi^x_{m,n}(x',y')\middle|g^x_{Mp_{i,j},p,q}(x',y')\right\rangle b^{Mp_{i,j}}_{x,p,q} \\ &\quad+\sum_{m,n}^{M,N}\sum_{p,q}^{P_x,Q_y}\left\langle g^x_{Mp_{k,l},p_0,q_0}(x,y)\middle|\Phi^x_{m,n}(x,y)\right\rangle \\ &\quad\times\Gamma^{xy}_{m,n}\left\langle \Phi^y_{m,n}(x',y')\middle|g^y_{Mp_{i,j},p,q}(x',y')\right\rangle b^{Mp_{i,j}}_{y,p,q} \\ &= \sum_{m,n}^{M,N}\sum_{p,q}^{P_x,Q_y}\Gamma xx^{Mpk,l}_{Mpi,j}(p,q,p_0,q_0)b^{Mi,j}_{x,p,q} \\ &\quad+\sum_{m,n}^{M,N}\sum_{p,q}^{P_x,Q_y}\Gamma xy^{Mpk,l}_{Mpi,j}(p,q,p_0,q_0)b^{M_{i,j}}_{y,p,q} \end{aligned} \tag{17.7}$$

where $(\Gamma uv^{Mpk,l}_{Mpi,j}(p,q,p_0,q_0))_{u,v=x,y}$ is the Green's function coupling the u-component of incident wave, relative to the local mode (p,q) on $Mp_{i,j}$, to v-component of reflected wave relative to local mode (p_0,q_0) on $Mp_{k,l}$.

The set of local basis functions in (17.5) being chosen as $g^{u=x,y}_{Mp_{i,j},p,q}(x,y)=(\delta a\times \delta b)^{-\frac{1}{2}}e^{j\frac{2\pi px}{\delta a}}e^{j\frac{2\pi qy}{\delta b}}$, without loss of generality, the Green's functions $(\Gamma uv^{Mp_{k,l}}_{Mp_{i,j}}$ $(p,q,p_0,q_0))_{u,v=x,y}$ can be expressed in the following form:

$$\begin{aligned} \left(\Gamma uv^{Mpk,l}_{Mpi,j}(p,q,p_0,q_0)\right)_{u=v=x} &= (N_x\times N_y)^{-1} \\ \times\sum_{m,n}^{N_x\times P,N_y\times Q} & e^{j2\pi(k-i)\frac{m}{N_x}}e^{j2\pi(l-j)\frac{n}{N_y}}\times Sc_{m,n}(p,q,p_0,q_0)R_{m,n}(k_0L) \end{aligned} \tag{17.8}$$

where the product k_0L of the wave number k_0 and the transverse dimension L is a scale parameter. $R_{m,n}$ depends on the modal reflection coefficients defined in (17.6) and Sc_{mn} is given by:

$$Sc_{m,n}(p,q,p_0,q_0) = Sc\left(\frac{m}{N_x}-p\right)Sc\left(\frac{m}{N_x}-p_0\right)\times Sc\left(\frac{n}{N_y}-q\right)Sc\left(\frac{n}{N_y}-q_0\right) \tag{17.9}$$

with Sc denoting the cardinal sine function. It is essential to underline that the coupling $(\Gamma uv^{M_{k,l}}_{M_{i,j}}(p,q,p_0,q_0))_{u,v=x,y}$ between modes (p,q) and (p_0,q_0) mainly depends on the relative differences $|k-i|$ and $|l-j|$.

17.3.2.2 Sub-Domain Couplings: Coupling Between Higher and Lower Order Modes

Traditionally, coupling is defined between sources through specified excitation modes. The concept of coupling between modes on different macro-pixels (composed of micro-pixels) can be understood as a generalization of the classical coupling between localized sources. In Fig. 17.2, a multi-grid Green's function is considered to evaluate the coupling between macro-pixel of order (k, l) and macro-pixel of order (i, j) through fundamental and higher order modes versus a normalized distance $|i–k|$ or $|j–i|$. The simulation domain is composed of 32×32 macro-pixels, each macro-pixel comprises 128 micro-pixels. The coupling resulting from the fundamental modes of the two macropixels is seen dominant by more than one decade in comparison with the higher order contributions. In Fig. 17.3 the parameters p, q represent orders of local modes to the macro-pixel of order

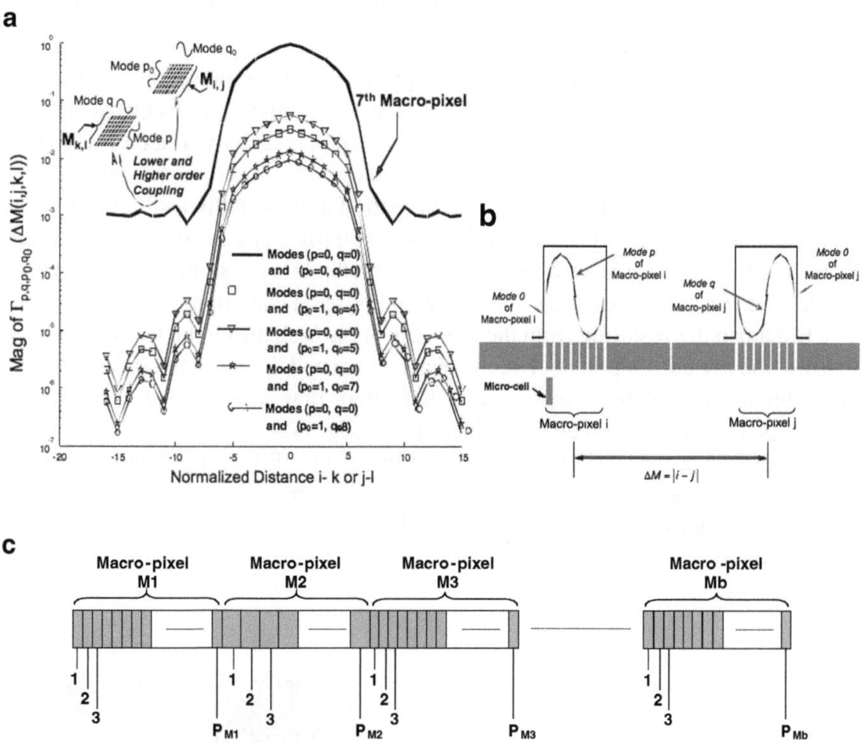

Fig. 17.3 Coupling between macro-pixel of order (i, j) and macro-pixel of order (k, l) through fundamental mode on macro-pixel of order (i, j) and higher order modes on macro-pixel of order (k, l) versus a normalized distance $\Delta M(i, j) = |i - j|$ (**a**), dominant coupling between fundamental modes (**b**). Schematic cross-section view of Fig. 17.2, (taking a centered numbering for the macro-pixels) illustrating coupling between local fundamental and higher order modes of two macro-pixels, *from* [18] (*c*)

(k, l) and parameters p_0, q_0 designate orders of local modes on the macro-pixel of order (i, j). p and p_0 refer to harmonics in the x direction; q and q_0 refer to harmonics in the y direction. It is observed from the curves of Fig. 17.3 that, given a reference macro-pixel, the most dominant coupling arises with an area not exceeding an optimal number of neighboring macro-pixels: this optimal number is found to be around 7. This means that for distances between the macro-pixel source and observation sub-domain (another macro-pixel) greater than one-seventh of the wavelength only a reduced number of terms are sufficient to accuralety compute the discrete Green's functions.

17.4 Application of Electromagnetic Partitioning Analysis to Component, Function Bloc and System Level Integration

17.4.1 Application to Component Level and Function-Bloc Scale

The proposed EM segmentation methodology is applied to representative structures where inductive couplings is an important aspect: a coupled 2-turns Octagonal and 2-turns 8-shaped inductor integrated on low cost high resistivity Silicon PICS (*Passive Integration Connecting Substrate*) technology. A simplified two-metal layer cross-section of the substrate stack is composed of a 650 μm thick silicon substrate with a conductivity of 0.1 S/m, covered by a 0.50 μm thick insulating SiO_2 layer. With this test-case carrier EM segmentation methodology is applied at function-block level in order to investigate critical couplings and ways to reduce their effects. The EM attributes resulting from the topology of the 8-shaped inductor are expected to lower the mutual coupling between the octagonal and the 8-shaped inductor. Such attributes are considered when dealing with floor-plan at chip level e.g., for the relative position of VCO blocks as illustrated in Fig. 17.4.

To apply the proposed partitioning methodology two different test-case structures represented in Figs. 17.6 and 17.7. Each test-case carrier is investigated based on the following steps. The first step is the partitioning of the physical layout topology into sub-partitions to analysis separately using electromagnetic divide-and-conquer methodology.

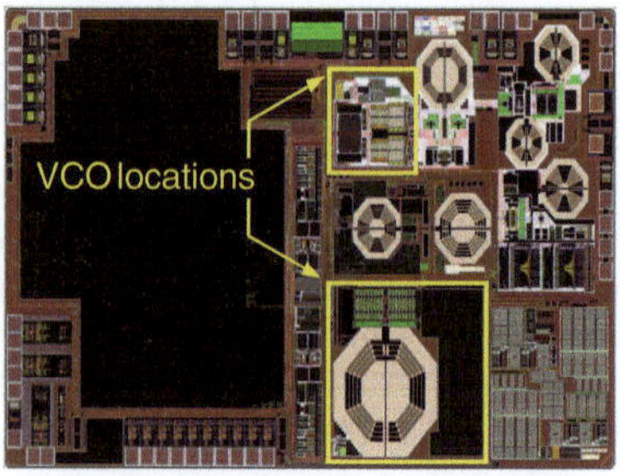

Fig. 17.4 Illustration of Chip floor-plan for optimal positioning of inductive modules

This initial step requires proper analysis and expertise of the specific structures in order to define appropriate sub-partition frontiers. Once the sub-partitions are defined, excitations are associated to the different partitions to extract the associated S-parameter models. The resulting sub-partitions are derived in terms of multi-port elements, the size of the multi-ports being determined by the number of used port excitations. The excitation of the different partitions implies the definition of internal ports with local grounding strategy [6]. Tradeoffs between low CPU time and accuracy over a broadband frequency response impose optimal partitioning strategy. The second step is combining the extracted S-parameters multi-ports with the appropriate connections in the framework of circuit analysis environment in order to synthesize the global response of the whole structure (Fig. 17.5). At this step the global (merging of the local ground connections into one single net) and local grounding strategies (keeping the local ground pins separate) can be studied in order to evaluate limits of equi-potential assumptions.

Concerning the time domain simulations SPICE/SPECTRE models are preferred to S-parameters based models to ensure smoother convergence and better DC behavior.

When SPICE models are not extracted natively broadband SPICE extractions (BBS) should be considered for time-domain simulations. A particular attention should be paid to passivity and power conservation.

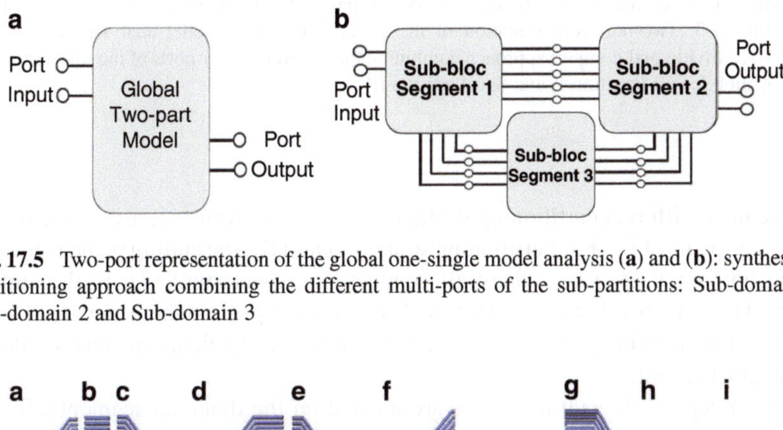

Fig. 17.5 Two-port representation of the global one-single model analysis (**a**) and (**b**): synthesized partitioning approach combining the different multi-ports of the sub-partitions: Sub-domain 1, Sub-domain 2 and Sub-domain 3

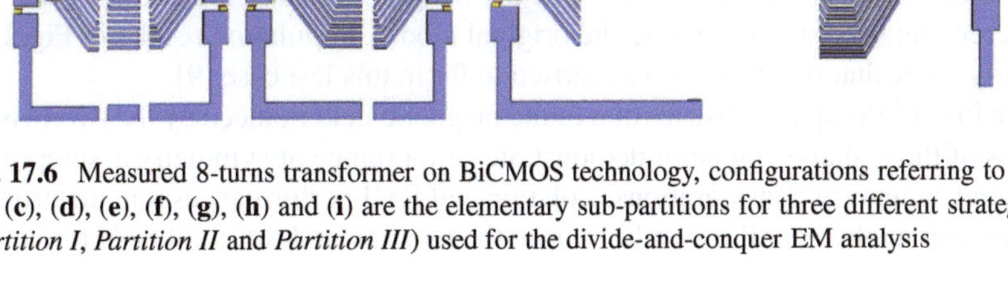

Fig. 17.6 Measured 8-turns transformer on BiCMOS technology, configurations referring to (**a**), (**b**), (**c**), (**d**), (**e**), (**f**), (**g**), (**h**) and (**i**) are the elementary sub-partitions for three different strategies (*Partition I*, *Partition II* and *Partition III*) used for the divide-and-conquer EM analysis

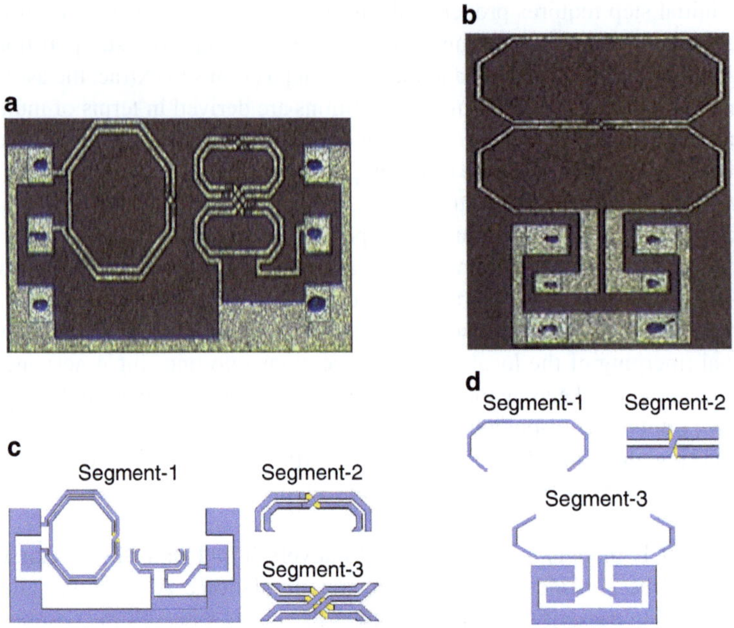

Fig. 17.7 Top views of measured coupled inductors on PICS technology (**a**), configurations referring to (**b**), (**c**) and (**d**) are the elementary sub-partitions for divide-and-conquer investigations. Fig. 17.3. Two-port representation of the global one-single model analysis (**a**) and (**b**): synthesized divide-and-conquer approach combining the different multi-ports of the sub-partitions: Sub-Partition I, Sub-Partition II and Sub-Partition III

The three different partitioning strategies for the transformer test-case are shown in Figs. 17.6 and 17.7. For partitioning strategy II and III, the cutting plane is located in the horizontal sections, where all conductors are perpendicular to the cutting plane. This ensures that in the EM analysis of each piece, where port feed lines are added at the cutting plane, the current flow and thus the fields are very similar to the original circuit.

For strategy I, the cutting planes are located on the diagonal segments. In this case, the cutting plane is not perpendicular to the conductor segments, and the port feed lines in the piecewise analysis point into other directions than the conductors in the original circuit. This means that the fringing fields at the cutting plane are different between the pieces and the original model. Simulation results in Fig. 17.8 clearly show that partitioning was carried to far in this last case [9].

In Fig. 17.9b, spatial distribution of the magnetic field induced by the two twisted loops of the 8-shaped inductor demonstrates importance of symmetry assumptions. Beyond symmetry considerations, isolation and EMI performances remain strongly dependent on the coupling resulting from the twisting and the number of turns.

Fig. 17.8 Comparison between full-wave single global model, three different partitioning strategies and measurement for the 8-turns transformer

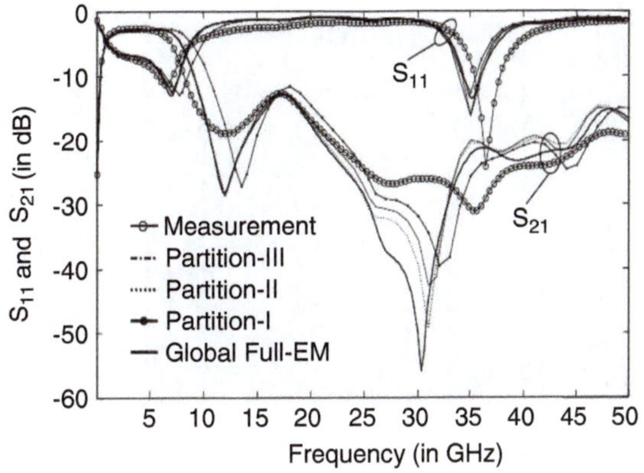

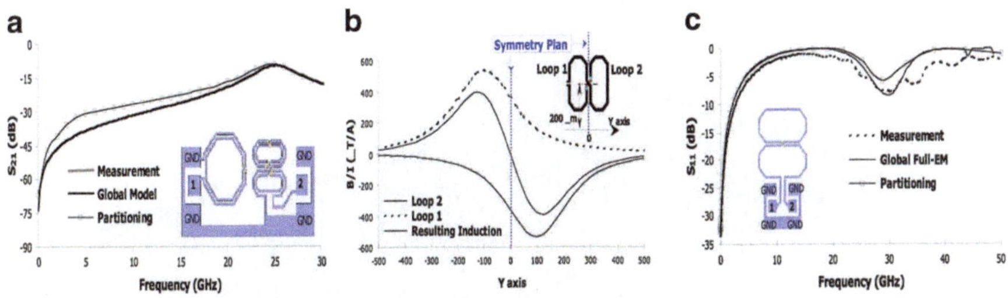

Fig. 17.9 S21 (**a**) parameter for the coupled 2-turns Octagonal and 2-turns 8-shaped inductor on PICS technology: comparison between full-wave single global model, one partitioning strategy (Partitioning label), and measurement, and S11-parameter (**b**) for a single 8-shape inductor. Illustration of magnetic field (induction) distribution for the 8-shaped inductance along the Y axis, at 200 μm from the loops' center (**c**)

Figure 17.9a–c compare the performances against frequency of the structure composed of coupled 2-turns Octagonal and 2-turns 8-shaped inductors in Fig. 17.7 with the structure of single 8-shaped inductor, which show acceptable correlations.

17.4.2 Application of EM Partitioning to Chip-Package-Board System: Concept of Chip-Package-Board Co-Design and Co-Simulation

17.4.2.1 Carrier Description and Flow Considerations

A SiP module composed of 3 dies stacked in a land-grid-arrays package mounted on a printed-circuit-board (PCB), shown in Fig. 17.10a–c is chosen to demonstrate the proposed co-simulation methodologies.

The three dies consist of one passive high resistivity substrate carrier using in-house low-cost process (passive integration connecting substrate) and two active

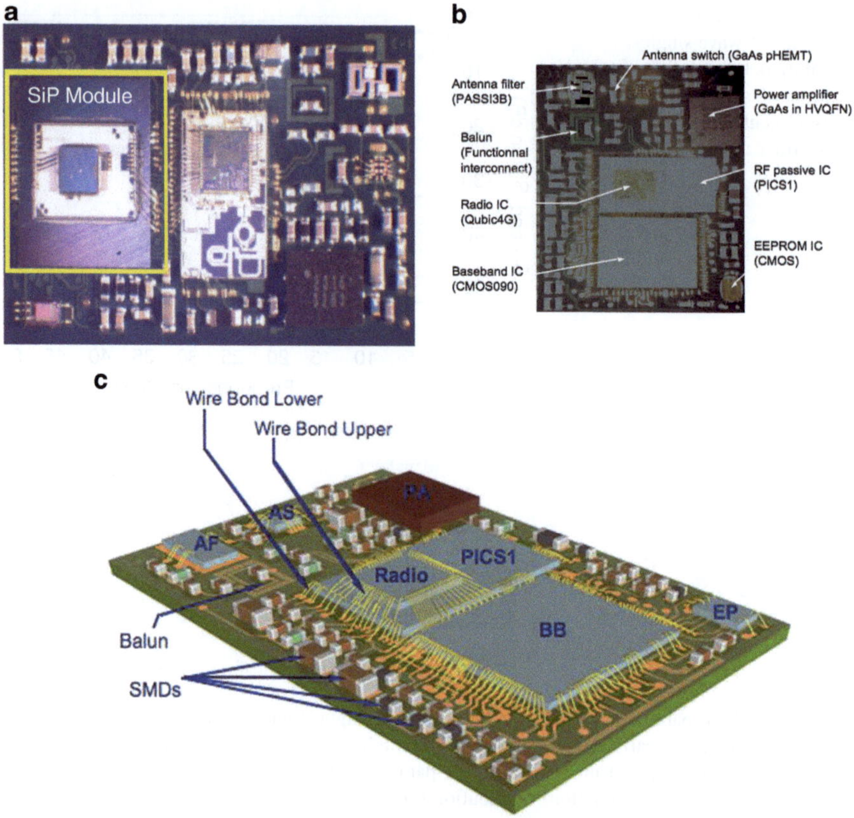

Fig. 17.10 Photograph of the SiP carrier, WLAN SiP Carrier developed by NXPSemiconductors used in this paper as a case study (**a**), 3D virtual prototyping model showing bond wired multi-chip modules (**b**), RF-passives(PICS) and SMD components (**c**)

digital and analog dies respectively with CMOS-90nm and BiCMOS technologies. The package 4-layer laminate substrate is designed with Cadence SiP (silicon-package-board, version 15.7 and 16.0). The PCB was designed using Mentor Graphics Expedition. The unified chip-package-board model is assembled using different SiP tooling suites, namely Ansoft, CST, Sigrity and Optimal-Apache [17]. For the purpose of chip-package-board co-design and co-simulation a multi-level path (delimited in Fig. 17.11a,b with a blue border line), which starts from an RF-input on the PCB, traverses the LGA and ends on the analog IC die at the input of the LNA (low-noise-amplifier), is selected.

Two co-simulation methodologies are investigated and compared both in full-wave and quasi-static assumptions, namely:

1. *Segregated methodology* where the selected complete chip-package-board path, shown in Fig. 17.11, is partitioned into 3 portions (chip-portion, package-portion

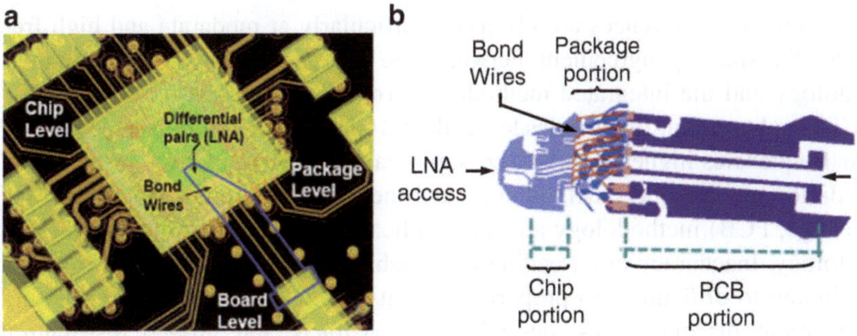

Fig. 17.11 Selected Multi-level complete chip-package-board (**a**) path for investigation of partitioning divide-and-conquer approach limitations (**b**)

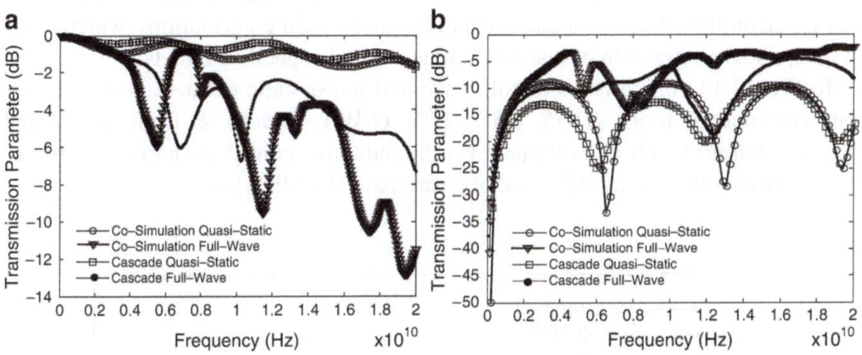

Fig. 17.12 (**a**) Comparisons of segregated and integrated methodologies for IC-package-board Co-Simulation in reference to a complete path shown in Fig. 17.2 for insertion loss, and (**b**) transmission loss. *These results from* [4] *extend full-wave analysis presented in with quasi-static results*

and board-portion) and the resulting S-parameters are cascaded to deduce the overall frequency-domain responses.

2. *Integrated methodology* where the complete chip-package-board is treated as a unified single model.

The first co-simulation methodology (segregated methodology) is based on solution cascading techniques. As a result, it requires special attentions in the full-wave analysis, as the partitioning approach is very sensitive to ground return-path settings and to the definition of excitation ports and associated numerical de-embedding. In the quasi-static analysis, the notion of ports is replaced by the concept of "sources" and "sinks" most suitable for nodal-based representations. In Fig. 17.12, full-wave and quasi-static solutions from the segregated methodology and results from the integrated methodology are compared.

While the location of the resonant frequencies are relatively about the same between full-wave and quasi-static solutions for the insertion and transmission losses, significant differences are observed particularly at moderate and high frequencies. Satisfactory agreement between the segregated (divide-and-conquer) methodology and the integrated methodology (one single model for IC, package and PCB) is obtained at frequencies lower than 2 GHz with a relative error less than 8%. At frequencies higher than 2 GHz, significant differences between the segregated (divide-and-conquer) methodology and the integrated (one single model for IC, package, PCB) methodology are seen, indicating limitations of the segregated methodology. Inaccuracy of cascading approaches at moderate and high frequencies is linked to difficulties to properly account for higher order effects resulting from different discontinuities, de-embedding artifacts and ground distribution [4–6]. Accuracy of the segregated (divide-and-conquer) methodology can be improved by introducing residual multi-ports between the different segmentation portions to account for the missing coupling effects between IC, package, and PCB [2].

The root limitation of quasi-static assumption with respect to full-wave analysis lies in uncoupled analysis of electric and magnetic field contributions, which leads to independent extraction of resistive, inductive and capacitive effects.

In Fig. 17.13 comparisons of full-wave and quasi-static extractions (using various commercial tools: *HFSS, Q3D, TPA, O-Wave, PackSi-E* from *Ansoft* and *Apache-Optimal*) with measurement for the inductive contribution of on-chip portion is presented showing influence of return path shielding [6].

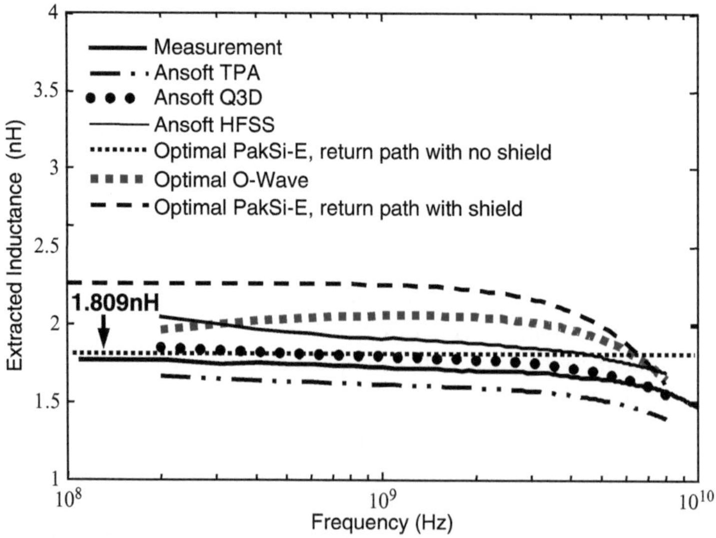

Fig. 17.13 Comparisons of Full-wave and Quasi-static inductance extraction with on-chip Silicon Measurement (chip portion) (data are from [4])

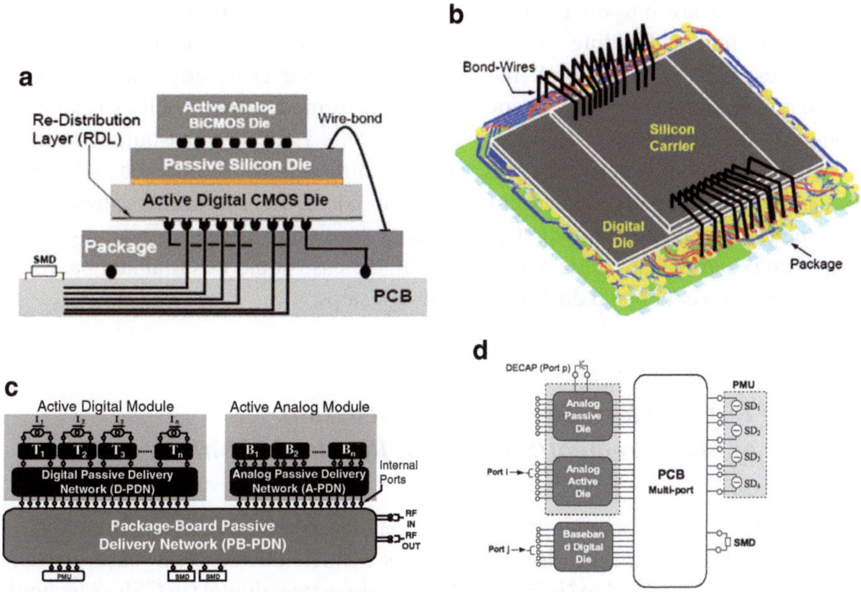

Fig. 17.14 Unified chip-package-board model, cross-section view of the SiP module mounted on board (**a**), and 3D view of the package with die-stack (**b**). Schematic representation of coupled active digital-analog modules and passive multi-ports (IC-level, package/board levels). A-PDN = analog-PDN, D-PDN = digital-PDN, PB-PDN = package-board-PDN (**c**), test-bench multi-port schematic representation (**d**)

17.5 Analog-Digital Co-Simulation Methodology

17.5.1 Concept of Power-Signature for Co-Simulation of Digital and Analog Modules Through Multiports

The proposed Co-Simulation methodology is articulated around three principal approaches: digital high-speed dynamic switching power activity modeling, analog active behavioural model representation and Electromagnetic extraction of generalized multi-port passive delivery network (PDN). The analog and digital active models are coupled through passive embedding environments. The passive embedding environments are referenced in Fig. 17.14 for the digital module, for the Analog blocks and the package/board following generalized multiport topology/functionality-based partitioning. The coupled quasi-static and full-wave electromagnetic-based partitioning approach proposed in [2–6] is applied to accurately derive the different multi-ports. The resulting multi-ports are iteratively coupled to macro-model representations of the digital and analog modules.

A concept of power-signature [5] is introduced to characterize high-speed digital modules switching activities. In Fig. 17.14c, two-port matching elements

$(T_1, T_2, \ldots T_n)$ are introduced to couple current switching sources for the digital module to their embedding PDN multi-ports. The analog blocks $(B_1, B_2, \ldots B_n)$ are represented by behavioral models or transistor-level descriptions. The proposed power-signature concept captures in a macro-model, a multi-clock frequency domains, the switching activity of standard cells (gates), input-output buffers, decoupling capacitors and memories. The switching activity is extracted at the level of the input/output pads of the digital active module. Different stimuli scenarios are considered to emulate both worst-case activity (assuming all gates switching at the same time) and realistic-case activity. A realistic case could be emulated based on timing analysis conducted during initial design steps.

17.5.2 Test Carriers Description, Discussion of Measurement/Simulation Results and Validation

To investigate the proposed global active-passive co-simulation methodology, a satellite-TV-system real-world NXP-Philips-Semiconductors test case carriers including two active analog (in BiCMOS technology) and digital (in CMOS technology) dies packaged and reported on test-board is considered. Different integration configurations are studied: the first configuration where the two active dies are separately reported on the test-board (case of Fig. 17.15b) and the second configuration where the two dies are stacked in a single system-in-package which is reported on test-bord (case of Fig. 17.15a). In the single package option two variants (with and without active SiP module) are introduced to analyze the coupling between the active SiP and the package-board at the interfacing junctions.

The measured average power of the active digital die is 140 mW. The considered digital die comprises 4 principal clock domains at 20, 60, 120 and 480 MHz. Assuming a constant voltage supply for each frequency domain, within a certain margin, at the delivery PMU (power management unit) sources, current switching activity waveforms are extracted in the time domain as depicted in Fig. 17.16a, up to 50 nS.

To evaluate important requirement for current waveform models concerning IR-drop analysis capability current derivatives are extracted based on numerical

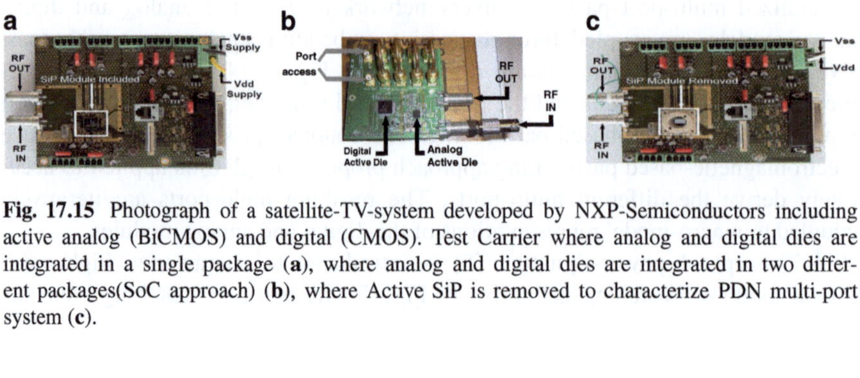

Fig. 17.15 Photograph of a satellite-TV-system developed by NXP-Semiconductors including active analog (BiCMOS) and digital (CMOS). Test Carrier where analog and digital dies are integrated in a single package (**a**), where analog and digital dies are integrated in two different packages(SoC approach) (**b**), where Active SiP is removed to characterize PDN multi-port system (**c**).

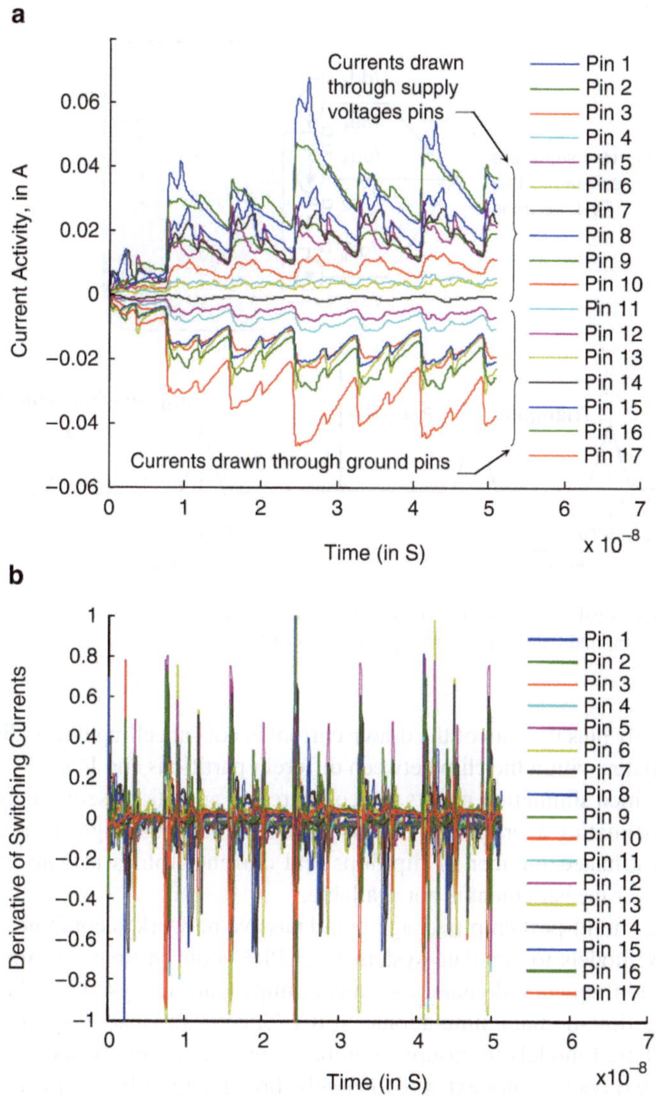

Fig. 17.16 Extracted current signatures behavior against time (**a**) and associated normalized derivatives for 17 IOs (Onput/Output pins) (**b**)

derivation from Fourier transform expansion as shown in Fig. 17.16b. In classical approaches based on triangular waveform assumptions (shown in Fig. 17.17b) the derivative of current models is not properly calculated since it leads to a pulse-like representation (Dirac type of discontinuity). It is essential to notice the polarity attribute of the extracted current activity waveforms in the time-domain (Fig. 17.16a): negative polarity refers to ground pins while positive polarity is associated to supply pins (drawing currents convention). It is observed for a pair of

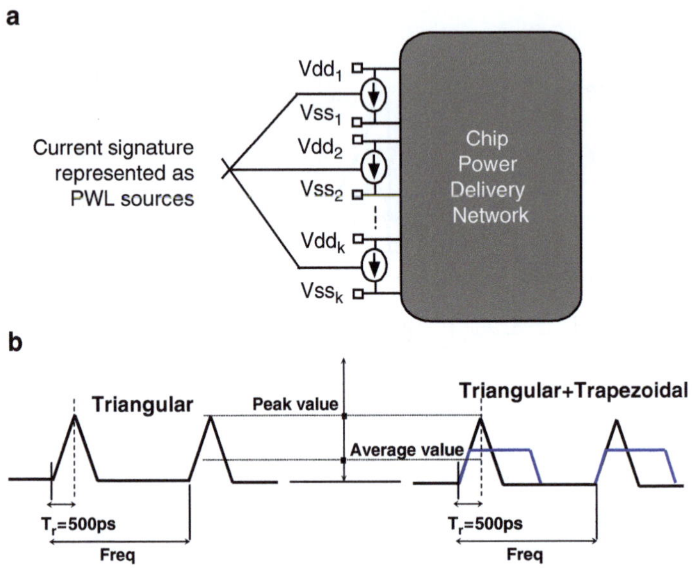

Fig. 17.17 Classical triangular or combined triangular-trapezoidal waveform profile for current activity profile (**a**) and CPM multi-port representation (**b**)

ground/supply pins the sum of the drawn current is not cancelling due to distribution of ground paths within the chip between different partitions and IOs.

The obtained simulation results are compared to analytical, semi-analytical calculations estimating average power consumption of a digital high speed active die based on estimated number of flip-flops and current profiles at gate level when average power measurement is not available.

Extracted multi-port chip-package-board passive network is combined with digital activity models to simulate system level PI/SI both in time-domain (for SSN analysis) and frequency-domain (for transfer impedance analysis) simulation. Concerning the time-domain simulations, SPICE/Spectre models are preferred to S-parameter based models to ensure smoother convergence and better DC behavior. When Spice model are not extracted natively, broadband SPICE extractions (BBS) are considered. A special attention should be paid to passivity and power conservation to guarantee stability and causality. Two use-models are investigated for the digital baseband switching activity model: a simplified first-order switching current activity profile defined by a triangular/trapezoidal waveform, where the amplitude is calculated to match the average power obtained from the measurements. For this simple model the rise and fall times are deduced from the technology-node information, relatively to 90 nm CMOS.

A more complex digital activity model that takes into account dynamic attributes through statistical analysis is extracted using Apache Redhawk CPM solution. In Fig. 17.17b a representation of a triangular waveform profile and a CPM multi-port model are shown respectively. The CPM consists of a Piecewise Linear (PWL)

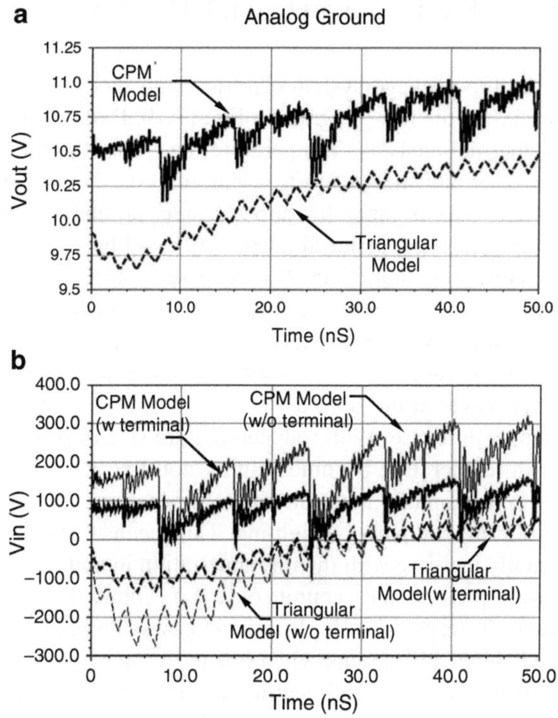

Fig. 17.18 Illustration of voltage fluctuations due to digital switching activity at Analog die output (**a**) and input (**b**) acces

compression sources and a passive parasitic network describing the digital on-chip delivery network. The influence digital activity switching on power/ground bounces as well as noise fluctuations at sensitive RF inputs (LNA access) of the analog active die is studied. For the simulation of decoupling capacitors impact on PI/SI, impedances in frequency-domain analysis are calculated at different locations. In Fig. 17.18 typical time-domain voltage fluctuations of power-supply nodes and signal input/output voltages both for triangular waveform activity profile and CPM model are shown.

Significant changes in the noise fluctuations are observed, which demonstrate importance of proper activity modeling for digital dies. Such fluctuations necessitate proper estimation of on-chip decoupling capacitance to reduce SSN impact on PI/SI. In Fig. 17.18 the voltage fluctuations on RF signal input/out are evaluated in two configurations: with and without 150 Ω equivalent termination for the LNA input impedance. To evaluate the accuracy of constant voltage supply assumption at PMU level the fluctuations of the voltage waveform are measured. Figure 17.18a represents the variation against time up to 50 ns of supply voltage and relative fluctuations with 206 mV margin value.

To investigate coupling effects at system level, including interferences between analog and digital dies, S-parameters of two port configuration are measured in reference to RFin and RFout access ports in Fig. 17.15. Both configurations in Fig. 17.15a, b have been measured and modeled based PDN extractions.

17.6 Electromagnetic and Thermal Co-Analysis

The chip-package-system electrical and thermal co-analysis flow could be represented by the flow diagram in Fig. 17.19. Detailed power map for chips is the input to thermal analysis. Temperature map from thermal analysis is used for material property adjustment in electrical analysis. Power analysis will generate adjusted power map as input for another thermal analysis. With elevated temperature on chips, electrical material properties could be very different than that at room temperature, e.g., more than 40% increase in electrical resistivity. This will significantly affect values of electrical parameters in functional prediction and leakage power dissipation which is the main source for elevated temperature.

For advanced 90 nm process, leakage power amounts more than 30% of the total power at elevated temperature. Leakage power is in general increasing with temperature rise. With reduced feature size in process, the leakage power increase with temperature becomes highly non-linear, i.e., quadratic or even more drastic. Also, the smaller the scale of the process, e.g., 65, 45, or 32 nm, the more leakage power in total power weighting. This means that the chip heating power will significantly increase with the miniaturization in manufacturing process. So chip heating power prediction is accurate only if it is through power-electrical-thermal co-design loop. In response to high leakage power on chip, one of the approach is partially switching- off circuits when not in use, e.g., clock gating and power gating. This will lead to "uneven" heating and temperature distribution on chip. For accurate leakage current prediction, material temperature-dependency must be considered. Hence, local heating in thermal prediction becomes important, in order to generate correct temperature map on chip which will affect subsequent power-electrical simulation. Other thermal impacts to package functions and reliability include effects of rising temperature to timing and cross-talk analysis. Electro-migration increases exponentially with temperature and will affect reliability of metal/via in IC when there is higher local temperature. Uniform power assumption will no longer provide accurate results in temperature prediction for nanometer design today.

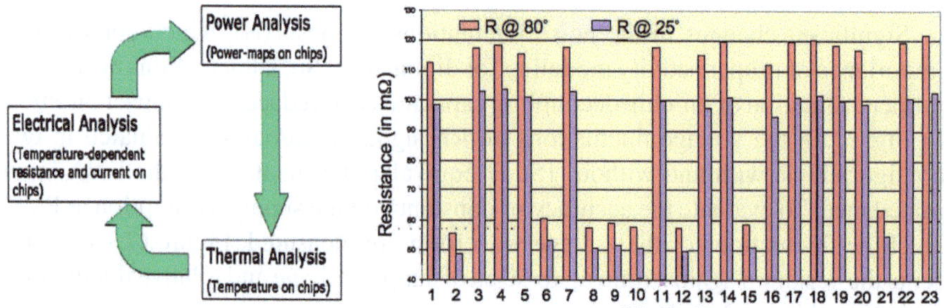

Fig. 17.19 IC/package/system co-analysis flow (**a**). Influence of temperature SiP-module design (**b**). (from Ref [4])

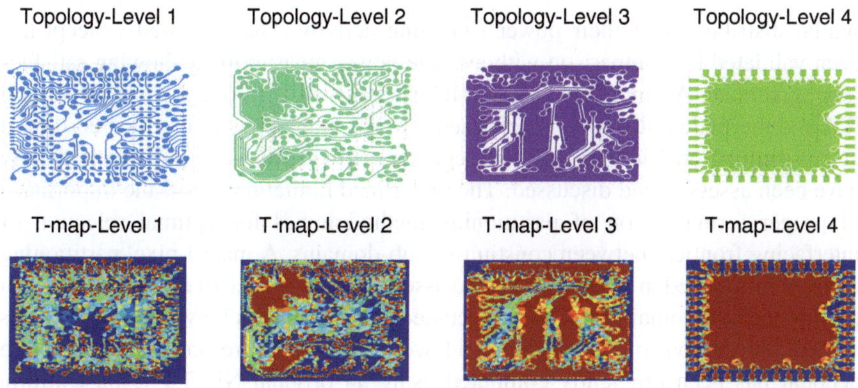

Fig. 17.20 Illustration of Temperature maps (T-map) on the 4-metal layers of the package: the top four pictures are the metal patterns and the lower contour-maps are the associated temperature distribution (from ref [4])

To illustrate the effect of temperature dependency of electrical material property on electrical parameters, electrical potential field on wire bonds for the SiP-module is calculated and wire-bond resistances are extracted (Fig. 17.19). Significant changes in wire resistance due to temperature changes, from 20°C to 80°C, can be found for all the wires. Hence, the temperature effects on electric properties in conductors should not be overlooked in IC-package-board design process. Figure 17.20 shows the automated metal layer modeling to reflect realistic metal distribution.

Top pictures are the original CAD design. The lower pictures represent temperature distributions on the metal layers. Colors from red to blue represent copper contents from 100% to nil. The temperature maps show hot (red) and cold (blue) zones on the metal layer. It is apparent that temperature map on chip will be affected by the degree of detailness on substrate metals.

17.7 Conclusion

In this contribution, EM and Thermal Co-analysis for chip, package and board co-design and co-simulation has been presented. Segregated (divide-and-conquer) and integrated (IC, package, PCB in one single model) methodologies have been applied to real-world SiP carrier applications. Results with both full-wave and Quasi-static assumptions are obtained and compared. Importance of return path settings as well as sensitivity of shielding options have been demonstrated. A global distributed co-simulation methodology for concurrent/simultaneous analysis of passive and active parts have been proposed and applied to two different real-world test carrier modules. An original power-signature concept for analog-digital cosimulation has been introduced to model high-speed digital modules temporal and spatial distribution of their power switching activity. The proposed concept has been validated by comparison with average power measurement showing satisfactory agreement. An integrated model (IC, package, PCB in one model) combining EM and thermal simulations is proposed towards multi-physics oriented co-design and co-simulation. Limitations of the segregated (divide-and-conquer) methodology have been assessed and discussed. The underlined limitations show the importance of proper formalization of partitioning methodologies for optimal selection of interfacing frontiers between constitutive sub-domains. A macro-pixel partitioning concept introduced in TWF approach to assess the importance of couplings between partitioned sub-domains has been discussed. The couplings between macro-pixels where contributions of both higher and lower order modes are accurately taken into account has been efficiently computed using an original NUFFT (Non-Uniform Fast Fourier Transform) offering the use of non-uniform multi-grid space stepping. Combination of proposed partitioning methodologies with energy considerations using power-waves based formulations opens possibilities of multiphysics modeling and analysis for coupling different energy-domains in one unified environment (using physics-based equivalent circuit models derivations fulfilling passivity and causality preservations).

Acknowledgements The authors would like to thank professor Henri Baudrand for fruitful discussions during the elaboration of the concept of Macro-Pixel domain-decomposition approach.

References

1. R.R. Tummala, SOP: What is it and why? A new microsystem-integration technology Paradigm-Moore's law for system integration of miniaturized convergent systems of the next decade. Trans. Adv. Packaging **27**(2), 241–247 (2004)
2. J. Mao, G. Fitzgerald, A. Kuo, S. Wane, Coupled analysis of quasi-static and full-wave solution towards IC, package and board co-design. in IEEE 2007 Electrical Performance of Electronic Packaging, 111–114 (2007)
3. S. Wane, Partition and global methodologies for IC, package and board co-simulation in SiP applications. in European Microwave Integrated Circuit Conference Proceedings, EuMIC 2007, 451–454 (2007)
4. S. Wane, A. Kuo, Electromagnetic and thermal co-analysis for distributed co-design and co-simulation of chip, package and board, in Proceedings of the IEEE-RFIC 2008 Radio Frequency Integrated Circuit Symposium Digest, 471–474 (2008)
5. S. Wane, G. Boguszewski, Global digital-analog co-simulation methodology for powerand signal integrity aware design and analysis. in *Proceedings of the 38th European Microwave Conference, Amsterdam*, Oct. 2008
6. S. Wane, D. Bajon, Partition-recomposition methodology for accurate electromagnetic analysis of SiP passive circuitry. in proceedings of EURONCON 2007, the International Conference on "Computer as a Tool", 15–23
7. J. Mao, B. Archambeault, J.L. Drewniak, T.P. Van Doren, Estimating DC power bus noise. in *IEEE International Symposium on EMC, Minneapolis*, Minnesota, USA, Aug. 19–23, 2002
8. P. Russer, D. Bajon, S. Wane, N. Fichtner, Overview and status of numerical electromagnetic field simulation methods applied to integrated circuits. in *IEEE Topical Meeting on Silicon Monolithic Integrated Circuits in RF Systems,* SiRF 2009, Jan. 2009

9. S. Wane, J.C. Rautio, V. Mühlhaus, Topological and Functional Partitioning in EM Analysis: Application to Wafer-Level Chip-Scale Harmonic Filters, in Proceedings of the IEEE MTT-S International Microwave Symposium Digest, June 2009, 1621–1624
10. M. Leone, V. Ricchiuti, G. Antonini, A. Orlandi, Measurement and modeling of noise current spectrum for large ASICs. in *IEEE 7th Workshop on Signal Propagation on Interconnects*, 2003
11. A. Raghunathan, S. Dey, N.K. Jha High-level macro-modeling and estimation techniques for switching activity and power consumption in *IEEE Transactions on VLSI Systems* Aug 2003 pp. 538–557
12. D.I. Cheng, K.-T. Cheng, D.C. Wang, M. Marek-Sadowska, A new hybrid methodology for power estimation. in *Proceedings of the Design Automation Conference*, June 1996, pp. 439–444
13. F.N. Najm, M.Y. Zhang, Extreme delay sensitivity and the worst-case switching activity in VLSI circuits. in *Proceedings of the Design Automation Conference*, June 1995, pp. 623–627
14. Z. Chen, K. Roy, A power macro-modeling technique based on power sensitivity. in *Proceedings of the Design Automation Conference*, June 1998, pp. 678–683
15. M. Barocci, L. Benini, A. Bogliolo, B. Ricco, G. De Micheli, Lookup table power macro-models for behavioral library components. in *Proceedings of the IEEE Alessandro Volta Memorial Workshop on Low-Power Design*, Mar. 1999, pp. 173–181
16. S. Mallat, A theory for multi-resolution signal decomposition: The wavelet representation. IEEE Trans. Pattern Anal. Mach. Intell. **11**(7), 674–693 (1989)
17. http://www.apachedesignsolutions.com/
18. S. Wane, D. Bajon, Derivation of multi-grid discrete and analytic green's functions free of poles in terms of transverse waves. in *IEEE MTT-S International Microwave Symposium Digest*, San Francisco, USA, June 2006, 1041–1044
19. Q.H. Liu, N. Nguyen, An accurate algorithm for non-uniform fast Fourier transforms (NUFFTs). IEEE Microw. Guid. Wave Lett. **8**, 18–20 (1998)
20. T.J. Cui, W.C. Chew, Fast evaluation of Sommerfeld integrals for EM scattering and radiation by three-dimensional buried objects. IEEE Geosci. Remote Sensing, **37**, 887–900 (1999)
21. S. Wane, D. Bajon, Power waves formulation of oscillation conditions: Avoidance of bifurcation modes in cross-coupled VCO architectures, in Proceedings of IEEE Topical Meeting on Silicon Monolithic Integrated Circuits in RF Systems proceedings, SiRF 2010, 2010, 176–179
22. S. Wane, D. Bajon, Attempts for bridging circuits geometrical topologies with equivalent network architecture synthesis, in Proceedings of International Conference on Electromagnetics in Advanced Applications, ICEAA'09, 2009, 705–708
23. R.F. Harrington, *Time-Harmonic Electromagnetic Fields* (McGraw-Hill, New York, 1961)
24. V.I. Okhmatovski, A.C. Cangellaris, A new technique for the derivation of closed-form electromagnetic green's functions for unbounded planar layered media. IEEE Trans. Antennas Propag. **50**(7) (2002)
25. C. Cangellaris, V.I. Okhmatovski, Novel closed-form green's function in shielded planar layered media. IEEE Trans. Microw. Theory Tech. **48**(12) (2000)
26. Q.-H. Liu, W.C. Chew, Application of the conjugate gradient fast Fourier Hankel transfer method with an improved fast Hankel transform algorithm. Radio Sci. **29**, 1009–1022 (1994)
27. R. Coifman, V. Rokhlin, S. Wandzura, The fast multipole method for the wave equation: a pedestrian prescription. IEEE Antennas Propag. Mag. **35**(3) (1993)
28. E. Darve, P. Havé, A fast multipole method for Maxwell equations stable at all frequencies. Phil. Trans. R. Soc. Lond. A **362**, 603–628 (2004)
29. B. Hu, W.C. Chew, E. Michielssen, J. Zhao, Fast inhomogeneous plane wave algorithm for the fast analysis of two-dimensional scattering problems. Radio Sci. **34**(4), 759–772 (1999)
30. L.J. Jiang, W.C. Chew, Low-frequency fast inhomogeneous plane-wave algorithm (LF-FIPWA). Microw. Opt. Tech. Lett. **40**, 117–122 (2004)
31. S. Ohnuki, W.C. Chew, Error Analysis of the Fast Inhomogeneous Plane Wave Algorithm for 2-D Free Space cases. Tech Report, June 2003
32. B. Hu, Fast inhomogeneous plane wave algorithm for electromagnetic scattering problems. Thesis, University of Illinois at Urbana-Champaign, 2001

33. O. Ozgun, R. Mittra, M. Kuzuoglu, Parallelized characteristic basis finite element method (CBFEM-MPI) – A non-iterative domain decomposition algorithm for electromagnetic scattering problems. J. Comput. Phys. **228**, 2225–2238 (2009)
34. I. Sebestyén, Electric-field calculation for HV insulators using domain-decomposition method. IEEE Trans. Magn. **38**(2), 1213–1216 (2002)
35. D. Lacour, X. Ferrieres, P. Bonnet, V. Gobinand, J.C. Alliot, Application of multi-domain decomposition method to solve EMC problem on an aeroplane. Electron. Lett. **33**(23), 1932–1933 (1997)
36. Y.J. Lu, C.Y. Shen, A domain decomposition finite-difference method for parallel numeric implementation of time-dependent Maxwell's equations. IEEE Trans. Antennas Propag. **45**(3), 556–562 (1997)
37. M. Kaltenbacher, S. Reitzinger, J. Schöberl, Algebraic multigrid method for solving 3D nonlinear electrostatic and magnetostatic field problems. IEEE Trans. Magn. **36**(4), 1561–1564 (2000)
38. K. Watanabe, H. Igarashi, T. Honma, Comparison of geometric and algebraic multigrid methods in edge-based finite-element analysis. IEEE Trans. Magn. **41**(5), 1672–1675 (2005)
39. G. Haase, A parallel AMG for overlapping and nonoverlapping domain decomposition. Electron. Trans. Numer. Anal. **10**, 41–55 (2000)
40. W.L. Briggs, V.E. Henson, S.F. Mc Cormick, *A Multigrid Tutorial*, 2nd edn. (SIAM, Philadelphia, PA, 2000)
41. R. Courant, D. Hilbert, *Methods of Mathematical Physics*, vol. II. Partial Differential Equations. (Springer, New York, 1962)
42. S. Wane, D. Bajon, H. Baudrand, A new full-wave hybrid differential-integral approach for the investigation of multilayer structures including non-uniformly doped diffusions. IEEE Trans. Microw. Theory Tech. **53** (2005), 200–214
43. N. Fichtner, S. Wane, D. Bajon, P. Russer, Interfacing the TLM and the TWF method using a diakoptics approach in *2008 IEEE MTT-S Int. Microwave Symposium Digest.* Atlanta, USA, June 2008, pp. 57–60
44. M.C Longtin, S. Din-Kow, J. Silvestro, Z. Cendes, Domain decomposition and distributed analysis for large microwave structures. in *2006 IEEE MTT-S International Microwave Symposium Digest* June 2006, pp. 1053–1056
45. M. Krumpholz, B. Bader, P. Russer, On the theory of discrete TLM Green's functions in three-dimensional TLM. IEEE Trans. Microw. Theory Tech. **43**, 1472–1482 (1995)
46. P. Russer, *Electromagnetics, Microwave Circuit and Antenna Design for Communications Engineering*, 2nd edn. (Artech House, Boston, 2006)
47. D. Bajon, S. Wane, Concept of marco-pixel formulation using non-uniform Fourier transform. in *25th Annual Review of Progress in Applied Computational Electromagnetics*, Monterey, USA, 8–12 Mar. 2009
48. S. Wane, D. Bajon, H. Baudrand, A congruent compact-cell approach for global EM analysis of multi-scale integrated circuits. in *IEEE International Microwave Symposium Digest*, Long Beach CA, USA, June 2005
49. P. Russer, M. Mongiardo, L.B. Felsen, Electromagnetic field representations and computations in complex structures III: network representations of the connection and subdomain circuits. Int. J. Numer. Model. Electron. Networks Devices Fields **15**, 127–145 (2002)
50. L.B. Felsen, M. Mongiardo, P. Russer, Electromagnetic field representations and computations in complex structures I: complexity architecture and generalized network formulation. Int. J. Numer. Model. Electron. Networks Devices Fields **15**, 93–107 (2002)
51. L.B. Felsen, M. Mongiardo, P. Russer, Electromagnetic field representations and computations in complex structures II: alternative Green's functions. Int. J. Numer. Model. Electron. Networks Devices Fields **15**, 109–125 (2002)
52. M. Mongiardo, C. Tomassoni, P. Russer, Generalized network formulation: Applicatioto flange–mounted radiating waveguides. IEEE Trans. Antennas Propag. **55**(6), 1667–1678 (2007)

Chapter 18
Parallel TLM Procedures for NVIDIA GPU

Poman So

18.1 Introduction

Massively parallel computing technology has undergone a paradigm shift in recent years. The driving force behind this change is the need for better graphics hardware for personal computers. The latest graphics processors from ATI, Intel and NVIDIA have advanced multi-processor hardware to support popular graphics interfaces such as DirectX and OpenGL. These new graphics processing units (GPU) employ the *Single Instruction Multiple Data* (SIMD) computing model which enables all processors in the GPU to work simultaneously on a vast amount of data using identical instructions. This approach is very suitable for performing graphics operations because all pixels in an image require identical transformation and mapping instructions.

The SIMD computing model, which revolutionized the GPU industry, is making its way into mainstream computing. Matrix operations, which are at the core of many computer graphics algorithms, are also found in many linear algebra routines. Moreover numerical procedures that require identical instructions to be executed on a large amount of data are suitable candidates for execution on the SIMD hardware in advanced GPUs. However, developing parallel algorithms for GPU hardware is not straightforward. The task is further complicated by the lack of a good software development kit (SDK) that encapsulates the hardware details in a software model.

As of the writing of this article, ATI has released a *Stream Computing* SDK [1] whereas NVIDIA has released a new version of their *Compute Unified Device Architecture* (CUDA) SDK [2]. In addition to that, NVIDIA is working on an Open Computing Language (OpenCL) for programming GPU hardware [3]. In this paper, a TLM engine implemented using the CUDA SDK is presented.

P. So
Computational Electromagnetics Research Laboratory, Department of Electrical and Computer Engineering, University of Victoria, Victoria, BC, Canada V8W 3P6
e-mail: pso@ece.uvic.ca

S. Lindenmeier and R. Weigel (eds.), *Electromagnetics and Network Theory and their Microwave Technology Applications*, DOI 10.1007/978-3-642-18375-1_18,

18.2 Transmission Line Matrix Algorithms

Transmission Line Matrix (TLM) method has three fundamental operations: scattering, transfer and reflection of voltage impulses. Figure 18.1 depicts these operations using the two-dimensional shunt node TLM [4]; for the three-dimensional symmetrical condensed node (SCN) TLM [5], the operations depicted in Fig. 18.1 must be applied to voltage impulses in all polarizations. In a typical TLM simulation there are a large number of nodes to be processed. Most of these nodes require the same scattering and impulse transfer procedure. For the nodes with boundaries adjacent to them, voltage impulses are reflected back to the nodes instead of transferred to their neighbouring nodes. To execute these operations in parallel using NVIDIA GPUs, dedicated functions or subroutines, called *kernels* in the CUDA nomenclature, are needed.

A CUDA kernel consists of special instructions for execution on the GPU's multi-processors; details about CUDA enabled GPU architecture and its SDK are given in [6]. One of the challenges in designing kernels is to ensure processes are synchronized across all multi-processors. This is particularly difficult when the number of nodes in a TLM mesh is larger than the number of available multi-processors. In that situation, the TLM mesh of interest must be divided into small data-blocks for sequential processing by the GPU multi-processors. TLM scattering is inherently parallel since each node requires no extra data other than what is already contained in the local node structure. TLM impulse transfer requires voltage pulses from adjacent nodes. Exchanging voltage pulses between neighbouring nodes internal to a data-block can be straightforwardly handled with some CUDA functions; however exchanging impulses at the boundaries across adjacent blocks requires special attention because the CUDA SDK does not provide any built-in mechanism for imposing synchronism across data-blocks.

18.3 Implementation of TLM Using the CUDA SDK

The easiest way to port an existing program to run on NVIDIA GPU is to implement some time-critical portions of the program using *C for CUDA*. Computer code written in *C for CUDA* can be cross-linked with C/C++ modules. The core of our

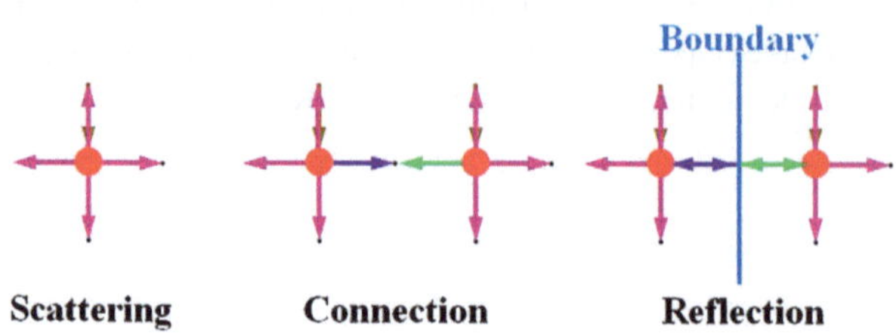

$$[V]^r_{k+1} = [S] \times [V]^i_k \qquad [V^*]^i_{k+1} = [C] \times [V]^r_{k+1} \qquad [V]^i_{k+1} = [\Gamma] \times [V]^r_{k+1}$$

Fig. 18.1 Schematic diagram that illustrates the fundamental operations in TLM

GPU TLM code consists of two CUDA kernels. The first kernel executes scattering, boundary condition, and local impulse interchange and the results are stored internally in the data-block, Fig. 18.2. For electromagnetic signal to propagate across data-blocks voltage pulses at the data-block boundaries must be moved over to the adjacent blocks; this is handled by the second kernel, Fig. 18.3.

Since CUDA kernel code can only access memory in the GPU, TLM voltage impulses and boundary information in the host CPU must be duplicated in the GPU; Fig. 18.4 depicts the data structure used for that purpose. Together with the CUDA's grid block models, a large TLM mesh could be mapped to a grid which consists of blocks of threads. Each block in the grid has an index associated with it. In the same way, each thread in a block has an index associated with it. The dimensions and indices are available via four internal variables: gridDim, blockDim, blockIdx and threadIdx. The technique for using these variables is illustrated in Fig. 18.5.

Using the code segment in Fig. 18.5, each thread in the active data-block can execute the TLM procedure on the node associated with the thread index specified by the CUDA runtime variables. Figure 18.6 depicts the SCN procedure implemented in *C for CUDA*; the code looks no different from standard $C/C++$ code because the per-thread attribute of the local variables is not explicitly shown in the code segment.

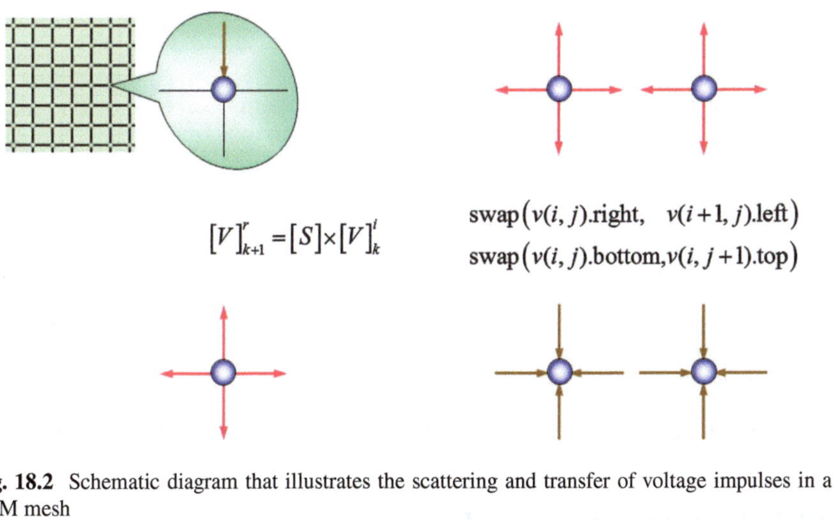

Fig. 18.2 Schematic diagram that illustrates the scattering and transfer of voltage impulses in a TLM mesh

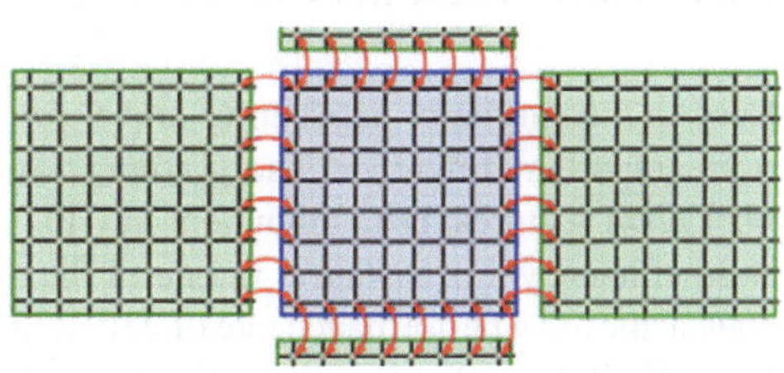

Fig. 18.3 Schematic diagram that illustrates the transfer of TLM voltage impulses across data blocks that made up of the TLM mesh

```
// Data structure for storing TLM voltage impulses and
// boundary information in the GPU memory
struct __align__(16) Link_Line
{
    float v1;
    float v2;
    float v3;
    float v4;
    float v5;
    float v6;
    float v7;
    float v8;
    float v9;
    float v10;
    float v11;
    float v12;
    float RefCoeff;        // Reflection Coeficient of Boundary
    float Bndry_Dir;       // Boundary Direction: none=0
                           //   x+ve = 1, x-ve = 2,
                           //   y+ve = 3, y-ve = 4,
                           //   z+ve = 5, z-ve = 6.
};

__device__ __shared__ Link_Line V_struct[BLOCK_SIZE_Z][BLOCK_SIZE_Y][BLOCK_SIZE_X];
```

Fig. 18.4 *C for CUDA* data structure for storing air filled SCN node

```
// Grid Dimensions
int bxDim = gridDim.x;             // # of blocks in x-dir of grid
int byDim = gridDim.y;             // # of blocks in y-dir of grid

// Block Dimensions
int block_x_dim = blockDim.x;     // # of nodes in the x-direction
int block_y_dim = blockDim.y;     // # of nodes in the y-direction
int block_z_dim = blockDim.z;     // # of nodes in the z-direction

// Block index (current block ID)
int bx = blockIdx.x;
int by = blockIdx.y;

// Thread index (used to map to shared memory)
int tx = threadIdx.x;
int ty = threadIdx.y;
int tz = threadIdx.z;
```

Fig. 18.5 *C for CUDA* source code excerpted from the author's TLM program

18.4 Data Alignment and Process Synchronization

Despite the similarity between *C* and *C for CUDA*, Figs. 18.4 and 18.6 do reveal some important difference. The __align__(16) and __syncthreads() constructs are *C for CUDA* features. Data alignment in GPU memory has a direct impact on kernel code efficiency. For instance, in one of the author's earlier implementation [7] the code achieved a performance rate of 210 MegaNodes/sec on the GeForce 8800 GTX Ultra NVIDIA GPU. By reorganizing the data alignment in the GPU memory, an

```
//////////////////////
//Scattering

float v1 = V_struct[tz][ty][tx].v1;
float v2 = V_struct[tz][ty][tx].v2;
float v3 = V_struct[tz][ty][tx].v3;
float v4 = V_struct[tz][ty][tx].v4;
float v5 = V_struct[tz][ty][tx].v5;
float v6 = V_struct[tz][ty][tx].v6;
float v7 = V_struct[tz][ty][tx].v7;
float v8 = V_struct[tz][ty][tx].v8;
float v9 = V_struct[tz][ty][tx].v9;
float v10 = V_struct[tz][ty][tx].v10;
float v11 = V_struct[tz][ty][tx].v11;
float v12 = V_struct[tz][ty][tx].v12;

V_struct[tz][ty][tx].v1  = 0.5* ( v2 + v3 + v9  - v11);
V_struct[tz][ty][tx].v2  = 0.5* ( v1 + v6 - v10 + v12);
V_struct[tz][ty][tx].v3  = 0.5* ( v1 + v4 + v8  - v12);
V_struct[tz][ty][tx].v4  = 0.5* ( v3 + v5 - v7  + v11);
V_struct[tz][ty][tx].v5  = 0.5* ( v4 + v6 - v8  + v10);
V_struct[tz][ty][tx].v6  = 0.5* ( v2 + v5 + v7  - v9);
V_struct[tz][ty][tx].v7  = 0.5* (-v4 + v6 + v8  + v10);
V_struct[tz][ty][tx].v8  = 0.5* ( v3 - v5 + v7  + v11);
V_struct[tz][ty][tx].v9  = 0.5* ( v1 - v6 + v10 + v12);
V_struct[tz][ty][tx].v10 = 0.5* (-v2 + v5 + v7  + v9);
V_struct[tz][ty][tx].v11 = 0.5* (-v1 + v4 + v8  + v12);
V_struct[tz][ty][tx].v12 = 0.5* ( v2 - v3 + v9  + v11);

__syncthreads();
```

Fig. 18.6 SCN scattering procedure in *C for CUDA*

execution rate of 350 MegaNodes/sec has been attained. *C for CUDA* has many software switches for manipulating data alignment in the GPU memory. It is thus necessary to make use of the features to avoid fragmented memory allocation. It was found that by using the previously reported implementation, 10% of the processing time is spent on computation and 90% is spent on transferring data between memory banks. In order to reduce elapsed time on data transfer, a new memory mapping method which ensures contiguous and coalescent data in GPU memory has been developed, Fig. 18.7. When a multiprocessor is ready to process the next data-block, it reads it from GPU global memory. Each data-block in global memory is organized to be contiguous. Furthermore, to achieve memory coalescence, each node structure within each data-block is organized such that each node address is aligned to the nearest 16 byte address. A consequence of the coalesced memory model is a 33% increase in memory usage.

The improved memory organization scheme has brought an overall speed improvement of 67%, which is a substantial improvement of overall performance and justifies the increase in memory usage. Figure 18.8 illustrates the speed of

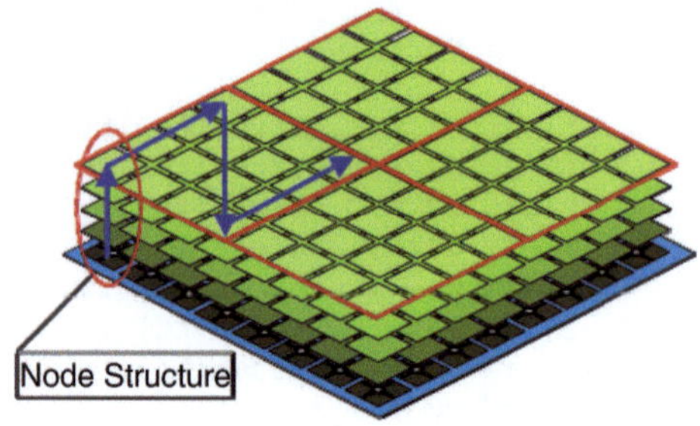

Fig. 18.7 Contiguous memory model

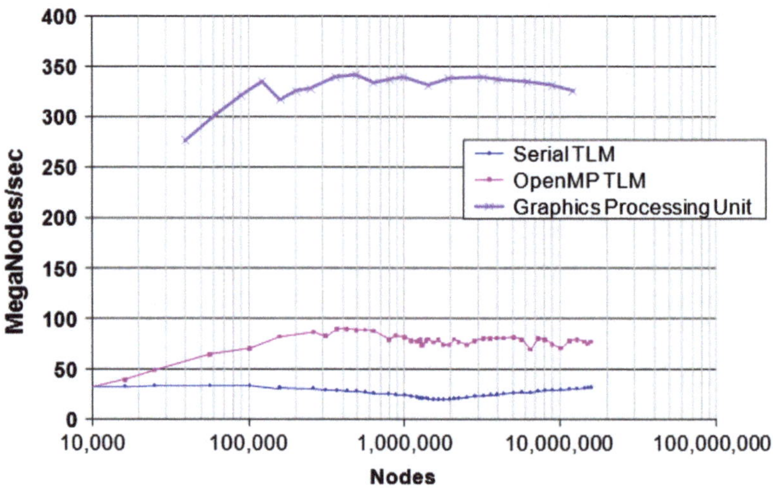

Fig. 18.8 Comparison of performance of three TLM implementations

execution of the GPU based TLM code, as well as a serial TLM adaptation and an OpenMP [7] TLM adaptation (4cpu) over varied mesh sizes.

18.5 Validation of Algorithm

Using the techniques mentioned above a three-dimensional GPU based TLM program has been developed. The program has achieved a 47 MegaNodes/sec simulation speed on NVIDIA FX 5600 GPUs. A WR-28 waveguide band-pass filter was modelled using the program and with a commercially available TLM simulation application. Simulation runs of both methods were measured and charted in Fig. 18.9. The TLM simulation application was able to engage one, two, three or four CPUs of the workstation. It can be seen that the GPU TLM routine outperforms the commercially available TLM application by 7.1 times in the single CPU mode, and 3.2 times in the 4-CPU mode.

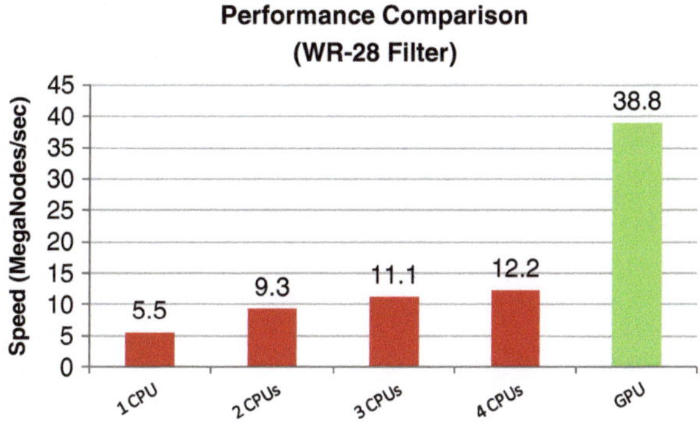

Fig. 18.9 Comparison of performance of three TLM implementations

18.6 Conclusion

A massively parallel three-dimensional TLM programs have been successfully designed and implemented for the NIVIDIA CUDA enabled GPUs. It is found that the SIMD computing paradigm is suitable for implementing time-domain computational electromagnetic methods such as TLM. The three-dimensional TLM program described in this paper has reached a 47 MegaNode/sec performance. Comparing to a commercially available SCN TLM simulation package running on a single CPU this *C for CUDA* implementation is 7.1 times faster. Technique for utilizing a cluster of four FX 5600 GPUs [8] is under investigation. Since memory transfer is always the bottleneck in multi-processor environment, this would also be one of the challenges in porting field simulation programs to the multi-GPU hardware.

References

1. ATI Stream Computing User Guide, April 2009
2. NVIDIA CUDA Programming Guide Version 2.2, 4/2/2009
3. The OpenCL Specification, version 1.0, revision 33, 2/4/2009
4. W.J.R. Hoefer, The transmission-line matrix method – theory and applications. IEEE Trans. Microw. Theory Techn. **MTT-33**(10), 882–893 (1995)
5. P.B. Johns, A symmetrical condensed node for the TLM method. IEEE Trans. Microw. Theory Techn. **35**(4), 370–377 (1987)
6. F.V. Rossi, P.P.M. So, N. Fichtner, P. Russer, Massively parallel two-dimensional tlm algorithm on graphics processing units. in *IEEE International Microwave Symposium*, Atlanta, GA, June 2008, pp. 153–156
7. ClusterInABox Quad (Q30) Product Information, http://www.acceleware.com
8. OpenMP Application Program Interface, OpenMP Architecture Review Board, V3.0, May 2008, http://www.OpenMP.org

Chapter 19
Stability Enhancement of Digital Predistortion Through Stationary Iterative Methods to Solve System of Equations

Xin Yu, Georg Fischer, and Andreas Pascht

19.1 Introduction

In wireless communication system the power amplifiers (PAs) are nonlinear in nature causing not only spectral regrowth but also in-band distortions [1]. In order to fulfill the linearity requirements of PAs one can simply back off the input signal, which results in degradation of PA efficiency. Another choice is the PA linearization techniques, which is introduced to compensate the nonlinearity of PAs. Through linearization techniques the spectral regrowth can be suppressed and the EVM (error vector magnitude) can be reduced simultaneously. Furthermore, one is able to push the operating points of PAs closer to their deep saturation region, so that the requirements of high efficiency and high linearity can be fulfilled at the same time.

The digital predistortion (DPD) is one of the popular linearization techniques because of its good properties concerning overall module efficiency, implementation effort and adaptation possibilities. Any PA exhibits some dynamic deviations from its static characteristics, which degrades the static DPD performance. Such deviation effects are known as "memory effects". The principle of the memory effects compensation is that the digital predistortion also need to have memory effects. There are several ways to introduce memory effects in the base-band model of PAs [2–6,9]. The most common architecture of DPD is the indirect learning architecture as shown in Fig. 19.1. The parameters estimation can be done offline (e.g. in DSP) by solving an over-determined system of equations by using Least Squares (LS) Method [7]. One critical problem in practice is the stability of the DPD system. The stability of a DPD system is completely dependent on the condition number of its system of equations. The DPD system becomes more and more instable with

X. Yu (✉) and A. Pascht
Alcatel-Lucent Bell Labs, Stuttgart, CO 70435, Germany
e-mail: xin.xy.yu@alcatel-lucent.com, andreas.pascht@alcatel-lucent.com

G. Fischer
University of Erlangen-Nuernberg, Erlangen-Nuernberg, CO 91054, Germany
e-mail: fischer@lte.e-technik.uni-erlangen.de

S. Lindenmeier and R. Weigel (eds.), *Electromagnetics and Network Theory and their Microwave Technology Applications*, DOI 10.1007/978-3-642-18375-1_19,

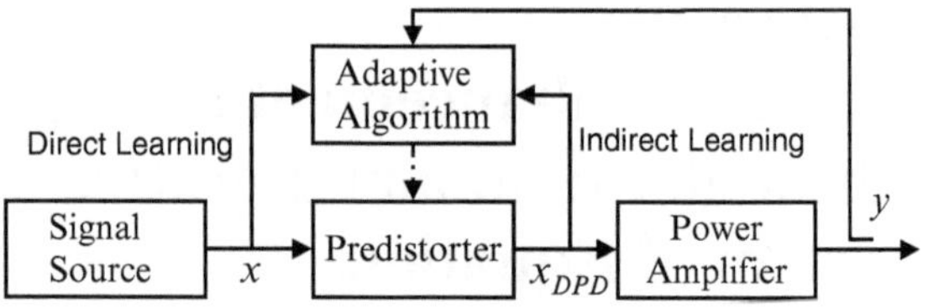

Fig. 19.1 Direct and indirect learning architectures of DPD system

increasing complexity of the DPD algorithm, e.g. when high order polynomials and memory compensation are in use.

Some techniques are introduced to increase the stability of DPD systems, i.e. to reduce the matrix condition number. For example the orthogonal technique [8] is often utilized to construct an orthogonal matrix for reduction of condition number. However, the orthogonal technique can only alleviate this problem of stability. In practice the orthogonal polynomial DPD with memory compensation still has a matrix with a large condition number. Furthermore, orthogonal technique is not always usable, because sometimes it is difficult to construct an orthogonal matrix for some DPD algorithms.

In this chapter the first-order and second-order stationary iterative methods for calculation of DPD parameters are investigated. The main idea is to solve the DPD system of equations not by means of LS method, but by means of iterative methods, so that the problem of calculating a inverse matrix of an ill-conditioned matrix can be avoided. The DPD parameters are computed only by matrix-vector multiplications and vector additions in the stationary iterative methods. Furthermore, if an identity matrix is used as a preconditioning matrix for stationary iterative methods, one can even avoid all divisions in the parameter estimation process. Section 19.2 presents a conventional DPD system of a polynomial model based on indirect learning architecture and the Least Squares (LS) Method for calculation of DPD coefficients. In Sect. 19.3 the stability estimation of DPD system with respect to matrix condition number is introduced. A discussion on the first-order and second-order stationary iterative methods and their experimental results are given in Sects. 19.4 and 19.5, respectively. Finally the conclusion is drawn in Sect. 19.6.

19.2 Digital Predistortion System

19.2.1 Direct and Indirect Learning Architechtures

The base-band model of a transmitter with DPD system is shown in Fig. 19.2. Let us denote the base-band input of PA (output of digital predistorter) by X_{DPD} and the corresponding base-band output of PA by Y. The indirect and direct learning architectures reflect two special cases of a DPD system. The direct learning architecture compares the original signal X and the normalized feedback signal Y to minimize the difference between these two signals. In contrast the indirect learning

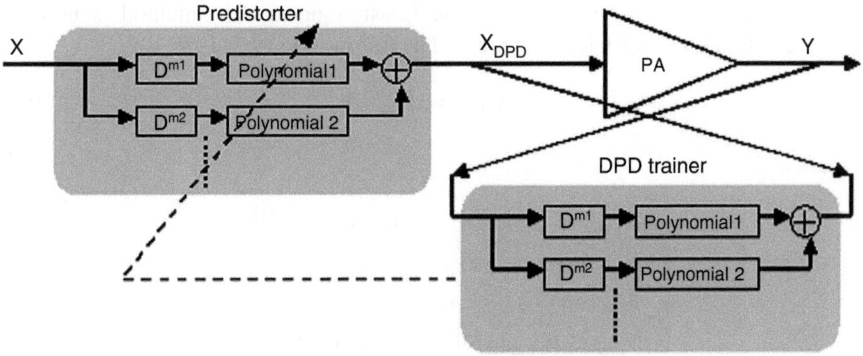

Fig. 19.2 Indirect learning DPD system with memory compensation

architecture derives the inverse input–output relationship of the PA by using the predistorted signal X_{DPD} and feedback signal Y. Thus the DPD characterized by the inverse PA behaviour is capable of compensating the nonlinearities of the PA.

19.2.2 Conventional Polynomial DPD Model

Just like PA modeling we need a nonlinear model to construct a digital predistorter. In order to compensate the memory effects of PAs one needs to have a digital predistorter with memory structure. The most general base-band model of PAs with memory effects is the Volterra series [6]. However, the large number of coefficients of the Volterra series makes it difficult to be applied in practice. The relative simple and effective digital predistorter model is the memory polynomial model [2], which can be viewed as a compromise between the DPD performance and DPD complexity. The memory polynomial DPD model can be described by

$$x_{DPD}[n] = \sum_{q=0}^{m} \sum_{k=1}^{l} a_{k,q} \cdot \Gamma_{k,q}(y[n]) \tag{19.1}$$

$$\Gamma_{k,q}(y[n]) = \left|y[n-d_q]\right|^{(k-1)} \cdot y[n-d_q] \tag{19.2}$$

where the y and X_{DPD} denote the output and input signal of PA, respectively. The $a_{k,q}$ presents the coefficient of DPD model. The k and q indicate the order of the polynomial and the length of memory effects, respectively. In other words the memory DPD system is constructed with parallel taps, one main tap and several memory taps, as shown in Fig. 19.2. The memory taps with individual delayed input signals are utilized to compensate the memory effects of PAs.

By indirect learning architecture the Least Squares (LS) method is used to calculate the DPD coefficients, expressed as follows in (19.3, 19.4)

$$X_{DPD} = M_y \cdot A \tag{19.3}$$

where

$$\begin{aligned}
X_{DPD} &= [x_{DPD}[0], \ldots, x_{DPD}[N-1]]^T \\
A &= [a_{1,0}, \ldots, a_{k,0}, \ldots, a_{1,q}, \ldots, a_{k,q},]^T \\
M_y &= [V_{1,0}, \ldots, V_{k,0}, \ldots, V_{1,q}, \ldots, V_{k,q}] \\
V_{k,q} &= [\Gamma_{k.q}(y[0]), \ldots, \Gamma_{k.q}(y[N])]^T \\
A &= \left(M^H \cdot M\right)^{-1} \cdot M^H \cdot X_{DPD}
\end{aligned} \tag{19.4}$$

The matrix M_y is constructed based on output signal Y of PAs. The $(.)^H$ and $(.)^{-1}$ denote the conjugate transpose and inverse transpose, respectively. In practice the expression (19.3) is an over-determined system of equations with dozen of unknowns (DPD coefficients) and thousands of equations.

19.3 The Estimation of Stability of DPD System

An indicator for the stability of a DPD system is the condition number of the matrix M_y. The condition number $cond(M_y)$ is the ratio of the largest singular value of matrix M_y to the smallest, described by the expression (19.5). One should consider the condition number as the rough rate, at which the solution of the system of equations may change with respect to a change in vector X, given in the expression (19.6). Thus with a large condition number even a small error in X may result in a large error in solution A.

$$cond(M_y) = \frac{\text{largest singular value of } M_y}{\text{smallest singular value of } M_y} \tag{19.5}$$

$$\frac{\|\Delta A\|}{\|A + \Delta A\|} \leq cond(M_y) \frac{\|\Delta X\|}{\|X\|} \tag{19.6}$$

To reduce the matrix condition number, the orthogonal technique is often utilized to construct an orthogonal matrix. However, the orthogonal technique can only alleviate this problem of high condition number. In Figs. 19.3 and 19.4 the condition numbers of the conventional and orthogonal polynomial for memoryless and memory DPD are illustrated considering a three carrier UMTS signal, respectively. The x-axis indicates the highest order of the polynomial. In the case of memoryless DPD very small condition numbers for orthogonal polynomial are obtained. In Fig. 19.4 for memory DPD the polynomial and orthogonal polynomial used in simulation have a degree of 5 and the x-axis indicates the number of the memory taps.

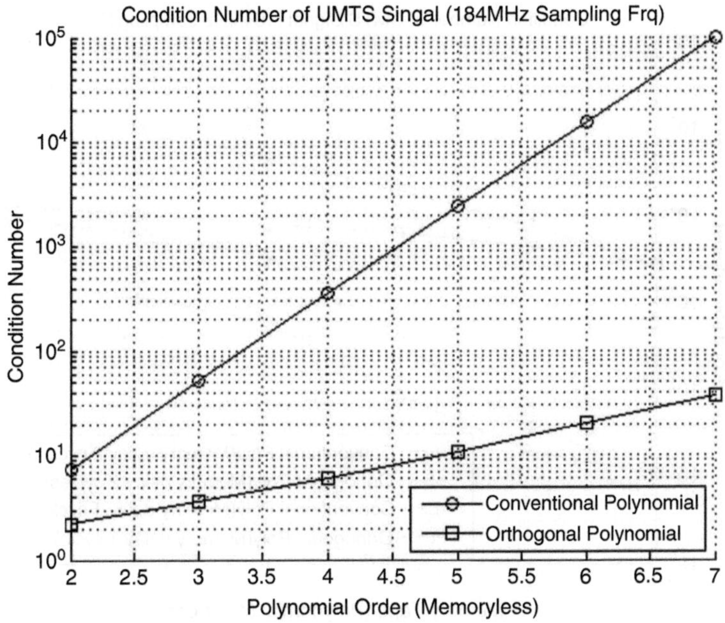

Fig. 19.3 Condition numbers of memoryless DPD for a three carrier UMTS signal

The condition numbers for orthogonal polynomial are much smaller in comparison to conventional polynomial. But the absolute values of the condition number of orthogonal polynomial are still large.

In practice the matrix M_y has a condition number normally ranging from several hundreds to several thousands for memory DPD, depending on different signals, DPD structures and DPD algorithms. The error range ΔA is dependent on the product of the condition number $cond(M_y)$ and the relative error ΔX of vector X. Assuming that only vector X in the matrix equation has a quantization error of $2^{-17}(\Delta X = 2^{-17})$, with the condition number $cond(M_y) = 150$ and $cond(M_y) = 1{,}000$ one would observe the error range of the DPD coefficients $\Delta A = 2^{-12}$ and $\Delta A = 2^{-7}$, respectively. The bad case with large error can cause the divergence of the DPD system. The case with small error results in. Furthermore, the spectral fluctuation shows up during updating new DPD coefficients. It can be deduced from the results above, that the stability of DPD system is still a major problem in practice.

19.4 DPD performance variation in terms of ACP suppression and EVM reduction

The basic idea described in this section is to estimate the DPD coefficients by iterative methods. Because iterative methods for solving linear system of equations embody a quite different approach from direct methods, the inversion of an

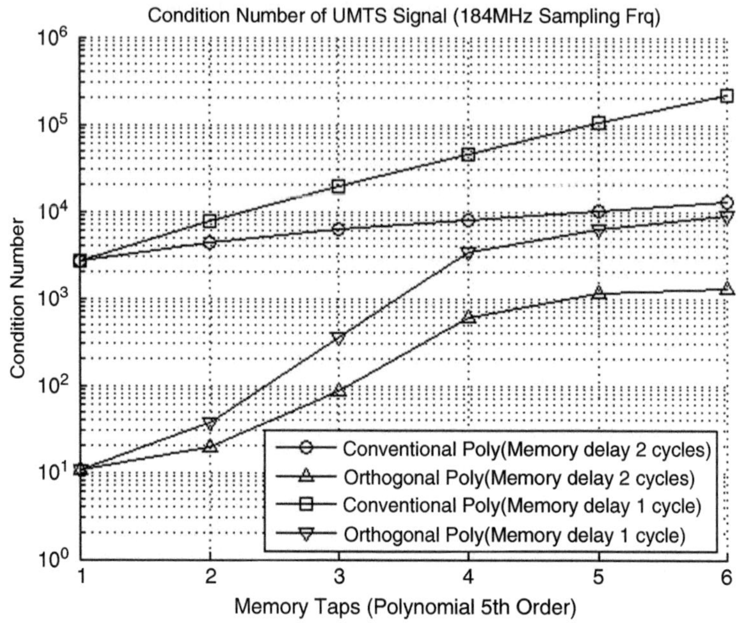

Fig. 19.4 Condition numbers of memory DPD for a three carrier UMTS signal

ill-conditioned matrix in the expression (19.3) can be avoided during the estimation of DPD parameters. Iterative methods attempt to solve the system of equations by approaching the solution step by step with an initial guess. The first order and second order stationary iterative methods are investigated. More information about stationary methods can be found in [10].

19.4.1 The First-Order Stationary Iterative Methods

To solve a square system of linear equations (19.7) with same number of equations and unknowns the first-order iterative methods start with an initial guess A^0. The residual R^k can be calculated with the current parameter A^k according to the expression (19.8). With a properly selected preconditioning nonsingular matrix N the correction $D^{(k+1)}$ in (19.9) is computed at stage k and then added to current parameter A, described by the expression (19.10). τ_k is a fixed parameter for first-order stationary iterative methods, which has an effect on the convergence rate of first-order iterative methods. The whole process would be repeated until a break condition is fulfilled. A crucial task in the construction of an efficient iterative method is the choice of preconditioning matrix N and convergence parameter τ_k, because at each iteration step a new linear system (19.9) with preconditioning matrix N and parameter τ_k must be solved. The computational cost of this linear system must be relative small but still effective for the increase of convergence rate. The selection of proper preconditioning matrix N and parameter τ_k is discussed in the following paragraphs.

$$M_{matrix} \cdot A = X_{vector} \quad (19.7)$$
$$R^k = M_{matrix} \cdot A^k - X_{vector} \quad (19.8)$$
$$N \cdot D^{(k+1)} = -\tau_k R^k \quad (19.9)$$
$$A^{(k+1)} = A^k + D^{(k+1)} = A^k - \tau_k (N^{-1} \cdot R^k) \quad (19.10)$$

However, there is still a hindrance to apply iterative methods to solve the DPD coefficients, because the system of equations for DPD is a over determined system of equations. At first the over-determined system of equations must be modified to become a square system of equations with the same number of equations and unknowns without information lost compared to the original system of equations.

To overcome this problem one can multiply the system of equations with a Hermitian transpose of the matrix M_y, given in the expression (19.11). Consequently, one is able to solve the regular linear system of equations in the expression (19.11) by iterative methods.

$$M_y \cdot A = X; \Rightarrow (M_y^H \cdot M_y) \cdot A = M_y^H \cdot X \quad (19.11)$$

One of the disadvantages of the first-order iterative methods is that the rate of convergence may be slow or even diverge, depending on the selected preconditioning matrix N and parameter τ_k. In subsequent paragraphs we present some possible choices of preconditioning matrix N and their optimal parameter τ_k for maximal convergence speed.

19.4.1.1 Basic Iterative Method Without Single Division

As mentioned before, one needs to solve the linear system (19.9) with matrix N at each iteration step. If we use an identity matrix as the preconditioning matrix N, the linear system of (19.9) becomes the simplest linear system with minimal computational cost. Furthermore, there is no single division in the parameter estimation process, which can cause numerical instabilities, if the denominator is extreme small. This iterative method starts also from the modified equation (19.11) with an initial guess of parameter A. The parameters are estimated only by multiplications and additions in expression (19.13). The optimal parameter τ_k can be derived according to the expression (19.14). The λ_{min} and λ_{max} are the minimal and maximal eigenvalues of matrix $(N^{-1}*(M_y^{H}*M_y))$.

$$N \cdot D^{(k+1)} = -\tau \cdot R^k; N \text{ is an identity matrix} \quad (19.12)$$
$$A^{(k+1)} = A^k + D^{(k+1)} = A - \tau \cdot R^k; \quad (19.13)$$
$$\tau_{opt} = 2/(\lambda_{\min} + \lambda_{\max}); \quad (19.14)$$

To stop iterative methods we have two criteria available, the maximal difference between two successive sets of DPD parameters A^{k+1} and A^k is smaller than a predefined value ε (19.15a), or the maximal difference (residual) of equation (19.11) is smaller than a predefined value ε (19.15b).

$$\max\left|A^{(k+1)} - A^{(k)}\right| < \varepsilon \tag{19.15a}$$

$$\max\left|\left(M_y^H \cdot M_y\right) \cdot A^{(k+1)} - M_y^H \cdot X\right| < \varepsilon \tag{19.15b}$$

19.4.1.2 Jacobi Method

For basic iterative method we need to calculate the extreme eigenvalues, which results in high computational cost. In this paragraph we present another first-order iterative method by the name of Jacobi Method, by which we do not need to know the extreme eigenvalues. In this method the matrix $(M_y^{H*} M_y)$in expression (19.11) is decomposed in three matrices L, D and U, that represent the diagonal, strictly lower triangular and strictly upper triangular parts, respectively. We take the diagonal part D as preconditioning matrix N and choose the parameter τ_k equal one, so that we do not need extreme eigenvalues to find optimal τ_k. The convergence condition is that the spectral radius of iteration matrix $(D^{-1}(L + U))$ need to be smaller than one. The single parameter a_j^k comprised in the vector A^k is computed according to the expression (19.18). The benefit of the Jacobi Method is already mentioned, that we save the computational cost for calculation of eigenvalues. If the matrix $(M_y^{H*} M_y)$ is strictly diagonal dominant, we even do not need to proof the convergence condition. But in practice Jacobi Method is not suitable for digital predistortion, because the transmitted signals are almost random signals. We cannot guarantee the convergence of Jacobi Method.

$$\left(M_y^H \cdot M_y\right) = L + D + U \tag{19.16}$$

$$L = \begin{vmatrix} 0 & 0 & 0 \cdots \\ \widetilde{y}_{21} & 0 & 0 \cdots \\ \widetilde{y}_{31} & \widetilde{y}_{32} & 0 \cdots \\ \vdots & \vdots & \vdots \cdots \end{vmatrix}; U = \begin{vmatrix} 0 & \widetilde{y}_{12} & \widetilde{y}_{13} & \cdots \\ 0 & 0 & \widetilde{y}_{23} & \cdots \\ 0 & 0 & 0 & \cdots \\ \vdots & \vdots & \vdots & \cdots \end{vmatrix};$$

$$D = \begin{vmatrix} \widetilde{y}_{11} & 0 & 0 & \cdots \\ 0 & \widetilde{y}_{22} & 0 & \cdots \\ 0 & 0 & \widetilde{y}_{33} & \cdots \\ \vdots & \vdots & \vdots & \cdots \end{vmatrix}; \tag{19.17}$$

$$a_i^{k+1} = \frac{1}{\widetilde{y}_{ii}} \cdot \left(- \sum_{\substack{j=1 \\ j \neq i}}^{n} \widetilde{y}_{ij} \cdot a_j^k + \widetilde{x}_i \right) \tag{19.18}$$

19.4.1.3 Gauss–Seidel Method

The Gauss–Seidel Method is derived from Jacobi Method. The basic idea is to use the most up-to-date parameter value a_j^{k+1} for calculation of the remainder a_{j+n}^{k+1}. In Gauss–Seidel Method the matrix $(D\text{-}L)$ is used as preconditioning matrix N and the single parameter a_j^k is computed according to the expression (19.19).

$$a_i^{k+1} = \frac{1}{\widetilde{y}_{ii}} \cdot \left(- \sum_{j=1}^{i-1} \widetilde{y}_{ij} \cdot a_j^{k+1} - \sum_{j=i+1}^{n} \widetilde{y}_{ij} \cdot a_j^k + \widetilde{x}_i \right) \tag{19.19}$$

The convergence condition of Gauss–Seidel Method is more relaxed compared to Jacobi Method. If the matrix $(M_y^{H*} M_y)$ in the expression (19.11) is positive-definite, the Gauss–Seidel Method is convergent [11]. The product of the matrixes M_y^H and M_y is proven to be nonnegative definite (see appendix) independent of DPD algorithms and test signals and hence the Gauss–Seidel Method was always convergent in simulations and hardware tests.

19.4.2 The Second-Order Stationary Iterative Methods

A second-order, or two-step iterative, method is defined by

$$A^{(k+1)} = A^k + D^{(k+1)} = \alpha_k \cdot A^k + (1 - \alpha_k) \cdot A^{k-1} - \tau_k \cdot (N^{-1} \cdot R^k) \tag{19.20}$$

where α_k and τ_k are constant for all k by stationary iterative methods. By second-order iterative methods we calculate the new set of coefficients on the base of last two sets of coefficients. Similar to the first order stationary iterative methods, the crucial task in the construction of an efficient second-order iterative method is the choice of preconditioning matrix N and convergence parameters α_k and τ_k. For a certain preconditioning matrix N the optimal convergence parameters α_k and τ_k are given in expressions (19.21, 19.22). The λ_{min} and λ_{max} are the minimal and maximal eigenvalues of matrix $(N^{-1*}(M_y^{H*} M_y))$ [10].

$$\alpha_{opt} = 1 + \left(\frac{1 - \sqrt{\lambda_{\min}/\lambda_{\max}}}{1 + \sqrt{\lambda_{\min}/\lambda_{\max}}} \right)^2 \tag{19.21}$$

$$\tau_{opt} = 2/(\lambda_{\min} + \lambda_{\max}); \tag{19.22}$$

In comparison to first-order stationary iterative methods the second-order stationary iterative methods have higher convergence speed generally. In simulation and hardware test the convergence speeds of second-order stationary iterative methods are almost tens of times higher than first-order stationary iterative methods with same preconditioning matrix N and optimal convergence parameters.

Among the stationary iterative methods the Gauss–Seidel Method is a good choice in comparison to other first-order and second-order stationary iterative methods. The main advantage of the Gauss–Seidel Method is that there is no need for calculation of extreme eigenvalues. The convergence speed of the Gauss–Seidel Method is also relative high, which depends on the condition number of $(M_y^{H*} M_y)$.

19.4.3 Simulation Results

Generally, the computational cost of iterative methods is relative high. The number of iterations for iterative method is strongly dependent on the convergence rate, the break conditions and initial values. To accelerate the computation, the old DPD coefficients can be used as the initial values for new adaptation. Furthermore, we can relax the break conditions to achieve a low number of iteration steps or just limit the maximal number of iterations.

The simulation results for DPD performance vs. maximal number of iterations are presented in Fig. 19.5. A base-band PA model with Volterra series structure is used as test object and compensated by orthogonal polynomial DPD mentioned in Sect. 19.2. The initial value of the Gauss–Seidel Method is neutral, i.e. the DPD has a gain of 1 and phase shift of 0. The old DPD coefficients would not be used as the initial values for next adaptation. We can see, that the ACP is more suppressed

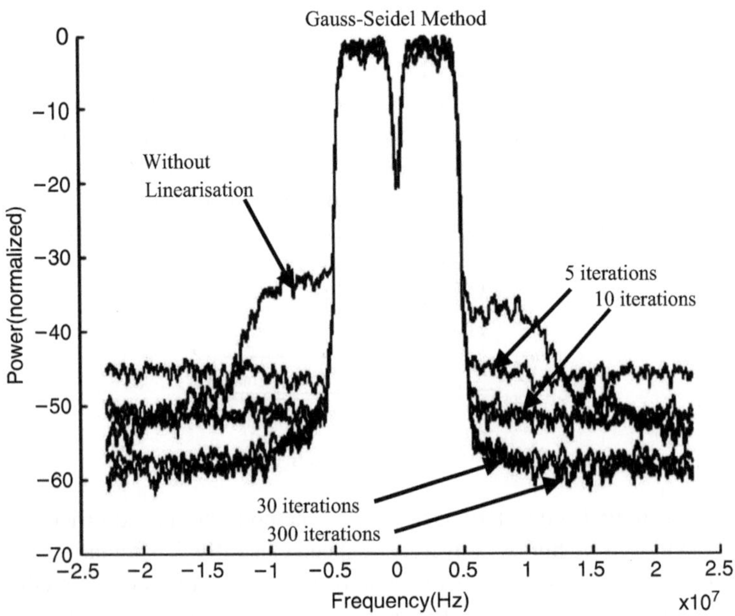

Fig. 19.5 DPD performance vs. maximal number of iterations for Gauss–Seidel Method

with increased number of iterations. However, the difference of DPD performance between the Gauss–Seidel Method with 30 iterations and 300 iterations is almost invisible.

The same simulation is carried out for DPD performance vs. maximal residual of the modified linear matrix equation (19.11). The smaller the maximal residual is, the better performs the DPD. It is quit evident, that the break-conditions of maximal residual and number of iterations correlate with each other, i.e. a small residual value indicates high number of iteration and vice versa. The optimal tradeoff between the break-conditions in consideration of computational cost and DPD performance can be drawn in practice by trying. It is worthwhile to point out, that a over stringent break condition can result in high computational cost instead of visible improvement of DPD performance As shown in Fig. 19.6 the DPD exhibits similar ACP suppression for the maximal residual of 0.001 and 0.0001. But the computational cost for the maximal residual of 0.0001 is much higher than it for the maximal residual of 0.001.

Furthermore, if we really need the complete new coefficients to replace the old ones when the old coefficients perform good ACP reduction and EVM correction. In practice the really utilized coefficients are a compromise of the old and new coefficients, described by (19.23).

$$a_{use,new} = \xi \cdot a_{use,old} + (1 - \xi) \cdot a_{new} \tag{19.23}$$

with $\xi < 1$.

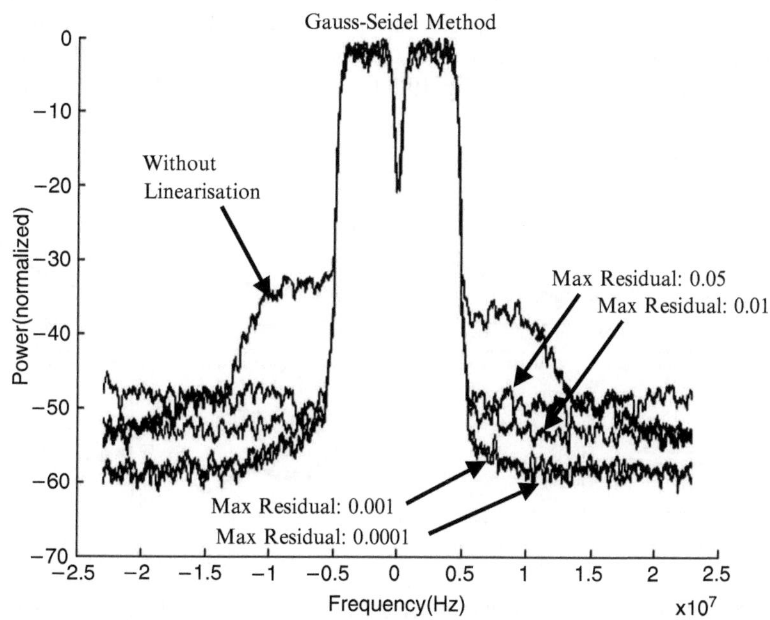

Fig. 19.6 DPD performance vs. maximal residual for Gauss–Seidel Method

In LS method the new coefficients are computed without consideration of the old coefficients, so that the deviation between the old and new coefficients maybe very large, which causes strong spectral fluctuations during updating new coefficients. On the contrary, the iterative methods need an initial value (old coefficients) to start calculation. In addition we have complete control to restrict the difference of old and new coefficients through the number of iterations. As a consequence, by properly chosing the maximal number of iterations the DPD coefficients solved by iterative method are much more stable in comparison to these coefficients solved by LS method. Furthermore, one can adapt the break conditions for iterative method according to matrix condition or other aspects, so that a balance between stability, computational cost and DPD performance can be optimised even during the operating time.

19.5 Measurements

The DPD system with memory compensation was implemented in a high speed Field Programmable Gate Array (FPGA). A Doherty power amplifier with the average output power of 50W was tested with different UMTS test signals pre-clipped to 6dB peak to average ratio. The orthogonal polynomial algorithm is utilized to describe the inverse model of this PA. The inverse characteristics (parameters a_j^k) are computed in Matlab by both LS method and stationary iterative methods.

The DPD performance concerning ACP and EVM suppression is almost same for LS method and iterative methods. However, the DPD by using iterative methods exhibits much more stable performance than the DPD by using LS method. The DPD characteristic curves of main tap and first memory tap are illustrated in Fig. 19.7 for a one carrier and in Fig. 19.8 for a three carrier UMTS test signal. For

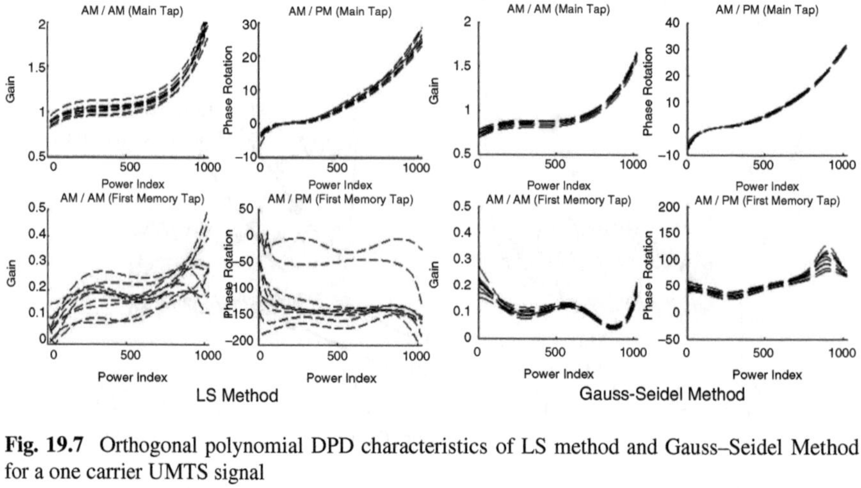

Fig. 19.7 Orthogonal polynomial DPD characteristics of LS method and Gauss–Seidel Method for a one carrier UMTS signal

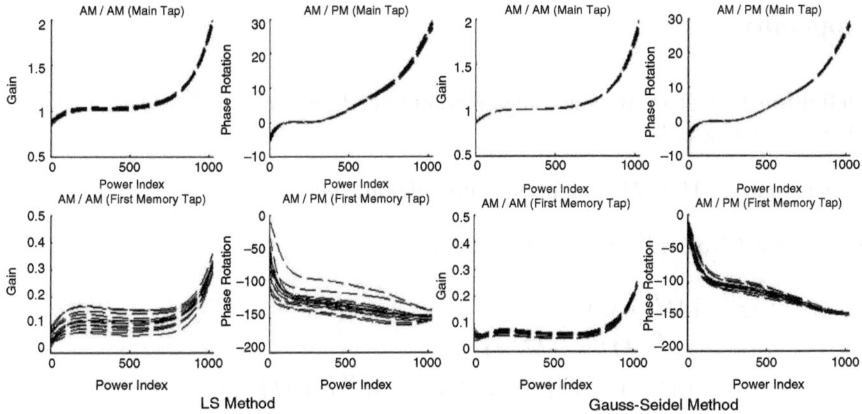

Fig. 19.8 Orthogonal polynomial DPD characteristics of LS method and Gauss–Seidel Method for a three carrier UMTS signal

LS method all curves of first memory tap are far-scattered, especially in AM/PM characteristic. This results not only in strong spectral fluctuation (up to 20 dB in ACP region) of the PA output during the coefficients updating (new curves) but also in instability of the whole system. For the Gauss–Seidel Method (iterative method) the curves of each parameter-set are bundled up. There is only slight spectral fluctuation (under 5 dB in ACP region) of PA output during parameter updates. The spectral fluctuation depends strongly on the number of iterations, so that we can just limit the maximal number of iterations to guarantee even smoother updating of new coefficients. Other iterative methods exhibit similar test results.

19.6 Conclusion

The parameters of DPD system solved by stationary iterative methods are much more stable in comparison to these solved by the LS method without performance degradation, because the inversion of an ill-conditioned matrix can be avoided in iterative methods. One does not need any other process to guarantee smooth updating of new coefficients. Furthermore, the old parameters can be used as initial values for calculation of next DPD parameters, which can significantly reduce the computational cost. In the end the iterative methods offer us the possibility to control the computing process, e.g. stopping iterative methods in any time according to a specific requirement.

Appendix

Definition 1. Matrix M is nonnegative-definite, if and only if $z^H \cdot M \cdot z \geq 0$ with all $z \in C^n$ and $z \neq 0$.

The matrix $(M^{H*}M)$ is a nonnegative-definite matrix.

Proof. $z \in C^n and, z^H \cdot (M^H \cdot M) \cdot z \geq 0;$

$$\begin{aligned} & z^H \cdot (M^H \cdot M) \cdot z \\ & = (z^H \cdot M^H) \cdot (M \cdot z) \\ & = (M \cdot z)^H \cdot (M \cdot z) \text{ with } (z^H \cdot M^H) = (M \cdot z)^H \\ & = \sum_{i=1}^{n} |a_i|^2 \geq 0; \text{ with } (M_y \cdot z) = [a_1, a_2 \cdots a_n]^T \quad a_i \in C \end{aligned}$$

References

1. S.C. Cripps, Some topics in PA nonlinearity. *Advanced Techniques in RF Power Amplifier Design* (Artech House, Norwood, MA, 2002), pp. 73–111
2. L. Ding, G.T. Zhou, D.R. Morgan, Z. Ma, J.S. Kenney, J. Kim, C.R. Giardina, Memory polynomial predistorter based on the indirect learning architecture in *Proceedings of the IEEE Global Telecommunication Conference*, Nov 2002, pp. 967–971
3. M. Schetzen, Nonlinear system modelling based on the wiener theory. Proc. IEEE **69**(12), 1557–1573 (1981)
4. K. Hyunchul, J.S. Kenney, Behavioral modeling of nonlinear RF power amplifiers considering memory effects. IEEE Trans. Microw. Theory Techn. **51**(12) (2003)
5. J. Vuolevi, T. Rahkonen, J. Manninen, Measurement technique for characterizing memory effects in RF power amplifiers. IEEE Trans. Microw. Theory Meas. **49**(8), 1383–1389 (2001)
6. A. Zhu, Behavioral modeling of RF power amplifiers based on pruned Volterra series. IEEE Microw. Wirelss Compon. Lett. **14**, 563–565 (2004)
7. L. Ding, Z. Ma, D.R. Morgan, M. Zierdt, J. Pastalan, A least-squares/newton method for digital predistortion of wideband signals. IEEE Trans. Commun. **54**(5), pp. 833–840 (2006)
8. R. Raich, H. Qian, G.T. Zhou, Digital baseband predistortion of nonlinear power amplifiers using orthogonal polynomials. in *Proceedings of the IEEE Interenational Conference Acoust., Speech, Signal Processing*, Apr. 2003, pp. 689–692
9. L. Ding, G.T. Zhou, D.R. Morgan, Z. Ma, J.S. Kenney, J. Kim, C.R. Giardina, A robust digital baseband predistorter constructed using memory polynomials. IEEE Trans. Commun. **52**(1), 159–165 (2004)
10. O. Axlsson, Basic iterative methods and their rates of convergence. *Iterative Solution Methods* (Cambrige University Press, NY, 1994), pp. 158–178
11. J. Gilbert, http://www.maths.lancs.ac.uk/~gilbert/m306b/node18.html. 1999

Chapter 20
Analysis of Complex Periodic Structures

Reinhold Pregla

20.1 Introduction

Efficient algorithms for the analysis of conventional periodic structures are described in [1, 2]. In this paper we present algorithms for the analysis of complex periodic structures. The algorithms can be used for structures in optics and for microwaves and millimeter waves. In Fig. 20.1a periodic waveguide structure is sketched in which the period sections are periodic structures themselves. For the analysis we will use the Floquets theorem twice. In modern microwave- and especially nano-technology circuits ring resonators and also circuits in photonic crystals might play an important role. Conventional ring resonators consist of a ring of a homogeneous (homogeneous in azimuthal direction) waveguide (e.g. a rib waveguide in optics or a microstrip waveguide in microwaves) or also a periodic structure. In principle, the wave propagation in azimuthal direction is analogous to that in straight direction. Therefore, also the analysis is similar. Further ring resonators with hyperperiodic waveguides can be analysed in the same manner. In Fig. 20.3 two ring resonators with equal size are coupled with each other. Figure 20.4 shows a ring resonator which is coupled to a straight waveguide (photonic or microstrip waveguide).

In this contribution will be shown how such complex structures can be analyzed with a similiar principle. This principle can also be used for other structures like bends and junctions in photonic crystals.

20.2 Analysis Algorithms

In this section we would like to present in detail the algorithms for the analysis of various devices which contain periodic structures.

R. Pregla
University of Hagen, 58084 Hagen, Germany
e-mail: r.pregla@Frnuni-Hagen.de

S. Lindenmeier and R. Weigel (eds.), *Electromagnetics and Network Theory and their Microwave Technology Applications*, DOI 10.1007/978-3-642-18375-1_20,

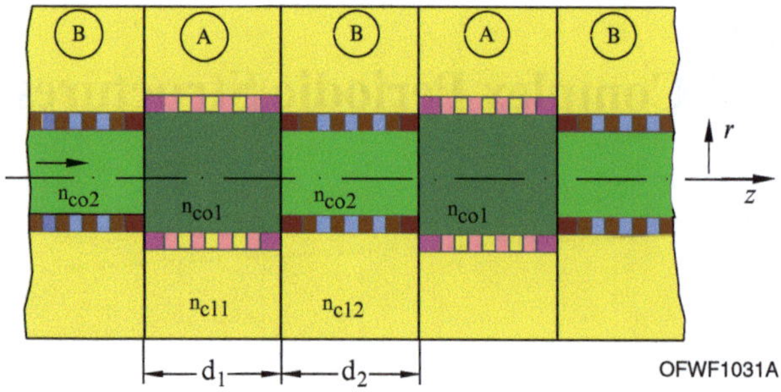

Fig. 20.1 Straight periodic waveguide: the sections are itself periodic

Fig. 20.2 Rib waveguide ring resonator of periodic sections

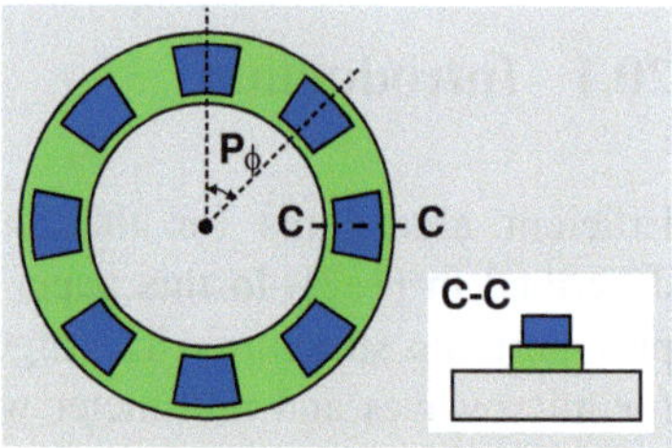

Fig. 20.3 Two coupled ring resonators of equal size

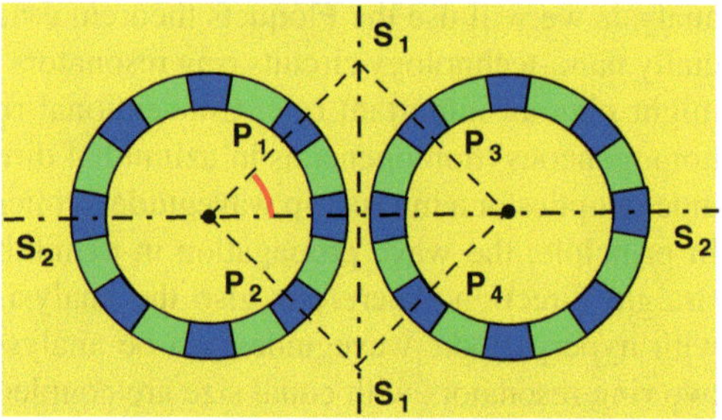

20.2.1 *Analysis of Super Grating Structures*

Generalized Transmission Line Equations in matrix notation as basis are used for efficient analysis algorithms with the Method of Lines (MoL) [3]. Such efficient algorithms were developed in the past for conventional periodic structures like fibre gratings [2, 3]. The Floquet's modes were obtained in a special way from half of the period by calculating the short and open circuit impedance matrices. Here it shall be shown how we can analyse periodic structures (Bragg gratings) where the sections of the periods are formed themselves as periodic structures (super periodic structures – see e.g. Fig. 20.1). In the first step we apply the above mentioned algorithm to the different periodic sections of our complex grating structure (see Fig. 20.5). We obtain new sections consisting of homogeneous Floquet transmission lines and some intermediate regions. These might come from the remaining parts in the periodes. For example, if we look at such a periodic part we see that we have the same

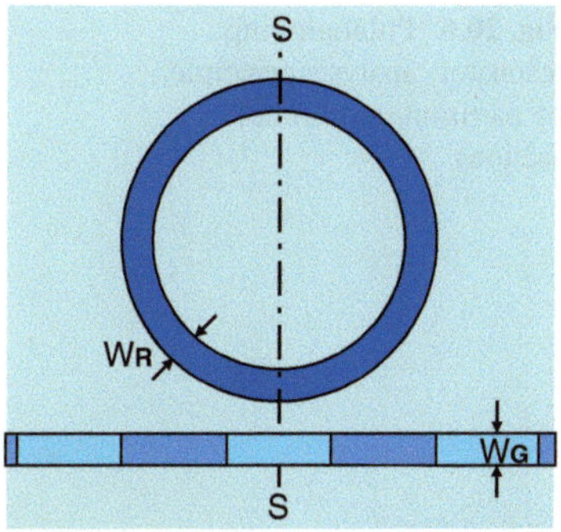

Fig. 20.4 Photonic ring resonator with coupled straight periodic waveguide

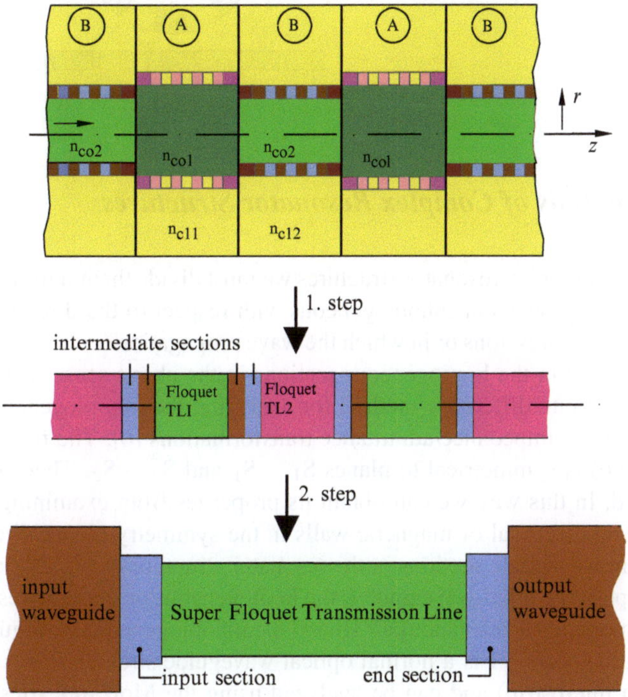

Fig. 20.5 Analysis steps for the supergrating

section at the beginning and at the end. Therefore, we have a remaining area besides the periodic part . We could also include additional sections here. After that we obtain a normal periodic structure. We can now replace this new structure again by an equivalent Floquet transmission line, the super Floquet transmission line. At the input and the output we have again additional sections which cannot be included into the periodic part of the structure. Now we can also introduce the concatenated input and output waveguides (which may be completely different to the other parts) and analyze the new structure in conventional way.

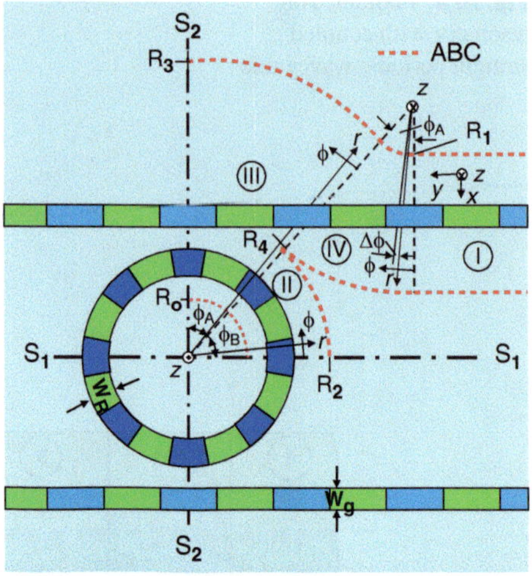

Fig. 20.6 Photonic ring resonator: analysis principle by partition into suitable sections

20.2.2 Analysis of Complex Resonator Structures

To analyze the complex resonator structures we must divide them into suitable parts, which are homogeneous or inhomogeneous with respect to the direction in which we use analytic expressions or in which the wave propagation occurs (see Fig. 20.6). The MoL is used in the homogeneous sections in the above sense. It is combined with a special finite difference method for analyzing the inhomogeneous parts. In both parts we use impedance/admittance transformations [6]. The resonator structure (Fig. 20.6) is symmetrical to planes $S_1 - S_1$ and $S_2 - S_2$. These symmetries are exploited. In this way we can obtain its properties from examining four cases by introducing electrical or magnetic walls in the symmetry planes. We have now four different waveguide sections in the resonator structure of Fig. 20.6: the homogeneous or periodic input waveguide I, the homogeneous or periodic resonator part II, the inhomogeneous resonator part III and the inhomogeneous waveguide part IV. The waveguide section I is a normal optical waveguide structure (or a microwave guide like a microstrip) and can be analyzed using the MoL in Cartesian coordinates. The resonator part II consists of concatenations of waveguide sections in azimuthal direction and is analyzed with the MoL in cylindrical coordinates and impedance/admittance transformation in ϕ-direction. In case of a periodic structure (with many periods) also Floquet's algorithm can be used. For the inhomogeneous parts III and IV we use impedance/admittance transformation with finite differences in cylindrical coordinates for the ϕ-direction [6,7] – even for the straight waveguide sections.

We start in the symmetry plane $S_2 - S_2$ (short or open circuited) with impedance transformation by FD in $-\phi$-direction and in the symmetry plane $S_1 - S_1$ (short or open circuited) with impedance transformation by the MoL in ϕ-direction. Therefore, we have to distinguish four cases. In the interface plane between the section III and sections II+IV we have to match the tangential fields (which is done by the impedances) to obtain the load impedance for the section IV (see Sect. 20.2.7). This impedance has then to be transformed through the section IV in $-\phi$-direction in cylindrical coordinates. The results is the load impedance of the input waveguide I.

20.2.3 Analysis of Bends and Junctions in Photonic Crystals

A bend and a Y-junction in photonic crystal waveguide are shown in Fig. 20.7. Because of the symmetry plane S − S in the waveguide bend we can obtain the properties from examining two cases by introducing an electrical or a magnetic wall in this symmetry plane. In symmetry plane of the Y-junction we introduce an electric or an magnetic wall depending on the mode which we would like to analyze. The periodic waveguide structures in section I and III are analyzed with the help of Floquets theorem [1, 2]. In section IIa or II we use impedance/admittance transformation with finite differences in cylindrical coordinates for the ϕ-direction [6]. For matching optimization the columns in the regions II should displaced in the given direction (→). The basic equations that we use in cylindrical coordinate for propagation in ϕ-direction are given in the book [5].

20.2.4 Symmetrical 4-Port: Analysis by Subdivision into Substructures

The general 4-port in Fig. 20.8 has two symmetry planes: I-I and II-II. By using symmetric and antimetrical fields at the ports – symbolized by amplitudes (1) and

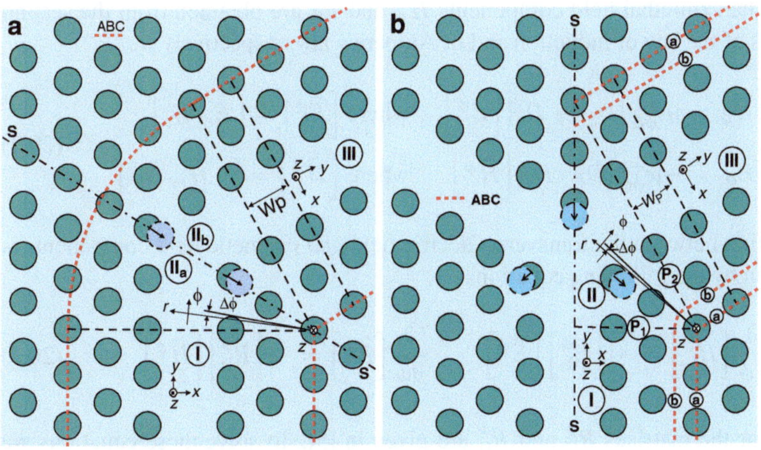

Fig. 20.7 Photonic crystal structures: partitioning into suitable sections: (**a**) sharp bend and (**b**) Y-junction. → Displacement for impedance matching

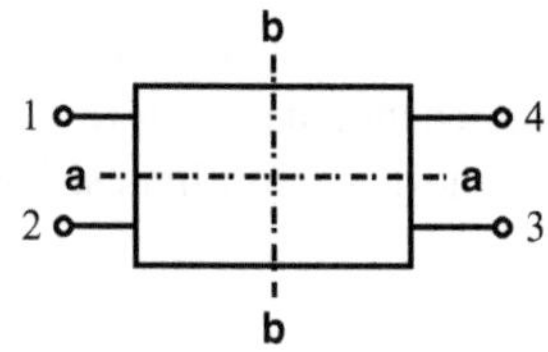

Fig. 20.8 Subdivision of the ring resonator in Fig. 20.6a

(-1) – we obtain electric (e) and magnetic (m) walls at the symmetry planes. The values A_i, $(i = 1, 2, 3, 4)$ are the amplitudes at port i. S_{11} is the scattering parameter at port 1. The scattering parameters S_{i1} (from port 1 to the other ones) are obtained from the four different parameters S_{11}^{uv}, u, v = e, m in the following way

a - a	b - b	A_1	A_2	A_3	A_4	S_{11}
m	m	1	1	1	1	S_{11}^{mm}
e	m	1	−1	1	−1	S_{11}^{em}
m	e	1	1	−1	−1	S_{11}^{me}
e	e	1	−1	−1	1	S_{11}^{ee}

$$\begin{aligned} S_{11} &= 0.25(S_{11}^{\mathrm{mm}} + S_{11}^{\mathrm{em}} + S_{11}^{\mathrm{me}} + S_{11}^{\mathrm{ee}}) \\ S_{21} &= 0.25(S_{11}^{\mathrm{mm}} - S_{11}^{\mathrm{em}} + S_{11}^{\mathrm{me}} - S_{11}^{\mathrm{ee}}) \\ S_{31} &= 0.25(S_{11}^{\mathrm{mm}} + S_{11}^{\mathrm{em}} - S_{11}^{\mathrm{me}} - S_{11}^{\mathrm{ee}}) \\ S_{41} &= 0.25(S_{11}^{\mathrm{mm}} - S_{11}^{\mathrm{em}} - S_{11}^{\mathrm{me}} + S_{11}^{\mathrm{ee}}) \end{aligned} \tag{20.1}$$

20.2.5 GTL Equations for ϕ-Direction

We summarize here the GTL-equations in cylindrical coordinates because the analysis should be performed in these coordinates in ϕ-direction in many sections of our structures. The material parameters ν_{rr}, $\nu_{\phi\phi}$, ν_{zz} and the off-diagonal elements ν_{rz} and ν_{zr} should be functions of r and z only. The remaining off-diagonal elements in the material tensors are zero. (The general case of material parameters is described in [5].) The azimuthal field components $\widetilde{H}_\phi$ and E_ϕ are obtained from the second equations of the law of induction and of Ampere's law, respectively

$$\begin{aligned} \widetilde{H}_\phi &= \mathrm{j}\mu_{\phi\phi}^{-1}\left[D_{\bar{z}} \;\; -D_{\bar{r}}\right]\left[\widehat{E}^\phi\right] \quad \text{where } \left[\widehat{E}^\phi\right] = \left[E_r, E_z\right]^{\mathrm{t}} \\ E_\phi &= -\mathrm{j}\epsilon_{\phi\phi}^{-1}\left[D_{\bar{r}} \;\; D_{\bar{z}}\right]\left[\widehat{H}^\phi\right] \quad \text{where } \left[\widehat{H}^\phi\right] = \left[-\widetilde{H}_z, \widetilde{H}_r\right]^{\mathrm{t}} \end{aligned} \tag{20.2}$$

The relation between the transverse electric field and magnetic field components is now given by the following equations

$$\frac{\partial}{\partial\phi}\left[\widehat{H}^\phi\right] = -\mathrm{j}\left[R_{\mathrm{E}}^\phi\right]\left[\widehat{E}^\phi\right] \qquad \frac{\partial}{\partial\phi}\left[\widehat{E}^\phi\right] = -\mathrm{j}\left[R_{\mathrm{H}}^\phi\right]\left[\widehat{H}^\phi\right] \tag{20.3}$$

Details for the matrices R_{E}^ϕ and R_{H}^ϕ are given in [5]. To solve these equations we must perform discretization in the cross-section (r, z-plane). Figure 20.9 shows the discretization of a rib waveguide as example.

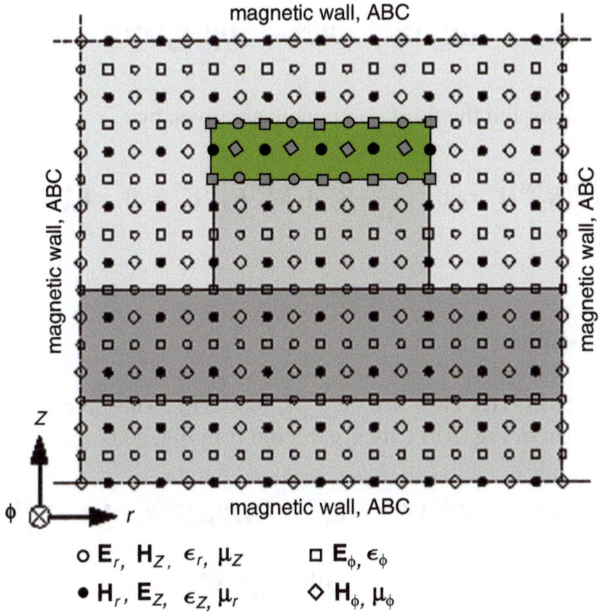

Fig. 20.9 Example for a 2-D discretization of the cross section (e.g. for the waveguide in the ring resonator in Fig. 20.2)

20.2.6 2D-Case

We now give the equations for the 2D case in detail. With $D_z = 0$ we obtain from the general equations in [5] separate expressions for the two polarizations.

20.2.6.1 TM$_\phi$-Case

The components in the TM$_\phi$-case are: E_r, H_z and E_ϕ. They are discretized in r-direction as shown in Fig. 20.10 for e.g. region II in Fig. 20.6.

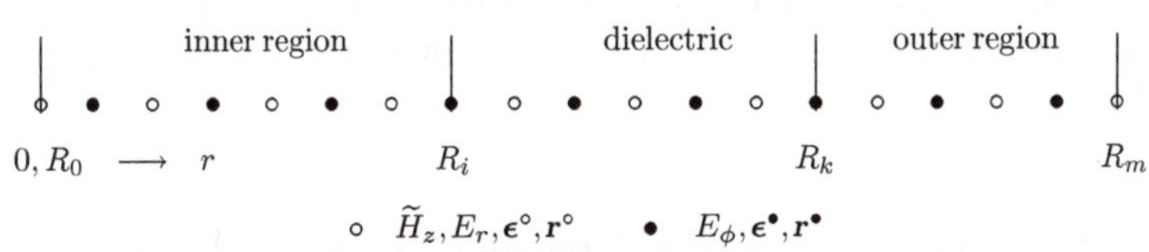

Fig. 20.10 Discretization scheme in case of TM$_\phi$ – modes. The subdivisions are schematic

In discretized form we obtain the following GTL -equations

$$\frac{\partial}{\partial \phi}\widetilde{\mathbf{H}}_u^\circ = -\mathrm{j}\boldsymbol{R}_\mathrm{E}^\mathrm{TM}\mathbf{E}_r^\circ \qquad \boldsymbol{R}_\mathrm{E}^\mathrm{TM} = \boldsymbol{\epsilon}_{rr}^\circ \overline{\boldsymbol{r}}_n^\circ \qquad \widetilde{\mathbf{H}}_u^\circ = -\widetilde{\mathbf{H}}_z^\circ \tag{20.4}$$

$$\frac{\partial}{\partial\overline{\phi}}\mathbf{E}_r^{\circ} = -\mathrm{j}\boldsymbol{R}_{\mathrm{H}}^{\mathrm{TM}}\widetilde{\mathbf{H}}_u^{\circ} \qquad \boldsymbol{R}_{\mathrm{H}}^{\mathrm{TM}} = -\boldsymbol{D}_{\overline{r}}^{\circ\mathrm{t}}\boldsymbol{\epsilon}_{\phi\phi}^{\bullet-1}\overline{\boldsymbol{r}}_n^{\bullet}\boldsymbol{D}_{\overline{r}}^{\circ} + \boldsymbol{\mu}_{zz}^{\circ}\overline{\boldsymbol{r}}_n^{\circ} \tag{20.5}$$

The wave equations and the third component are given by

$$\frac{\partial^2}{\partial^2\overline{\phi}}\widetilde{\mathbf{H}}_z^{\circ} + \boldsymbol{R}_{\mathrm{E}}^{\mathrm{TM}}\boldsymbol{R}_{\mathrm{H}}^{\mathrm{TM}}\widetilde{\mathbf{H}}_z^{\circ} = 0 \qquad \frac{\partial^2}{\partial^2\overline{\phi}}\mathbf{E}_r^{\circ} + \boldsymbol{R}_{\mathrm{H}}^{\mathrm{TM}}\boldsymbol{R}_{\mathrm{E}}^{\mathrm{TM}}\mathbf{E}_r^{\circ} = 0 \qquad \mathbf{E}_{\phi}^{\bullet} = -\mathrm{j}\boldsymbol{\epsilon}_{\phi\phi}^{\bullet-1}\boldsymbol{D}_{\overline{r}}^{\circ}\widetilde{\mathbf{H}}_u^{\circ} \tag{20.6}$$

20.2.6.2 TE$_\phi$-Case

The components in TE$_\phi$-case are: E_z, H_r and H_ϕ. They are discretized in r-direction as shown in Fig. 20.11 for e.g. region II in Fig. 20.6.

$$\frac{\partial}{\partial\overline{\phi}}\widetilde{\mathbf{H}}_r^{\bullet} = -\mathrm{j}\boldsymbol{R}_{\mathrm{E}}^{\mathrm{TE}}\mathbf{E}_z^{\bullet} \qquad \boldsymbol{R}_{\mathrm{E}}^{\mathrm{TE}} = -\boldsymbol{D}_{\overline{r}}^{\bullet\mathrm{t}}\boldsymbol{\mu}_{\phi\phi}^{\circ-1}\overline{\boldsymbol{R}}_n^{\circ}\boldsymbol{D}_{\overline{r}}^{\bullet} + \boldsymbol{\epsilon}_{zz}^{\bullet}\overline{\boldsymbol{R}}_n^{\bullet} \tag{20.7}$$

$$\frac{\partial}{\partial\overline{\phi}}\mathbf{E}_z^{\bullet} = -\mathrm{j}\boldsymbol{R}_{\mathrm{H}}^{\mathrm{TE}}\widetilde{\mathbf{H}}_r^{\bullet} \qquad \boldsymbol{R}_{\mathrm{H}}^{\mathrm{TE}} = \boldsymbol{\mu}_{rr}^{\bullet}\overline{\boldsymbol{R}}_n^{\bullet} \tag{20.8}$$

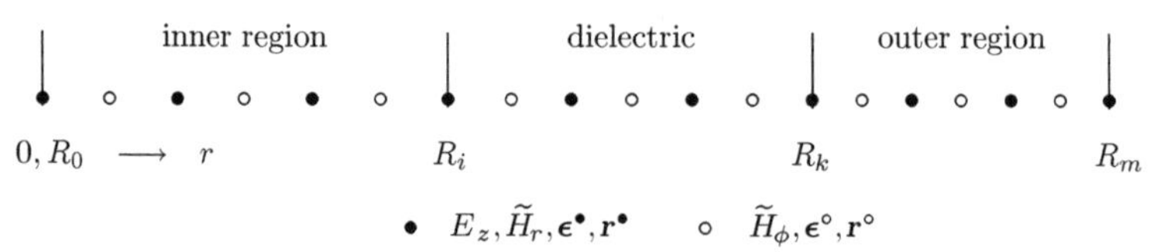

Fig. 20.11 Discretization scheme for TE$_\phi$ – modes. The subdivisions are schematic

The wave equations and the third component are given by

$$\frac{\partial^2}{\partial^2\overline{\phi}}\widetilde{\mathbf{H}}_r^{\circ} + \boldsymbol{R}_{\mathrm{E}}^{\mathrm{TE}}\boldsymbol{R}_{\mathrm{H}}^{\mathrm{TE}}\widetilde{\mathbf{H}}_r^{\circ} = 0 \qquad \frac{\partial^2}{\partial^2\overline{\phi}}\mathbf{E}_z^{\circ} + \boldsymbol{R}_{\mathrm{H}}^{\mathrm{TE}}\boldsymbol{R}_{\mathrm{E}}^{\mathrm{TE}}\mathbf{E}_z^{\circ} = 0 \qquad \widetilde{\mathbf{H}}_{\phi}^{\circ} = -\mathrm{j}\boldsymbol{\mu}_{\phi\phi}^{\circ-1}\boldsymbol{D}_{\overline{r}}^{\bullet}\mathbf{E}_z^{\bullet} \tag{20.9}$$

We assume the relative permeability being equal to 1. To develop an Finite Difference impedance/admittance transformation algorithm [6, 7] we use the expressions in (20.3).

20.2.7 *Matching at Interface Between Concatenated Regions*

In this subsection we would like describe especially the matching at the interface between the regions III and II+IV in Fig. 20.6. We mark the interfaces in the following way: region III at $\phi^{\mathrm{III}} = 0$ by → A, region II at $\phi^{\mathrm{II}} = \phi_{\mathrm{B}}$ by → B and region IV at $\phi^{\mathrm{IV}} = \phi_{\mathrm{A}}$ by → C.

Matching of the electric and magnetic fields (vectors of the discretized fields) results in the following equations ($\mathrm{U} = \mathrm{A}, \mathrm{B}, \mathrm{C}$)

$$\begin{aligned}\mathbf{E}_\mathrm{U} &= \boldsymbol{Z}_\mathrm{U}\widetilde{\mathbf{H}}_\mathrm{U} \\ \widetilde{\mathbf{H}}_\mathrm{U} &= \boldsymbol{Y}_\mathrm{U}\mathbf{E}_\mathrm{U}\end{aligned} \qquad \text{where} \qquad \mathbf{E}_\mathrm{A} = \begin{bmatrix}\mathbf{E}_\mathrm{B} \\ \mathbf{E}_\mathrm{C}\end{bmatrix} \qquad \widetilde{\mathbf{H}}_\mathrm{A} = \begin{bmatrix}\widetilde{\mathbf{H}}_\mathrm{B} \\ \widetilde{\mathbf{H}}_\mathrm{C}\end{bmatrix} \tag{20.10}$$

We subdivide the matrix $\boldsymbol{Y}_\mathrm{A}$ corresponding to the regions B and C and may write by using (20.10)

$$\widetilde{\mathbf{H}}_\mathrm{A} = \begin{bmatrix}\widetilde{\mathbf{H}}_\mathrm{B} \\ \widetilde{\mathbf{H}}_\mathrm{C}\end{bmatrix} = \boldsymbol{Y}_\mathrm{A}\mathbf{E}_\mathrm{A} = \begin{bmatrix}\boldsymbol{Y}^\mathrm{A}_\mathrm{BB} & \boldsymbol{Y}^\mathrm{A}_\mathrm{BC} \\ \boldsymbol{Y}^\mathrm{A}_\mathrm{CB} & \boldsymbol{Y}^\mathrm{A}_\mathrm{CC}\end{bmatrix}\begin{bmatrix}\boldsymbol{Z}_\mathrm{B}\widetilde{\mathbf{H}}_\mathrm{B} \\ \mathbf{E}_\mathrm{C}\end{bmatrix} \tag{20.11}$$

By calculating $\widetilde{\mathbf{H}}_\mathrm{B}$ from the first equation in this system and introducing this quantity into the second equation of the system we obtain

$$\widetilde{\mathbf{H}}_\mathrm{C} = \left(\boldsymbol{Y}^\mathrm{A}_\mathrm{CB}\boldsymbol{Z}_\mathrm{B}\left(\boldsymbol{I} - \boldsymbol{Y}^\mathrm{A}_\mathrm{BB}\boldsymbol{Z}_\mathrm{B}\right)^{-1}\boldsymbol{Y}^\mathrm{A}_\mathrm{BC} + \boldsymbol{Y}^\mathrm{A}_\mathrm{CC}\right)\mathbf{E}_\mathrm{C} \tag{20.12}$$

and therefore the following admittance matrix at interface part C

$$\boldsymbol{Y}_\mathrm{C} = \boldsymbol{Y}^\mathrm{A}_\mathrm{CC} - \boldsymbol{Y}^\mathrm{A}_\mathrm{CB}\left(\boldsymbol{Y}^\mathrm{A}_\mathrm{BB} - \boldsymbol{Y}_\mathrm{B}\right)^{-1}\boldsymbol{Y}^\mathrm{A}_\mathrm{BC} \tag{20.13}$$

Alternative by using

$$\mathbf{E}_\mathrm{A} = \begin{bmatrix}\mathbf{E}_\mathrm{B} \\ \mathbf{E}_\mathrm{C}\end{bmatrix} = \boldsymbol{Z}_\mathrm{A}\widetilde{\mathbf{H}}_\mathrm{A} = \begin{bmatrix}\boldsymbol{Z}^\mathrm{A}_\mathrm{BB} & \boldsymbol{Z}^\mathrm{A}_\mathrm{BC} \\ \boldsymbol{Z}^\mathrm{A}_\mathrm{CB} & \boldsymbol{Z}^\mathrm{A}_\mathrm{CC}\end{bmatrix}\begin{bmatrix}\boldsymbol{Y}_\mathrm{B}\mathbf{E}_\mathrm{B} \\ \widetilde{\mathbf{H}}_\mathrm{C}\end{bmatrix} \tag{20.14}$$

we obtain the following impedance matrix at interface part C

$$\boldsymbol{Z}_\mathrm{C} = \boldsymbol{Z}^\mathrm{A}_\mathrm{CC} - \boldsymbol{Z}^\mathrm{A}_\mathrm{CB}\left(\boldsymbol{Z}^\mathrm{A}_\mathrm{BB} - \boldsymbol{Z}_\mathrm{B}\right)^{-1}\boldsymbol{Z}^\mathrm{A}_\mathrm{BC} \tag{20.15}$$

Because r and ϕ in region IV are in opposite direction to those in II and III we obtain e.g. the load impedance Matrix $\boldsymbol{Z}^{IV}_\mathrm{B}$ for region IV by multiplication with the exchange matrix $\boldsymbol{J}$ (rotated identity matrix)

$$\boldsymbol{Z}^{IV}_\mathrm{B} = \boldsymbol{J}\boldsymbol{Z}_\mathrm{C}\boldsymbol{J} \qquad \boldsymbol{Y}^{IV}_\mathrm{B} = \boldsymbol{J}\boldsymbol{Y}_\mathrm{C}\boldsymbol{J} \tag{20.16}$$

20.2.8 Impedance/Admittance-Transformation in ϕ-Direction

We perform the analysis with impedance/admittance transformation. Therefore, as usual we develop the transformation formulas from the GTL-equations (20.3). In the inhomogeneous sections (in propagation direction) we obtain from these GTL-equations (in 2D-case (20.4), (20.5) and (20.7), (20.8)) an FD impedance/admittance transformation algorithm analogous to [6,7]. The (20.3) can be combined to

$$\frac{\mathrm{d}}{\mathrm{d}\overline{u}}\widehat{\mathbf{F}} = \widehat{\boldsymbol{Q}}\widehat{\mathbf{F}} \qquad \widehat{\boldsymbol{Q}} = \begin{bmatrix} \boldsymbol{0} & -\mathrm{j}\boldsymbol{R}_{\mathrm{H}} \\ -\mathrm{j}\boldsymbol{R}_{\mathrm{E}} & \boldsymbol{0} \end{bmatrix}\widehat{\mathbf{F}} \qquad \widehat{\mathbf{F}} = \begin{bmatrix} \widehat{\mathbf{E}} \\ \widehat{\mathbf{H}} \end{bmatrix} \qquad \overline{u} = \overline{r}_0\phi \tag{20.17}$$

To describe the impedance/admittance transformation in u-direction by an FD algorithm we replace the first derivatives with respect to $\overline{u}$ by central differences between planes A and B and the right side by an arithmetic mean value and obtain:

$$\frac{\mathrm{d}}{\mathrm{d}\overline{u}}\widehat{\mathbf{F}}_{\mathrm{S}} = \widehat{\boldsymbol{Q}}\widehat{\mathbf{F}} \longrightarrow \frac{\widehat{\mathbf{F}}_{\mathrm{B}} - \widehat{\mathbf{F}}_{\mathrm{A}}}{\Delta\overline{u}} = \widehat{\boldsymbol{Q}}(\overline{u}_m)\frac{\widehat{\mathbf{F}}_{\mathrm{B}} + \widehat{\mathbf{F}}_{\mathrm{A}}}{2} \qquad \overline{u}_m = 0.5(\overline{u}_{\mathrm{A}} + \overline{u}_{\mathrm{B}}) \tag{20.18}$$

The subscripts A and B mark cross-sections A and B for which we would like to calculate the fields. $\Delta\overline{u}$ is the normalized azimuthal distance between them. With (20.18) we approximate the differential equation between the two cross-sections. Assuming this cross-section is centered between A and B then the approximation of the left side in (20.18) is of second order accuracy. If a cross-section is not centered between A and B a third term must be added on the left side of the equation. The right side of the approximate equation in (20.18) is only of first order (or linear) approximation. To obtain the same order of accuracy we must use three terms [6]. Now we define $\widehat{\boldsymbol{Q}}_m, \boldsymbol{R}_{\mathrm{E}}^m$ and $\boldsymbol{R}_{\mathrm{H}}^m$ as

$$\widehat{\boldsymbol{Q}}_m = 0.5\Delta\overline{u}\widehat{\boldsymbol{Q}}(\overline{u}_m) \qquad \boldsymbol{R}_{\mathrm{E}}^m = 0.5\Delta\overline{u}\boldsymbol{R}_{\mathrm{E}}(\overline{u}_m) \qquad \boldsymbol{R}_{\mathrm{H}}^m = 0.5\Delta\overline{u}\boldsymbol{R}_{\mathrm{H}}(\overline{u}_m) \tag{20.19}$$

and obtain from (20.18)

$$\widehat{\mathbf{F}}_{\mathrm{A}} = \left(\widehat{\boldsymbol{I}} + \widehat{\boldsymbol{Q}}_m\right)^{-1}\left(\widehat{\boldsymbol{I}} - \widehat{\boldsymbol{Q}}_m\right)\widehat{\mathbf{F}}_{\mathrm{B}} \qquad \widehat{\mathbf{F}}_{\mathrm{B}} = \left(\widehat{\boldsymbol{I}} - \widehat{\boldsymbol{Q}}_m\right)^{-1}\left(\widehat{\boldsymbol{I}} + \widehat{\boldsymbol{Q}}_m\right)\widehat{\mathbf{F}}_{\mathrm{A}} \tag{20.20}$$

or

$$\widehat{\mathbf{F}}_{\mathrm{A}} = \begin{bmatrix} \boldsymbol{I} & -\mathrm{j}\boldsymbol{R}_{\mathrm{H}}^m \\ -\mathrm{j}\boldsymbol{R}_{\mathrm{E}}^m & \boldsymbol{I} \end{bmatrix}^{-1}\begin{bmatrix} \boldsymbol{I} & \mathrm{j}\boldsymbol{R}_{\mathrm{H}}^m \\ \mathrm{j}\boldsymbol{R}_{\mathrm{E}}^m & \boldsymbol{I} \end{bmatrix}\widehat{\mathbf{F}}_{\mathrm{B}} \tag{20.21}$$

$$\widehat{\mathbf{F}}_{\mathrm{B}} = \begin{bmatrix} \boldsymbol{I} & \mathrm{j}\boldsymbol{R}_{\mathrm{H}}^m \\ \mathrm{j}\boldsymbol{R}_{\mathrm{E}}^m & \boldsymbol{I} \end{bmatrix}^{-1}\begin{bmatrix} \boldsymbol{I} & -\mathrm{j}\boldsymbol{R}_{\mathrm{H}}^m \\ -\mathrm{j}\boldsymbol{R}_{\mathrm{E}}^m & \boldsymbol{I} \end{bmatrix}\widehat{\mathbf{F}}_{\mathrm{A}} \tag{20.22}$$

The inversion of the matrix on the left side can be simplified. For matrix expressions like that we have

$$\begin{bmatrix} a\boldsymbol{I} & \boldsymbol{A} \\ \boldsymbol{B} & a\boldsymbol{I} \end{bmatrix} \longrightarrow \begin{bmatrix} a\boldsymbol{I} & \boldsymbol{A} \\ \boldsymbol{B} & a\boldsymbol{I} \end{bmatrix}^{-1} = \begin{bmatrix} a\boldsymbol{I} & -\boldsymbol{A} \\ -\boldsymbol{B} & a\boldsymbol{I} \end{bmatrix}\begin{bmatrix} (a^2\boldsymbol{I} - \boldsymbol{A}\boldsymbol{B})^{-1} & \boldsymbol{0} \\ \boldsymbol{0} & (a^2\boldsymbol{I} - \boldsymbol{B}\boldsymbol{A})^{-1} \end{bmatrix} \tag{20.23}$$

with $a = 1$ in our case. The order of the matrix product on the right side can also be changed. Therefore, we obtain instead of the (20.22) for transformation from port B to port A or from port A to port B, respectively

$$\widehat{\mathbf{F}}_{\mathrm{A}} = \begin{bmatrix} (\boldsymbol{I}+\boldsymbol{R}_{\mathrm{H}}^{m}\boldsymbol{R}_{\mathrm{E}}^{m})^{-1} & \boldsymbol{0} \\ \boldsymbol{0} & (\boldsymbol{I}+\boldsymbol{R}_{\mathrm{E}}^{m}\boldsymbol{R}_{\mathrm{H}}^{m})^{-1} \end{bmatrix} \begin{bmatrix} (\boldsymbol{I}-\boldsymbol{R}_{\mathrm{H}}^{m}\boldsymbol{R}_{\mathrm{E}}^{m}) & \mathrm{j}2\boldsymbol{R}_{\mathrm{H}}^{m} \\ \mathrm{j}2\boldsymbol{R}_{\mathrm{E}}^{m} & (\boldsymbol{I}-\boldsymbol{R}_{\mathrm{E}}^{m}\boldsymbol{R}_{\mathrm{H}}^{m}) \end{bmatrix} \widehat{\mathbf{F}}_{\mathrm{B}} \quad (20.24)$$

$$\widehat{\mathbf{F}}_{\mathrm{B}} = \begin{bmatrix} (\boldsymbol{I}+\boldsymbol{R}_{\mathrm{H}}^{m}\boldsymbol{R}_{\mathrm{E}}^{m})^{-1} & \boldsymbol{0} \\ \boldsymbol{0} & (\boldsymbol{I}+\boldsymbol{R}_{\mathrm{E}}^{m}\boldsymbol{R}_{\mathrm{H}}^{m})^{-1} \end{bmatrix} \begin{bmatrix} (\boldsymbol{I}-\boldsymbol{R}_{\mathrm{H}}^{m}\boldsymbol{R}_{\mathrm{E}}^{m}) & -\mathrm{j}2\boldsymbol{R}_{\mathrm{H}}^{m} \\ -\mathrm{j}2\boldsymbol{R}_{\mathrm{E}}^{m} & (\boldsymbol{I}-\boldsymbol{R}_{\mathrm{E}}^{m}\boldsymbol{R}_{\mathrm{H}}^{m}) \end{bmatrix} \widehat{\mathbf{F}}_{\mathrm{A}} \quad (20.25)$$

In shorter form we may write for these two equation systems

$$\begin{bmatrix} \mathbf{E}_{\mathrm{A}} \\ \mathbf{H}_{\mathrm{A}} \end{bmatrix} = \begin{bmatrix} \boldsymbol{V}_{\mathrm{E}} & \mathrm{j}\boldsymbol{X}_{\mathrm{H}} \\ \mathrm{j}\boldsymbol{Y}_{\mathrm{E}} & \boldsymbol{V}_{\mathrm{H}} \end{bmatrix} \begin{bmatrix} \mathbf{E}_{\mathrm{B}} \\ \mathbf{H}_{\mathrm{B}} \end{bmatrix} \qquad \begin{bmatrix} \mathbf{E}_{\mathrm{B}} \\ \mathbf{H}_{\mathrm{B}} \end{bmatrix} = \begin{bmatrix} \boldsymbol{V}_{\mathrm{E}} & -\mathrm{j}\boldsymbol{X}_{\mathrm{H}} \\ -\mathrm{j}\boldsymbol{Y}_{\mathrm{E}} & \boldsymbol{V}_{\mathrm{H}} \end{bmatrix} \begin{bmatrix} \mathbf{E}_{\mathrm{A}} \\ \mathbf{H}_{\mathrm{A}} \end{bmatrix} \quad (20.26)$$

At the ports A and B we may write the relations

$$\mathbf{E}_{\mathrm{A,B}} = \mathrm{j}\boldsymbol{X}_{\mathrm{A,B}}\mathbf{H}_{\mathrm{A,B}} \qquad \mathbf{H}_{\mathrm{A,B}} = \mathrm{j}\boldsymbol{Y}_{\mathrm{A,B}}\mathbf{E}_{\mathrm{A,B}} \quad (20.27)$$

The impedance/admittance transformation between them is then given by

$$\boldsymbol{X}_{\mathrm{A}} = (\boldsymbol{V}_{\mathrm{E}}\boldsymbol{X}_{\mathrm{B}}+\boldsymbol{X}_{\mathrm{H}})\,(\boldsymbol{V}_{\mathrm{H}}-\boldsymbol{Y}_{\mathrm{E}}\boldsymbol{X}_{\mathrm{B}})^{-1} \quad (20.28)$$

$$\boldsymbol{Y}_{\mathrm{A}} = (\boldsymbol{V}_{\mathrm{H}}\boldsymbol{Y}_{\mathrm{B}}+\boldsymbol{Y}_{\mathrm{E}})\,(\boldsymbol{V}_{\mathrm{E}}-\boldsymbol{X}_{\mathrm{H}}\boldsymbol{Y}_{\mathrm{B}})^{-1} \quad (20.29)$$

$$\boldsymbol{X}_{\mathrm{B}} = (\boldsymbol{V}_{\mathrm{E}}\boldsymbol{X}_{\mathrm{A}}-\boldsymbol{X}_{\mathrm{H}})\,(\boldsymbol{V}_{\mathrm{H}}+\boldsymbol{Y}_{\mathrm{E}}\boldsymbol{X}_{\mathrm{A}})^{-1} \quad (20.30)$$

$$\boldsymbol{Y}_{\mathrm{B}} = (\boldsymbol{V}_{\mathrm{H}}\boldsymbol{Y}_{\mathrm{A}}-\boldsymbol{Y}_{\mathrm{E}})\,(\boldsymbol{V}_{\mathrm{E}}+\boldsymbol{X}_{\mathrm{H}}\boldsymbol{Y}_{\mathrm{A}})^{-1} \quad (20.31)$$

For open (OC), or alternatively, short (SC) circuiting the ports A and B we obtain

$$\text{SC:}\quad \boldsymbol{X}_{\mathrm{B}} = \boldsymbol{0} \longrightarrow \boldsymbol{X}_{\mathrm{A}} = \boldsymbol{X}_{\mathrm{H}}\boldsymbol{V}_{\mathrm{H}}^{-1} \qquad \boldsymbol{X}_{\mathrm{A}} = \boldsymbol{0} \longrightarrow \boldsymbol{X}_{\mathrm{B}} = -\boldsymbol{X}_{\mathrm{H}}\boldsymbol{V}_{\mathrm{H}}^{-1} \quad (20.32)$$

$$\text{OC:}\quad \boldsymbol{Y}_{\mathrm{B}} = \boldsymbol{0} \longrightarrow \boldsymbol{Y}_{\mathrm{A}} = \boldsymbol{Y}_{\mathrm{E}}\boldsymbol{V}_{\mathrm{E}}^{-1} \qquad \boldsymbol{Y}_{\mathrm{A}} = \boldsymbol{0} \longrightarrow \boldsymbol{Y}_{\mathrm{B}} = -\boldsymbol{Y}_{\mathrm{E}}\boldsymbol{V}_{\mathrm{E}}^{-1} \quad (20.33)$$

20.3 Numerical Results

Results for the reflectivity $|S_{11}|^2$ and transmittivity $|S_{21}|^2$ for the fibre Bragg grating in Fig. 20.1 – analyzed with the described algorithms – are shown in Fig. 20.12a, b for two different period sections and $M = 200$ periods [4].

For the analysis of resonator structures we first would like to check our algorithm for impedance/admittance transformation with finite differences in cylindrical coordinates. To do this we use a straight film waveguide as in Fig. 20.13b for which we can easily calculate the eigenmodes by the MoL. Between the ports P_1 and P_2 we have the symmetry plane S where we may again introduce an electric or a magnetic wall. We obtain the whole behavior of the device from the input impedances of these two cases. The section between the port P_1 and the symmetry plane S is divided into two inhomogeneous cylindrical sections S_1 and S_2. Then, we transform the impedances from the symmetry plane trough these sections by the described algorithm to the port 1. From the input impedances we can calculate all other quantities.

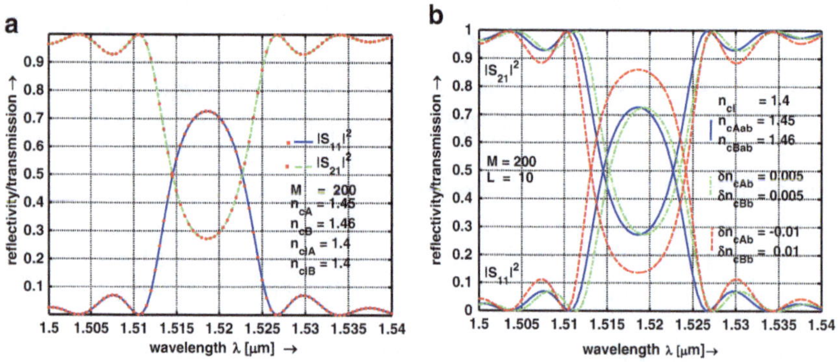

Fig. 20.12 Reflectivity of the Fibre Bragg-gratings in Fig. 20.1 as function of the wavelength with symmetrical periods of homogeneous (**a**) and periodic (**b**) sections. $L = 10$ is the number of periods in the sections

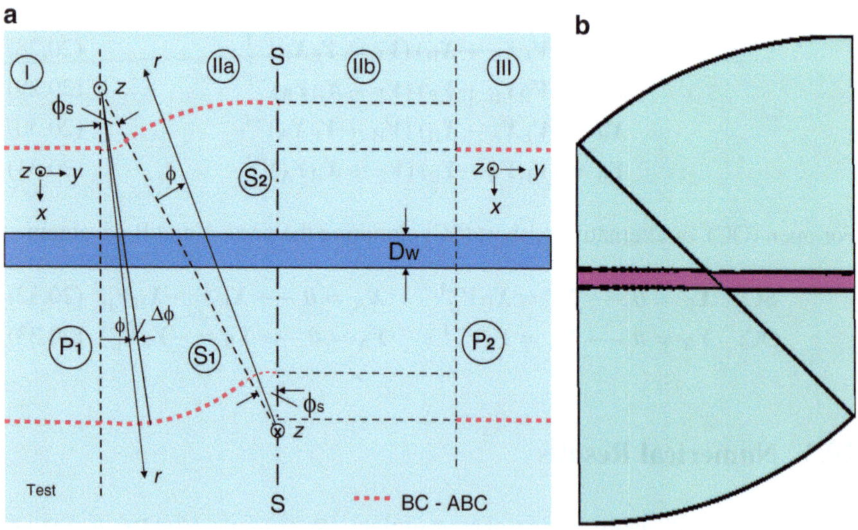

Fig. 20.13 (**a**) Test structure: propagation analysis between ports P_1 and P_2 by cylindrical FD algorithm. (**b**) Structure constructed by permittivity matrices

By introducing the field of the fundamental mode at the port P_1 we obtain the field at the output port P_2. These fields are shown in Fig. 20.14. The output fields are in best agreement with the input fields. Therefore, the distributions are not distinguishable. For these calculations we have divided the sections S_1 and S_2 each into 50 subsection and we used 296/297 discretization points. The curves for reflectance and transmittances are shown in Fig. 20.15. The numerical reflections are small enough for practical purposes. The parameters S_{11} and S_{21} are for the fundamental mode only. Since we excite higher order modes as well, the relation $|S_{11}|^2 + |S_{21}|^2 = 1$ is usually not fulfilled even in lossless structures.

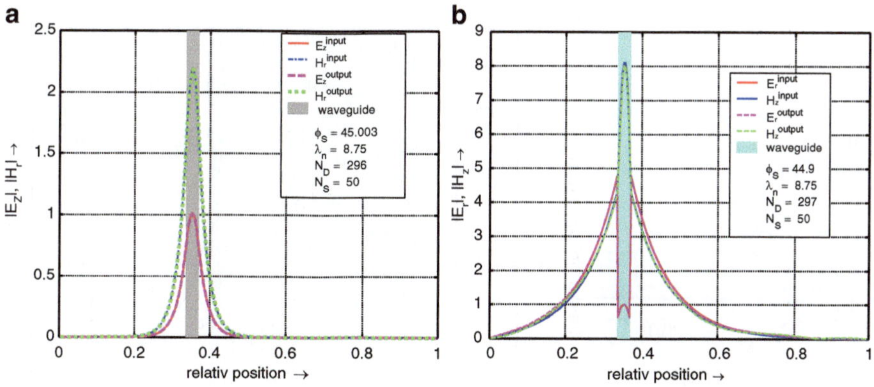

Fig. 20.14 Field distribution at input and output ports for the test structure: (**a**) TE (**b**) TM

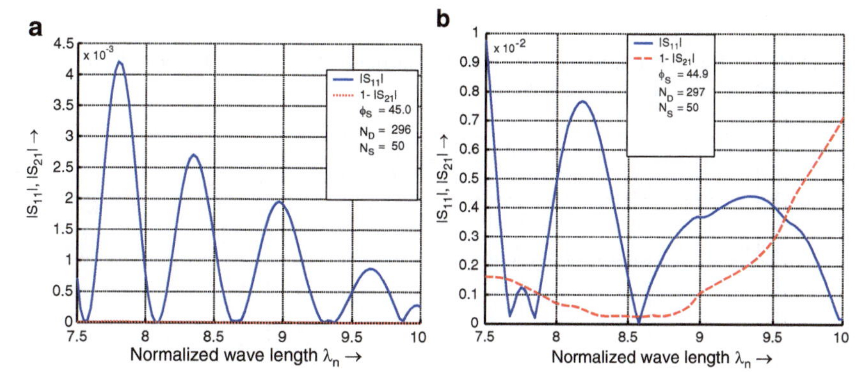

Fig. 20.15 Scattering parameters between input and output ports for the test structure: (**a**) TE (**b**) TM

Results for the field distribution in the ring part II and the scattering parameters $|S_{\mathrm{ik}}|^2$ for the structure in Fig. 20.6 – analyzed with the algorithms described – are shown in Fig. 20.16a, b and in Fig. 20.17 [8]. The ring resonator and the coupled waveguides are assumed to be homogeneous 2-D structures (i.e. non-periodic).

20.4 Other Structures

The described combination of the MoL with the impedance/admittance transformation in different coordinate systems can also be used for the analysis of many other structures like the mirror in Fig. 20.18a [9]. In sections II and III the special FD algorithms in Cartesian- and in section IV in cylindrical-coordinates will be used. An other example is the output of a coupler with the connecting lines in Fig. 20.18b.

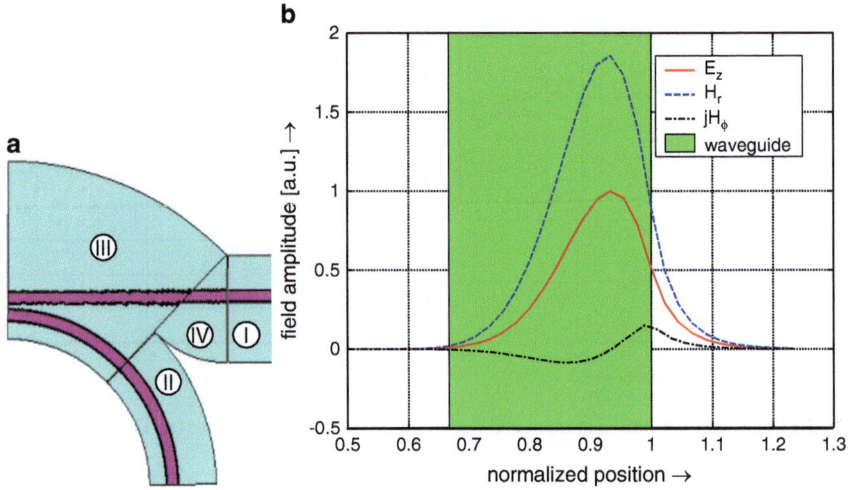

Fig. 20.16 (**a**) Resonator structure constructed by permittivity matrices (**b**) Field distribution in the homogeneous part II of the resonator in Fig. 20.6

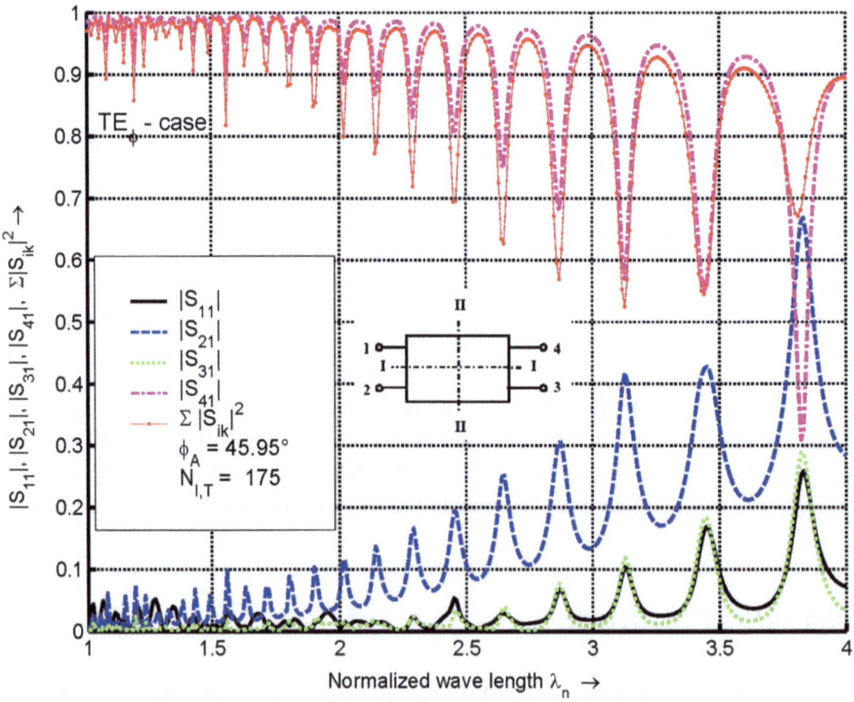

Fig. 20.17 Scattering parameters between the ports of the structure in Fig. 20.6 as function of the normalized wavelength

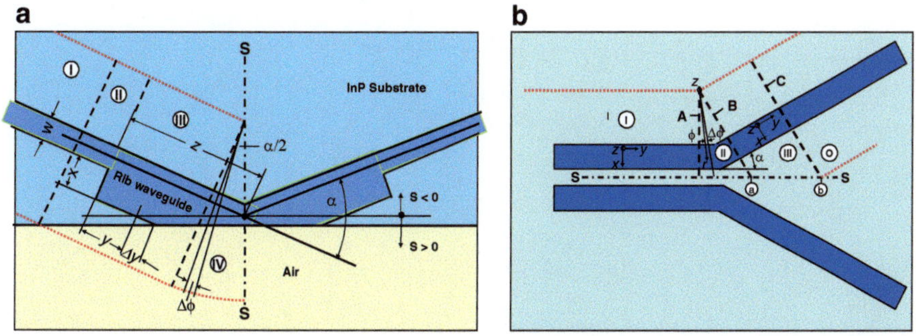

Fig. 20.18 Geometry and principle for the analysis of the (**a**) Self-aligned waveguide total internal reflection(TIR) mirror according to [9]. (**b**) Output of a microstrip or rib waveguide coupler with connecting lines

In section II the FD algorithm in cylindrical-coordinates will be used. For sections III, however, we use the special FD algorithm in Cartesian-coordinates.

Acknowledgements The author would like to acknowledge S. F. Helfert for the help in preparing this paper.

References

1. R. Pregla, Efficient modeling of periodic structures. Int. J. Electron. Commun. (AEÜ), **57**(3), 185–189 (2003)
2. R. Pregla, Analysis of gratings with symmetrical and unsymmetrical periods. in *6th International Conference on Transparent Optical Networks*, Wrozlaw, July 2004
3. R. Pregla, Modeling of optical waveguide structures with general anisotropy in arbitrary orthogonal coordinate systems. IEEE J. Sel. Top. Quantum Electron. **8**(6), 1217–1224 (2002)
4. R. Pregla, Analysis of complex photonic structures. in *6th International Conference on Transparent Optical Networks*, Roma, June 2007
5. R. Pregla, *Analysis of Electromagnetic Fields and Waves – The Method of Lines* (Wiley, Chichester, 2008)
6. R. Pregla, Modeling of optical waveguides and devices by combination of the method of lines and finite differences of second order accuracy. Opt. Quantum Electron. **38**, 3–17 (2006)
7. R. Pregla, Analysis of microwave structures by combination of the method of lines and finite differences. *MIKON 2006*, 08.-09.04.2006, Cracow
8. R. Pregla, Analysis of general optical resonator structures coupled by straight waveguides. in *6th International Conference on Transparent Optical Networks*, ICTON, Athens, 22–26 June 2008
9. J. Ctyroky et al., Modelling of self-aligned total internal reflection waveguide mirrors: an interlaboratory comparison. *Opt. Quantum Electron.* **27**, 935–942 (1995)

Chapter 21
Macromodeling in Finite Differences

Lukasz Kulas and Michal Mrozowski

21.1 Introduction

Photonic crystals, metamaterials and electromagnetic band gap structures are among those problems where fine geometrical details play an essential role in shaping the response of an electromagnetic system. In order to design and investigate circuits involving such materials new computational tools have to be developed. Traditional and versatile techniques of computational electromagnetics such as finite differences with Yee's mesh or finite elements, are capable of handling complex geometries encountered in new materials. However, they do not perform well in problems involving objects of different scale. In these methods, large systems of linear equations arise and to solve them one often has to use iterative algorithms whose convergence depends on a norm of a matrix that results from the discretization of Maxwell's equations. This norm increases as the mesh becomes finer. In consequence, if one uses a very dense mesh to resolve fine geometrical details, the convergence rate drops drastically. This is true even if a dense mesh is used only in a small region inside the computational space. As a rule of thumb, one may say that, when the mesh density increases locally by the factor of k, the solution time increases at least k times. A similar problem occurs in the time domain formulation e.g. in the Finite Difference Time Domain scheme. Due to the explicit character of the time marching scheme the increasing norm entails the reduction of the time step that is proportional to the refinement factor. As a result, resolving geometrical details by refining a mesh requires a higher number of iterations to reach the steady state than it would have to be performed in standard ten-cells-for-a-wavelength computations.

L. Kulas (✉) and M. Mrozowski
Gdansk University of Technology, Faculty of Electronics, Telecommunications and Informatics, Department of Microwave and Antenna Engineering, WiComm Centre of Excellence,
ul. Narutowicza 11/12, 80–233, Gdansk, Poland
e-mail: luke@eti.pg.gda.pl, m.mrozowski@ieee.org

S. Lindenmeier and R. Weigel (eds.), *Electromagnetics and Network Theory and their Microwave Technology Applications*, DOI 10.1007/978-3-642-18375-1_21,

Macromodeling is a technique that can be used to overcome the shortcomings of the finite difference or finite element schemes related to high mesh density. A macromodel is a technique of encapsulating a complex behaviour of a system in a form of a compact matrix-valued transfer function. Macromodeling is widely used in VLSI design to characterize interconnects, but this technique remains virtually unexplored in computational electromagnetics. Macromodels for finite difference schemes employing Yee's grid can be constructed by applying the concept of reduced order modelling to Maxwell's grid equations in a selected region of the computational space covered with a very fine mesh [1–4]. Macromodels are then incorporated into a standard mesh formulation. Model order reduction applied to the equations defined for a fine mesh eliminates most of the internal state variables and, at the same time, reduces the operator norm. Consequently, iterative solvers converge faster and what is even more important, longer time steps can be used in time domain schemes. Indeed several papers have demonstrated [1, 3, 5] that macromodels operate with a much longer time step than the standard scheme with an equivalent spatial resolution.

Usually, model order reduction is applied only to the equations defined for a fine mesh. Model order reduction techniques eliminate most of the internal state variables, so when a macromodel is incorporated into the standard mesh algorithm, highly accurate results are obtained with low memory and CPU cost [3]. Up till now, macromodels have been demonstrated for various grid based methods including 2-dimensional FEM [6], 1D, 2D, 3D FDTD [1, 3, 5, 7, 8] and 2D, 3D FDFD [5, 8, 9]. This chapter reviews progress in this area that has been achieved at the computational electromagnetics group at Gdansk University of Technology.

21.2 Creation of a Macromodel in Finite Difference Equations

To understand the concept of a macromodel in the context of Finite Difference algorithms, let us consider a coarse 2D Yee's mesh [10] having E_x, E_y and H_z field components. Let us assume that a certain area inside the mesh is refined to increase the accuracy locally. This means that the original coarse mesh is removed and then the area is discretized using finer mesh as shown in Fig. 21.1.

Model order reduction techniques have originally been developed for approximating the transfer function of a dynamical system around a certain frequency of interest. Therefore a set of first order differential equations that will be subjected to reduction has to be written in the Laplace s-domain. For the time domain algorithms, like FDTD, this can be readily achieved by starting from Maxwell's grid equations [11] rather than a traditional leap-frog scheme. Maxwell's grid equations in the Laplace s-domain ($s = j\omega = \frac{\partial}{\partial t}$) written for a finely meshed region take up the following form:

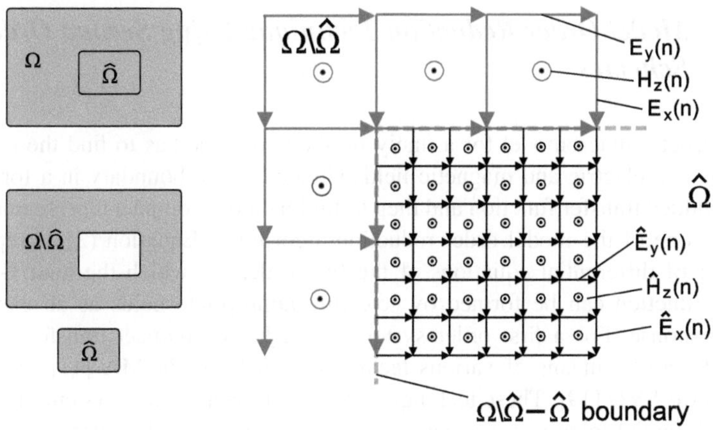

Fig. 21.1 Two-dimensional computational domain Ω with a local subdomain $\widehat{\Omega}$ and corresponding 2D Yee's mesh used in Finite Differences for TE polarization (mesh refinement factor equals 3)

$$\begin{aligned}\widehat{\mathbf{e}}_b + \widehat{\mathbf{R}}_E\widehat{\mathbf{e}} &= -s\widehat{\mathbf{D}}_\mu\widehat{\mathbf{h}} \\ \widehat{\mathbf{R}}_H\widehat{\mathbf{h}} &= s\widehat{\mathbf{D}}_\varepsilon\widehat{\mathbf{e}}\end{aligned} \tag{21.1}$$

where $\widehat{\mathbf{R}}_E, \widehat{\mathbf{R}}_H = \widehat{\mathbf{R}}_E^T$ and $\widehat{\mathbf{D}}_\mu, \widehat{\mathbf{D}}_\varepsilon$ are discrete curl operators and diagonal material matrices, respectively and the hat denotes a matrix or a vector defined on the fine mesh (FM). Similar equations can be formed for the coarse mesh (CM):

$$\begin{aligned}\mathbf{R}_E\mathbf{e} &= -s\mathbf{D}_\mu\mathbf{h} \\ \mathbf{h}_b + \mathbf{R}_H\mathbf{h} &= s\mathbf{D}_\varepsilon\mathbf{e}\end{aligned} \tag{21.2}$$

Boundary vectors $\widehat{\mathbf{e}}_b$ and $\mathbf{h}_b$ link the two regions by providing the boundary conditions at the CM-FM interface.

Note that meshes with different densities are employed. It is obvious then that interpolation of field values between grids has to be performed. The interpolation techniques for coupling Yee's grids with different densities is a research subject of its own but many proposed schemes use a linear interpolation [12–16]. For linear interpolation the algorithm for computing $\widehat{\mathbf{e}}_b$ and $\mathbf{h}_b$ can be written in a matrix form:

$$\widehat{\mathbf{e}}_b = \widehat{\mathbf{B}}_E\mathbf{I}_E\mathbf{L}_E \cdot \mathbf{e} \qquad \mathbf{h}_b = \mathbf{B}_H\mathbf{I}_H\widehat{\mathbf{L}}_H \cdot \widehat{\mathbf{h}} \tag{21.3}$$

where $\mathbf{L}_E$, $\widehat{\mathbf{L}}_H$ choose CM and FM fields to interpolate, $\mathbf{I}_E$ and $\mathbf{I}_H$ are matrices containing interpolation coefficients and $\widehat{\mathbf{B}}_E$, $\mathbf{B}_H$ place interpolated fields in FM and CM, respectively.

21.2.1 Model Order Reduction Technique Using Second Order Schemes

To construct a macromodel for a finely meshed area one has to find the relationship between electric and magnetic fields situated at the boundary in a form of a matrix-valued transfer function and then to find its more compact representation by applying one of the model order reduction algorithms. Equation (21.1) represent a system of differential equations of the first order for which the matrix-valued transfer function can be formed by selecting appropriate fields as an excitation and a response. For a first order system the reduced compact transfer function can be found by means of various techniques, such as, PRIMA [17], PVL [18] or Laguerre-SVD [19]. These techniques, however, operate on a system of the first order differantial equations and require inversion (or rather a LU factorisation) of the system matrix. For grid based methods this may become a problem in three dimensions as the number of the equations in system (21.1) may be high. To increase the efficiency of model order reduction process one has to reduce the number of equations by eliminating the electric or magnetic field from Maxwell's equations. With a proper normalisation this transforms (21.1) to a symmetric and (semi) positive definite system of the second order. PRIMA, PVL or Laguerre-SVD can not handle such systems, however other techniques like ENOR [20], SAPOR [21] or SMOR [22] are available for second order systems. These algorithms operate on half as many variables as PRIMA or PVL and also involve Cholesky decomposition which makes them much faster and therefore preferable when macromodels for grid based methods are constructed. Tests carried out in [23] showed that there are no significant differences in the accuracy obtained with all there second order MOR schemes i.e. ENOR, SAPOR, SMOR. For this reason, the ENOR algorithm will be used in all examples presented in this chapter.

The matrix-valued transfer function appropriate for reduction using the second order reduction scheme ENOR can be obtained by transforming (21.1) into the form:

$$\widehat{\mathbf{h}}_M = \widehat{\mathbf{L}}^T \left(\frac{1}{s} \widehat{\Gamma} + s \widehat{\mathbf{C}} \right)^{-1} \widehat{\mathbf{B}} \cdot \widehat{\mathbf{e}}_M = \widehat{\mathbf{H}}(s) \cdot \widehat{\mathbf{e}}_M \tag{21.4}$$

where $\widehat{\Gamma} = \widehat{\mathbf{R}}_E \widehat{\mathbf{D}}_\varepsilon^{-1} \widehat{\mathbf{R}}_E^T$ and $\widehat{\mathbf{C}} = \widehat{\mathbf{D}}_\mu$ are symmetric positive semidefinite matrices, $\widehat{\mathbf{L}}^T = \widehat{\mathbf{L}}_H$ is a matrix choosing FM's magnetic fields. Subscript M refers to the excitation and response of the macromodel, so $\widehat{\mathbf{h}}_M$ denotes the system response and $\widehat{\mathbf{e}}_M$ is a vector containing fine mesh excitation fields. Matrix $\widehat{\mathbf{B}} = -\widehat{\mathbf{B}}_E$ places excitation fields in the system in such a way that they form the boundary vector $\widehat{\mathbf{e}}_b = -\widehat{\mathbf{B}} \cdot \widehat{\mathbf{e}}_M$ in (21.1).

ENOR applied to the system defined by (21.4) produces an orthonormal basis $\widehat{\mathbf{V}}$ which is then used to find the fine grid reduced matrix-valued transfer function $\widehat{\mathbf{H}}_m(s)$. This function is defined as

$$\widehat{\mathbf{h}}_M = \widehat{\mathbf{L}}_m^T \left(\frac{1}{s} \widehat{\Gamma}_m + s\widehat{\mathbf{C}}_m \right)^{-1} \widehat{\mathbf{B}}_m \cdot \widehat{\mathbf{e}}_M = \widehat{\mathbf{H}}_m(s) \cdot \widehat{\mathbf{e}}_M \tag{21.5}$$

where $\widehat{\mathbf{L}}_m = \widehat{\mathbf{V}}^T \widehat{\mathbf{L}}$, $\widehat{\Gamma}_m = \widehat{\mathbf{V}}^T \widehat{\Gamma} \widehat{\mathbf{V}}$, $\widehat{\mathbf{C}}_m = \widehat{\mathbf{V}}^T \widehat{\mathbf{C}} \widehat{\mathbf{V}}$, $\widehat{\mathbf{B}}_m = \widehat{\mathbf{V}}^T \widehat{\mathbf{B}}$. The reduced matrix-valued transfer function has the size $m \times m$, where $m = p \cdot q$, p is the number of system's input ports and q is the model order. $\widehat{\mathbf{H}}_m(s)$ approximates the fine grid transfer function $\widehat{\mathbf{H}}(s)$ in a limited frequency band which depends on q.

The most compact representation of the matrix-valued transfer function is obtained when the matrices selecting fields also perform interpolation. In other words $\mathbf{L}^T = \mathbf{I}_H \cdot \widehat{\mathbf{L}}_H$ and $\mathbf{B} = -\widehat{\mathbf{B}}_E \cdot \mathbf{I}_E$ have to be defined as choosing and excitation matrices. The mesh transfer function describing the relation between CM fields $\mathbf{e}_M$ and $\mathbf{h}_M$ at the CM-FM interface becomes

$$\mathbf{h}_M = \mathbf{L}^T \left(\frac{1}{s} \widehat{\Gamma} + s\widehat{\mathbf{C}} \right)^{-1} \mathbf{B} \cdot \mathbf{e}_M = \mathbf{H}(s) \cdot \mathbf{e}_M \tag{21.6}$$

21.3 *Finite Difference Frequency Domain* Analysis Using Macromodels

Having Maxwell's grid equations for a fine mesh written in a reduced form (21.6) one can incorporate them into the Finite Difference Frequency Domain (*FDFD*) formulation. For the example from Fig. 21.1 the computational domain consists of two regions. One region is covered by the Coarse Mesh (CM) and described using (21.1), while in the second one by the Fine Mesh (FM) with corresponding equations given by (21.2).

In order to plug a macromodel into the *FDFD* equations it is convenient to construct global Finite Difference operators in the s-domain in such a way that they describe fine and coarse mesh relations (21.1), (21.2) together with the interface dependence (21.3). This results in the following form of Maxwell's grid equations:

$$\begin{bmatrix} \mathbf{R}_E & 0 \\ \mathbf{S}_E & \widehat{\mathbf{R}}_E \end{bmatrix} \begin{bmatrix} \mathbf{e} \\ \widehat{\mathbf{e}} \end{bmatrix} = -s \begin{bmatrix} \mathbf{D}_\mu & 0 \\ 0 & \widehat{\mathbf{D}}_\mu \end{bmatrix} \begin{bmatrix} \mathbf{h} \\ \widehat{\mathbf{h}} \end{bmatrix} \tag{21.7}$$

$$\begin{bmatrix} \mathbf{R}_H & \mathbf{S}_H \\ 0 & \widehat{\mathbf{R}}_H \end{bmatrix} \begin{bmatrix} \mathbf{h} \\ \widehat{\mathbf{h}} \end{bmatrix} = s \begin{bmatrix} \mathbf{D}_\varepsilon & 0 \\ 0 & \widehat{\mathbf{D}}_\varepsilon \end{bmatrix} \begin{bmatrix} \mathbf{e} \\ \widehat{\mathbf{e}} \end{bmatrix} \tag{21.8}$$

where $\mathbf{R}_E$, $\mathbf{R}_H = \mathbf{R}_E^T$ and $\mathbf{D}_\mu$, $\mathbf{D}_\varepsilon$ are discrete curl operators and diagonal material matrices, the hat denotes local grid operators and fields. Matrices $\mathbf{S}_E = \widehat{\mathbf{B}}_E \mathbf{I}_E \mathbf{L}_E$ and $\mathbf{S}_H = \widehat{\mathbf{B}}_H \mathbf{I}_H \mathbf{L}_H$ are responsible for the coupling between the fields at the coarse to fine mesh grid interface. Equations (21.7) and (21.8) are a compact s-domain version of relations derived in [16].

To incorporate a macromodel into grid equations one has to begin with macromodel equations (21.6) that give the relationship between electric and magnetic fields on the perimeter of the coarse mesh in the form of a matrix-valued transfer function. Rather than using the transfer function directly one can apply projection to the discrete operators (21.7), (21.8). To this end, we left-multiply the second group of equations in (21.7) by $\widehat{\mathbf{V}}^T$:

$$\begin{bmatrix} \mathbf{R}_E & 0 \\ \widehat{\mathbf{V}}^T \mathbf{S}_E & \widehat{\mathbf{V}}^T \widehat{\mathbf{R}}_E \end{bmatrix} \begin{bmatrix} \mathbf{e} \\ \widehat{\mathbf{e}} \end{bmatrix} = -s \begin{bmatrix} \mathbf{D}_\mu & 0 \\ 0 & \widehat{\mathbf{V}}^T \widehat{\mathbf{D}}_\mu \end{bmatrix} \begin{bmatrix} \mathbf{h} \\ \widehat{\mathbf{h}} \end{bmatrix} \tag{21.9}$$

Projecting $\widehat{\mathbf{h}}$ on $\widehat{\mathbf{V}}$ and taking the advantage of the orthogonality of $\widehat{\mathbf{V}}$ we get the projected set of (21.7) and (21.8) in the form:

$$\begin{bmatrix} \mathbf{R}_E & 0 \\ \widehat{\mathbf{V}}^T \mathbf{S}_E & \widehat{\mathbf{V}}^T \widehat{\mathbf{R}}_E \end{bmatrix} \begin{bmatrix} \mathbf{e} \\ \widehat{\mathbf{e}} \end{bmatrix} = -s \begin{bmatrix} \mathbf{D}_\mu & 0 \\ 0 & \widehat{\mathbf{V}}^T \widehat{\mathbf{D}}_\mu \widehat{\mathbf{V}} \end{bmatrix} \begin{bmatrix} \mathbf{h} \\ \widehat{\mathbf{h}}_m \end{bmatrix} \tag{21.10}$$

$$\begin{bmatrix} \mathbf{R}_H & \mathbf{S}_H \widehat{\mathbf{V}} \\ 0 & \widehat{\mathbf{R}}_H \widehat{\mathbf{V}} \end{bmatrix} \begin{bmatrix} \mathbf{h} \\ \widehat{\mathbf{h}}_m \end{bmatrix} = s \begin{bmatrix} \mathbf{D}_\varepsilon & 0 \\ 0 & \widehat{\mathbf{D}}_\varepsilon \end{bmatrix} \begin{bmatrix} \mathbf{e} \\ \widehat{\mathbf{e}} \end{bmatrix} \tag{21.11}$$

with

$$\widehat{\mathbf{h}}_m = \widehat{\mathbf{V}}^T \widehat{\mathbf{h}} \tag{21.12}$$

Using the above set of equations one can easily solve any deterministic or eigenproblem written in a form of *FDFD* equations. For instance, algebraic transformations of (21.10), (21.11) and (21.12) lead to the *FDFD* eigenproblem with a reduced number of FM magnetic field samples:

$$\begin{bmatrix} \mathbf{R}_E \mathbf{D}_\varepsilon^{-1} & 0 \\ \widehat{\mathbf{V}}^T \mathbf{S}_E \mathbf{D}_\varepsilon^{-1} & \widehat{\mathbf{V}}^T \widehat{\mathbf{R}}_E \widehat{\mathbf{D}}_\varepsilon^{-1} \end{bmatrix} \begin{bmatrix} \mathbf{R}_H & \mathbf{S}_H \widehat{\mathbf{V}} \\ 0 & \widehat{\mathbf{R}}_H \widehat{\mathbf{V}} \end{bmatrix} \begin{bmatrix} \mathbf{h} \\ \widehat{\mathbf{h}}_m \end{bmatrix} = \omega^2 \begin{bmatrix} \mathbf{D}_\mu & 0 \\ 0 & \widehat{\mathbf{V}}^T \widehat{\mathbf{D}}_\mu \widehat{\mathbf{V}} \end{bmatrix} \begin{bmatrix} \mathbf{h} \\ \widehat{\mathbf{h}}_m \end{bmatrix} \tag{21.13}$$

Solving the above eigenvalue problem one will receive resonant angular frequencies (ω) together with the coarse mesh magnetic field samples $\mathbf{h}$ and the vector containing projected fine mesh field samples $\mathbf{h}_m$. To calculate the remaining vectors $\widehat{\mathbf{h}}$, $\mathbf{e}$ and $\widehat{\mathbf{e}}$ one has to modify (21.8) and (21.12) to obtain the following formulas:

$$\widehat{\mathbf{h}} = \widehat{\mathbf{V}}\widehat{\mathbf{h}}_m \tag{21.14}$$

$$\begin{bmatrix} \mathbf{e} \\ \widehat{\mathbf{e}} \end{bmatrix} = \frac{1}{j\omega} \begin{bmatrix} \mathbf{D}_\varepsilon^{-1} & 0 \\ 0 & \widehat{\mathbf{D}}_\varepsilon^{-1} \end{bmatrix} \begin{bmatrix} \mathbf{R}_H & \mathbf{S}_H \\ 0 & \widehat{\mathbf{R}}_H \end{bmatrix} \begin{bmatrix} \mathbf{h} \\ \widehat{\mathbf{h}} \end{bmatrix} \tag{21.15}$$

To verify the effectiveness of the *FDFD*-macromodel eigenproblem a few resonant frequencies of *TE* modes of a rectangular resonator with a wedge (see Fig. 21.2) were calculated. It was assumed, that there is no field variation across the z direction

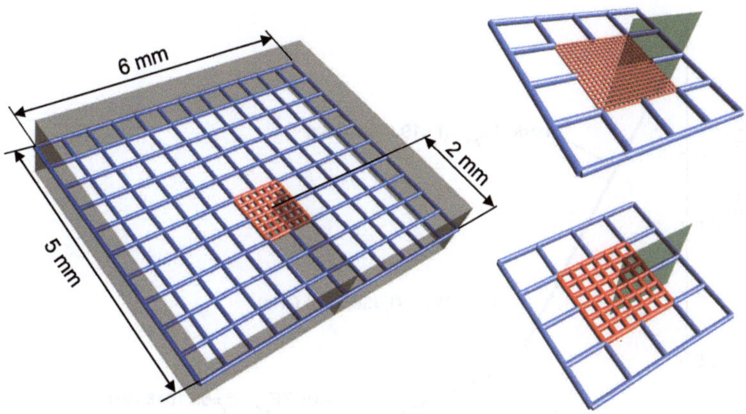

Fig. 21.2 A resonator with a wedge used in tests of reduced eigenproblem and refined local mesh (refinement factor $k = 3$ and $k = 9$)

and a 2D *FDFD* eigenproblem formulation was used with a uniform mesh size ($\Delta x = \Delta y = 0.5\,\text{mm}$). To reduce the errors due to the field singularity at the wedge tip a macromodel was created for a small area ($2\Delta x \times 2\Delta y$) containing the wedge's tip.

To show the results of the *FDFD* algorithm with a macromodel incorporated into the mesh relative errors of computations versus the refinement factor used for the local mesh around the wedge tip were plotted in Fig. 21.3. The reference values, were calculated by extrapolation of the results for progressively refined meshes. When the macromodel is used in the *FDFD* scheme, significant improvement due to the macromodel is visible for all modes.

As indicated in the introduction, the mesh refinement entails an increase of the norm of the system matrix, which in turn slows down the convergence of iterative matrix solvers. As discussed in [5], macromodels can be constructed in such a way that the norm of the projected system is of the same order as the norm of the matrix formed for the main coarse grid. As a result, no convergence penalty is incurred. This effect is clearly visible in Table 21.1 which shows the size of the problem and the number of iterations of an eigenvalue solver for a standard *FDFD*, the *FDFD* with subgridding and the *FDFD* with macromodels incorporated into the mesh according to the scheme proposed herein. In each of these cases the computations were carried out for the main grid density $\Delta x = \Delta y = 0.1\,\text{mm}$ and the first *TE* mode. In all instances a similar improvement in terms of accuracy was obtained (the relative error decreased from approximately 0.8% for $k = 1$ to about 0.09% for $k = 9$). As for the convergence, the scheme with the macromodel converges at the same rate regardless of the refinement factor, while the convergence of the standard *FDFD* and the formulation involving subgridding deteriorates significantly as k increases.

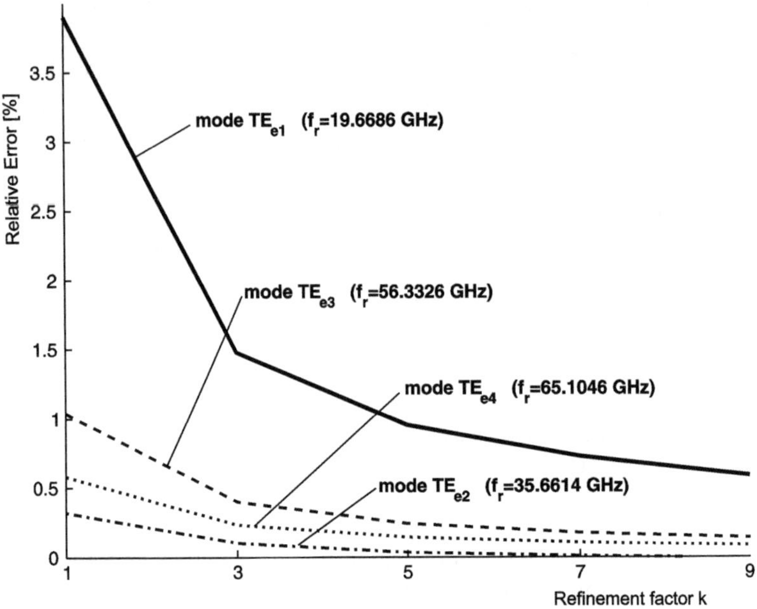

Fig. 21.3 Relative errors versus the refinement factor for the first 4 *TE* even modes for the macromodel incorporated into *FDFD* method (see text for explanation)

Table 21.1 Comparison of the problem size and the number of iterations needed for convergence for standard, subgridded and reduced eigenproblem formulations

k	$EIG_{standard}$		$EIG_{subgr.}$		$EIG_{macro.}$	
	size	iter.	size	iter.	size	iter.
1	3000	447	3000	447	3000	447
3	27000	2140	3032	1073	3000	460
5	75000	2918	3096	1577	3000	470
7	147000	3542	3192	2421	3000	460
9	243000	4159	3320	3629	3000	438

21.4 *Finite Difference Time Domain* **Analysis Using Macromodels**

Although a macromodel is created in the frequency domain and the example given so far was related to eigenvalue analysis using the *F*DFD method, the same macromodel can be used in time domain. Several techniques have been proposed to incorporate macromodels [1, 3, 5] in time domain algorithms. An elegant scheme requiring only a slight reformulation of time domain iterations has been proposed in [8]. The advantage of this approach is that the scheme does not require one to perform any operations on macromodel's matrices and does not need averaging in time to preserve synchronism of electric and magnetic field updates.

The procedure goes as follows. After the second order system is created (21.6) for the refined region one applies the ENOR algorithm to generate the basis $\widehat{\mathbf{V}}$. This basis is then used to project the first order system of Maxwell's grid equations for the fine mesh (21.1). Since vectors forming basis $\widehat{\mathbf{V}}$ are orthogonal, using the reduced vector of magnetic field samples representing the internal states of the macromodel $\widehat{\mathbf{h}}_m$ (21.12) and formulas (21.3), equation (21.1) can be written as:

$$\underbrace{\widehat{\mathbf{V}}^T\widehat{\mathbf{B}}_E\mathbf{I}_E}_{-\mathbf{B}_m}\underbrace{\mathbf{L}_E\mathbf{e}}_{\mathbf{e}_M}+\widehat{\mathbf{V}}^T\widehat{\mathbf{R}}_E\widehat{\mathbf{e}}=-s\widehat{\mathbf{V}}^T\widehat{\mathbf{D}}_\mu\widehat{\mathbf{V}}\widehat{\mathbf{h}}_m$$
$$\widehat{\mathbf{R}}_H\widehat{\mathbf{V}}\widehat{\mathbf{h}}_m=s\widehat{\mathbf{D}}_\varepsilon\widehat{\mathbf{e}} \tag{21.16}$$

and

$$\mathbf{h}_M=\mathbf{L}_m^T\widehat{\mathbf{h}}_m=\mathbf{I}_H\widehat{\mathbf{L}}_H\widehat{\mathbf{V}}\widehat{\mathbf{h}}_m \tag{21.17}$$

To derive the iterative scheme we first eliminate nonprojected vector $\widehat{\mathbf{e}}$ by substituting the second row of (21.16) into the first one as follows:

$$-\mathbf{B}_m\mathbf{e}_M+\frac{1}{s}\underbrace{\widehat{\mathbf{V}}^T\widehat{\mathbf{R}}_E\widehat{\mathbf{D}}_\varepsilon^{-1}\widehat{\mathbf{R}}_H\widehat{\mathbf{V}}}_{\widehat{\Gamma}_m}\cdot\widehat{\mathbf{h}}_m=-s\underbrace{\widehat{\mathbf{V}}^T\widehat{\mathbf{D}}_\mu\widehat{\mathbf{V}}}_{\widehat{\mathbf{C}}_m}\cdot\widehat{\mathbf{h}}_m \tag{21.18}$$

We now rewrite (21.18) as follows

$$-s\widehat{\mathbf{V}}^T\mathbf{B}_m\mathbf{e}_M+\widehat{\Gamma}_m\cdot\widehat{\mathbf{h}}_m=-s^2\widehat{\mathbf{C}}_m\cdot\widehat{\mathbf{h}}_m \tag{21.19}$$

Transforming it back to time domain and applying the time discretization one obtains the following scheme

$$\begin{aligned}\widehat{\mathbf{h}}_m^{n+1}&=2\widehat{\mathbf{h}}_m^n-\Delta t^2\widehat{\mathbf{C}}_m^{-1}\widehat{\Gamma}_m\widehat{\mathbf{h}}_m^n-\widehat{\mathbf{h}}_m^{n-1}\\&\quad+\Delta t\widehat{\mathbf{C}}_m^{-1}\mathbf{B}_m\left(\mathbf{e}_M^{n+0.5}-\mathbf{e}_M^{n-0.5}\right)\end{aligned} \tag{21.20}$$

were boundary fields $\mathbf{e}_M$ are updated at each iteration using the macromodel's memory. Vectors forming basis $\widehat{\mathbf{V}}$ are orthogonal and when the permeability is constant, matrix $\widehat{\mathbf{D}}_\mu$ is diagonal with a constant element. As a result $\mathbf{C}_m=\widehat{\mathbf{V}}^T\widehat{\mathbf{D}}_\mu\widehat{\mathbf{V}}$ is also a diagonal matrix.

An iterative update algorithm for a macromodel can easily be incorporated into the regular *FDTD* scheme. Applying the time discretization to (21.2) and taking into account (21.3), (21.17) and (21.20), the *FDTD* scheme containing macromodels can be written as:

- Apply the excitation and boundary conditions
- $\mathbf{e}^{n+0.5}=\mathbf{e}^{n-0.5}+\Delta t\mathbf{D}_\varepsilon^{-1}\mathbf{R}_H\mathbf{h}^n+\Delta t\mathbf{D}_\varepsilon^{-1}\mathbf{B}_H\mathbf{h}_M^n$
- $\mathbf{h}^{n+1}=\mathbf{h}^n-\Delta t\mathbf{D}_\mu^{-1}\mathbf{R}_E\mathbf{e}^{n+0.5}$
- $\mathbf{e}_M^{n+0.5}=\mathbf{L}_E\mathbf{e}^{n+0.5}$

- Calculate $\widehat{\mathbf{h}}_m^{n+1}$ from (21.20)
- $\mathbf{h}_M^{n+1} = \mathbf{L}_m^T \widehat{\mathbf{h}}_m^{n+1}$
- Iterate

Since the model order reduction reduces the norm related to fine grid the same time step can be used throughout the computational space without increasing the overall CPU time needed to complete the calculations.

Additionally, system matrices $\widehat{\mathbf{C}}_m$ and $\widehat{\Gamma}_m$ can be diagonalized [2], so each iterations are carried out very fast. After diagonalization, the scheme given by (21.20) reduces to:

$$\widehat{\mathbf{h}}_m^{n+1} = \left(2\widehat{\mathbf{I}}_m - \Delta t^2 \widehat{\Lambda}_m\right) \widehat{\mathbf{h}}_m^{n} - \widehat{\mathbf{h}}_m^{n-1} + \Delta t \mathbf{B}_m \left(\mathbf{e}_M^{n+0.5} - \mathbf{e}_M^{n-0.5}\right) \tag{21.21}$$

where instead of $\widehat{\mathbf{C}}_m$ and $\widehat{\Gamma}_m$ (m^2 nonzero entries in each matrix), new diagonal matrices $\widehat{\mathbf{I}}_m$ (identity matrix) and $\widehat{\Lambda}_m$ (m nonzero entries) were introduced.

21.4.1 Stability of the FDTD*-Macromodel Scheme*

For the time domain iteration to be useful, one has to establish if the scheme is stable. The simplest way to do this is to perform an analysis of the resulting scheme for a resonator having a high quality factor excited by a Gaussian pulse. While this approach is often used in practice it only allows one to test the stability of the resulting algorithm and gives no guidelines useful in the development of stable and accurate transitions between subdomains.

A much better way to verify *FDTD*-macromodel scheme's stability is to use mathematical approach [13, 24, 25]. This involves construction of global operators for both subdomains and creation of coupling matrices $\widehat{\mathbf{S}}_E$ and $\mathbf{S}_H$ (21.3), which contain information on how the field samples at the $\Omega \backslash \widehat{\Omega} - \widehat{\Omega}$ boundary provide boundary conditions for schemes operating in subdomains $\Omega \backslash \widehat{\Omega}$ and $\widehat{\Omega}$.

Using a compact form of Finite Difference discrete operators (21.7 and 21.8) one can easily form stability conditions for the combined *FDTD*-macromodel scheme [24, 25]. The overall algorithm is stable if time discretization step Δt fulfills the *Courant* stability condition for discretization steps in both subdomains and coupling matrices are transpose of each other, i.e.:

$$\widehat{\mathbf{S}}_E = \mathbf{S}_H^T \tag{21.22}$$

The above equation can be generalized to the case where a scaling factor α occurs in relation (21.22) Let us assume that:

$$\widehat{\mathbf{S}}_E = \alpha \mathbf{S}_H^T \tag{21.23}$$

where $\alpha \neq 0$. Using (21.23) one can write (21.7) and (21.8) as:

$$\begin{bmatrix} \mathbf{R}_E & \mathbf{0} \\ \alpha \mathbf{S}_H^T & \widehat{\mathbf{R}}_E \end{bmatrix} \begin{bmatrix} \mathbf{e} \\ \widehat{\mathbf{e}} \end{bmatrix} = -s \begin{bmatrix} \mathbf{D}_\mu & \mathbf{0} \\ \mathbf{0} & \widehat{\mathbf{D}}_\mu \end{bmatrix} \begin{bmatrix} \mathbf{h} \\ \widehat{\mathbf{h}} \end{bmatrix} \tag{21.24}$$

$$\begin{bmatrix} \mathbf{R}_H & \mathbf{S}_H \\ \mathbf{0} & \widehat{\mathbf{R}}_H \end{bmatrix} \begin{bmatrix} \mathbf{h} \\ \widehat{\mathbf{h}} \end{bmatrix} = s \begin{bmatrix} \mathbf{D}_\varepsilon & \mathbf{0} \\ \mathbf{0} & \widehat{\mathbf{D}}_\varepsilon \end{bmatrix} \begin{bmatrix} \mathbf{e} \\ \widehat{\mathbf{e}} \end{bmatrix} \tag{21.25}$$

Introducing new variables $\widetilde{\widehat{\mathbf{e}}} = \frac{1}{\sqrt{\alpha}}\widehat{\mathbf{e}}$, $\widetilde{\widehat{\mathbf{h}}} = \frac{1}{\sqrt{\alpha}}\widehat{\mathbf{h}}$ and using some basic operations one can easily transform (21.24), (21.25) into:

$$\begin{bmatrix} \mathbf{R}_E & \mathbf{0} \\ \widetilde{\widehat{\mathbf{S}}}_E & \widehat{\mathbf{R}}_E \end{bmatrix} \begin{bmatrix} \mathbf{e} \\ \widetilde{\widehat{\mathbf{e}}} \end{bmatrix} = -s \begin{bmatrix} \mathbf{D}_\mu & \mathbf{0} \\ \mathbf{0} & \widehat{\mathbf{D}}_\mu \end{bmatrix} \begin{bmatrix} \mathbf{h} \\ \widetilde{\widehat{\mathbf{h}}} \end{bmatrix} \tag{21.26}$$

$$\begin{bmatrix} \mathbf{R}_H & \widetilde{\mathbf{S}}_H \\ \mathbf{0} & \widehat{\mathbf{R}}_H \end{bmatrix} \begin{bmatrix} \mathbf{h} \\ \widetilde{\widehat{\mathbf{h}}} \end{bmatrix} = s \begin{bmatrix} \mathbf{D}_\varepsilon & \mathbf{0} \\ \mathbf{0} & \widehat{\mathbf{D}}_\varepsilon \end{bmatrix} \begin{bmatrix} \mathbf{e} \\ \widetilde{\widehat{\mathbf{e}}} \end{bmatrix} \tag{21.27}$$

where $\widetilde{\mathbf{S}}_H = \sqrt{\alpha}\mathbf{S}_H$ and $\widetilde{\widehat{\mathbf{S}}}_E = \sqrt{\alpha}\widehat{\mathbf{S}}_E = \sqrt{\alpha}\mathbf{S}_H^T$ are new coupling matrices. It is readily seen that $\widetilde{\mathbf{S}}_H = \widetilde{\widehat{\mathbf{S}}}_E^T$, which makes (21.23) a general stability condition.

Because the above approach allows one to verify stability of the *FDTD*-subgridding scheme by examining the symmetry of global operator matrices, operator-based approach is much more meaningful than the experimental one. One should note however, that if global operators containing coupling matrices are not constructed with care, stability condition (21.23) will not be observable in terms of condition (21.23), even if the algorithm is stable. IIn such a case, one should use the stability verification method that employs *reciprocity principle* [26], which examines the relations between coupling matrices in global operators.

21.5 *Finite Differences* Analysis Using Nested Macromodels

The efficiency of model order reduction diminishes when the refinement factors are high – this is because creation of a macromodel involves a solution of the grid equations (via LU decomposition) in the refined mesh. The size of the problems that can be handled efficiently by the ENOR algorithm is limited to about tens of thousands of unknowns. This requirement imposes a limit on the applicability of MOR techniques. To overcome the limitation of maximum number of unknowns in MOR methods and to achieve high refinement factors one can apply nested macromodels [9].

The starting point are Maxwell's grid equations with an embedded macromodel derived from a system of unknowns having projected fine mesh operators (21.10), (21.11).

The concept of nested macromodels will be explained for a two level case involving nested meshes from a structure shown in the upper-right corner in Fig. 21.4. A cross section of the structure is covered with coarse mesh and the areas around wedge tips are refined. This fine grid is further refined in the center, so that three meshes are formed. Discretized Maxwell's equations for three nested meshes around one tip can be written as follows:

$$\begin{bmatrix} \mathbf{R}_E & 0 \\ \mathbf{S}_E & \begin{bmatrix} \widehat{\mathbf{R}}_E & 0 \\ \widehat{\mathbf{S}}_E & \widehat{\widehat{\mathbf{R}}}_E \end{bmatrix} \end{bmatrix} \begin{bmatrix} \mathbf{e} \\ \begin{bmatrix} \widehat{\mathbf{e}} \\ \widehat{\widehat{\mathbf{e}}} \end{bmatrix} \end{bmatrix} = -s \begin{bmatrix} \mathbf{D}_\mu & 0 \\ 0 & \begin{bmatrix} \widehat{\mathbf{D}}_\mu & 0 \\ 0 & \widehat{\widehat{\mathbf{D}}}_\mu \end{bmatrix} \end{bmatrix} \begin{bmatrix} \mathbf{h} \\ \begin{bmatrix} \widehat{\mathbf{h}} \\ \widehat{\widehat{\mathbf{h}}} \end{bmatrix} \end{bmatrix} \tag{21.28}$$

$$\begin{bmatrix} \mathbf{R}_H & \mathbf{S}_H \\ 0 & \begin{bmatrix} \widehat{\mathbf{R}}_H & \widehat{\mathbf{S}}_H \\ 0 & \widehat{\widehat{\mathbf{R}}}_H \end{bmatrix} \end{bmatrix} \begin{bmatrix} \mathbf{h} \\ \begin{bmatrix} \widehat{\mathbf{h}} \\ \widehat{\widehat{\mathbf{h}}} \end{bmatrix} \end{bmatrix} = s \begin{bmatrix} \mathbf{D}_\varepsilon & 0 \\ 0 & \begin{bmatrix} \widehat{\mathbf{D}}_\varepsilon & 0 \\ 0 & \widehat{\widehat{\mathbf{D}}}_\varepsilon \end{bmatrix} \end{bmatrix} \begin{bmatrix} \mathbf{e} \\ \begin{bmatrix} \widehat{\mathbf{e}} \\ \widehat{\widehat{\mathbf{e}}} \end{bmatrix} \end{bmatrix} \tag{21.29}$$

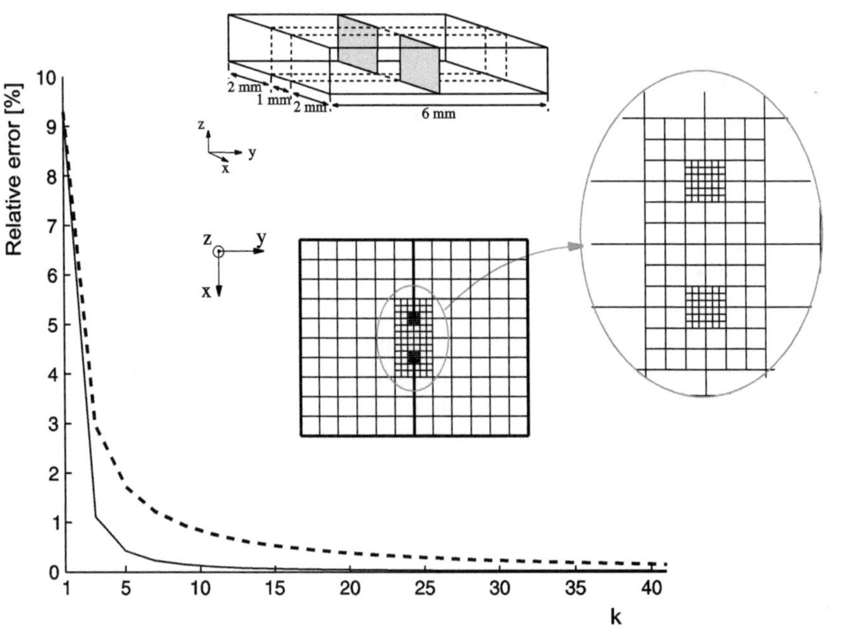

Fig. 21.4 Relative error in first TE resonant frequency for a two-wedge rectangular resonator versus mesh refinement for a single (*dashed*) and nested (*solid*) macromodel

where the double hat is used to denote the operators and fields for the finest mesh. In the above equations the fields defined on the innermost mesh are decoupled from the fields of the main grid. Accordingly, the projection of these fields does not affect the fields or operators pertaining to the coarse mesh. We may therefore create a macromodel for this area by projecting suitable matrices and fields in the manner similar to that seen in (21.10), (21.11). To this end one uses a basis $\widehat{\widehat{\mathbf{V}}}$ generated by ENOR for the innermost mesh. Projection of (21.28) gives then

$$\begin{bmatrix} \mathbf{R}_E & 0 \\ \mathbf{S}_E & \begin{bmatrix} \widehat{\mathbf{R}}_E & 0 \\ \widehat{\widehat{\mathbf{V}}}^T \widehat{\mathbf{S}}_E & \widehat{\widehat{\mathbf{V}}}^T \widehat{\widehat{\mathbf{R}}}_E \end{bmatrix} \end{bmatrix} \begin{bmatrix} \mathbf{e} \\ \begin{bmatrix} \widehat{\mathbf{e}} \\ \widehat{\widehat{\mathbf{e}}} \end{bmatrix} \end{bmatrix} = -s \begin{bmatrix} \mathbf{D}_\mu & 0 \\ 0 & \begin{bmatrix} \widehat{\mathbf{D}}_\mu & 0 \\ 0 & \widehat{\widehat{\mathbf{V}}}^T \widehat{\widehat{\mathbf{D}}}_\mu \widehat{\widehat{\mathbf{V}}} \end{bmatrix} \end{bmatrix} \begin{bmatrix} \mathbf{h} \\ \begin{bmatrix} \widehat{\mathbf{h}} \\ \widehat{\widehat{\mathbf{V}}}^T \widehat{\widehat{\mathbf{h}}} \end{bmatrix} \end{bmatrix} \tag{21.30}$$

A similar result is obtained for (21.29).

Once the projection for the innermost mesh has been carried out, a new basis is generated for the outer mesh, but this time the basis $\widehat{\mathbf{V}}$ has to be generated using the matrices obtained as the result of projection carried out at the previous level. The final result is

$$\begin{bmatrix} \mathbf{R}_E & 0 \\ \widehat{\mathbf{V}}^T \mathbf{S}_E & \widehat{\mathbf{V}}^T \begin{bmatrix} \widehat{\mathbf{R}}_E & 0 \\ \widehat{\widehat{\mathbf{V}}}^T \widehat{\mathbf{S}}_E & \widehat{\widehat{\mathbf{V}}}^T \widehat{\widehat{\mathbf{R}}}_E \end{bmatrix} \end{bmatrix} \begin{bmatrix} \mathbf{e} \\ \begin{bmatrix} \widehat{\mathbf{e}} \\ \widehat{\widehat{\mathbf{e}}} \end{bmatrix} \end{bmatrix} = -s \begin{bmatrix} \mathbf{D}_\mu & 0 \\ 0 & \widehat{\mathbf{V}}^T \begin{bmatrix} \widehat{\mathbf{D}}_\mu & 0 \\ 0 & \widehat{\widehat{\mathbf{V}}}^T \widehat{\widehat{\mathbf{D}}}_\mu \widehat{\widehat{\mathbf{V}}} \end{bmatrix} \widehat{\mathbf{V}} \end{bmatrix} \begin{bmatrix} \mathbf{h} \\ \widehat{\mathbf{V}}^T \begin{bmatrix} \widehat{\mathbf{h}} \\ \widehat{\widehat{\mathbf{V}}}^T \widehat{\widehat{\mathbf{h}}} \end{bmatrix} \end{bmatrix} \tag{21.31}$$

$$\begin{bmatrix} \mathbf{R}_H & \mathbf{S}_H \widehat{\mathbf{V}} \\ 0 & \begin{bmatrix} \widehat{\mathbf{R}}_H & \widehat{\mathbf{S}}_H \widehat{\widehat{\mathbf{V}}} \\ 0 & \widehat{\widehat{\mathbf{R}}}_H \widehat{\widehat{\mathbf{V}}} \end{bmatrix} \widehat{\mathbf{V}} \end{bmatrix} \begin{bmatrix} \mathbf{h} \\ \widehat{\mathbf{V}}^T \begin{bmatrix} \widehat{\mathbf{h}} \\ \widehat{\widehat{\mathbf{V}}}^T \widehat{\widehat{\mathbf{h}}} \end{bmatrix} \end{bmatrix} = s \begin{bmatrix} \mathbf{D}_\varepsilon & 0 \\ 0 & \begin{bmatrix} \widehat{\mathbf{D}}_\varepsilon & 0 \\ 0 & \widehat{\widehat{\mathbf{D}}}_\varepsilon \end{bmatrix} \end{bmatrix} \begin{bmatrix} \mathbf{e} \\ \begin{bmatrix} \widehat{\mathbf{e}} \\ \widehat{\widehat{\mathbf{e}}} \end{bmatrix} \end{bmatrix} \tag{21.32}$$

The whole procedure can easily be generalized to more levels and several regions of nested meshing.

The performance of nested macromodels can be shown by computation of the resonant frequency of the first *TE* mode of a resonator (shown in Fig. 21.4). The 2D *FDFD* eigenproblem formulation was used with a uniform mesh size ($\Delta x = \Delta y =$ 0.5 mm). Macromodels formed for Yee's mesh around each wedge tip were nested.

Table 21.2 Comparison of macromodels (see text for explanation)

Error (%)	Number of variables for reduction	
	Single macromodel	Nested macromodel
1	324 (9)	40 & 36 (3,3)
0.3	1444 (19)	104 & 100 (5,5)
0.15	6724 (41)	328 & 324 (9,9)
0.03	33124 (91)	3368 & 3364 (29,29)

The outer macromodel covered area $2\Delta x \times 2\Delta y$ (refinement factor k) and the inner macromodel covered area $2\frac{\Delta x}{k} \times 2\frac{\Delta y}{k}$ (refinement factor k).

Results for nested macromodels, presented in Fig. 21.4 and Table 21.2, show an obvious advantage of nested macromodels over a single macromodel. Table 21.2 compares the number of variables involved in each case for achieving the same level of error (reference value is 15.93 GHz [27]). Two numbers in the column for nested macromodels indicate the problem size at two reduction levels. The refinement factors of the corresponding meshes are given in brackets. A single macromodel scheme quickly reaches the point where the reduction becomes inefficient due to a large number of variables that have to be handled during the reduction. The multi-level technique involves a reduction of a small problem at each level and hence it is carried out fast.

Acknowledgements This work has been partially supported by ERO of the US Army under contract N62558-06-P-0103.

References

1. L. Kulas, M. Mrozowski, Reduced-order models in FDTD. IEEE Microw. Wireless Compon. Lett. **11**, 422–424 (2001)
2. P. Sypek, L. Kulas, M. Mrozowski, Low reflection macromodels for a stable *FDTD* scheme operating with highly refined local meshes. in *Proceedings Of International Microwave Symposium, IMS-2005*, June 2005, pp. 195–198
3. B. Denecker, F. Olyslager, L. Knockaert, D.D. Zutter, Generation of FDTD subcell equations by means of reduced order modeling. IEEE Trans. Antennas Propag. **51**, 1806–1817 (2003)
4. B. Denecker, F. Olyslager, D.D. Zutter, Z. Klinkenbusch, L. Knockaert, Efficient analysis of photonic crystal structures using a novel FDTD-technique. in *Proceedings of the IEEE Antennas and Propagation Society International Symposium*, vol. 4, June 2002, pp. 344–347
5. L. Kulas, M. Mrozowski, Reduced order models of refined yee's cells. IEEE Microw. Wireless Compon. Lett. **13**, 164–166 (2003)
6. Y. Zhu, A.C. Cangellaris, Macro-elements for efficient FEM simulation of small geometric features in waveguide components. IEEE Trans. Microw. Theory Tech. **48**, 2254–2260 (2000).
7. B. Denecker, F. Olyslager, L. Knockaert, D.D. Zutter, Automatic generation of subdomain models in 2-D FDTD using reduced order modeling. IEEE Microw. Guid. Wave Lett. **10**, 301–303 (2000)
8. L. Kulas, M. Mrozowski, A fast high-resolution 3-D finite-difference time-domain scheme with macromodels. IEEE Trans. Microw. Theory Tech. **52**, 2330–2335 (2004)
9. L. Kulas, M. Mrozowski, Multilevel model order reduction. IEEE Microw. Wireless Compon. Lett. **14**, 165–167 (2004)

10. K.S. Yee, Numerical solution of initial boundary value problems involving maxwell's equations in isotropic media. IEEE Trans. Antennas Propag. **14**, 302–307 (1966)
11. T. Weiland, Maxwells's grid equations. Frequenz **44**, 9–15 (1990)
12. L. Kulas, M. Mrozowski, A simple high-accuracy subgridding scheme. in *Proceedings of the 33rd European Microwave Conference*, Oct 2003, pp. 347–350
13. P. Thoma, T. Weiland, Numerical stability of finite difference time domain methods. IEEE Trans. Magn. **34**, 2740–2743 (1998)
14. M.W. Chevalier, R.J. Luebbers, V.P. Cable, FDTD local grid with material traverse. IEEE Trans. Antennas Propag. **45**, 411–421 (1997)
15. M. Okoniewski, E. Okoniewska, M.A. Stuchly, Three-dimensional subgridding algorithm for FDTD. IEEE Trans. Antennas Propag. **45**, 422–429 (1997)
16. O. Podebrad, M. Clemens, T. Weiland, New flexible subgridding scheme for the finite integration technique. IEEE Trans. Magn. **39**, 1662–1665 (2003)
17. A. Odabasioglu, M. Celik, L.T. Pileggi, PRIMA: passive reduced-order interconnect macromodeling algorithm. IEEE Trans. Comput. Aided Des. **17**, 645–654 (1998)
18. P. Feldmann, R.W. Freund, Efficient linear circuit analysis by Padé approximation via Lanczos process. IEEE Trans. Comput. Aided Des. **14**, 639–649 (1995)
19. L. Knockaert, D.D. Zutter, Laguerre-SVD reduced-order modeling. IEEE Trans. Microw. Theory Tech. **48**, 1469–1475 (2000)
20. B.N. Sheehan, *ENOR*: model order reduction of RLC circuits using nodal equations for efficient factorization. in *Proceedings of the IEEE 36th Design Automation Conference*, June 1999, pp. 17–21
21. Y. Su, J. Wang, X. Zeng, Z. Bai, C. Zhou, SAPOR: second-order Arnoldi method for passive order reduction of RCS circuits. in *IEEE/ACM International Conference Computer Aided Design, ICCAD-2004.*, 2004, pp. 74–79
22. H. Zheng, L. Pileggi, Robust and passive model order reduction for circuits containing susceptance elements. in *Proceedings of IEEE/ACM ICCAD 2002*, 2002, pp. 761–766
23. J. Przewocki, L. Kulas, M. Mrozowski, Digital system interconnects analysis using model order reduction methods. in *16th International Conference on Microwaves, Radar and Wireless Communications, MIKON-2006*, vol. 2, May 2006, pp. 577–580
24. M. Mrozowski, Stability condition for the explicit algorithms of the time domain analysis of Maxwell's equations. IEEE Microw. Guid. Wave Lett. **4**, 279–281 (1994)
25. L. Kulas, M. Mrozowski, Stability of the FDTD scheme containing macromodels. IEEE Microw. Wireless Compon. Lett. **14**, 484–486 (2004)
26. L. Kulas, M. Mrozowski, Reciprocity principle for stable subgridding in the finite difference time domain method. in *International Conference on "Computer as a Tool", EUROCON-2007*, Sept 2007, pp. 106–111
27. P. Przybyszewski, M. Mrozowski, A conductive wedge in Yee's mesh. IEEE Microw. Guid. Wave Lett. **8**, 66–68 (1998)

Chapter 22
Analysis of a Time-Space Periodic Filter Structure with Tunable Band-Pass Characteristic

Johannes A. Russer and Andreas C. Cangellaris

22.1 Introduction

Electromagnetic wave interaction with periodic structures is an extensively studied phenomenon [1–10]. A good overview on wave interaction in active and passive periodic structures is given in [11]. The periodic structure can arise from a crystalline structure of the medium supporting the electromagnetic wave, from thermal vibrations or standing acoustics waves in a resonator etc. Due to the distributed feedback arising from the periodic variation of the refraction index or the structure's shape, periodic structures can prevent waves at certain frequencies form propagation while guiding them at other frequencies thus exhibiting a pass- and stop-band behavior.

Analytic solutions to the wave equation in a medium changing periodically in space can be found with the help of Mathieu functions [12–14] and for the more general case of a time- and space-periodic media by use of Floquet's Theorem [15, 16]. For the latter case energy transfer between the electromagnetic wave and the wave exhibited by the varying material parameters may occur in form of a parametric amplification [17, 18]. The effect of the variation in time of the periodic structure on the bandgap characteristic can be utilized to create tunable filter structures.

For a solution using numerical techniques we have developed a finite difference time domain (FDTD) [19–23] formulation that allows for a mapping of curved moving boundaries onto a stationary rectangular grid on which the numerical computations for the electromagnetic fields are performed. Our objective is to validate the results obtained from this mapping approach with the help of analytic results.

J.A. Russer (✉) and A.C. Cangellaris
Department of Electrical and Computer Engineering, University of Illinois at Urbana-Champaign, Urbana, IL 61801, USA
e-mail: russer@illinois.edu, cangella@illinois.edu

S. Lindenmeier and R. Weigel (eds.), *Electromagnetics and Network Theory and their Microwave Technology Applications*, DOI 10.1007/978-3-642-18375-1_22,

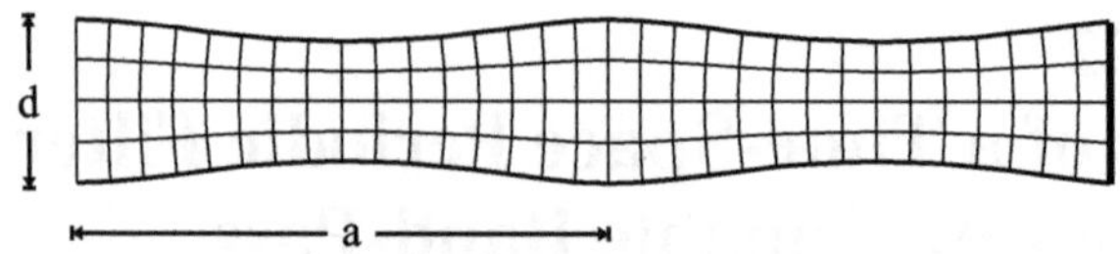

Fig. 22.1 Two sections of the corrugated parallel-plate waveguide

22.2 Formulation of the Problem

We consider a filter structure which is comprised of a parallel plate wave guide with perfect electrically conducting (PEC) corrugated walls and filled by dielectric with the permittivity ε_m. The variation of the plate separation is described by the function

$$d(t,z) = \frac{d_0}{\sqrt{1 + M\cos(\omega_p t - \beta_p z)}} \tag{22.1}$$

and depicted in Fig. 22.1. The spatial period in (22.1) is given by a and thus $\beta_p = 2\pi/a$. The problem is two-dimensional hence we use the reduced formulation in the FDTD mapping approach [22]. The variation of the plate separation modulates the capacitive loading of the waveguide and the impedance is proportional to the plate separation. The mapping method provides for a rigorous formulation for general boundary displacements. However, for this particular case the plate separation translates into an impedance modulation $Z = \sqrt{\frac{\mu}{\varepsilon_m}}\frac{d}{w}$, where w is the waveguide's width, which is equivalent to a one-dimensional transmission line with

$$\varepsilon(t,z) = \varepsilon_m + \Delta\varepsilon\cos(\omega_p t - \beta_p z)\,, \tag{22.2}$$

$\mu = \text{const.}$ and $M = \Delta\varepsilon/\varepsilon_m$. We will use this permittivity variation in the one-dimensional wave equation for our analytic reference solution.

22.2.1 *Analytic Approach*

For a one-dimensional transmission line Maxwell's equations yield, using ε from (22.2),

$$\frac{\partial^2 E_x}{\partial z^2} = \mu_0 \frac{\partial^2}{\partial t^2}(\varepsilon E_x)\,. \tag{22.3}$$

We follow [4–6] and choose the Floquet ansatz to solve (22.3)

$$E_x = E_0 \mathrm{e}^{-j(\omega t - \kappa z)} \sum_{n=-\infty}^{+\infty} a_n \mathrm{e}^{-jn(\omega_p t - k_p z)}\,. \tag{22.4}$$

This yields the following relation for neighboring Fourier coefficients of (22.4)

$$a_{n+1} + D_n a_n + a_{n-1} = 0 \tag{22.5}$$

with

$$D_n = \frac{2\varepsilon_m}{\Delta\varepsilon}\left[1 - \left(\frac{\kappa a + 2\pi n}{ka + 2\pi v n}\right)^2\right], \tag{22.6}$$

$v = v_p/v_0$, $v_p = \omega_p/\beta_p$, $v_0 = \omega/k$ and $k = \omega\sqrt{\mu\varepsilon_m}$. Finally, (22.5) is expanded into the recursive relation using continued fraction expansions

$$D_n - \cfrac{1}{D_{n-1} - \cfrac{1}{D_{n-2} - \cfrac{1}{\ddots}}} - \cfrac{1}{D_{n+1} - \cfrac{1}{D_{n+2} - \cfrac{1}{\ddots}}} = 0\,, \tag{22.7}$$

which is truncated for the numerical computation after a chosen value for n. Equation (22.7) is solved using a nonlinear Newton method [24] and yields a relation for $\kappa(\omega)$ respectively $\omega(\kappa)$. To ensure convergence of the above method v should be outside the interval specified by [5, 6]

$$\frac{1}{\sqrt{1 + \Delta\varepsilon/\varepsilon_m}} \leq v \leq \frac{1}{\sqrt{1 - \Delta\varepsilon/\varepsilon_m}}\,. \tag{22.8}$$

22.2.2 FDTD Mapping Approach

The FDTD mapping approach solves a modified set of Maxwell's equations on a rectangular computational grid, using Yee's staggered grid for E- and H-fields and the leap frog scheme [19, 20]. The original boundary value problem (BVP) with curved and time-varying boundaries is mapped on a static, rectangular reference domain [22, 23]. The mapping technique allows for an accurate implementation of moving curved boundaries while relaxing the need for staircase approximations, spatial oversampling and on-the-fly remeshing of the computational grid. The time-varying curved boundary is absorbed into a time dependent operator, which is a modified Nabla operator for Maxwell's equations. This concept is illustrated in Fig. 22.2, where the deformed configuration of the original BVP is denoted by the set of base vectors (x, y, z) and the mapped BVP in the reference domain by (α, β, γ). We will use the ˜ symbol to denote vector quantities cast into the equivalent BVP of the reference domain. We define a gradient matrix

$$\mathscr{G}(t) = \begin{bmatrix} \frac{\partial\alpha}{\partial x(t)} & \frac{\partial\beta}{\partial x(t)} & \frac{\partial\gamma}{\partial x(t)} \\ \frac{\partial\alpha}{\partial y(t)} & \frac{\partial\beta}{\partial y(t)} & \frac{\partial\gamma}{\partial y(t)} \\ \frac{\partial\alpha}{\partial z(t)} & \frac{\partial\beta}{\partial z(t)} & \frac{\partial\gamma}{\partial z(t)} \end{bmatrix} \tag{22.9}$$

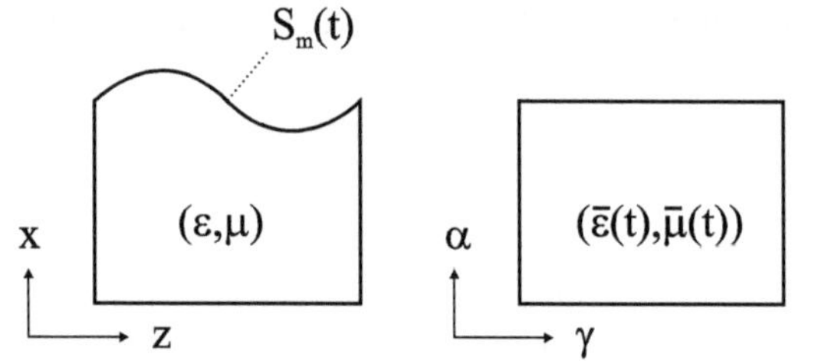

Fig. 22.2 Non-uniform, time-varying and reference configuration. (**a**) Deformed configuration (**b**) Reference configuration

and the Jacobian determinant

$$D_{x_1 x_2}^{\alpha_1 \alpha_2}(t) := \det \begin{bmatrix} \frac{\partial \alpha_1}{\partial x_1(t)} & \frac{\partial \alpha_1}{\partial x_2(t)} \\ \frac{\partial \alpha_1}{\partial x_1(t)} & \frac{\partial \alpha_2}{\partial x_2(t)} \end{bmatrix}. \tag{22.10}$$

Maxwell's equations in the reference domain are found to be

$$\frac{\partial}{\partial t}\tilde{\mathbf{E}} = \frac{1}{\varepsilon}\tilde{\nabla}(t) \times \tilde{\mathbf{H}} \tag{22.11}$$

$$\frac{\partial}{\partial t}\tilde{\mathbf{H}} = -\frac{1}{\mu}\tilde{\nabla}(t) \times \tilde{\mathbf{E}}. \tag{22.12}$$

where the modified operator $\tilde{\nabla}$ is defined as

$$\tilde{\nabla}(t)\times \equiv \mathscr{G}^{-1}(t)\tilde{\mathscr{D}}(t)\nabla\times \tag{22.13}$$

and

$$\tilde{\mathscr{D}}(t) = \begin{bmatrix} D_{yz}^{\beta\gamma}(t) & -D_{yz}^{\alpha\gamma}(t) & D_{yz}^{\alpha\beta}(t) \\ -D_{xz}^{\beta\gamma}(t) & D_{xz}^{\alpha\gamma}(t) & -D_{xz}^{\alpha\beta}(t) \\ D_{xy}^{\beta\gamma}(t) & -D_{xy}^{\alpha\gamma}(t) & D_{xy}^{\alpha\beta}(t) \end{bmatrix}. \tag{22.14}$$

22.3 Analysis

We have computed the bandpass characteristic using the above described methods for a transmission line with $\varepsilon_m = 2\varepsilon_0$, $\Delta\varepsilon_m = 0.6\varepsilon_0$ for (22.2) and hence $M = 0.3$ in (22.1). The period length is chosen $a = 4$ cm and the constant for the plate separation $d_0 = 1$cm, thus we find the plate separation varying between $d_{min} = 0.88d_0$ and $d_{max} = 1.2d_0$. The waveguide is excited below higher order cut-off frequencies, exciting only the fundamental, transversal electromagnetic (TEM) mode. The frequency f_{bg} of the stop-band can be estimated from the condition for the Bragg reflection [25] yielding maximum reflection for

$$f_{bg,n} = \frac{n}{2a\sqrt{\mu\varepsilon_m}}, \tag{22.15}$$

where n is the order of the stop-band, and hence $f_{bg,1} = 2.65\,\text{GHz}$, $f_{bg,2} = 5.30\,\text{GHz}$.

The frequency f' observed by a moving corrugation profile deviates from the actual frequency f according to the Doppler effect

$$f' = f\sqrt{\frac{1 - v_p/c}{1 + v_p/c}}, \tag{22.16}$$

with $c = 1/\sqrt{\mu\varepsilon_m}$. Hence, for the case where the propagation vector of the TEM wave and of the corrugation profile point in the same direction, the observed frequency f' is reduced. The Bragg condition will apply to the observed frequency f' and occur at a frequency f, which increases along with the velocity v_p. Thus the stop-bands shift upward in frequency. Conversely, it will shift to lower frequencies for an opposite directed motion of the corrugation profile.

Approximating the analytic solution in (22.7) numerically yields the dispersion relation graphed in Fig. 22.3 for $v_p = 1 \times 10^7$ m/s respectively $v_p = -1 \times 10^7$ m/s, with the wave number $\kappa = \kappa' - j\kappa''$. Whereas this solution is for an infinite structure we consider for the mapped FDTD method a waveguide section with 30 periods (see Fig. 22.1). The waveguide is excited at one side of the moving corrugated section with a modulated Gaussian pulse and sampled at the other side. The power spectrum of the input signal is centered at 4.7 GHz and the full width half maximum (FWHM) bandwidth is given as 2.9 GHz. The mesh spacing in numerical reference grid is 2.5 mm. We have computed the lower and upper edge of the

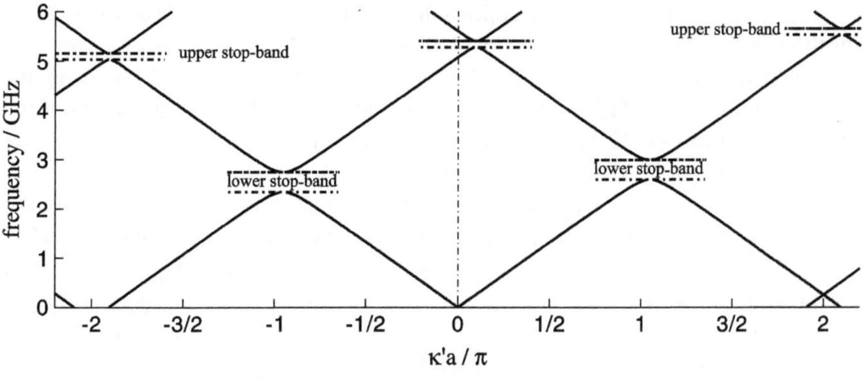

Fig. 22.3 Dispersion diagram

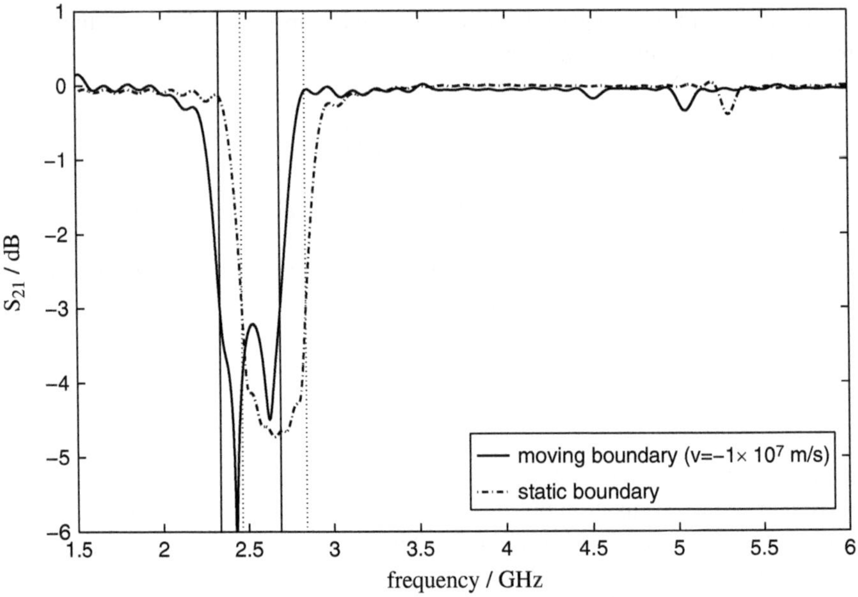

Fig. 22.4 Transmission coefficient $|S_{21}|$ for moving boundary with $v_p = -1 \times 10^7$ m/s and $v_p = 0$ m/s. *Vertical lines* mark the −3 dB frequency points

first stop-band of the fundamental mode, and the center frequencies of the first two stop-bands. For the FDTD method the lower and upper edge for the stop-band are specified as the frequency points where the ratio energy transfer drops to $S_{21}(f) = -3$ dB. The S_{21} parameter of the waveguide for $v_p = -1 \times 10^7$ m/s, $v_p = 0$ m/s, and $v_p = 1 \times 10^7$ m/s are shown in Figs. 22.4 and 22.5. For the solution of (22.7) at $v_p = 0$ we find κ'' at the center of the lower and upper stop-band to be $\kappa''_{bg,1} = 1.78\,\mathrm{m}^{-1}$ and $\kappa''_{bg,2} = 5.9\,\mathrm{m}^{-1}$. The difference in κ'' is reflected in different attenuation levels of the two stop-bands in the plots of Figs. 22.4 and 22.5. However, these values for κ'' are applicable to an infinite structure. The frequencies for the stop-band are presented in Table 22.1 with results obtained from an approximation using the Bragg condition (22.15) and the Doppler effect (22.16), using the analytic approach for TEM waves of (22.7), and the FDTD implementation of the mapped equations in (22.12). The values obtained for the stop-band frequencies are in good agreement for the different methods.

Furthermore, we have excited the waveguide with a sinusoidal in the spectrum of the pass-band and we have computed the S_{21} parameter using the mapping FDTD. The results for the input signal at $f = 4.2$ GHz are plotted in Fig. 22.6. For the corrugation profile moving with $v_p = 1 \times 10^7$ m/s we observe intermodulation products at very low power levels at frequencies separated by integer multiples of $f_p = \omega_p/(2\pi) = \beta_p v_p/(2\pi) = 250$ MHz away from the excitation frequency, as suggested in Floquet's ansatz (22.4).

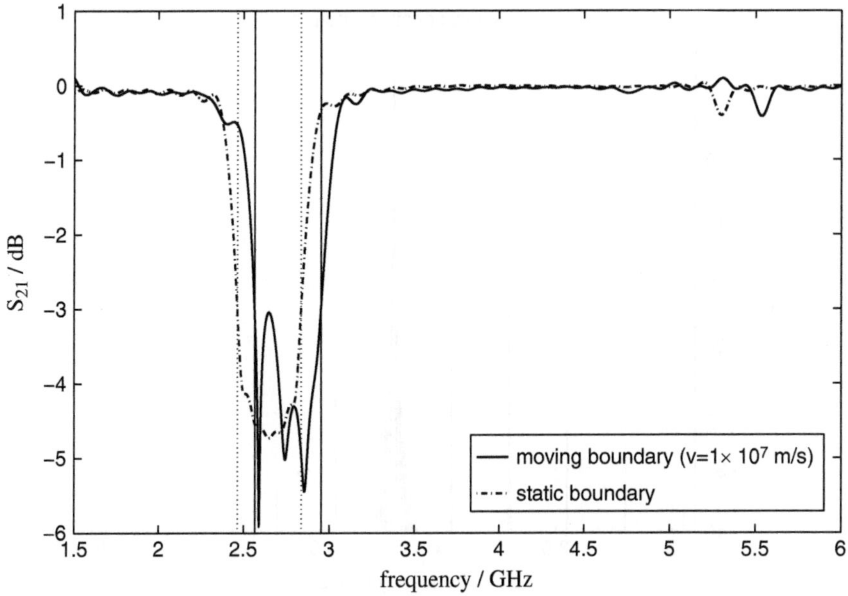

Fig. 22.5 Transmission coefficient $|S_{21}|$ for a moving boundary with $v_p = 1 \times 10^7$ m/s and $v_p = 0$ m/s. *Vertical lines* mark the −3 dB frequency points

Table 22.1 Stop-band frequencies of a time-space periodic waveguide with its profile depicted in Fig. 22.1, with results obtained from the Doppler shift approximation, analytic solution and mapped FDTD method

	Frequency *lower stop-band*/GHz						
	Doppler	Analytic			FDTD		
v/ms^{-1}	approx.	Lower edge	Upper edge	Center	Lower edge	Upper edge	Center
-1×10^7	2.53	2.34	2.74	2.54	2.34	2.69	2.51
0	2.65	2.47	2.87	2.67	2.46	2.84	2.65
1×10^7	2.78	2.59	2.99	2.79	2.57	2.95	2.76
	Frequency *upper stop-band*/GHz						
	Doppler	Analytic			FDTD		
v/ms^{-1}	approx.	Lower edge	Upper edge	Center	Lower edge	Upper edge	Center
-1×10^7	5.06	5.03	5.16	5.09	–	–	5.04
0	5.30	5.29	5.41	5.35	–	–	5.30
1×10^7	5.56	5.53	5.66	5.59	–	–	5.54

22.4 Conclusion

A time-space periodic filter structure has been analyzed for the tunability of its band-gaps. We have implemented a time dependent mapping approach for Maxwell's equations with moving curved boundaries. We have compared the results to an analytic solution existing for the infinite extended time-space periodic structure.

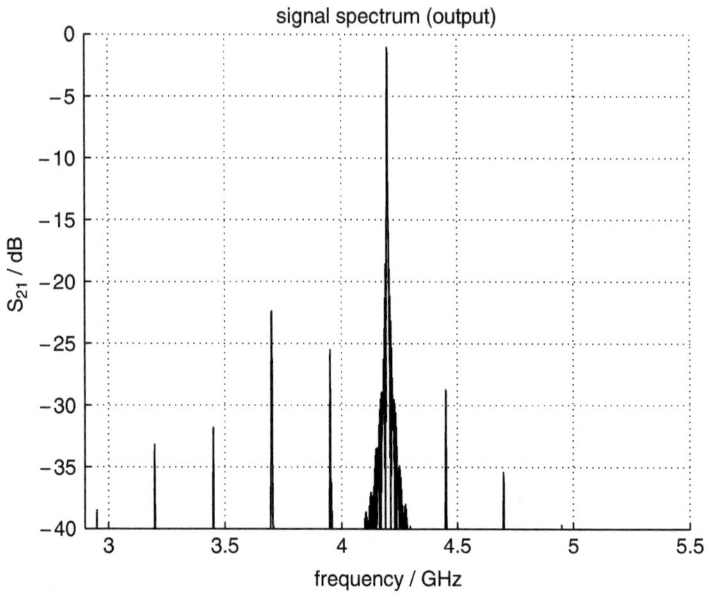

Fig. 22.6 Transmission coefficient $|S_{21}|$ for a moving boundary with $v_p = 1 \times 10^7$ m/s, excited at $f = 4.2$ GHz

The results support prior investigations of the time dependent mapping approach as an accurate method [23]. These results add to the benefit of the mapping FDTD approach. This approach can be used for filters with more complex structures where analytic approximations are not available. The benefits include an accurate and natural implementation of boundary conditions for curved boundaries, the relaxation of spatial oversampling requirements, and the relaxation of on-the-fly re-meshing [22, 23].

Acknowledgements This material is based upon work supported in part by the U.S. Army Research Office as a Multi-disciplinary University Research Initiative on Standoff Inverse Analysis and Manipulation of Electronic Systems under grant number W911NF-05-1-0337.

References

1. L. Brillouin, *Wave Propagation in Periodic Structures*. (Dover, New York, 1953)
2. J.C. Slater, Interaction of waves in crystals. Rev. Mod. Phys. **30**(1)
3. A. Oliner, A. Hessel, Guided waves on sinusoidally-modulated reactance surfaces. Antennas Propag. IRE Trans. **7**(5), 201–208 (1959)
4. J.-C. Simon, Action of a progressive disturbance on a guided electromagnetic wave. Microw. Theory Techn. IRE Trans. **8**(1), 18–29 (1960)
5. A. Hessel, A. Oliner, Wave propagation in a medium with a progressive sinusoidal disturbance. Microw. Theory Techn. IRE Trans. **9**(4), 337–343 (1961)

6. E. Cassedy, A. Oliner, Dispersion relations in time-space periodic media: Part i-stable interactions. Proc. IEEE **51**(10), 1342–1359 (1963)
7. T. Tamir, H. Wang, A. Oliner, Wave propagation in sinusoidally stratified dielectric media. Microw. Theory Techn. IEEE Trans. **12**(3), 323–335 (1964)
8. C. Yeh, K. Casey, Z. Kaprielian, Transverse magnetic wave propagation in sinusoidally stratified dielectric media. Microw. Theory Techn. IEEE Trans. **13**(3), 297–302 (1965)
9. L. Matekovits, G. Colome, M. Orefice, Propagation of electromagnetic waves in a sinusoidally modulated dielectric substrate. Antennas Wireless Propag. Lett. IEEE **6**, 207–210 (2007)
10. P. Russer, *Electromagnetics, Microwave Circuit and Antenna Design for Communications Engineering*, 2nd edn. (Artech House, Boston, 2006)
11. C. Elachi, Waves in active and passive periodic structures: A review. Proc. IEEE **64**(12), 1666–1698 (1976)
12. É. Mathieu, Mémoire sur le mouvement vibratoire d'une membrane de forme elliptique. Journal des Mathématiques Pures et Appliquées **13**, 137–203 (1868)
13. N.W. McLachlan, *Theory and Application of Mathieu Functions*. (Clarendon, Oxford, 1951)
14. P.M. Morse, H. Feshbach, *Methods of Theoretical Physics – Part 1*. (McGraw-Hill, New York, 1953)
15. G. Floquet, Sur les équations différentielles linéaires à coefficients périodiques. *Annales Scientifiques de L'É.N.S.* **2**(12), 47–88 (1883)
16. R.E. Collin, *Field Theory of Guided Waves*. (IEEE, New York, 1991)
17. J. Manley, H. Rowe, Some general properties of nonlinear elements – Part I. General energy relations. Proc. IRE **44**(7), 904–913 (1956)
18. G.N. Burlak, N.Y. Kotsarenko, S.V. Koshevaya, Interaction of electromagnetic and acoustic waves in solids. Russ. Phys. J. **24**(8), 732–742 (1981)
19. K.S. Yee, Numerical solution of initial boundary value problems involving Maxwell's equations in isotropic media. IEEE Trans. Antennas Propagat. **14**, 302–307 (1966)
20. A. Taflove, S. Hagness, *Computational Electrodynamics: The Finite-Difference Time-Domain Method*. (Artech House, Boston, 2005)
21. J.A. Russer, P.S. Sumant, A.C. Cangellaris, A Lagrangian approach for the handling of curved boundaries in the finite-difference time-domain method. in *IEEE MMT-S International Microwave Symposium*, pp. 717–720. June 2007
22. J.A. Russer, P.S. Sumant, A.C. Cangellaris, Modeling of curved boundaries in the finite-difference time-domain method using a Lagrangian approach. Springer Proc. Phys. **121**, 55–68 (2008)
23. J.A. Russer, A.C. Cangellaris, An efficient methodology for the modeling of electromagnetic wave phenomena in domains with moving boundaries. in *IEEE MMT-S International Microwave Symposium*, pp. 157–160. June 2008
24. C.T. Kelley, *Solving Nonlinear Equations with Newton's Method*. (Society for Industrial and Applied Mathematics, Philadelphia, 2003)
25. M. Born, E. Wolf, *Principles of Optics*, 7th edn. (Cambridge University Press, Cambridge, 2002)

Autobiography

The Impossible Takes Longer

Peter Russer

That which hath been is now;
and that which is to be hath already been;
Ecclesiastes 3;15

1 Speak, Mnemosyne

The outpouring of good wishes that I received from so many friends, colleagues, former and present students at the celebration event and symposium on occasion of my retirement was overwhelming. This encourages me to give in the following a very personal account of my life and my career. I will try to give credit to all who have fostered me and enriched my path of life through their love, friendship and collaboration.

I don't have to emphasize that it is a delicate task to write an autobiographical text. Diving into the waters of mnemosyne for the treasures of memory, capturing them in the drift-net of language and reason the essential may slip away. Going back to my early childhood in remembrance of things past, islands of memory are surfacing, showing myself as a little boy with my parents at the countryside where woods were lovely, dark, and deep and the meadows and the heaven were bright. The solitariness of early recollections is in a peculiar contrast to the wide extension of the subjective time scale into the past. The origin is empty and infinite. The mind is setting up, creating space and time, and begins to order experience and thoughts in images and language and assembles the constituents of imagination and conceptual thinking.

I was born in 1943 in Vienna where I grew up in the Schottenfeldgasse. Looking at an old family photograph from 1950 gives a sense of how time passed by

P. Russer
Institute for Nanoelectronics, Technische Universität München, Arcisstrasse 21, 80333 Munich, Germany
e-mail: russer@tum.de

S. Lindenmeier and R. Weigel (eds.), *Electromagnetics and Network Theory and their Microwave Technology Applications*, DOI 10.1007/978-3-642-18375-1,

Fig. 1 With my parents and my sister in 1950

and world has changed (Fig. 1). The picture shows my mother Theresia, my father Eduard and my beloved sister Herta, who, 14 years older than I, has been like a second mother to me and passed away too early in 2006. Images are not a mere illustration but carry their own deep meaning complementing what words can say. Absent on the picture but still present in the mind of the family was my brother Eduard. Born in 1906 he had studied chemistry at the University of Vienna and had graduated with Wolfgang Pauli senior on colloidal gold, today also called nano-gold, a suspension of nanometer-sized particles of gold in a fluid [1–4]. My brother Eduard and his wife Toni died in Spring 1944 in Leuna. Our overlap in this life time had been too short to give me any memory of him. Like a message in a bottle traveling through the time his library in art and philosophy, his photographs, and his photography equipment reached me and had some impact on me in my youth.

Between the age of six to ten I spent summer vacations with my parents in eastern Styria. Since my father was already retired we could stay many weeks there. For one or two weeks we were joined by my sister joined who already was working and had less vacations. Collecting bugs and chasing butterflies together with my father and with the children of the village was one of my favorite occupations there. Preparing and mounting the prey of these subtle pursuits yielded a quite presentable collection over the years.

Railways and trains captivated me in this age. As a young engineer, between 1903 and 1905 my father was engaged in the railway tunnel construction in Slovenia. The 6.339 km long tunnel of Wochein (Bohijn) crosses the eastern foothills of the Julian alps from the north to south. I followed my father's stories from his experiences and adventures in the tunnel. The technical challenges in the construction of this tunnel is documented in a text also describing gorgeous details of geology and landscape in [5, pp. 232–242]. Let me give a short impression "*Der Wocheinertunnel, 6339 m lang, durchfährt in fast nordsüdlicher Richtung den Gebirgszug der östlichen Ausläufer der Julischen Alpen, welcher die Wasserscheide zwischen dem Adriatischen*

Fig. 2 My first electric toy train locomotive in 1949

und Schwarzen Meere bildet. Der Nordeingang liegt zunächst des Dorfes Wocheiner Feistritz auf der Meereshöhe von 525,4 m in dem flachgeneigten Vorlande der Kolba und dem breiten, landschaftlich herrlich schönen Savetal angesichts der imposanten Triglavgruppe. … Auf eine Länge von 1600 m dunkelgraublauen Tonmergel, mehr oder weniger feinkörnigen Sand mit kalkigem Bindemittel und Lettenlassen. Diese tertiären Ablagerungen sind in der folgenden an Dachsteinkalke angrenzenden Strecke mit Kalkgerölle konglomeratartig vermengt". I apologize for not translating this text. For me, the tunnel became a metaphor for the human quest for the intangible goal.

When I got my first electric toy train for Christmas in 1949 I was fascinated by the electric circuits for the train, illumination, and signals (Fig. 2). Since the toy train was directly connected to the 220 V DC power supply I got some sensual experience of electricity when touching the rails. This did not reduce my interest in this field. Already in my elementary school time it was clear to me that my later profession will have to do something with electricity. Another impulse fostering my interest in electrical things came from a drawer with radio components in the home of my aunt. My late uncle, like many people in the twenties and thirties of the past century, had built his radios himself and had left a rich collection of resistors, capacitors, inductor coils, and other miraculous things which I gained now.

After finishing the elementary school I attended from 1953 to 1961 the Realschule in the Neustiftgasse of the seventh district of Vienna. This type of school in Austria was a secondary school like the German Gymnasium with focus on mathematics and natural science. I had no difficulties in school but I cannot say that I loved to go to school since its main drawback was to keep me off from even more interesting things. However, looking back I acknowledge the excellent education the school and my teachers had provided. Especially, I would like to give credit to three of them, namely Erich Skalicky, who had given over the 8 years excellent courses in mathematics, Richard Tauber an impressive personality teaching French language, and Erich Liedl giving an inspiring course in literature. Over all the education in the Realschule furnished the students with a rich cultural background. I consider it a great fortune to have spent my childhood and youth in Vienna. This marvelous culturally vibrant city was still breathing the splendid plentifulness in art and science in the afterglow of its great epoch of intellectual movements from mid of the nineteenth century to the early twentieth century. Johnston's book on the Austrian mind gives some impression of this cultural heritage [6].

At the age of twelve I developed a strong interest in radio techniques and built my first radios with diodes and transistors. The OC 390, a popular germanium "high frequency" transistor at this time, had a cutoff frequency of 900 kHz and I had to spent my pocket money of a whole month for it. At this time I also got the "Kosmos" construction kit "Radiotechnik" from my father. Different from today's construction kits this one contained beech wood base plates, brass clamps with milled screws, inductors made from silk isolated wire on cardboard, rotary capacitors, a small galena crystal for assembling a detector (Fig. 3), an induction coil, a low voltage vacuum tetrode, and the excellent instruction book written by Wilhelm Fröhlich [7]. This enabled a wireless link with an induction coil transmitter and a coherer receiver, the latter made from iron swarf in a glass tube between magnetized knitting needles, a technology from the time of Karl Ferdinand Braun [8]. These experiments were successful, however my mother was not amused about her magnetic knitting needles.

The scientist's impetus has the same origin as the child's playing aptitude. In his magnum opus "Homo Ludens", a study in the play-elements in culture, the Dutch cultural anthropologist Johan Huizinga has named the sympathy and the solemn emotion generated by the game the holy gravity of the play [9, 10]. Like all culture,

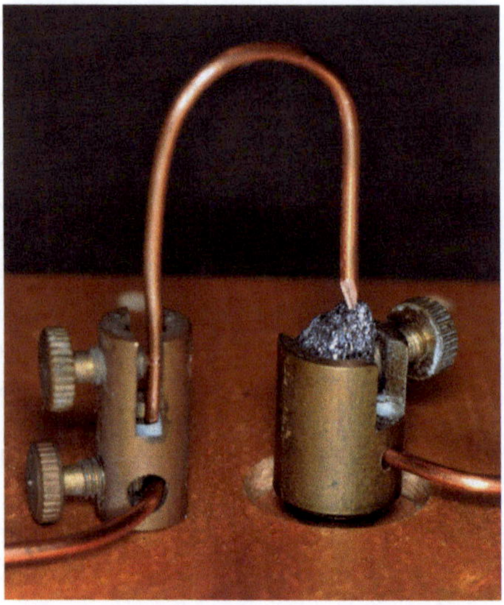

Fig. 3 Galena crystal detector assembled from the Kosmos construction kit

science emanates from the spirit of the play. I kept during my life the joy in playing and in playful combination of the building blocks of imagination.

2 At the Technische Universität Wien

I began my university studies in electrical engineering at the Technische Universität Wien in Fall 1961. Some extraordinarily excellent academic teachers have highly impressed me. Concerning the first 2 years I have to mention Rudolf Inzinger who gave a brilliant course in mathematics, in equal measure clear and profound. I also would like to mention the excellent courses in theoretical physics, namely in electrodynamics and thermodynamics, held by Otto Hittmair which were mandatory for electrical engineers in the sixties. After the intermediate diploma I had chosen a specialization in Communications Engineering. Günther Kraus covering communications engineering and Herbert W. König representing high frequency engineering were outstanding and highly influential academic teachers and scientists. I also attended the four–term courses in theoretical physics held by Theodor Sexl and Walter Thirring at the University of Vienna.

In 1966, Dieter Schuöcker offered me to do a diploma thesis on microwave amplification utilizing the quasiparticle tunnel effect in superconducting tunnel junctions. Superconducting tunnel junctions made by superconductors of different energy gap parameters on both sides of the tunnel barrier exhibit a current–voltage characteristics with a region of negative differential resistance. I investigated the amplification and noise properties of superconducting quasiparticle tunnel heterojunctions [11] in a theoretical work based on the microscopic theory of superconductivity from Bardeen, Cooper and Schrieffer [12].

At the beginning of the year 1968 Professor Hans Pötzl who held the chair of Physical Electronics at the Technische Universität Wien offered me a position as a research associate. In the sixties Hans Pötzl gave the course on semiconductor devices at the Technische Universität Wien. His area of research was focused on transport phenomena in semiconductors. Hans Pötzl was an extraordinary personality. Being scientifically brilliant, highly cultured, modest and kind he impressed everyone. Figure 4 shows Hans Pötzl among his coworkers.

When Hans Pötzl read my Diplom-Ingenieur thesis on the quasiparticle tunnel effect, he suggested to me to work on the AC Josephson effect and to investigate Josephson junctions and their applications for microwave detection and mixing. He got familiar with this matter during a sabbatical stay with Theodore Van Duzer at Berkeley and he was strongly interested in it. At our first discussion he said to me that he possibly would not be able to supervise my thesis as intensively as usual since the topic was a little separate from the main direction of his research. Nevertheless I joyfully accepted and all in all I think I have learned a lot from Hans Pötzl and feel deep gratitude to him.

The Josephson effect is the phenomenon of a supercurrent flowing between two weakly coupled superconductors where the weak coupling is achieved via an insulating tunnel barrier or a narrow bridge [14–16]. One interesting property of

Fig. 4 At the Institute for Physical Electronics: Ernst Bonek, Konrad Frank, Mrs. Lindner, Hans Pötzl, Erwin Hochmair, Ditmar Kranzer, Franz Seifert, and Peter Russer (from *left to right*)

Josephson junctions is that a DC voltage can be applied across the junction under maintenance of the superconducting state and quantum phase coherence over the junction. Application of a DC voltage V_0 yields an AC current with the frequency $f_0 = 2e_0V_0/h$, proportional to the applied voltage, where e_0 is the electron charge and h is Planck's constant. This phenomenon is called the AC Josephson effect. The ratio of frequency to applied voltage is 483.6 GHz/mV. Under microwave irradiation the Josephson oscillations synchronize to the irradiated microwave and the DC voltage–current characteristics exhibit constant voltage steps at voltages corresponding to the frequency of the incident radiation and their harmonics [17]. The step height depends on the amplitude of the incident radiation and our idea was to explore the potential of this effect for the realization of sensitive microwave detectors.

I performed my experimental investigations in 1968 at the Ludwig Boltzmann Institute for Solid State Physics in Vienna, where I had in time intervals of several weeks access to liquid helium. Helium was very expensive at this time and the institute stocked only a small quantity of it. The evaporating helium had to be collected for re-liquifying. Figure 5a shows the coaxial resonator used for the experiments with tantalum/niobium Josephson point junctions. On the bottom of the inner conductor of the coaxial resonator a niobium whisker was fixed. By a differential screw the niobium whisker could be moved vertically and brought into contact with a tantalum plate fixed on the bottom of the coaxial resonator. Microwaves were coupled into the resonator from an X-band steel waveguide via a coupling pin. The measurements were made in a liquid helium glass cryostat which was embedded in a liquid nitrogen cryostat. The differential screw allowed the variation of the pressure of the whisker during the measurements. On days when I got the ration of liquid helium I started at eight in the morning, assisted by two diploma students, with the preparation of the probe and then we filled and cooled down the cryostats, first the

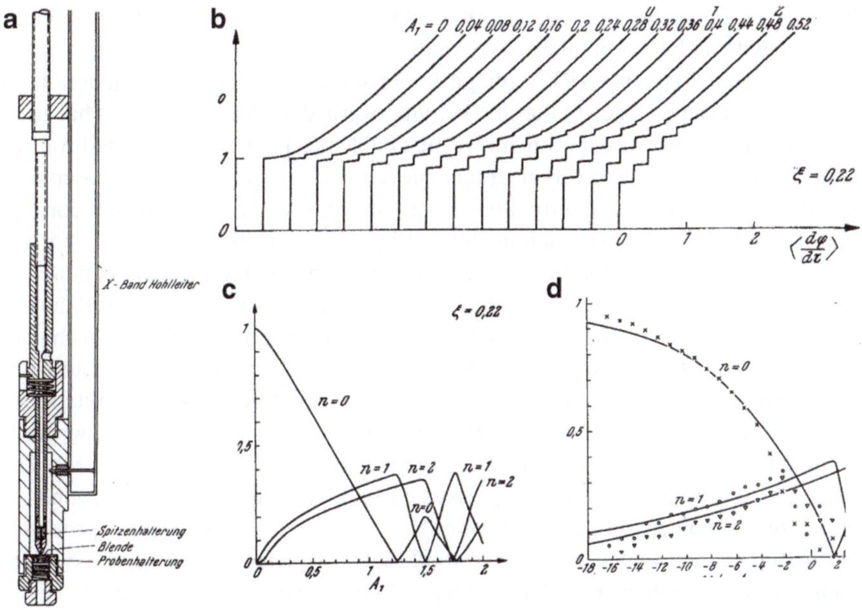

Fig. 5 Investigation of the AC Josephson effect: (**a**) cross sectional view of the coaxial resonator, (**b**) analog computation of the DC characteristics of the Josephson junction under microwave irradiation, (**c**) computed step height dependence on microwave amplitude, (**d**) comparison of theoretical and experimental values for the first three steps [13]

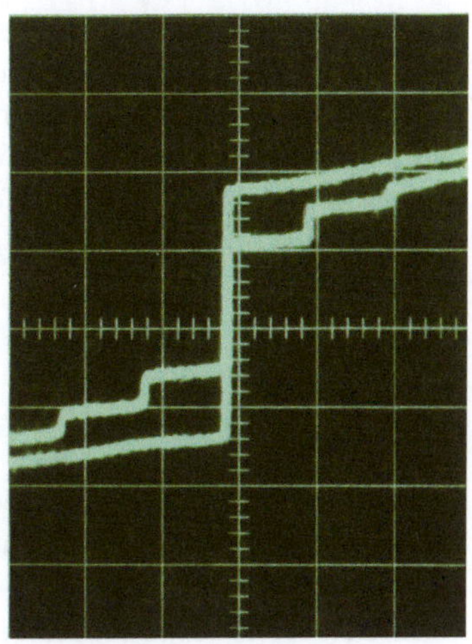

Fig. 6 DC voltage–current characteristics of a tantalum/niobium Josephson point junction with and without microwave irradiation at 10 GHz, *horizontal*: 20 μV/div, *vertical*: 500 μA/div [18]

liquid nitrogen cryostat and then the liquid helium cryostat. The filling of the liquid helium cryostat took several hours in order to minimize the evaporation during the filling. Usually we could start the measurements at eight in the evening and if we were successful in not to damaging the junction while varying the parameters we could continue the measurements until four in the morning. Figure 6 shows the oscilloscope screen-shot of the DC voltage tantalum/niobium Josephson point junction with and without microwave irradiation [13, 18].

For a radio frequency (RF) voltage impressed into the Josephson junction an RF amplitude dependence of the step height governed by Bessel functions has been predicted. However, the experimental results published in literature deviated considerably from this. For me it was clear that due to the low impedance of the Josephson junction we can impress a current but not a voltage. Using a simple model consisting of an ideal Josephson junction shunted by a linear resistor accounting for the normal conducting quasiparticle current flowing in parallel to the superconducting Josephson current I could give for the first time a quantitatively correct explanation of the influence of the microwave radiation on the step structure of the DC characteristics of Josephson junctions [13, 19, 20]. Figure 5b shows the DC characteristics for different values of the normalized impressed microwave current amplitude A_1 and for a value of 0.22 of the normalized parameter $\xi = hf_1/2e_0RI_{max}$, where R is the quasiparticle resistance and I_{max} the maximum DC Josephson current. The value $\xi = 0$ corresponds to an impressed current whereas $\xi = \infty$ corresponds to an impressed voltage. For the steps of order $n = 1, 2, 3$ the computed dependence of the step height from the microwave amplitude is depicted in Figs. 5c, d shows the comparison between measured and computed step height dependence on a logarithmic scale of incident microwave amplitude.Further work on the Josephson effect concerned the derivation of general energy relations for Josephson junctions governing the application of Josephson junctions as detectors, mixers and parametric amplifiers [21, 22].

Later, when working at other places I returned from time to time to engage with the Josephson effect. In a theoretical work from 1977, I proposed a DC pumped Josephson traveling wave amplifier [23, 24]. In 1983, I derived the generally covariant sine-Gordon equation for arbitrarily shaped large-area Josephson junctions, and I investigated the dynamics of rotating ring-shaped Josephson junctions with respect to possible applications for inertial rotation sensing [25]. Further work on the Josephson effect is discussed in Sect. 5.10.

3 Youth

In 1969 an important change took place in my life when I met Hilde Heimerl. Figure 7 shows both of us in that year. We imagined the magic of living together and married in July 1970.

In October 1971 we moved to Ulm, a small city at the Danube with the Gothic cathedral which has the tallest steeple in the world. There, Hilde and I spent ten happy years from 1971 to 1981. Our three children were born in Ulm, Martin in 1972, Andrea in 1974, and Johannes in 1977. Since 1974 we lived at a beautiful place on a hill rising from the danube, in close walking distance to the center of the city as well as to my work place. In every season we loved to hike – there is no English equivalent for the German word *wandern* – all together through the

Fig. 7 Hilde and I in 1970

surroundings, strolling through the valleys, meadows and woods of the Swabian Alps.

All that is now long ago and the memory of the past has crystallized over the depth of the years as the treasure of remembrance of blissful times of pure happiness. Our faces are transient. The time regained, is what has been preserved in images and words.

4 At the AEG–Telefunken Research Institute in Ulm

After finishing my PhD thesis at the Institute for Physical Electronics of the Technische Universität Wien I joined the research group of Berthold G. Bosch at the AEG–Telefunken Research Institute in Ulm. My task has been to develop the electronic circuits for broadband fiber optic communications. From November 1971 to the end of 1980 I have been with the Research Institute, where I worked on fiber optic communication, broadband solid-state electronic circuits, statistical noise analysis of microwave circuits, laser modulation, and fiber optic gyroscopes.

4.1 Optical Fiber Communication

The availability of coherent optical sources after the invention of the laser [26, 27] greatly stimulated the research in optical communications since the high optical carrier frequencies in the order of some 10^{14}Hz yields a high available bandwidth. However, the breakthrough for the idea of optical communications came with the concept of fiber optical communications. Based on a patent of Manfred Börner who has been department head in the AEG–Telefunken research institute Ulm since the sixties, in 1967 research towards high bit rate optical fiber communications has been started at AEG–Telefunken [28, 29]. The idea has been to use a directly modulated

GaAs based double hetero-structure semiconductor injection laser as the optical transmitter, a monomode quartz fiber as the optical transmission medium and a photo diode as the optical receiver. Similar proposals came at the same time from Charles K. Kao and George Hockham in England [30], and from Alain Werts [31] in France. The expectation has been to realize by this way low cost fiber optical transmission systems with Gbit/s transmission rates [32–34].

At the end of 1971, the chances to realize broad-band optical fiber communication could have been considered to be rather discouraging since the lifetime of GaAs semiconductor injection lasers has been in the order of minutes under continuous wave room temperature operation conditions and the attenuation of optical fibers has had to be expressed in dB per meter. In spite of these adverse conditions every endeavor has been made at AEG–Telefunken to push forward the research in fiber optical communications. In the optical communications research group of Stefan Maslowski around 30 people performed research and development covering all components required for fiber optical communications [35–41]. From the members of this research group I would like to mention Günther Arnold, Joachim Guttmann, Oskar Krumpholz, Peter Marschall, Ewald Schlosser, Hans-Peter Vollmer, Edgar Weidel, Claus Wölk, and since 1976 also Klaus Petermann. The topics included material technology and structuring of semiconductor injection lasers, photo diodes, optical fiber technology, and related topics. In 1972 Berthold G. Bosch left the research institute and I joined together with my laboratory the fiber optics group of Stefan Maslowski.

My main task was to develop experimental broadband fiber optic communication systems achieving gigabit per second (Gbit/s) rates. At this time the direct modulation of semiconductor injection lasers at bit rates of several hundred megabit per second (Mbit/s) was a considerable challenge. Figure 8 shows me investigating

Fig. 8 At the experimental investigation of the direct modulation of a semiconductor injection laser with a bit rate of 500 Mbit/s in the year 1972

the direct modulation of a semiconductor injection laser at 500 Mbit/s. Since at the mid of the seventies modulation amplitudes in the order of 100 mA were required for direct modulation of semiconductor injection lasers it was not possible to realize modulation amplifiers for gigabit rates with transistors. The problem could be solved with the step recovery diode amplifier [42, 43]. Furthermore, monolithic circuits for such high bit rates have not been available. Together with my research group members Johann Gruber, Michael Holz, Peter Marten, Reinhard Petschacher, and Siegfried Schulz, I developed electronic components for digital fiber optic transmitters and receivers with bit rates from several hundred Mbit/s up into the Gbits/s range. All high speed components were realized in thin film hybrid integrated technology using silicon bipolar transistors, Schottky diodes, and step recovery diodes. In particular, drivers and multiplexers suitable for direct laser modulation were developed for use in the transmitter units. A demultiplexer using fast hybrid integrated emitter coupled logic (ECL) gates for 1 Gbit/s pulse code modulation signals has been realized in 1977 [44] and a demultiplexer and clock regenerator circuit was developed for optical receivers [45]. The technicians of the group, Siegfried Neumann and Roman Sobkowiak, gave valuable assistance in the fabrication of the circuits. Together with Johann Gruber, Peter Marten, and Reinhard Petschacher, from the Nachrichtentechnische Gesellschaft (NTG), I received the NTG award 1979 for the publication *"Electronic circuits for high bit rate fiber optic communication systems"* [45].

The development of hybrid integrated circuits for signal processing in the Gbit/s region yielded worldwide the first realization of an optical fiber transmission link for 1 Gbit/s [45–49]. Figure 9 shows me with the laboratory setup of the 1 Gbit/s fiber optic communications link. The cable reel contains the 1.6 km long cable of the fiber

Fig. 9 With an early high-bit-rate fiber optic link

optic test link. In 1979 also an experimental 280 MBit/s fiber optic communication link based on monolithic integrated ECL circuits was realized [50, 51].

4.2 Dynamics of Semiconductor Injection Lasers

One major problem to be solved in order to facilitate high bit rate fiber optical communication was the direct modulation of semiconductor injection lasers. Under direct modulation at high frequencies semiconductor lasers exhibit nonlinear relaxation oscillations. It already had been shown experimentally that sinusoidal modulation of semiconductor lasers is possible up into the GHz range. However, the direct modulation of semiconductor lasers with a bit pattern in the Gbit/s range had not been realized at this time.

In 1973, I demonstrated together with Siegfried Schulz the direct modulation of a semiconductor injection laser at 2.3 Gbit/s with low bit pattern dependence. This result was an essential precondition for the realization of broadband digital optical fiber communication links and remained unsurpassed by other research groups until the end of the seventies [52]. In the review papers [53,54] an overview of the state of the art in direct modulation of semiconductor injection lasers has been given. When doing the first gigabit modulation experiments in 1973, we initially had to build a bit pattern generator for this bit rate since Gbit/s bit pattern generators have not been commercially available at this time. This problem has been solved by converting the 460 MHz signal of a radio frequency generator into a narrow pulse train and after power splitting, variable delay, and switched recombination, two different 5 bit words at 2.3 Gbit/s could be generated. By comparing the two modulation signals we were able to check to what extent bit pattern effects occurred. The 2.3 Gbit/s originated from the circumstance that the only available old radio frequency power generator in the laboratory did not provide sufficient output power beyond 460 MHz in spite of its specification up to 500 MHz. A consequence of this has been that in the following years I often have been asked at conferences whether the German Post is planning broadband fiber optic communications at 2.3 Gbit/s.

In the years from 1975 to 1977, I have performed investigations on the improvement of the spectral and modulation behavior of injection lasers by coherent light injection. The first papers [55, 56] published in 1975 have shown the improvement of the modulation behavior by light injection theoretically. In [56] the improvement of the PCM modulation behavior of injection lasers has been demonstrated. In the German Patent DE2514140 [57], submitted on March 29th, 1975 also several methods of laser coupling, including the application of an optical isolator have been proposed. In [58] an integrated structure of two laterally coupled injection lasers is proposed. The US Patent 4,101,845 [59] is based on the German patents [57, 58]. The paper [60] contains the experimental investigation of coherent light injection on injection laser modulation behavior. An extended version of this work has been published in [61].

4.3 Thermal Noise Analysis

In 1975 the semiconductor division of AEG–Telefunken in Heilbronn asked for support for the development of a low-noise silicon monolithic integrated broadband amplifier with 1 GHz bandwidth. Together with Herbert Hillbrand, I developed the mathematical tools needed for the noise analysis and optimization of microwave and millimeterwave circuits by combining methods of circuit analysis and the representation of noise signals using correlation spectra [62–65]. The methods have been applied successfully for the computer aided design of monolithic integrated differential amplifier with 1 GHz bandwidth [66]. Subsequently, the methods developed in this project have been widely adopted by software developers and are now incorporated in all leading computer aided design (CAD) programs. Later on, I extended this work together with Stefan Müller to the S-Parameter analysis of linear noisy networks with general topology [67–70]. In 1994, I have been elevated to the Fellow Grade of the IEEE *for fundamental contributions to noise analysis and low-noise optimization of linear electronic circuits with general topology.*

4.4 The Invention of the SiGe Hetero-Bipolar Transistor

The state of the art of today's silicon based semiconductor devices allows the realization of circuits with operating frequencies beyond 200 GHz. The availability of ecologically friendly low-cost high frequency semiconductor devices opened the door for consumer applications in communication technology and sensorics up into the millimeter wave region. A key element of the silicon based high frequency semiconductor electronics, is the silicon-germanium based hetero-bipolar transistor (SiGe HBT). A bipolar transistor with an emitter of wider energy gap than the base was already mentioned explicitly in William Shockley's original patent [71]. The hetero-junction bipolar transistor however was proposed for the first time by Alfons Hähnlein from the Fernmeldetechnische Zentralamt (FTZ) in Darmstadt, the research institute of the German Federal Post Office [72]. Hähnlein's German patent DE 1021488 with the title *"Halbleiter-Kristallode der Schichtenbauart"* (semiconductor cristallode with layer design) has been filed February 19th, 1954 and issued on July 10th, 1958 [72, 73]. In his patent Alfons Hähnlein described a bipolar transistor for which the emitter layer exhibits a higher band gap than the basis layer, with the special feature that the base layer is doped higher than the emitter layer. In the second claim of the patent, Alfons Hähnlein proposed Si as the emitter material, and Ge as the base material. In July 1954, Herbert Kroemer submitted a paper in which he formulated the idea of wide-gap emitter design [74]. He presented the theory of the wide-band emitter transistor in detail in 1957 [75, 76]. However, at this time the technology for the realization of this transistor was not available.

In the mid of the seventies at the AEG–Telefunken research institute, Erich Kasper has grown one-dimensional SiGe superlattices with periods ranging from 10 to 80 nanometers on Si substrates by means of ultra high vacuum epitaxy [77].

The reason has been the quest for an artificial silicon based optical semiconductor. In early 1978 I met Alfons Hähnlein in Darmstadt who told me about his broad-band emitter transistor patent from 1954. I discussed this idea with Erich Kasper and we concluded that his ultra high vacuum epitaxy technology would be suitable to realize the broad-band emitter transistor if we could cope with the lattice mismatch problem. The solution has been the double hetero-structure. In the invention submitted to the German Patent office on April 30th, 1977 and disclosed by the German Patent office on December 21st, 1978 (Disclosure P 27 19 464, "*Verfahren zur Herstellung von Bipolartransistoren*"), Erich Kasper and I proposed for the first time a double hetero-structure bipolar transistor on the basis of a mono-crystalline silicon germanium mixed crystal system and specified precise dimensioning rules and technological fabrication procedures [78]. Figure 10 shows the schematic of the double hetero-structure transistor which was proposed in this patent. According to this disclosure, by application of ultra-high vacuum technology to a mono-crystalline silicon substrate (1), first an n/p silicon layer (2) is grown as the collector. Then a thin p/n silicon-germanium mixed crystal layer (3) with a thickness less than 200 nm is grown to form the base of the transistor. On this layer the silicon emitter layer (4) is grown. This has been an essential step to reduce the lattice mismatch.

At the time when we made this invention Erich Kasper and coworkers already had grown SiGe superlattices with their highly developed silicon germanium ultra high vacuum epitaxy equipment at the AEG–Telefunken Research Institute. We had the technological means to realize the SiGe HBT [77]. However, we could not persuade our company to pursue the project.

The first realized SiGe HBT has been reported in literature by IBM researchers more than 10 years after our invention [79, 80]. Many people thought the idea was of value only for a few exotic niche applications. In his paper on the early history of IBM's SiGe mixed signal technology David L. Harame stated "This is a story about how a small group of people persuaded a large digital computer manufacturer to invest in a new unproven technology for telecommunication applications in a field which the company knew little about. It is a success story, as SiGe technology has now become the only BiCMOS technology in development in IBM and is in the roadmaps of every major telecommunication company" [81].

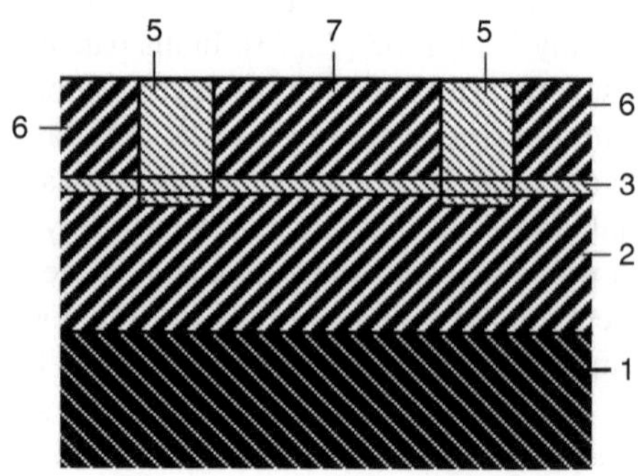

Fig. 10 Schematic of the SiGe HBT as proposed in the disclosure [78]. The numbers correspond to various layers used to fabricate the transistor

4.5 *Optical Fiber Gyroscopes*

In 1978, we started research work on fiber-optic gyroscopes at the AEG–Telefunken research institute. The fiber-optic gyroscope uses the interference of two light waves propagating in a ring interferometers along a fiber coil in opposite directions for inertial rotation sensing. Based on a general relativistic effect the propagation time of the two counter-propagating light waves becomes different when the fiber coil rotates around its axis with respect to the inertial frame. The sensitivity of the gyroscope was limited by noise due to Rayleigh backscattering of the light wave in the fiber. One day Konrad Böhm, when investigating the temperature dependence of the experimental setup, placed a fan on the vibration isolated table supporting the setup. The oscilloscope screen immediately showed a dramatic decrease of the system noise. The explanation was found soon. The vibrations of the fan reduced the time coherence of the light so that the interference of backscattered light yielded a broad noise signal spectrum for which only a small part overlapped with the signal spectrum. We could show that the noise can be reduced either by introducing a phase modulation into the fiber ring or by the use of a low-coherence source. By this way we could increase the sensitivity of fiber gyros by more than one order of magnitude compared with the state of the art at this time [82–84].

5 At the Technische Universität München

In 1980, I was appointed Full Professor and Ordinarius of the Institute of High Frequency Engineering of the Technische Universität München as of January 1st, 1981. At first I started to develop new courses. A four-term course in High Frequency Engineering comprised electromagnetic fields, waveguides, antennas, active linear, nonlinear and noisy circuits. Courses in Optical Communications and Quantum Electronics and also an introductory course covering the Fundamentals of Information Technology followed. For all courses I wrote lecture notes which were published and distributed by the institute. For two courses I also wrote textbooks. The book on fundamentals of information theory appeared in 1988 [85]. I introduced for the very first time the exterior differential calculus in the teaching of applied electromagnetics. Exterior calculus can considerably simplify the formulation of Maxwell's theory and its applications. For the three term electromagnetics course I wrote the textbook "Electromagnetics, Microwave Circuit and Antenna Design for Communications Engineering" which appeared in 2003 and in a considerably extended second edition in 2006 [86, 87]. The exterior differential calculus developed by Élie Cartan [88] is based on the algebraic structures introduced by Hermann Günter Grassmann in his book "*Die lineale Ausdehnungslehre, ein neuer Zweig der Mathematik*", published in 1844 [89]. Exterior differential calculus has simple and concise rules for computation. Its objects have a clear geometrical significance and the geometrical laws of electromagnetics assume a simple and elegant form [90–95].

Today mathematicians consider exterior differential calculus to be the most suitable framework for geometrical analysis and field theory.

Since the eighties, I also gave a course on Optical Communications, dealing with the fundamentals of optical fiber communications and a course on Quantum Electronics, treating the quantum theoretical foundations of the interaction of electromagnetic radiation and matter. After 2005, I also treated superconducting and semiconducting quantum devices in this course. Since the name "Quantum Electronics" is already occupied for the physics dealing with the interactions of electrons in matter with photons, I have chosen the name "Quantum Nanoelectronics" for the course.

Over the years I have graduated more than 400 students and supervised and graduated 60 PhD students. The diploma and PhD students were embedded in our research projects and were guided in this way for the scientific work. Ten of my former students have became Professors themselves:

- Erwin Biebl, Technische Universität München
- Gerhard Fischerauer, Universität Bayreuth
- Josef Hausner, Ruhr-Universität Bochum
- Franz X. Kärtner, Massachusetts Institute of Technology, Cambridge, MA
- Stefan Lindenmeier, Universität der Bundeswehr, München
- Martin Rieger, University of Applied Sciences, Albstadt-Sigmaringen
- Sebastian Sattler, Universität Erlangen-Nürnberg
- Gerd Scholl, Universität der Bundeswehr, Hamburg
- Alejandro Valenzuela, University of Applied Sciences, Bonn-Rhein-Sieg
- Robert Weigel, Universität Erlangen-Nürnberg

Figure 11 shows some of them together with me at the Symposium on the occasion of my retirement on 8 October 2008.

I would like to thank some of my coworkers for their valuable assistance and support. Until 1986, Karl-Heinz Türkner and thereafter Gerhard Olbrich have served as Academic Directors. In this capacity they have contributed to research and teaching, and to the administration of the institute. In the fine mechanical workshop of the institute run until 2000 by Manfred Fuchs, Manfred Agerer, and Josef Franzisi, mechanical components of the highest precision were made. Manfred Fuchs who led the workshop passed away in 2000. Since then the workshop is lead by Manfred Agerer. I thank our technician Thomas Mittereder who did an excellent job in assembling electronic circuits and in serving our computer systems.

The last 30 years brought an increasing internationalization of the University. I have established numerous scientific collaborations with colleagues from European countries, North America, China, and Japan. Through activities in the European and International Microwave Communities, especially in the IEEE Microwave Society and in the European Microwave Association, I could establish scientific exchange and personal relations with colleagues all over the world.

Numerous colleagues spent research semesters at my institute, supported by the Deutsche Forschungsgemeinschaft, the German Academic Exchange Service (DAAD), or the Alexander von Humboldt Foundation. To provide an international

Fig. 11 Stefan Lindenmeier, Robert Weigel, Gerd Scholl, Peter Russer, Franz Kärtner Josef Hausner, Erwin Biebl and Gerhard Fischerauer (from *left to right*)

course program focused on education in radio frequency engineering I have put on the way the course "Master of Science in Microwave Engineering". The course started in the Winter term 2000/2001 and comprised three terms with lectures and one term dedicated to a master thesis. The students were coming from Bangladesh, Brasilia, Bulgaria, Cameroon, Canada, China, Czech Republic, Greece, India, Ireland, Israel, Korea, Nepal, New Zealand, Palestine, Russia, Turkey, Venezuela, Vietnam. Also in the courses held in German language a large number of students from other countries, especially European countries and the former Soviet Union, were participating in the last years.

In the years from 2002 nearly every year two visiting professors from North America or England stayed the whole summer term at the institute and gave courses within the Master of Science in Microwave Engineering program. I have to thank here the following colleagues who gave courses and also contributed to research projects:

- Andreas Cangellaris, University of Illinois, Urbana-Champaign, USA
- Christos Christopoulos, University of Nottingham, UK
- Wolfgang J. R. Hoefer, University of Victoria, Canada
- Steve Maas, University of California, Los Angeles, USA
- Zoya Popovic, University of Colorado, Boulder, USA
- Mohamed I. Sobhy, University of Kent, UK
- Emmanouil Tentzeris, Georgia Institute of Technology, Atlanta, USA
- Karl F. Warnick, Brigham Young University, Provo, Utah, USA
- Ke Wu, University of Montréal, École Polytechnique, Canada

Since 1990, I have a still ongoing research cooperation with Wolfgang Hoefer. He has drawn my attention to the transmission line matrix (TLM) method, a powerful method for numerical modeling of electromagnetic fields, which became one of my main research areas. We started our scientific cooperation during my research stay with him at the University of Ottawa from March to May 1990. Research stays of Wolfgang Hoefer in Munich and Berlin and of me in Victoria followed. Also our PhD students were involved in numerous joint publications. In 2008, Wolfgang Hoefer has been bestowed the Honorary Doctor Degree by the Faculty of Electrical Engineering and Information Technology at the Technische Universität München for "extraordinary scientific achievements in the theory of electromagnetic fields" [96].

An intensive and prolific scientific collaboration has taken place since 1991 with Leopold Felsen and Mauro Mongiardo. Ties with Leopold Felsen were initiated through his invited attendance of the "International Workshop on Discrete Time Domain Modeling of Electromagnetic Fields and Networks", which I have organized in Munich in October 1991. Over a 14 years period we have had a fruitful scientific cooperation together with Mauro Mongiardo. Our cooperation yielded numerous publications [97–100] and at the end the monograph "Electromagnetic Field Computation by Network Methods" [101]. Leopold Felsen has been an exceptional theoretician in electromagnetics and also a strong human character. To meet him has been a great encounter. In 2004, the Faculty of Electrical Engineering and Information Technology at the Technische Universität München bestowed him the Honorary Doctor degree for "extraordinary scientific achievements in the theory of electromagnetic fields". The contributions to a workshop organized in honor of Leopold Felsen are summarized in [102]. Leopold Felsen passed away on September 24th, 2005. We miss him.

Andreas Cangellaris came two times, together with his family, to Munich for whole summer terms. With Andreas I already had started cooperation in the area of TLM in 2001 in network–oriented modeling, complexity reduction and system identification techniques for electromagnetic systems [103]. Numerous joint research activities in that area followed. Ke Wu came together with his Family. Also Karl Warnick came two times together with his wife and their six children for a longer research stay to Munich. Karl Warnick and I worked together in the area of electromagnetics, especially on the application of exterior differential forms [104], and we have written a book on solving problems in electromagnetics applying exterior differential forms [105].

I also would like to mention the fruitful cooperation with the Moscow Aviation Institute (MAI) over the past two decades. The impetus came from Dmitriy Leonov, a student of the MAI, who visited me 1990 and expressed the desire to cooperate and to exchange students. In December 1990, I visited together with Jürgen Detlefsen and Gerhard Olbrich the MAI and on January 24th, 1991 a cooperation agreement between MAI and TUM was signed. From July 28th to August 4th, 1991 a first group of students, young scientists, and professors of the MAI visited the Institute for High Frequency Engineering of the TUM, and between 1991 and 2004 nine scientific exchange seminars were held, five in Munich and four in

Moscow. This exchange was funded by the DAAD. The first exchange scientists were Vitali Chtchekatourov staying in Munich from April 1998 to April 2001 and Ivan Daviditch staying in Munich during September 1998. With the visit of Vitali Chtchekatourov we started joint research in the application of system identification methods to numerical electromagnetics which has been extended considerably since 2003 by the cooperation with Yury Kuznetsov and Andrey Baev who have visited Munich since 2003 every year. We have focused our work on compact model generation for electromagnetic structures. On April 25th, 2007, I was awarded an honorary doctorate from the MAI.

In the list of my scientific partners I also have to give credit to Damienne Bajon from the Institut Supérieur de l'Aéronautique et de l'Espace (SUPAERO) in Toulouse, to Wen-Quan Che from the Nanjing University of Science and Technology, to Poman So from the University of Victoria, and to Ayhan Altintas from the Bilkent University in Ankara for productive scientific cooperation. I apologize to all colleagues with whom I have worked in the past 30 years and I have not mentioned here.

The scientific exchange also brought close private contacts with the partners and led to marvelous and enriching friendships, also between the families. During my 3-month research stay in Ottawa with Wolfgang Hoefer in 1990 my wife and our three children, Martin, Andrea, and Johannes were with me. The children went to school in Ottawa which has been a positive experience for them. Many colleagues visited us for a longer stay, together with their families in Munich. Our international scientific community also is a marvelous social and cultural network that enriched our lives in many ways.

Before I am going to give an overview over my research activities in the following decades I would like to make a general remark. In engineering sciences, research means to make the impossible possible. This distinguishes a research project from a development task. Naturally, the research plan has to be established upon a solid fundament of knowledge and experience and one must have a clear plan how to approach the goal initially looking intangible. The impossible takes longer but it is the only thing that pays the effort.

5.1 Electromagnetics

With increasing bandwidths and data rates of modern electronic circuits and systems, electromagnetic wave phenomena which in the past had to be considered only in the domain of the radio frequency engineering, are now becoming crucial in the design of analog and digital systems. Design, modeling and optimization of high-speed analog and digital electronic circuits and systems, photonic devices, antennas, radar and communications systems, require the application of advanced tools in computational electromagnetics. Methods of electromagnetic field computation and their application to circuits, components, antennas and systems developed to the central area of research in my institute. My areas of research in electromagnetics comprised analytic as well as numerical methods and also combinations and hybridization of these methods.

Compared to a network-oriented design, a field-oriented design of circuits and systems requires a tremendously higher computational effort. The availability of steadily increasing computing facilities has not reduced the demand for efficient methods of electromagnetic field computation. This is readily understandable especially in the highly competitive design of broadband and high-speed electronic components. The demands for volume, weight, and cost reduction foster a compact and complex design of electromagnetic structures yielding a high computational effort in electromagnetic modeling.

Applying electromagnetic field analysis to technical problems requires numerical computations in general. However, the numerical effort can be considerably reduced by analytic preprocessing of the problem. Analytic methods are less versatile than numerical ones and usually they are applicable to a special class of problems only. Therefore, when performing an electromagnetic design task the most appropriate method and design tool has to be chosen. If a certain class of design tasks has to be solved repeatedly, it pays to develop a specific method based on advanced analytic preprocessing. Furthermore, a profound knowledge of theoretical fundamentals and analytical methods of electromagnetic theory is an indispensable basis for the design engineer, even if he or she uses numerical design tools. In the following I give a brief overview over my research in the area of electromagnetics.

Together with Leopold Felsen and Mauro Mongiardo I investigated network methods for a systematic treatment of electromagnetic field representations in complex structures [97–101, 106–114]. The application of network methods has proven to be an efficient tool in electromagnetic problem formulation and solution. In the context of network methods I also investigated gyrator surfaces which are a field theoretical analogue to Tellegen's gyrator circuit in network theory[115].

Network methods based on mode matching, also called partial wave synthesis, are an efficient tool for electromagnetic field computation of all structures which can be segmented into substructures for which analytic field solutions are available. Jochen Kessler applied partial wave synthesis to model the electromagnetic properties of high-temperature superconducting coplanar waveguides [116–120]. This project was supported by Siemens. The work has been continued by Rolf Schmidt, who extended the scope to waveguide discontinuities [121, 122]. Later, Dzianis Lukashevich used these methods for the modeling of interconnect structures in monolithic integrated circuits [123–125]. He also introduced a hybrid mode matching–TLM method together with Borys Broido to model discontinuities and waveguide junctions [126–128]. Further work on mode matching has been done together with Leopold Felsen, Mauro Mongiardo, Roberto Sorrentino, and Cristiano Tomassoni [129–134]. Bruno Biscontini applied mode matching to cylindrical structures to model antenna arrays [135–139].

The transmission line matrix (TLM) method is a powerful method for the numerical modeling of electromagnetic structures in the time domain. First published by Johns and Beurle in 1971 [140], the TLM method has been further developed by Wolfgang Hoefer [141–147]. I started my research work on the TLM method during my research stay at the University of Ottawa in 1990 where I was visiting Wolfgang Hoefer. In TLM the electromagnetic field is modeled by wave pulses propagating in a mesh of transmission lines. The wave pulses are scattered in the mesh nodes. It is interesting to note that the TLM scheme shows similarities to the theoretical concept that Christian Huygens has presented in 1690 in his "Traité de la lumière" [148, p. 14] explaining light propagation by a model looking like a billiard game of small ether spheres.

The TLM method exhibits an excellent numerical stability and is also suitable for modeling of complex three-dimensional structures exhibiting lossy, dispersive, and nonlinear media. The TLM method is based upon the mapping of the electromagnetic field problem into a network problem. This makes the TLM method excellently suited for applying network oriented concepts for problem solution [103, 133, 149, 150]. During my visit in Ottawa I investigated together with Wolfgang Hoefer and Poman So the modeling of nonlinear active distributed circuits in TLM [151]. In Munich I continued the work on TLM together with Bertram Isele who developed the first in–house TLM simulator software at the Institute for High Frequency Engineering. This simulator software has been further developed over many years by Tobias Mangold, Wolfgang Dressel, and Petr Lorenz and resulted in the open source software YATSIM (Yet Another TLM Simulator) [152]. Bertram Isele applied TLM to model nonlinear dispersive active structures [153, 154], planar and coplanar circuits [155–157]. He also developed a technique together with Mohamed Sobhy and Christos Christopoulos for analyzing general electromagnetic structures including distributed regions and lumped non-linear sub-circuits, interfacing the TLM with the state space method [158].

When I headed the Ferdinand Braun Institute in Berlin from 1992 to 1995 (see Sect. 6) I also supervised a small group of students there, doing research work on electromagnetics. Members of this group were Bernhard Bader, Michael Krumpholz, Stefan Lindenmeier, and Monika Niederhoff. Michael Krumpholz investigated the theoretical foundations of the TLM method. We formulated the TLM scheme in Hilbert space and derived it from Maxwell's equations using the Method of Moments [159–166]. Bernhard Bader worked on the alternating transmission line matrix (ATLM) scheme [167–169]. Monika Niederhoff developed a full-vector beam-propagation method in which the discretization of Maxwell's equations is performed by finite integration and she applied it successfully to the modeling of laser diode structures [170–172]. Stefan Lindenmeier developed a hybrid dynamic-static finite-difference method for numerical modeling of electromagnetic fields [173–177]. This method improved the computational efficiency of the finite-difference scheme considerably by combining the dynamic full-wave analysis with a quasi-static approach. Structure details which require a spatial resolution far below the wavelength are treated by a quasi-static analysis. The mesh for the dynamic analysis can be coarse without degrading the computational accuracy. In 1996, Stefan Lindenmeier joined the Institute for High Frequency Engineering of the TUM. He worked on numerical electromagnetic methods for applications concerning electromagnetic compatibility, and microwave circuit and antenna design.

In continuation of the work of Stefan Lindenmeier who has introduced static subgridding to the finite-difference method [178], Wolfgang Dressel introduced static subgridding into the transmission line matrix method [179, 180].

Luca Pierantoni, coming from the Università Politecnica delle Marche in Ancona, joined the Institute for High Frequency Engineering from 1996 to 1998. Together with Stefan Lindenmeier he developed a hybrid finite-difference-integral equation (FDIE) method combining the versatility of the finite-difference method with the computational efficiency of the integral equation method [178, 181–185]. Hence, the FDIE method is excellently suited for the analysis of electromagnetic compatibility (EMC) problems. It allows the electromagnetic modeling of structures consisting of complex objects with large separation distance. Stefan Lindenmeier's Habilitation Thesis has been related to this area [186]. Rachid Khlifi developed a hybrid method combining the transmission-line matrix method and the time-domain method of moments [187–190]. The method is highly effective for the analysis of the interaction between complex electromagnetic structures separated by large free space intervals.

Martin Aidam derived the TLM scheme from Maxwell's equations by finite integration [191]. The focus of his work was on the investigation of wavelet methods in connection with finite-difference schemes for the solution of partial differential equations [192–194].

Wolfgang Hoefer and I investigated the generation of lumped element equivalent circuits of distributed microwave circuits on the basis of TLM simulations. Starting with a TLM analysis of a distributed multi-port circuit the impulse response functions for reflection and transmission between the ports are computed. The poles are extracted within a specified domain of the complex frequency plane after numerical Laplace-transformation of the impulse functions. From these poles canonical equivalent circuits representing the branches of the lumped element equivalent circuit are derived directly. In this manner the topology as well as the parameters of the lumped element equivalent circuit are determined [195, 196]. Tobias Mangold continued this work and developed a method for the automated extraction of lumped-element equivalent circuits for linear passive reciprocal distributed microwave circuits on the basis of the numerical data obtained from TLM simulation. The method yielded the lumped element equivalent circuit topology as well as parameter values while preserving circuit properties like reciprocity and passivity [197–200]. Tobias Mangold also applied the method to the modeling of multichip modules.

Vitali Chtchekatourov who came from the Moscow Aviation Institute has contributed system identification and spectral analysis methods to calculate the circuit parameters and to establish network models of distributed microwave circuits [201–204]. The were extracted in real-time during the running TLM simulation, and the simulation was terminated when the approximation accuracy was adequate. By this way the computation time could be reduced considerably.

Fabio Coccetti introduced a system identification and Prony's method based algorithm, for the computation and prediction of time-domain transient response of passive distributed circuits [205]. With this approach he could synthesize a lumped element equivalent circuit modeling the distributed circuit over a wide frequency band. In his PhD thesis he investigated the application of system identification to full-wave time domain characterization of microwave and millimeter wave passive structures [206].

In a successful long–term cooperation with Yury Kuznetsov and Andrey Baev from the Moscow Aviation Institute the application of system identification methods to the extraction of lumped element and delay line models from wide-band transfer functions of complex three-dimensional electromagnetic structures has been investigated systematically [150, 207–216].

With Andreas Cangellaris from the University of Illinois at Urbana-Champaign I have an ongoing research cooperation in numerical electromagnetics since 2001. Model order reduction became a principal area of our joint research. The basic idea of the model order reduction is to reduce the order of a large linear system of equations before solving it. Dzianis Lukashevich has investigated together with Andreas Cangellaris the application of model order reduction to the transmission line matrix scheme by applying Krylov subspace methods and using the basic Arnoldi and non-symmetric Lanczos algorithms [217–219]. A novel scattering-symmetric Lanczos algorithm, which is faster and requires less memory in comparison to the conventional non-symmetric Lanczos algorithm has been proposed in [220, 221]. A further improvement has been achieved by the introduction of a second projection of the TLM system in order to extract only those eigenvalues and associated eigenstates that are the most influential on the system response in the frequency band of interest [222, 223]. Dzianis Lukashevich and Fabio Coccetti combined the application of model order reduction and system identification to TLM [224]. They also applied a fast multipole method (FMM) to the model order reduction (MOR) for the fast and efficient treatment of large scattering problems [225, 226].

Petr Lorenz, together with José Vagner Vital and Bruno Biscontini proposed a high-throughput transmission line matrix (HT-TLM) system, capable of performing high-performance computing of complex electromagnetic structures in grid environments [227–229].

Martin Aidam and Jürgen Rebel investigated the accuracy and the convergence of the symmetrical condensed node–transmission line matrix scheme [230, 231]. In his PhD thesis Jürgen Rebel investigated the foundations of the TLM method [232].

Marcello de Sousa and José Vagner Vital together with Leonardo de Menezes from the Universídade de Brasilia applied the two dimensional transmission line matrix power flow (TLMPF) method to model the ultra wide band system coverage [233, 234]. A similar approach has been applied by Uwe Siart, Susanne Hofmann, and Nikolaus Fichtner [235, 236].

Petr Lorenz developed a method for the modeling of discrete and modal sources in the transmission line matrix (TLM) method by means of connection networks. Discrete sources are modeled with connection networks based on parallel and series adaptors of wave digital filters (WDFs). Modal sources are modeled with an ideal transformer network [237].

MEMS (micro electro mechanical system) radio frequency switches exhibit low insertion loss, high linearity and exhibit low power consumption for control. Wolfgang Dressel, Fabio Coccetti, Vitali Chtchekatourov and Larissa Vietzorreck worked on the electromagnetic modeling of MEMS components [238–242]. Also three-dimensional silicon structures have been modeled [243]. For the modeling of complex three-dimensional structures the computational effort could be reduced considerably by introducing a static sub-gridding [180].

Together with Damienne Bajon from the Institut Supérieur de l'Aéronautique et de l'Espace (SUPAERO) in Toulouse and Sidina Wane from NXP Semiconductors in Caen, Nikolaus Fichtner and I investigated the application of numerical electromagnetic field simulation methods to integrated circuit design [244]. Several modeling approaches including hybrid methods and global methodologies were discussed. In this context we also investigated a combination of the TLM method and the transverse wave formulation (TWF) method for efficient modeling of multi-scale and multilayered planar structures [245, 246].

One challenging area in electromagnetics are metamaterials. Metamaterials are structured artificial materials with properties not occurring in nature [247]. Left–handed metamaterials are artificial electromagnetic structures exhibiting special properties like negative permeability, negative permittivity and negative refractive index. The name left-handed metamaterials is due to the circumstance that the vectors of the electric field, the magnetic field and phase velocities form a left-hand oriented trihedron. Together with Michael Zedler I investigated three–dimensional metamaterials. We have shown that the transmission line matrix scheme provides a fundamental theoretical framework for the finding and exploration of three-dimensional metamaterial structures [248–250]. Michael Zedler, Uwe Siart, and I have shown that space-discretizing numerical schemes can be considered the unifying framework behind metamaterials [251]. This work on metamaterials has been continued with Christophe Caloz from the École Polytechnique, Montréal [252–255] and George Eleftheriades from the University of Toronto [256].

Working at the German Aerospace Center (DLR) in Oberpfaffenhofen on his PhD thesis, Ali Eren Culhaoglu performed analytic investigations of left–handed metamaterials. The concept of the perfect lens, made of left–handed metamaterial allows to overcome the diffraction limit and sub-wavelength imaging became possible. A full wave analysis of a three dimensional, finite and impedance matched metamaterial lens was performed and the impact of the aperture size on the imaging quality was analyzed in [257, 258].

On 26th October 1991, Leopold Felsen and Wolfgang Hoefer – they were in Munich to attend a workshop I had organized – were visiting us in our home. My younger son Johannes, in the age of 13, liked to design simple computer games. Wolfgang, observing this, said to Johannes: "If you can do this you could also program a TLM code" and he explained him the two-dimensional TLM scheme. Since it has been a sunny afternoon I went together with Leopold and Wolfgang through the English Garden. When we came back Johannes had finished the mathematical core of the 2D-TLM simulator. In the following weeks he designed the user interface and the graphics, demonstrating the propagation and scattering of the wave pulses as a pinball game. This simulator has been very useful as an educational tool to demonstrate to the students how TLM works. Wolfgang proposed to publish this simulator, wrote the paper together with Johannes and presented it at the International Conference on Computation in Electromagnetics in London [259].

5.2 *Circuits*

Electronic noise occurs due to random fluctuations of electrons. It is unavoidable in electronic systems and yields undesired perturbations of the information carrying signals. Methods for optimization of the signal-to-noise ratio in electronic devices, circuits and systems therefore are of great importance. Based on previous work at AEG-Telefunken, I continued my work on noisy linear circuits together with Martin Rieger and Stefan Müller [62–66]. By combining methods of circuit analysis and the representation of noise signals using correlation spectra we developed the mathematical tools needed for the analysis and optimization of microwave and millimeter-wave circuits. We developed computer algorithms permitting the modeling of multi-port circuits containing internal noise sources [67–70, 260]. Subsequently, these algorithms have been widely adopted by software developers and are now incorporated in all leading CAD programs for linear circuit analysis. We also developed the commercial CAD program SANA for the analysis of linear microwave circuits under consideration of the noise properties.

Microwave oscillators are key components for signal generation and signal conversion in many applications, especially in wireless communications and sensorics. They became a major research topic at my institute. An oscillator is an autonomous system generating a harmonic oscillation of definite amplitude and frequency. It has to fulfill operating requirements concerning output power, frequency stability, low phase noise, low costs, and low power consumption and in some cases also frequency tunability. All these requirements can be fulfilled by monolithic integrated oscillators. The design of monolithic integrated oscillators requires advanced computer aided design methods applicable to complex equivalent circuit structures.

Franz Kärtner investigated the noise behavior of oscillators described in time domain by a set of nonlinear ordinary differential equations with intrinsic noise sources [261, 262]. In his work he gave for the first time a general definition of amplitude and phase noise. Martin Schwab applied the multiple shooting algorithm for the solution of the cyclic boundary value problem of oscillators and created a powerful tool for the modeling of complex microwave oscillators [263,264]. Werner Anzill applied perturbation theory to simulate the noise behavior of free-running microwave oscillators and together with Roland Bulirsch and Oskar von Stryk from the Mathematics Department of the TUM he developed a time domain phase noise analysis method [265–267]. Marion Filleböck applied a continuation method to deal with the start-up problem in the large-signal analysis of oscillators and for the computation of the tuning characteristics of microwave oscillators [268–270].

Our theoretical activities on oscillator modeling have been the basis for numerous microwave oscillator design projects. Josef Hausner designed dielectric resonator oscillators [271]. A low-phase-noise hybrid 2 GHz oscillator with acoustic surface transverse wave delay lines as frequency–determining elements has been designed by Ludwig Eichinger, Bernd Fleischmann and Robert Weigel [272, 273]. Ralf Klieber, Roland Ramisch, Alejandro Valenzuela, and Robert Weigel worked on microwave oscillators with coplanar high-temperature superconducting resonators [274]. Together with Werner Anzill and Gerhard Olbrich, Tilman Felgentreff investigated up-conversion of generation–recombination noise to oscillator phase noise in AlGaAs-GaAs-HEMT oscillators. Volker Güngerich investigated broad–band tunable GaAs-MESFET microwave oscillators [275–279]. The contributions of Robert Wanner to the design of monolithic integrated millimeterwave oscillator will be discussed in Sect. 5.6.

Josef Hausner has carried out the very ambitious project to design and realize a tunable Bragg-type distributed feedback microwave resonator. The resonator is formed by a transmission line space periodically loaded with varactor diodes. A tunable periodic superstructure is superimposed on the transmission line by periodically DC biasing of the varactor diodes. With this resonator configuration, tuning bandwidths from 400 MHz to 4 GHz were achieved [280, 281].

In connection with our engagement in the area of microwave oscillators I co-founded in 1986 together with my coworkers Karl-Heinz Türkner and Gerhard Olbrich and commercial partners the company WORK Microwave in Holzkirchen. The company started its activities with the development of microwave oscillators and frequency synthesizers and today, it is developing microwave components and systems. Karl-Heinz Türkner, Gerhard Olbrich, and I left the company little more than a year after foundation.

Another challenging research project has been the development of demultiplexer circuit for an fiber optical receiver for 43 Gbit/s in a joint research project with the Siemens Information and Communications Network Division, started in 2001. Jung Han Choi developed a Si Schottky diode sampling demultiplexer and realized it in hybrid thin-film technology [282–284]. In the case of monolithic integration this demultiplexer circuit would be viable for much higher bit rates.

The increasing number of frequency bands and services in wireless communications yields a demand for front-end circuits with a wide frequency tuning range. Mahmoud Al-Ahmad worked at Siemens on capacitive piezoelectric tuning elements. With these tuning elements he was able to realize wide-band tunable filters [285–287].

5.3 Medical Electronics

In 1988, Dr. K.G. Riedel from the University Ophthalmic Clinic in Munich contacted me in the matter regarding the development of a microwave hyperthermia system for thermo-radiotherapy of malignant choroid melanoma. The malignant choroidal melanoma is an eye cancer arising from the blood-vessel layer choroid beneath the retina. The therapeutic effect of heat as an adjunct to irradiation is an efficient method in oncology. Intraocular malignant tumors offer excellent conditions for heat applications since tumor volumes are small and heat can be locally generated to the tumor through the overlying sclera. It was shown that hyperthermia in addition to irradiation may allow for radiation dose reduction which may be followed by a decreased irradiation induced mortality rate [288]. Dr. Riedel became acquainted with the hyperthermia treatment of eye tumors during a research stay in the United States and was convinced of this method. At that time, however, industrial equipment for treatment of eye tumors was not available.

Together with Karl-Heinz Türkner I developed a microwave hyperthermia system exclusively dedicated to the treatment of intra-ocular tumors. The system used a calotte shaped applicator matched to the shape of the eye and a microprocessor controlled 2.45 GHz generator with 5 W maximum output power. Temperatures between 40°C and 45°C and duration times of treatment between 1 and 60 min could be chosen [289]. The medical application of the hyperthermia system developed at the TUM is discussed in [288].

5.4 Optics and Acoustco-Optics

In the eighties we worked on acousto-optic spectrometers. An acousto-optic spectrometer is based on the diffraction of a laser light beam at an ultrasonic wave. A piezoelectric transducer, modulated by a radio frequency signal, applies an acoustic wave to a crystal. The acoustic wave propagating through the crystal modulates the crystal's refractive index, yielding a propagating Bragg grating. The angular distribution of the diffracted light beam represents the spectral distribution of the radio frequency signal. Focusing the deflected beam on a linear photodetector array yields the electrical signal representation of the spectrum. Such acousto-optic spectrometers are interesting for the surveillance of broad radio frequency spectra. Adalbert Bandemer developed an acousto-optic time and frequency domain Bragg cell signal analyzer [290, 291].

In a number of subsequent projects we have investigated the application of planar acousto-optic Bragg deflectors. Planar acousto-optic deflection occurs when a surface acoustic wave propagates in a planar optical waveguide producing a variation of the refractive index due to the photo-elastic effect. Robert Weigel and Kimon Anemogiannis have investigated planar acousto-optic interactions in lithium niobate [292, 293]. Erwin Biebl and Kimon Anemogiannis have developed novel methods for experimental characterization of arbitrarily anisotropic piezoelectric substrates and applied these methods to the determination of so far unknown constants of proton-exchanged lithium niobate[294, 295].

Adalbert Bandemer has investigated non–linearities in single–mode fibers. His calculations of cross talk due to stimulated Raman scattering yields a severe limitation of the performance of fiber optic wavelength-multiplexing systems [296]. Robert Osborne has constructed an all-fiber sub-picosecond Raman ring laser [297]. Furthermore Robert Osborne has investigated nonlinear pulse propagation and the generation and amplification of Stokes radiation in a single-mode fiber theoretically [298, 299].

5.5 Surface Acoustic Waves

Surface acoustic wave (SAW) devices are key devices in modern communications. Modern mobile phone technology only became feasible due to the availability of low cost SAW filters with low insertion loss. Cooperation with Siemens in the area of surface acoustic wave devices started as early as 1981. At my institute the projects have been supervised at the beginning by Gerhard Olbrich and later by Robert Weigel.

Gerd Scholl, Andreas Christ, Werner Ruile, and Robert Weigel worked on efficient design tools for SAW-resonator filters on the basis of a combination of the coupling-of-modes formalism and the transmission-matrix approach. This allowed to create exact and computationally efficient analysis and synthesis CAD tools for the design of SAW-resonator filters [300–302]. Kimon Anemogiannis designed a novel, 900-MHz SAW microstrip antenna-duplexer for use in mobile radio systems [303] and a microstrip front-end circuit in the low GHz range for applications in time division multiple access systems [304]. In 1991, Erwin Biebl demonstrated the feasibility of the combination of SAW and microstrip technologies for the development of low-cost mobile radio units [305]. Design, fabrication and performance of a low-loss SAW microstrip front-end circuit at 1.7 GHz for applications in time division multiple access (TDMA) systems has been investigated by Hans Meier, Erwin Biebl, and Robert Weigel [306]. Hans Meier also analyzed the propagation and reflection characteristics of leaky surface acoustic waves (LSAW) under periodic metal grating structures [307, 308]. This has been the basis of the developed sophisticated CAD tools at Siemens for the design of LSAW based filters. Ulrike Rösler investigated propagation, reflection and coupling of LSAWs on $LiTaO_3$ applying the Finite Element Method (FEM) [309].

Andreas Holm developed a nondestructive high-resolution technique for the optical detection of the phase and amplitude of high frequency surface acoustic waves. The test setup incorporated a mode-locked picosecond laser, harmonic mixing, and coherent detection, and it allows the measurement of the surface wave field and the direct determination of the phase velocity [310–312].

5.6 SIMMWICs and Silicon Based Millimeterwave Devices

Since 1984 I have done research work on monolithic millimeterwave integrated circuits (SIMMWICs). This work has been done in cooperation with the microwave electronics group of the AEG–Telefunken Research Institute which later has merged into the Daimler Research Center. This group in Ulm first has been headed by

Erich Kasper and later, after Erich Kasper moved to the University of Stuttgart, by Johann-Friedrich Luy. Arye Rosen from the RCA David Sarnoff Research Center in Princeton has been the first who has suggested the use of silicon as the substrate for millimeter-wave monolithically integrated circuits [313, 314]. Referring to the work of Arye Rosen Erich Kasper proposed to me to work together in this area. In 1984, together with Josef Büchler I started to work on this project. When we began this work there was the unanimous opinion in the professional community, that silicon would be completely inappropriate as the base material for integrated millimeterwave circuits. Soon, we realized together with Erich Kasper and his group integrated millimeterwave circuits in silicon technology, like planar transmitters and receivers for frequencies up to 100 GHz and with integrated antenna structures [315–324]. Planar passive circuits also have been investigated. In 1994, I edited together with Johann-Friedrich Luy the book "Silicon–Based Millimeterwave Devices" which gives an overview over the state of the art of silicon-based millimeterwave technology at this time [322]. Erich Biebl also joined the SIMMWIC project and later continued it with his own research group [323]. Today, silicon and silicon-germanium-based monolithic integrated millimeter-wave circuits allow the realization of sensing and communication systems with operating frequencies up into the millimeter-wave range and are the basis for millimeter-wave consumer applications in communication technology and automotive technology [325].

Robert Wanner designed fully monolithically integrated millimeterwave oscillators in SiGe HBT technology [326–330]. Integrated millimeterwave oscillators are basic components for radar sensors in vehicular technology. The monolithic integrated circuits were fabricated at Infineon. In [329] a monolithically integrated J-band push-push oscillator tunable between 275.5 and 279.6 GHz has been presented. For his thesis [330], Robert Wanner received the Joseph Ströbl award.

Investigations of the resonance phase transistor (RPT) resulted in the first experimental verification of the power gain of the RPT beyond its transit frequency [329–332]. The RPT is a SiGe hetero bipolar transistor in which current amplification is achieved far beyond the transit frequency due to coherent carrier transport in the base region. This allows for a transistor design with a higher base width for a given operating frequency yielding an increase of the radio frequency output power by one order of magnitude.

Hristomir Yordanov modeled multi-conductor transmission line interconnect structures in integrated circuits using Schwarz-Christoffel mapping and solved the multi-conductor transmission line equations in frequency domain. The resulting frequency response was used to compute the pulse distortion and the crosstalk effect in an on-chip digital bus [333]. Based on this results, together with Josef A. Nossek and Michel Ivrlač, the crosstalk effects in bus systems were investigated [334]. Furthermore, Hristomir Yordanov worked on wired and wireless inter-chip and intra-chip communication [335–339]. In this project the utilization of the electronic circuit ground planes as radiating elements for the integrated antennas was investigated. This yields optimal usage of chip area, since the antennas share the same metallization structure as the circuits.

5.7 Microwave Applications of Superconductors

After my appointment at the Technical University of Munich started, I took the opportunity to occupy myself again with the Josephson effect. Martin Rieger and Josef Büchler investigated theoretically the microwave frequency conversion in Josephson junctions [340–342].

At the end of the eighties we have investigated in cooperation with Siemens microwave applications of high-temperature superconductors. Soon after the discovery of high temperature superconductivity by Johannes Bednorz and Karl Müller in 1986 [343], Siemens started research activities concerning the application of high temperature superconducting thin films for low-loss microwave circuits. Theoretic investigations of the high-frequency behavior of planar high-temperature superconducting circuits have been started together with Jochen Kessler who worked on his PhD at the TUM and with Roland Dill from Siemens [116, 117]. Coplanar waveguide structures have been investigated using a partial wave synthesis taking into account the complex conductivity of the high temperature superconducting material. Micrometer transmission line dimensions were considered in the frequency range up to 100 GHz. Roland Ramisch, Alejandro Valenzuela, and Robert Weigel investigated passive and active circuits with high-temperature superconductors [274, 344, 345].

Roland Ramisch and Gerhard Olbrich developed a superconducting chirp filter using a niobium-on-silicon shielded microstrip technology. The chirp filter had a dispersive time delay of 26 ns and a 3.4-GHz bandwidth centered at 4.7 GHz [346]. Such chirp filters are interesting components for spread spectrum systems.

High-temperature superconductors allow the realization of high-Q planar resonators and hence the realization of microwave oscillators with low phase noise. Ralf Klieber, Roland Ramisch, Robert Weigel, Martin Schwab, and Alejandro Valenzuela, together with Roland Dill from Siemens developed GaAs MESFET oscillators stabilized by high-temperature-superconducting coplanar resonators, operating at 77 K [274, 347, 348].

5.8 Antennas and Wireless Communications

Throughout the years numerous projects dealt with antennas and wireless communications, comprising electromagnetic design as well as system considerations. The sizes of the antennas ranged from below 1 mm in the case of integrated on-chip antennas to several meters.

Bruno Biscontini developed an efficient design and optimization method for cylindrical multilayer conformal antennas. The approach is based on the integral equation method in combination with the method of moments [136, 137, 139]. This work has been performed for Rohde & Schwarz to create a design tool for ship antennas. Christoph Ullrich investigated the radiation of a linear antenna placed in the rear window of a car. To compute the field in the aperture he applied the method of moments. For far-field corrections he used uniform theory of diffraction [349, 350]. He computed the field in the aperture by the method of moments (MoM). Then, the resulting far-field is corrected using the Uniform Theory of Diffraction. This work has been performed at INI.TUM, the competence center of the TUM in Ingolstadt for cooperation with AUDI. Libo Huang designed a tunable receiver antenna for the digital video broadcast band from 462 to 696 MHz [351–353]. The project started with Siemens, then went to BENQ. After the crash of BENQ Libo Huang could finish his work at the TUM with support of the Werner von Siemens Foundation.

Stefan Lindenmeier, Gerhard Olbrich, and I, together with Johann-Friedrich Luy from Daimler, developed an extremely compact multifunctional antenna for the application in terrestrial radio services like GSM 900 MHz, DCS 1800 MHz as well as for satellite radio services like GPS 1575 MHz. At the terrestrial frequency bands the antenna exhibits omnidirectional radiation characteristics in the horizontal plane for vertically polarized waves whereas at the frequency bands for the satellite radio services the antenna exhibits a radiation characteristic with a vertical main lobe and circular polarization [354, 355]. At the European Microwave Week 2003 we received an innovation award for this work.

Robert Wanner investigated a bidirectional active antenna for vehicular and mobile applications. Active field compensation is performed using a shielding electrode inserted between the antenna electrode and the ground electrode and hence, the electrical antenna height is increased substantially. This allows the realization of flat conformal antennas for vehicular and mobile applications [356].

Direction-of-arrival (DOA) estimation plays a role for computing beamforming vectors in smart antennas. Smart antennas are antenna arrays which, in combination with signal processing algorithms, can track mobile stations. This allows multiple use of frequency channels in mobile communications. A wide-band DOA estimation method for wide-band smart antennas based on frequency-domain frequency-invariant beam-formers (FDFIB) has been developed by Tuan Do-Hong. By appropriately designing the weights for frequency-domain beam-formers at different frequencies, the frequency-invariant beam-patterns are obtained [357–359].

Together with Karl Warnick, I studied the noise penalty caused by mutual coupling of antenna elements in an antenna array [360, 361]. In this work a matching condition for minimizing the receiver noise temperature over multiple beams was formulated and we investigated the noise performance of arrays for multiple input – multiple output (MIMO) communications.

A serious problem in monolithic integration of antennas is the high chip-area requirement of antenna structures which would considerably enhance the costs of chips with integrated antennas. In [336–338] the use of the electronic circuit ground planes as radiating elements for the integrated antennas has been proposed. This allows an optimum utilization of the chip area. Michel Ivrlač, Josef Nossek, Hristomir Yordanov, and I have shown the applicability of isotropic radiators in antenna array modeling [362]. Although isotropic antennas do not exist, their application in theory is legitimate, since they yield a correct antenna coupling and qualitatively correct analysis of antenna gain.

5.9 Electromagnetic Interference

In the year 2000, together with Florian Krug, I started to investigate time-domain measurement methods for electromagnetic interference (EMI) [363–367]. At this time commercial EMI measurement systems used heterodyne receivers which slowly scanned the frequency spectrum. The measurement of the EMI emission of an object under test in the frequency range typically took 45 min. We developed a time domain electromagnetic interference measurement system that uses ultra high-speed analog-to-digital converters and real-time digital signal processing systems to enable ultra fast tests and measurements for electromagnetic compliance that fulfill the demand for measurements of today's complex electronic equipment and systems.

My first application for project funding was rejected since one of the reviewers considered the project goal to be intangible and the second one classified it as a project in signal theory and not a project in EMI. If a project goal is said to be intangible I consider it as a challenge and, hence, a real research project. I also could not convince the industry to engage in this area. One of the great advantages of German universities is that professors have a number of scientific coworkers, independent of project funding. This allows to launch projects without dedicated support. And this was what I have done in this case. Nine month after the denial of support Florian Krug received the 2002 Best Student Paper Award of the IEEE Electromagnetic Compatibility Society for the paper "Ultra-fast broadband EMI measurement in time-domain using FFT an periodogram" [363] at the IEEE International Symposium on Electromagnetic Compatibility in Minneapolis.

In 2004, Stephan Braun realized a first time-domain EMI measurement system for the frequency range from 30 MHz to 1 GHz in [368–372]. The system performs the calculation of the spectrum via the fast Fourier transform (FFT) and a simultaneous evaluation of the spectrum under the peak, average, and root-mean-square detector mode. In [373–377] the suitability for full compliance measurements has been demonstrated. With the time-domain EMI measurement system described in [378, 379] a reduction of the measurement time by a factor of 8000 was achieved. Applying three parallel analog-to-digital converters a multi-resolution system was realized that fulfills the international EMC standards CISPR 16-1-1 [380]. Stephan Braun received the 2006 Best Student Paper Prize at the 17th International Zurich Symposium in Singapore [375] and for his PhD thesis the 2007 E.ON Future Award [381].

In November 2007, I founded together with my scientific coworkers Stephan Braun and Arnd Frech a spin-off company: the GAUSS INSTRUMENTS GmbH. We have chosen this company name since signal processing in our systems is based on the fast Fourier transform, which for the first time has been described in Carl Friedrich Gauss' publication "*Theoria interpolationis methodo nova tractata*" [382]. With the presentation of the first time domain electromagnetic interference measurement system, the TDEMI® –1G system at the EMC Zurich 2007 Conference in Munich, we started to establish a further growing product family which covers a wide range of the demands of modern EMC testing. A major success was the order of the VDE to equip the new test center in Offenbach with our time domain electromagnetic interference measurement systems. At the opening event of the VDE Test Center in Offenbach on June 10th, 2008 the guests witnessed an impressive demonstration of the capabilities of our systems. Manufacturers use the time domain electromagnetic interference measurement system especially for the emission measurement of intermittent signals from devices such as microwave ovens and electric actuators in cars. The developed methods also introduce new concepts of analysis including phase spectra, short-time spectra, statistical evaluation, and FFT-based time-frequency analysis methods.

Ambient cancellation techniques in time-domain for full compliance EMI measurements are investigated in [383, 384]. In this system two channels are fed from two broad-band antennas, where the first antenna is receiving predominantly the EMI radiated from the device under test and a second antenna receives predominantly the ambient noise. These techniques allow fast measurements of electromagnetic interference in the time-domain at open area test sites.

5.10 Quantum Nanoelectronics

Since 1900 quantum physics has revolutionized step by step our knowledge of physics and enabled technology as we know it today. In a first step quantum theory brought the understanding of the properties of atoms, molecules and solids [385]. Besides its implications on technology quantum theory also changed our cognition and the concept of physical reality. Our imagery thinking is properly adapted to the concepts of classical physics. Quantum theory, however, often conflicts with our habitual structures of thinking and often yields results which seem to be paradox or even contradictory. Bernard d'Espagnat stated in the preface of his book "On Physics and Philosophy" that "*trying to understand what contemporary physics is truly about unavoidably raises philosophical problems*" [386].

By mid of the twentieth century quantum theoretically based understanding of the properties of solids gave rise to the onset and prodigious development of semiconductor electronics. The quantum theory of radiation yielded the invention of the laser and fostered the development of quantum electronics and modern lightwave technology. However, u to a few years ago scientists and engineers dealing with electronic and optoelectronic devices have not been confronted with the strangeness of quantum theory. Due to the circumstance that in today's electronic and optoelectronic devices large numbers of electrons and photons are manipulated, most of the phenomena can be described in terms of classical models. This is going to change now with the ongoing miniaturization in electronics. Nanoelectronic devices with structure dimensions down to the atomic scale will finally allow to control single electrons and single photons [387–389]. Such devices will be governed essentially by quantum mechanical laws. Representing information by quantum mechanical states will provide a tremendous increase of computational power of future quantum computers compared to classical computers. Quantum mechanics will play a crucial role in future electronics for the understanding of devices circuits and systems.

Superconducting nanoelectronic Josephson devices exhibit a considerable potential for application in future RF electronics [389]. The Josephson effect allows generation, detection, mixing, and parametric amplification of high frequency signals up into the THz region and also quantum information processing. For these applications superconducting devices may be the most promising candidates in future since Josephson devices exhibit extremely small size and small energy consumption. Since I already worked on the Josephson effect in Vienna I have never lost my interest in this field.

In 1990, together with Franz Kärtner I have shown the possibility of generating squeezed quantum states (i.e. two-photon coherent states) by a DC pumped degenerate parametric Josephson junction oscillator [390, 391]. Squeezed quantum states, also called two-photon coherent states, are a generalization of the well-known quantum mechanical minimum uncertainty states [392, 393]. Referring to [391], Paternostro discussed the possibility to transfer entanglement from a two–mode squeezed state generated by Josephson junctions to a pair of quantum bits (qubits) [394]. A qubit is the unit quantum information in quantum computing [395–397]. Distinct from a classical bit, the qubit cannot only assume the states '0' and '1' but also any quantum superposition of these states.

Quantum computing offers interesting perspectives for the simulation of complex physical systems. Taking into account that the real world obeys quantum laws Richard Feynman argued that a real simulation of the physical world should be possible where the computer is doing the same as nature [398, 399]. Such a computer, mapping the laws of the physical world, should be reversible and should be built by quantum mechanical elements. In 1985, David Deutsch for the first time has given a fully quantum mechanical model for the theory of quantum computation [400]. Detailed treatments of quantum computing are given in [395–397]. In a quantum computer the problem to be simulated is mapped into a quantum mechanical system. The program is represented by a quantum mechanical Hamilton operator. The prospects and challenges for implementing a quantum computer using Josephson junctions have been discussed in [394, 401–405].

The tremendous potential of quantum computing is due to the utilization of quantum-mechanical phenomena such as quantum parallelism and entanglement. Quantum information processing essentially is a consequence of the famous work that Albert Einstein, Boris Podolsky, and Nathan Rosen have published in 1935 [406]. They postulated that every element of the physical reality must have a counterpart in the physical theory. As a consequence of this work we have to drop either the assumption of physical reality or the assumption of physical locality. For Einstein this has been an argument against quantum theory. However, today most physicists have the tendency to drop the assumption of physical realism and to keep physical locality. The nonclassical correlations between quantum systems is the potential of many strange quantum phenomena like quantum cryptography, quantum teleportation, and quantum computing [407].

Together with Siddharta Sinha, an excellent master course student of 2008, I developed quantum computing algorithm for electromagnetic field simulation on the basis of the transmission line matrix (TLM) method [408]. The Hilbert space formulation of TLM allows us to obtain a time evolution operator for the TLM method, which can then be interpreted as the time evolution operator of a quantum system, thus yielding a quantum computing algorithm. Furthermore, the quantum simulation is done within the framework of the quantum circuit model of computation. Our aim has been to address the design problem in electromagnetics – given an initial condition and a final field distribution, find the structures which satisfy these. Quantum computing offers us the possibility to solve this problem from first principles. Using quantum parallelism it will be possible to simulate a large number of electromagnetic structures in parallel in time and then try to filter out the ones which have the required field distribution.

The modeling, design, and optimization of complex physical, technical, biological, and economic systems will be one of the major future applications of quantum computing. However, there are still major problems unsolved. Although quantum parallelism in principle allows to model all possible structures in parallel, there still remains the problem of enhancing the contribution of the desired solutions in the quantum superposition of the solutions for all possible problems. A possible solution could be a coefficient-booster module, applying oracles based on nonlinear quantum mechanics as suggested by Daniel Abrams and Seth Lloyd [409].

6 At the Ferdinand Braun Institute in Berlin

In Spring 1992, I received from the Senate of Berlin the offer to take over the management of the Ferdinand Braun Institut für Höchstfrequenztechnik (FBH) in Berlin Adlershof as the founding director. For me this was a very interesting challenge. The FBH originated from the departments of two former Central Institutes of the Academy of Sciences of the former German Democratic Republic. These have been the Department of GaAs-Electronics in the Central Institute of Electron Physics and the Department of Optoelectronics in the Central Institute of Optics and Spectroscopy. In accordance with the reunification agreement between the two German states of October 3rd, 1990, the Academy of Sciences of the former German Democratic Republic was dissolved on December 31st, 1991. Based on a recommendation of the Wissenschaftsrat (Council of Science and Humanities), i.e. the scientific advisory board to the German Federal Government, the FBH was reestablished on January 1st, 1992. In accordance with the Senate of Berlin and the Federal Ministry of Research and Technology (BMFT) it was recommended that the FBH should become a center of excellence, especially in the research areas of metal organic vapor phase epitaxy and (MOVPE) and computer aided design (CAD). The competence developed in that areas should drive research and development of novel and innovative devices and circuits in microwave technology and optoelectronics.

The foundation committee whose members were highly qualified scientists from universities and companies was chaired by Günter Weimann from the Walter Schottky Institute of the TUM. A person whose importance for the conservation of the institute in a turbulent time cannot be put too high, was Rudolf Gründler. He was an outstanding scientist and an excellent manager who, with his critical attitude, did not climb to higher positions in the past regime. As the Assistant Director of the institute he supported me with great competence and strong engagement. Unfortunately Rudolf Gründler died in a car accident in early 1995.

My appointment in Berlin was for 3 years, during which I held positions in Berlin and in Munich. Finally, I obtained the prospective appointment for a permanent position as the director of the Ferdinand Braun Institute in connection with a position as a Full Professor at the Technische Universität Berlin. The decision I had to make was not easy, since the offered positions have been attractive and I also felt well in Berlin. The determining factor for my decision to decline this offer and to go back to Munich has been, that leading a large institute is primarily a management task and leaves little freedom in personal engagement in research.

Today, the Ferdinand Braun Institute, as a centre of competence for III-V compound semiconductors is doing research on innovative technologies for innovative applications in the fields of microwaves and optoelectronics.

7 Retired

Retirement, which came upon me on October 1st, 2008, usually is defined as the point where a person stops professional activity completely. Wolfgang A. Herrmann, the president of the Technische Universität München bestowed on me the TUM Emeriti of Excellence Award. The selected Emeriti of Excellence receive research-possibilities, take an active part in academic teaching, and are provided with organizational and financial support for their activities. Hence, this honorable status gave a good support for the continuation of my scientific work. Paolo Lugli offered me a room and also working places for my PhD students at the Institute for Nanoelectronics. My research projects are funded by the Deutsche Forschungsgemeinschaft and the Bayerische Forschungsstiftung. I am working on network methods in electromagnetic field modeling [133, 133, 410] and nanoelectronic topics [389, 411–414].

I am still supervising several PhD students. Arnd Frech worked on time-domain electromagnetic interference measurement techniques in the presence of ambient noise [383, 384], Nikolaus Fichtner worked on a dissertation on the hybridization of the transmission-line-matrix method with the integral equation method for the analysis of electromagnetic coupling [415–417], and Hristomir Yordanov worked on wired and wireless inter-chip and intra-chip communication [337–339]. These works have been finished in 2010. Three further PhD students are continuing their work. Christian Hoffmann is working on a broadband time–domain electromagnetic interference measurement system for measurements up to 18 GHz [418–420]. Hassan Slim is also working on EMI measurement systems [421]. Farooq Mukhtar is working on network methods in electromagnetic field modeling [215, 216, 422].

Since May 2010 my son Johannes holds a position as a Postdoctoral Research Fellow at the Institute of Nanoelectronics and I am also happy to work with him. Johannes brought experience in multi-physics modeling [423, 424] and electromagnetic interference modeling [425–430] from his stay with Andreas Cangellaris in Urbana / Champaign. We now are working together in the area of network modeling [150, 216, 429]

I could continue my international cooperations. In 2008 and 2009, I visited Andreas Cangellaris and my son Johannes, at the University of Illinois at Urbana / Champaign. During my research stay at the University of Illinois at Urbana Champaign in December 2009 I have been invited to participate the Graduation Ceremony where I congratulate my son Johannes (Fig. 12). For this ceremony I have taken the gown I received from the Moscow Aviation Institute in 2007. In 2009 I stayed for 3 month with Damienne Bajon at the Institut Supérieur de l'Aéronautique et de l'Espace in Toulouse. In 2010 I hosted for 3 month Yury Kuznetsov from the Moscow Aviation Institute and we have worked together on system identification methods applied to equivalent circuit model synthesis [150, 215, 216].

In acatech – the German Academy of Science and Engineering – I am leading the project group "Nanoelectronics" in which the potential of nanoelectronic developments are described and assessed. The availability and utilization of nanoelectronic research, development and production potential is necessary to ensure the continued strong performance of the German information and communications industry. The questions are: What opportunities do nanoelectronics offer with respect to improvement in efficiency and the development of new technologies? What is the current state of research and what are the research requirements in science and industry? What are the implications for action and what recommendations can be made for policy makers, industry and science?

In 2009, I was appointed "European Microwave Lecturer" by the European Microwave Society to give presentations on the topic "Network Methods in Electromagnetic Field Computation". This brings me to many places to give my presentations on this topic and to have discussions with colleagues. I am editing the

Fig. 12 Participating the Graduation Ceremony at the University of Illinois at Urbana Champaign on 19 December 2009 and congratulating my son Johannes

European Microwave Book series which is going to appear at Cambridge University Press.

In September 2009, at the European Microwave Conference in Rome, I received the Distinguished Service Award from the European Microwave Association, and on November 8th, 2010 I have been awarded the Golden Ring of Distinction, of the VDE – the German Association for Elcctrical, Electronic and Information Technologies – for achievements in the area of microwave engineering.

As I continue to pursue my research interests, to launch new projects, to work together with young researchers, and to have scientific exchange with colleagues at home and abroad, I am enjoying my life as a retired professor.

8 Coda

For now we see through a mirror in an enigma,
but then face to face.
Now I know in part,
but then I shall know as also I was fully known.
Corinthians 13;12

On December 16th, 1808 Goethe wrote from Jena to Friedrich August Wolf: *"Ich hatte mir manches zu arbeiten vorgesetzt, daraus nichts geworden ist, und manches getan, woran ich nicht gedacht habe; das heisst also ganz eigentlich das Leben leben."* – "I had planned to work on several things, which has become nothing, and I have done some things, which I have not planned to do, that is to say quite truly to live the life."[431, p. 533]. I like the serenity of this assessment.

A comprehensive personal review touches on existential explorations of the question of being such as

Where do we come from?
What are we?
Where are we going?

Paul Gauguin has paraphrased these questions in a striking image (Fig. 13). Leopold Felsen has loved this painting and considered it as a profound expression of the human condition [432].

In all reasoning of daily life, including scientific work, we take the world of phenomena as the reality. This mental pattern is reasonable and justified by the tremendous success of western science and technology. However, our fundamentals of knowledge are built upon a fragile conceptual groundwork. The only knowledge we have from the world we obtain via our mind. Spinoza's epistemology which is based on the ontology of the substance, is an illuminative contribution to the body-mind problem. Spinoza introduces the notion that thought (*cogitatio*) and extension (extensio) are attributes of the same universal substance [433]. Mind is a mode of thinking whereas the robust material world of appearances is represented by modes

Fig. 13 Paul Gauguin, 1897, Museum of Fine Arts, Boston, Mass. USA – Where do we come from? What are we? Where are we going?

of extension. Spinoza's idea of unity of substance conforms to the concepts of modern physics at least better than Cartesian dualism [386, p. 272].

The tremendous progress in neuroscience will not change anything, since the distinction between thought and extension is a matter of categories. Thomas Nagel proposed: "*Consciousness should be recognized as a conceptually irreducible aspect of reality that is necessarily connected with other equally irreducible aspects – as electromagnetic fields are irreducible to but necessarily connected with the behavior of charged particles and gravitational fields with the behavior of masses, and vice versa*" [434]. Colin McGinn argued that the "*mind–body problem brings us bang–up against our capacity to understand the world*" [435]. We see through a mirror in an enigma. This mirror is the last frontier.

In his first book "Über die vierfache Wurzel des Satzes vom zureichenden Grunde" Arthur Schopenhauer says *"Certain thoughts which wander about for a long time in our heads, belong to this sort of reflection: thoughts which come and go, now clothed in one kind of intuition, now in another, until they at last become clear, fix themselves in conceptions and find words to express them. Some, indeed, never find words to express them, and these are, unfortunately, the best of all: quæ voce meliora sunt, as Apuleius says"* [436, p. 113], [437, p. 133]. In the fictive letter *"Ein Brief"*, ostensibly written in 1603 by Lord Chandos to Francis Bacon, Hugo von Hofmannsthal reflects his distrust of language and dismisses the idea that language can describe the world [438]. Erwin Schrödinger has brought the issue to the point, saying that the attempt to express thoughts through communicable and noticeable words is like the task of the silkworm. The fabric receives its value only by shaping. At the light of day, the fabric solidifies and is no longer malleable [439, p. 55].

The elements of being are linked together in a strange way which may be expressed by the metaphor of Indra's net: Far away in the heavenly abode of the great god Indra, there is a wonderful net stretching out infinitely in all directions with a single glittering jewel in each eye of the net. Looking closely at one arbitrarily selected jewel we will discover that all the other jewels in the net are reflected in

Fig. 14 In the mountains

its surface, infinite in number, and each of the jewels reflected in this one jewel is also reflecting all the other jewels. [440, p. 2].

The world is mysterious. In Zarathustra's roundelay we listen "*The world is deep, and deeper than day can comprehend*" [441]. Prospero in the tempest: "*We are such stuff / As dreams are made on; and our little life / Is rounded with a sleep"* [442].

Hilde and I live happily in Munich. Our Children Martin, Andrea, and Johannes, Johannes' wife Moushumi and our grandson Aditya all are close to us. Throughout the year Hilde and I like to go for a walk in nature, either in the *Isarauen* or in the English Garden or to go out into the countryside. There we find happiness in absorption of the beauty and changing moods of nature. We are hiking and I am taking pictures attempting to preserve the impressions (Fig. 14).

When I am writing these lines the year again draws to a close. We were walking through the park. The soft light of the late afternoon made the leaves of the trees shine in an eternal gold. The evening falls into the twilight. The impression turns into remembrance.

References

1. Eduard Russer. *Zur Konstitution des kolloiden Goldes*. PhD thesis, Universität Wien, 1931.
2. Wolfgang Pauli and Eduard Russer. Die Konstitution des kolloiden Goldes. *Colloid & Polymer Science*, 58(1):22–44, 1932.
3. Wolfgang Pauli, Eduard Russer, and Erik Brunner. Aufbau und Eigenschaften der azidoiden Goldsole mit aufladenden Mischkomplexen. *Colloid & Polymer Science*, 72(1):26–35, 1935.
4. Wolfgang Pauli, Eduard Russer, and Paul Balog. Aufbau und zeitliche Reaktionen eines hochgereinigten Schwefelsols. *Helvetica Chimica Acta*, 27(1):585–612, 1944.

5. Josef Hannack. Tunnelbau. In Julius Derschatta Edler von Standhalt, editor, *Geschichte der Eisenbahnen der Österreichisch–Ungarischen Monarchie*, volume VI/2, pages 201–284. K.u.K. Hofbuchdruckerei & Verlagsbuchhandlung Karl Prohaska, 1908.
6. William M. Johnston. *The Austrian Mind: An Intellectual and Social History, 1848-1938*. University of California Press, 1983.
7. Wilhelm Fröhlich. *Radio-Technik in praktischen Versuchen: Ein Radio-Labor mit einem Lehrgang für Anfänger. Anleitungsbuch zum Kosmos-Baukasten Radiotechnik*. Franckh'sche Verlagsbuchhandlung, Stuttgart, 1951.
8. Peter Russer. Ferdinand Braun - a pioneer in wireless technology and electronics. In *Proc. European Microwave Conference, 2009. EuMC 2009.*, pages 547–554, September 2009.
9. Johan Huizinga. *Homo Ludens: Vom Ursprung der Kultur im Spiel*. Rowohlt, Reinbek, 1956.
10. Johan Huizinga. *Homo ludens: A study of the play-element in culture*. Taylor & Francis, 2003.
11. Peter Russer. Der Tunneleffekt bei Supraleitern. Diplomarbeit, Technische Universität Wien, 1967.
12. John Bardeen, Leon N. Cooper, and John R. Schrieffer. Microscopic theory of superconductivity. *Physical Review*, 106(1):162–164, February 1957.
13. Peter Russer. Untersuchungen am Wechselstrom-Josephsoneffekt (Investigations of the a.c. Josephson effect). *Acta Physica Austriaca*, 32(3-4):373–381, 1970.
14. Brian D. Josephson. Possible new effects in superconductive tunnelling. *Physics Letters*, 1(7):251–253, 1 July 1962.
15. Brian D. Josephson. Coupled superconductors. *Reviews of Modern Physics*, 36(1):216–220, January 1964.
16. Brian D. Josephson. The discovery of tunnelling supercurrents. *Reviews of Modern Physics*, 46(2):251–254, April 1974.
17. Sidney Shapiro. Josephson currents in superconducting tunneling: The effect of microwaves and other observations. *Physical Review Letters*, 11(2):8–82, July 1963.
18. Peter Russer. Untersuchungen des Josephsoneffektes. Dissertation, Technische Universität Wien, 1971.
19. Peter Russer. Influence of microwave radiation on current-voltage characteristic of superconducting weak links. *Journal of Applied Physics*, 43(4):2008–2010, April 1972.
20. Peter Russer and Hedayatollah Bayegan. Analog-computer studies on microwave mixing in superconducting weak links. *Proceedings of the IEEE*, 61(1):46–50, January 1973.
21. Peter Russer. Parametric amplification with Josephson junctions. *AEÜ Archiv der Elektrischen Übertragung*, 23(8):417–420, 1969.
22. Peter Russer. General energy relations for Josephson junctions. *Proceedings of the IEEE*, 59(2):282–283, February 1971.
23. Peter Russer. Ein gleichstromgepumpter Josephson-Wanderwellenverstärker (A direct-current pumped Josephson travelling-wave amplifier). *Wissenschaftliche Berichte AEG Telefunken*, 50:171–182, 1977.
24. Peter Russer. Circuit arrangement for amplifying high frequency electromagnetic waves. US Patent Nr. 4,132,956, filed Mar. 28, 1978, January 1979.
25. Peter Russer. Dynamics of accelerated Josephson junctions. *AEÜ Archiv der Elektrischen Übertragung*, 37:153–159, June 1983.
26. Theodore H. Maiman. Stimulated optical radiation in ruby. *Nature*, 187:493–494, 06 Aug. 1960.
27. Ali Javan, William R. Bennett, and Donald R. Herriott. Population inversion and continuous optical maser oscillation in a gas discharge containing a He-Ne mixture. *Physical Review Letters*, 6(3):106–110, February 1961.
28. Manfred Börner. Mehrstufiges Übertragungssystem für in Pulscodemodulation dargestellte Nachrichten. German Patent P 1 254 523, issued 12/21/1978, filed 30 April 1966.
29. Manfred Börner. Electro–optical transmission system using lasers. US Patent Nr. 3,845,293, filed September 28th, 1972, October 1974.
30. Charles K. Kao and George A. Hockham. Dielectric–fibre surface waveguides for optical frequencies. *Proceedings of the IEE*, 113:1151–1158, 7 July 1966.

31. Alain Werts. Propagation de la lumière cohérente dans les fibres optiques. *L'Onde Électrique*, 46:967–980, 1966.
32. Stefan Maslowski. Activities in fibre-optical communications in germany. *Optical and Quantum Electronics*, 5(4):275–284, July 1973.
33. Manfred Börner and Dietrich Rosenberger. Laser communication technology in germany. *IEEE Transactions on Communications*, 22(9):1305–1309, 1974.
34. Peter Russer. Introduction to optical communications. In M. J. Howes and D. V. Morgan, editors, *Optical Fibre Communications*, Chichester New York Brisbane Toronto, 1980. John Wiley.
35. Edgar Weidel. Light coupling from a junction laser into a monomode fibre with a glass cylindrical lens on the fibre end. *Optics Communications*, 12:93–97, September 1974.
36. K. Berchtold, Oskar Krumpholz, and J. Suri. Avalanche photodiodes with a gain-bandwidth product of more than 200 GHz. *Applied Physics Letters*, 26(10):585–587, May 1975.
37. Joachim Guttmann and Oskar Krumpholz. Location of imperfections in optical glass-fibre waveguides. *Electronics Letters*, 11(10):216–217, May 1975.
38. W. Eickhoff and Oskar Krumpholz. Determination of the ellipticity of monomode glass fibres from measurements of scattered light intensity. *Electronics Letters*, 12(16):405–407, 1976.
39. Peter Marschall, Ewald Schlosser, and Claus Wölk. New diffusion-type stripe-geometry injection laser. *Electronics Letters*, 15(1):38–39, 1979.
40. Klaus Petermann. Calculated spontaneous emission factor for double-heterostructure injection lasers with gain-induced waveguiding. *IEEE Journal of Quantum Electronics*, 15(7):566–570, 1979.
41. Günther Arnold, Klaus Petermann, and Ewald Schlosser. Spectral characteristics of gain-guided semiconductor lasers. *IEEE Journal of Quantum Electronics*, 19(6):974–980, 1983.
42. Peter Russer and Johann Gruber. Circuit arrangement for amplifying pulsed signals. US Patent Nr. 4,060,739, filed December 12th, 1975, November 1977.
43. Peter Russer and Johann Gruber. Hybrid integrierter Multiplexer mit Speicherschaltdioden für den Gbit/s-Bereich. *Wissenschaftliche Berichte AEG-Telefunken*, 48:55–60, 1975.
44. Reinhard Petschacher and Peter Russer. Demultiplexer using fast hybrid integrated ECL-gates for 1 Gbit/s pcm systems. *Proceedings of the 7th European Microwave Conference, Copenhagen*, pages 527–531, September 1977.
45. Johann Gruber, Peter Marten, Reinhard Petschacher, and Peter Russer. Electronic circuits for high bit rate fiber optic communication systems. *IEEE Transactions on Communications*, 26(7):1088–109, July 1978.
46. Peter Russer. Elektrische Bausteine für die breitbandige optische Nachrichtenübertragung. In *NTG Fachberichte "Neue Entwicklungen in der Nachrichtenübertragung," (München, 17–19. April 1978)*, München, April 17th–19th 1978. Nachrichtentechnische Gesellschaft.
47. Johann Gruber, Peter Marten, Reinhard Petschacher, Peter Russer, and Edgar Weidel. A 1Gbit/s fibre optic communication link. In *Proc. 4th European Conference on Optical Communication, Genova*, pages 556–563, September 12th–15th 1978.
48. Johann Gruber, Peter Marten, Reinhard Petschacher, Peter Russer, and Edgar Weidel. Digital fibre optic communications link for 1 Gbit/s. In *Proc. Laser 79 Optoelectronics Conference, Munich*, pages 305–308, July 1979.
49. Johann Gruber, Michael Holz, Reinhard Petschacher, Peter Russer, and Edgar Weidel. Digitale Lichtleitfaser–übertragungsstrecke für 1 Gbit/s. *Wissenschaftliche Berichte AEG-Telefunken*, 52:123–130, 1979.
50. E. Kremers, Peter Marten, Peter Russer, and H.J. Thomas. A 280 Mbit/s fibre optic communication link. In *Proc. 5th European Conference on Optical Communication*, pages 22.2.1–4, Amsterdam, September 17th–19th, 1978.
51. Michael Holz, E. Kremers, Peter Marten, and Peter Russer. Optischer Repeater für 280 Mbit/s. *Wissenschaftl. Berichte AEG–Telefunken*, 53:56–61, 1980.
52. Peter Russer and Siegfried Schulz. Direkte Modulation eines Doppelheterostrukturlasers mit einer Bitrate von 2,3 Gbit/s (direct modulation of a double heterostructure semiconductor injection laser at 2.3 Gbit/s). *AEÜ Archiv der Elektrischen Übertragung*, 27:193–195, 1993.

53. Günther Arnold, Peter Russer, and Klaus Petermann. Modulation behavior of semiconductor injection lasers. In H. Kressel, editor, *Topics in Applied Physics, Vol. 39, Optical Semiconductor Devices*, number 39 in Springer Series on Topics in Applied Physics, pages 213–242. Springer, Berlin, 1979.
54. Peter Russer and Günther Arnold. Direct modulation of semiconductor injection lasers. *IEEE Transactions on Microwave Theory and Techniques*, 30(11):1809–1821, November 1982.
55. Peter Russer. Modulation behaviour of injection lasers with coherent irradiation into their oscillating mode. *AEÜ Archiv der Elektrischen Übertragung*, 29:231–232, 1975.
56. Herbert Hillbrand and Peter Russer. Large signal P.C.M. behaviour of injection lasers with coherent irradiation into one of their oscillating modes. *Electronics Letters*, 11(16):372–374, August 7th, 1975.
57. Peter Russer. Verfahren zur Erzeugung mit hoher Bitrate modulierter kohärenter modenreiner Strahlung mit zwei optisch gekoppelten, getrennt voneinander ansteuerbaren Injektionslasern. German Patent DE2514140, Filed: September 30th, 1976, Issued April 6th, 1978, 29 March 1975.
58. Peter Russer. Laseranordnung. German Patent DE2548796, Priority data: 31 Oct. 1975, Filed: 35 May 1977, Issued 25 Oct 1984, 31 Oct. 1975.
59. Peter Russer. Method and arrangement for producing coherent mode radiation. US Patent 4,101,845, Priority data: 29 March 1975 and 31 Oct. 1975, Filed: 26, March 1976., 31 Oct. 1976.
60. Peter Russer, Günther Arnold, and Klaus Petermann. High–speed modulation of dhs lasers in the case of coherent light injection. In *Proc. 3rd European Conference on Optical Communication, Munich*, pages 139–141, September 14th–16th 1977.
61. Günther Arnold, Klaus Petermann, Peter Russer, and Franz-Josef Berlec. Modulation behaviour of double heterostructure injection lasers with coherent light injection. *AEÜ Archiv der Elektrischen Übertragung*, 32:128–136, 1978.
62. Herbert Hillbrand and Peter Russer. Rauschanalyse von linearen Verstärkernetzwerken. In *Nachrichtentechnische Fachberichte*, volume 51, pages 39–44, 1975.
63. Herbert Hillbrand and Peter Russer. An efficient method for computer aided noise analysis of linear amplifier networks. *IEEE Transactions on Circuits and Systems*, 23(4):235–238, April 1976.
64. Herbert Hillbrand and Peter Russer. correction to 'an efficient method for computer aided noise analysis of linear amplifier networks'. *IEEE Transactions on Circuits and Systems*, 23(11):691, November 1976.
65. Peter Russer and Herbert Hillbrand. Rauschanalyse von linearen Netzwerken. *Wissenschaftliche Berichte AEG Telefunken*, 49:127–138, 1976.
66. Herbert Hillbrand, Johann Gruber, Peter Russer, and K. Wörner. Computer aided design of a 1 GHz bandwidth monolithic integrated amplifier. In *Proc. 3rd European Solid State Circuits Conference, 1977, ESSCIRC '77.*, pages 122–124, September 1977.
67. Peter Russer and Stefan Müller. Noise analysis of linear microwave circuits. *International Journal of Numerical Modelling, Electronic Networks, Devices and Fields*, 3:287–316, 1990.
68. Peter Russer and Stefan Müller. Noise analysis of circuits with general topology and arbitrary representation. In *Proceedings of the 1992 Asia-Pacific Microwave Conference, APMC '92.*, pages 819–822, 1992.
69. Peter Russer and Stefan Müller. Noise analysis of microwave circuits with general topology. In *Microwave Symposium Digest, 1992., IEEE MTT-S International*, pages 1481–1484, 1992.
70. Peter Russer. Noise analysis of linear microwave circuits with general topology. *The review of radio science 1993–1996, Oxford, England,*, pages 887–890, 1996.
71. William Shockley. Circuit element utilizing semiconductive material. United States Patent 2,569,347, September 1951.
72. Alfons Hähnlein. Halbleiter-Kristallode der Schichtenbauart. German Patent DE 1 021 488, filed February 19th, 1954, July 1958.
73. Herbert Kroemer. Nobel lecture: Quasielectric fields and band offsets: teaching electrons new tricks. *Reviews of Modern Physics*, 73(3):783–793, 2001.

74. Herbert Krömer. Zur Theorie des Diffusions- und Drifttransistors - III Dimensionierungsfragen. *Archiv der Elektrischen Übertragung*, 8, July 1954.
75. Herbert Kroemer. Quasi-electric and quasi-magnetic fields in nonuniform semiconductors. *RCA Review*, 18:332–342, 1957.
76. Herbert Kroemer. Theory of a wide-gap emitter for transistors. *Proceedings of the IRE*, 45(11):1535–1537, November 1957.
77. Erich Kasper, H. Herzog, and H. Kibbel. A one-dimensional SiGe superlattice grown by UHV epitaxy. *Applied Physics A: Materials Science & Processing*, 8(3):199–205, November 1975.
78. Erich Kasper and Peter Russer. Verfahren zur Herstellung von bipolaren Hochfrequenztransistoren. German Disclosure P 27 19 464.5, issued 12/21/1978, filed 30 April 1977.
79. G.L. Patton, S.S. Iyer, S.L. Delage, S. Tiwari, and J.M.C. Stork. Silicon-germanium base heterojunction bipolar transistors by molecular beam epitaxy. *IEEE Electron Devices Letters*, 9(4):165–167, 1988.
80. S.S. Iyer, G.L. Patton, J.M.C. Stork, B.S. Meyerson, and D.L. Harame. Heterojunction bipolar transistors using si-ge alloys. *IEEE Transactions on Electron Devices*, 36(10):2043–2064, 1989.
81. D.L. Harame and B.S. Meyerson. The early history of ibm's sige mixed signal technology. *IEEE Transactions on Electron Devices*, 48(11):2555–2567, 2001.
82. Konrad Böhm, Peter Russer, Reinhard Ulrich, and Edgar Weidel. Fibre-optic rotation sensor. In *Proc. Symposium Gyro Technology (Deutsche Gesellschaft für Ortung und Navigation)*, pages 10.1–10.9., 1980.
83. Konrad Böhm, Peter Russer, Edgar Weidel, and Reinhard Ulrich. Low-noise fiber optic rotation sensing. *Optics Letters*, 6:64, 1981.
84. Klaus Petermann and Peter Russer. Ring interferometer, July 1985. U.S. Classification: 356/350; International Classification: G01B 902; G01C 1964.
85. Peter Russer. *Informationstechnik*. VCH, Weinheim, 1988.
86. Peter Russer. *Electromagnetics, Microwave Circuit and Antenna Design for Communications Engineering*. Artech House, Boston, 2003.
87. Peter Russer. *Electromagnetics, Microwave Circuit and Antenna Design for Communications Engineering*. Artech House, Boston, 2nd edition, 2006.
88. Élie Cartan. *Les systèmes différentielles extérieurs*. Hermann, Paris, 1945.
89. Hermann Grassmann and Lloyd C. Kannenberg. *A New Branch of Mathematics: The "Ausdehnungslehre" of 1844 and Other Works*. Open Court Publishing, Chicago, 1995.
90. Harley Flanders. *Differential Forms*. Academic Press, New York, 1963.
91. Theodore Frankel. *The Geometry of Physics*. Cambridge University Press, Cambridge, 1997.
92. Georges A. Deschamps. Electromagnetics and differential forms. *Proceedings of the IEEE*, 69(6):676–696, June 1981.
93. Friedrich W. Hehl and Yuri N. Obukov. *Foundations of Classical Electrodynamics*. Birkhäuser, Boston Basel Berlin, 2003.
94. Peter Russer. The geometry of electrodynamics. *European Microwave Journal*, 1(1):3—16, 2005.
95. Asim Egemen Yilmaz. Grassmann and his contributions to electromagnetics [Historical corner]. *IEEE Antennas and Propagation Magazine*, 52(4):186–193, August 2010.
96. Peter Russer and Uwe Siart, editors. *Time-Domain Methods in Modern Engineering Electromagnetics, A Tribute to Wolfgang J.R. Hoefer*, volume 121 of *Springer Proceedings in Physics*. Springer, 1 edition, 2008.
97. Leopold B. Felsen, Mauro Mongiardo, Peter Russer, G. Conti, and Cristiano Tomassoni. Waveguide component analysis by a generalized network approach. In *Proceedings of the 27th European Microwave Conference, Jerusalem*, pages 949–954, 1997.
98. Leopold B. Felsen, Mauro Mongiardo, and Peter Russer. Electromagnetic field representations and computations in complex structures I: Complexity architecture and generalized network formulation. *International Journal of Numerical Modelling, Electronic Networks, Devices and Fields*, 15:93–107, 2002.

99. Leopold B. Felsen, Mauro Mongiardo, and Peter Russer. Electromagnetic field representations and computations in complex structures II: Alternative Green's functions. *International Journal of Numerical Modelling, Electronic Networks, Devices and Fields*, 15:109–125, 2002.
100. Peter Russer, Mauro Mongiardo, and Leopold B. Felsen. Electromagnetic field representations and computations in complex structures III: Network representations of the connection and subdomain circuits. *International Journal of Numerical Modelling, Electronic Networks, Devices and Fields*, 15:127–145, 2002.
101. Leopold B. Felsen, Mauro Mongiardo, and Peter Russer. *Electromagnetic Field Computation by Network Methods*. Springer, Berlin, Heidelberg, New York, 2009.
102. Peter Russer and Mauro Mongiardo, editors. *Fields, Networks, Methods, and Systems in Modern Electrodynamics*. Springer, Berlin, 2004.
103. Peter Russer and Andreas C. Cangellaris. Network–oriented modeling, complexity reduction and system identification techniques for electromagnetic systems. *Proc. 4th Int. Workshop on Computational Electromagnetics in the Time–Domain: TLM/FDTD and Related Techniques, 17–19 September 2001 Nottingham*, pages 105–122, September 2001.
104. Karl F. Warnick and Peter Russer. Two, three and four-dimensional electromagnetics using differential forms. *Turkish Journal of Electrical Engineering and Computer Sciences*, 14(1):153–172, 2006.
105. Peter Russer. *Problem Solving in Electromagnetics, Microwave Circuit, and Antenna Design for Communications Engineering*. Artech House, Boston, 2006.
106. Mauro Mongiardo, Peter Russer, M. Dionigi, and Leopold B. Felsen. Waveguide step discontinuities revisited by the generalized network formulation. In *1998 International Microwave Symposium Digest, Baltimore, ML, USA*, pages 917–920, 1998.
107. Mauro Mongiardo, Peter Russer, M. Dionigi, and Leopold B. Felsen. Generalized networks for waveguide step discontinuities. *Proceedings of the 14th Annual Review of Progress in Applied Computational Electromagnetics ACES, Monterey*, pages 952–956, March 1998.
108. Mauro Mongiardo, Peter Russer, Cristiano Tomassoni, and Leopold B. Felsen. Analysis of n-furcation in elliptical waveguides via the generalized network formulation. In *1999 International Microwave Symposium Digest, Anaheim, CA, USA*, pages 27–30, 1999.
109. Mauro Mongiardo, Peter Russer, Cristiano Tomassoni, and Leopold B. Felsen. Analysis of n-furcation in elliptical waveguides via the generalized network formulation. *IEEE Transactions on Microwave Theory and Techniques*, 47(12):2473–2478, 1999.
110. Mauro Mongiardo, Peter Russer, Cristiano Tomassoni, and Leopold B. Felsen. Analysis of N–furcation in elliptical waveguides via the generalized network formulation. *1999 International Microwave Symposium Digest, Anaheim, CA, USA*, pages 27–30, June 1999.
111. Mauro Mongiardo, Peter Russer, Cristiano Tomassoni, and Leopold B. Felsen. Analysis of N–furcation in elliptical waveguides via the generalized network formulation. *IEEE Transactions on Microwave Theory and Techniques*, 47:2473–2478, December 1999.
112. Mauro Mongiardo, Peter Russer, Cristiano Tomassoni, and Leopold B. Felsen. Generalized network formulation analysis of the N–furcations application to elliptical waveguide. *Proc. 10th Int. Symp. on Theoretical Electrical Engineering, Magdeburg, Germany, (ISTET)*, pages 129–134, September 1999.
113. Peter Russer and Mauro Mongiardo. The application of network methods to distributed microwave circuit analysis. In *Microwaves, Radar and Wireless Communications. 2000. MIKON-2000. 13th International Conference on*, pages 189–200, 2000.
114. Peter Russer. Overview over network methods applied to electromagnetic field computation. In *ICEAA 2009, International Conference on on Electromagnetics in Advanced Applications*, pages 276–279, Torino, Italy, September 14th–18th, 2009.
115. Peter Russer. Electromagnetic properties and realisability of gyrator surfaces. In *Electromagnetics in Advanced Applications, 2007. ICEAA 2007. International Conference on*, pages 320–323, 2007.
116. Jochen Kessler, Roland Dill, Peter Russer, and Alejandro A. Valenzuela. Property calculations of a superconducting coplanar waveguide resonator. *Proceedings of the 20th European Microwave Conference, Budapest*, pages 798–903, September 1990.

117. Jochen Kessler, Roland Dill, and Peter Russer. Field theory investigation of high-tc superconducting coplanar waveguide transmission lines and resonators. *IEEE Transactions on Microwave Theory and Techniques*, 39(9):1566–1574, 1991.
118. Jochen Kessler, Roland Dill, and Peter Russer. Characterization of millimeterwave transmission lines on silicon substrates. In *Antennas and Propagation Society International Symposium, 1992. AP-S. 1992 Digest. Held in Conjuction with: URSI Radio Science Meeting and Nuclear EMP Meeting., IEEE*, pages 2296–2299 vol.4, 1992.
119. Jochen Kessler, Peter Russer, and Roland Dill. Modelling of miniaturized coplanar striplines based on $YBa_2Cu_3O_{7-x}$ thin films. In *Microwave Symposium Digest, 1992., IEEE MTT-S International*, pages 1127–1130, 1992.
120. Jochen Keßler. *Untersuchung hochtemperatursupraleitender planarer Wellenleiter mittels Partialwellenanalyse*. Dissertation, Technische Universität München, München, 1993.
121. Rolf Schmidt and Peter Russer. Modeling of cascaded coplanar waveguide discontinuities by the mode-matching approach. *IEEE Transactions on Microwave Theory and Techniques*, 43(12):2910–2917, 1995.
122. Rolf Schmidt. *Vollwellenanalyse von verlustbehafteten koplanaren Leitungen und Leitungsdiskontinuitäten*. Dissertation, Technische Universität München, München, 1996.
123. Dzianis Lukashevich, Larissa Vietzorreck, and Peter Russer. Numerical investigation of transmission lines and components in damascene technology. In *European Microwave Conference, 2002. 32nd*, pages 1–4, 2002.
124. Dzianis Lukashevich and Peter Russer. Full-wave analysis of transmission line structures in damascene technology. In *The 19th Annual Review of Progress in Applied Computational Electromagnetics ACES 2003, Monterey, California, USA*, pages 519–524, March 2003.
125. Dzianis Lukashevich and Peter Russer. Network-oriented models of transmission line structures in mmics. In *Silicon Monolithic Integrated Circuits in RF Systems, 2003. Digest of Papers. 2003 Topical Meeting on*, pages 178–181, 2003.
126. Dzianis Lukashevich, Borys Broido, Martin Pfost, and Peter Russer. The hybrid TLM-mm approach for simulation of mmics. In *European Microwave Conference, 2003. 33rd*, pages 339–342, 2003.
127. Borys Broido, Dzianis Lukashevich, and Peter Russer. Hybrid method for simulation of passive structures in rf-mmics. In *2000 Topical Meeting on Silicon Monolithic Integrated Circuits in RF Systems Digest, Garmisch, 26-28 April 2000*, pages 182–185, 2003.
128. Dzianis Lukashevich, Borys Broido, and Peter Russer. Using of transmission line matrix method and mode matching approach for simulation of MMICs. In *2003 International Microwave Symposium Digest, Philadelphia, PA, USA*, pages 993–996, 2003.
129. Mauro Mongiardo, Cristiano Tomassoni, and Peter Russer. Generalized network formulation: Application to flange—mounted radiating waveguides. *IEEE Transactions on Antennas and Propagation*, 55(6):1667–1678, 2007.
130. Mauro Mongiardo, Peter Russer, Roberto Sorrentino, and Cristiano Tomassoni. Spherical mode expansions for flange-mounted waveguide apertures. In *Electromagnetics in Advanced Applications, 2007. ICEAA 2007. International Conference on*, pages 41–44, 2007.
131. Mauro Mongiardo, Peter Russer, Roberto Sorrentino, and Cristiano Tomassoni. Spherical modal expansion for arrays of flange-mounted rectangular waveguides. In *Microwave Conference, 2007. European*, pages 32–35, 2007.
132. Cristiano Tomassoni, Mauro Mongiardo, Peter Russer, and Roberto Sorrentino. Rigorous computer-aided design of coaxial/circular antennas with semi-spherical dielectric layers. In *2008 IEEE MTT-S International Microwave Symposium Digest, Atlanta, GA, USA*, pages 975–978, 2008.
133. Peter Russer. Electromagnetic field computation by network methods. In *Proceedings of the 25th Annual Review of Progress in Applied Computational Electromagnetics, ACES, Monterey, CA*, Monterey, California, USA, March 8–12 2009.
134. Mauro Mongiardo, Cristiano Tomassoni, Peter Russer, and Roberto Sorrentino. Rigorous computer-aided design of spherical dielectric resonators for wireless non-radiative energy transfer. In *2009 IEEE MTT-S International Microwave Symposium Digest, June 7th–12th, Boston, MA, USA*, pages 1249–1252, June 7th–12th 2009.

135. Bruno Biscontini, Markus Burger, Franz Demmel, and Peter Russer. Dyadic Green's function for conformal antennas in multi layered cylindrical structures using generalized transmission lines. In *34th European Microwave Conference, Amsterdam, The Netherlands, 11.-15.10.2004*, pages 953–956, October 2004.
136. Bruno Biscontini, M. Burger, and Peter Russer. Network methods applied to multilayered cylindrical radiating structures. In Peter Russer and Mauro Mongiardo, editors, *Fields, Networks, Methods, and Systems in Modern Electrodynamics*, pages 129–142. Springer, Berlin, 2004.
137. Bruno Biscontini, S. Hamid, Franz Demmel, and Peter Russer. A novel antenna for ultra wide band (UWB) intelligent antenna systems. In *2006 International Microwave Symposium Digest, San Francisco, CA, USA*, pages 2023–2026, 2006.
138. Bruno Biscontini. *Network Methods Applied to Multilayered Cylindrical Radiating Structures*. Dissertation, Technische Universität München, München, 2006.
139. Bruno Biscontini, Uwe Siart, and Peter Russer. On the modeling of ultra wide band (UWB) radiating structures. In Peter Russer and Uwe Siart, editors, *Time-Domain Methods in Modern Engineering Electromagnetics*, Technische Universität München, 2007. Springer.
140. Peter B. Johns and R.L. Beurle. Numerical solution of 2-dimensional scattering problems using a transmission-line matrix. *Proceedings IEE*, 118(9):1203–1208, September 1971.
141. Wofgang J.R. Hoefer. The transmission line matrix method-theory and applications. *IEEE Transactions on Microwave Theory and Techniques*, 33:882–893, October 1985.
142. Wofgang J.R. Hoefer. The transmission line matrix (TLM) method. In Tatsuo Itoh, editor, *Numerical Techniques for Microwave and Millimeter Wave Passive Structures*, pages 496–591. John Wiley, New York, 1989.
143. Christos Christopoulos. *The Transmission-Line Modeling Method TLM*. IEEE Press, New York, 1995.
144. Peter Russer. The transmission line matrix method. In *Applied Computational Electromagnetics*, NATO ASI Series, pages 243–269. Springer, Berlin, 2000.
145. Christos Christopoulos and Peter Russer. Application of TLM to microwave circuits. In *Applied Computational Electromagnetics*, NATO ASI Series, pages 300–323. Springer, Berlin, 2000.
146. Christos Christopoulos and Peter Russer. Application of TLM to EMC problems. In *Applied Computational Electromagnetics*, NATO ASI Series, pages 324–350. Springer, Berlin, 2000.
147. Peter Russer. The transmission line matrix method. In Henri Baudrand, editor, *New Trends and Concepts in Microwave Theory and Technics*, pages 41–82. Research Signpost, Trivandrum, India, 2003.
148. Christiaan Huygens. *Traité de la lumière: où sont expliquées les causes de ce qui luy arrive dans la reflexion, & dans la refraction, et particulièrement dans l'étrange refraction du Cristal d'Islande*. Pierre Vander Aa, Leyden, 1690.
149. Peter Russer. Network methods applied to computational electromagnetics. In *Proceedings of the 9th International Conference on Telecommunication in Modern Satellite, Cable, and Broadcasting Services, 2009. TELSIKS '09.*, pages 329–338, 2009.
150. Johannes A. Russer, Yury Kuznetsov, and Peter Russer. Discrete-time network and state equation methods applied to computational electromagnetics. *Mikrotalasna Revija (Microwave Review)*, pages 2–14, July 2010.
151. Peter Russer, Poman P. M. So, and Wofgang J. R. Hoefer. Modeling of nonlinear active regions in TLM [distributed circuits]. *Microwave and Guided Wave Letters, IEEE*, 1(1):10–13, 1991.
152. Wolfgang Dressel, Bastian Lewke, and Fabio Coccetti. A TLM simulation package. 2004.
153. Wofgang J.R. Hoefer, Bertram Isele, and Peter Russer. Modelling of nonlinear active devices in TLM. In *Proceedings of the First International Conference on Computation in Electromagnetics*, pages 327–330, 1991.
154. Bertram Isele and Peter Russer. Modeling of nonlinear dispersive active elements, in TLM. In *Microwave Symposium Digest, 1992., IEEE MTT-S International*, pages 1217–1220, Albuquerque, New Mexico), 1–5 June 1992.

155. Bertram Isele, Hartmut Bender, Robert Weigel, Josef Hausner, and Peter Russer. Accurate characterization of microstrip filter and Hybrid-Ring coupler via an improved TLM method using variable and curved meshes. In *Proceedings of the 21st European Microwave Conference, Stuttgart*, pages 315–320, 1991.
156. Bertram Isele and Peter Russer. The modeling of coplanar circuits in a parallel computing environment. In *1996 International Microwave Symposium Digest, San Francisco, CA, USA*, pages 1035–1038, 1996.
157. Bertram Isele, Martin Aidam, and Peter Russer. TLM modeling of planar microwave circuits. In *Proceedings of the 26h European Microwave Conference, Prague*, pages 444–446, Prague, September 9th–12th, 1996.
158. Mohamed I. Sobhy, Essam A. Hosny, Peter Russer, Bertram Isele, and Christos Christopoulos. Interfacing the transmission line method (TLM) and state-space (ss) techniques to analyse general non-linear structures. In *Proceedings of the Second International Conference on Computation in Electromagnetics*, pages 299–302, 1994.
159. Peter Russer and M. Krumpholz. The Hilbert space formulation of the TLM method. *International Journal of Numerical Modelling, Electronic Networks, Devices and Fields*, 6(1):29–45, February 1993.
160. Michael Krumpholz, Peter Russer, Qi Zhang, and Wolfgang J.R. Hoefer. Field-theoretic foundation of two-dimensional TLM based on a rectangular mesh. In *1994 International Microwave Symposium Digest, San Diego, CA, USA*, pages 333–336, 1994.
161. Michael Krumpholz and Peter Russer. A field theoretical derivation TLM. *IEEE Transactions on Microwave Theory and Techniques*, 42(9):1660–1668, September 1994.
162. Michael Krumpholz and Peter Russer. TLM and Maxwell's equations. In *Proceedings of the Second International Conference on Computation in Electromagnetics*, pages 12–15, April 1994.
163. Michael Krumpholz and Peter Russer. On the dispersion in TLM and FDTD. *IEEE Transactions on Microwave Theory and Techniques*, 42(7):1275–1279, 1994.
164. Michael Krumpholz and Peter Russer. Two-dimensional FDTD and TLM. *International Journal of Numerical Modelling, Electronic Networks, Devices and Fields*, 7:141–153, April 1994.
165. Michael Krumpholz, Christian Huber, and Peter Russer. A field theoretical comparison of FDTD and TLM. *IEEE Transactions on Microwave Theory and Techniques*, 43(8):1935–1950, 1995.
166. Michael Krumpholz, L. Roselli, and Peter Russer. Dispersion characteristics of the TLM scheme with symmetrical super-condensed node. In *1995 International Microwave Symposium Digest, Orlando, FL, USA*, pages 369–372, 1995.
167. Peter Russer and Bernhard Bader. The alternating transmission line matrix (ATLM) scheme. In *1995 International Microwave Symposium Digest, Orlando, FL, USA*, pages 19–22, 1995.
168. Stefan Lindenmeier, Bernhard Bader, and Peter Russer. Investigation of various h-shaped antennas with an ATLM field-solver. In *1997 International Microwave Symposium Digest, Denver, CO, USA*, pages 1365–1368, 1997.
169. Bernhard Bader. *Untersuchung der Alternating-Transmission-Line-Matrix-Methode (ATLM) für die Zeitbereichsanalyse elektromagnetischer Felder*. Dissertation, Technische Universität München, München, 1997.
170. Monika Niederhoff, Wolfgang Heinrich, and Peter Russer. The finite-integration beam-propagation method (FIBPM). In *1995 International Microwave Symposium Digest, Orlando, FL, USA*, pages 483–486, 1995.
171. Monika Niederhoff, Wolfgang Heinrich, and Peter Russer. Three-dimensional modelling of high-power laser diodes based on the finite integration beam propagation method. In *1996 International Microwave Symposium Digest, San Francisco, CA, USA*, pages 1429–1432, 1996.
172. Monika Niederhoff. *Feldberechnung in Hochleistungslaserdioden*. Dissertation, Technische Universität München, München, 1996.

173. Stefan Lindenmeier, Peter Russer, and Wolfgang Heinrich. Hybrid dynamic-static finite-difference approach for MMIC design. *1996 International Microwave Symposium Digest, San Francisco, CA, USA*, 44:197–200, June 1996.
174. Stefan Lindenmeier, Wolfgang Heinrich, and Peter Russer. A fast magneto-static field simulation for the incorporation into a hybrid dynamic-static finite-integral algorithm. In *Proceedings of the 26h European Microwave Conference, Prague*, pages 447–451, 1996.
175. Stefan Lindenmeier. *Finite Differenzen–Methoden zur Modellierung planarer Hochfrequenzschaltungen*. Dissertation, Technische Universität München, München, 1996.
176. Stefan Lindenmeier and Peter Russer. Design of planar circuit structures with an efficient magneto-static field solver. In *1997 International Microwave Symposium Digest, Denver, CO, USA*, pages 1807–1810, June 1997.
177. Stefan Lindenmeier and Peter Russer. Design of planar circuit structures with an efficient magneto-static field solver. In *1997 International Microwave Symposium Digest, Denver, CO, USA*, pages 1807–1810, 1997.
178. Stefan Lindenmeier, Luca Pierantoni, and Peter Russer. Hybrid space discretizing-integral equation methods for numerical modeling of transient interference. *IEEE Transactions on Electromagnetic Compatibility*, 41(4):425–430, 1999.
179. Wolfgang Dressel. *Modellierung von elektromagnetischen Strukturen mit Hilfe der TLM–Methode*. Dissertation, Technische Universität München, München, 2005.
180. Wolfgang Dressel and Peter Russer. TLM modelling of electromagnetic structures using static sub-griddings. In *Proceedings of the 16th International Conference on Microwaves, Radar & Wireless Communications, MIKON 2006*, pages 707–710, 2006.
181. Luca Pierantoni, Stefan Lindenmeier, and Peter Russer. A combination of integral equation method and FD/TLM method for efficient solution of emc problems. In *Microwave Conference and Exhibition, 1997 27th European*, pages 937–942, 1997.
182. Federigo Alimenti, F. Tiezzi, Roberto Sorrentino, Stefan Lindenmeier, Luca Pierantoni, and Peter Russer. Accurate analysis and modeling of slot coupled patch antennas by the TLM–IE and the FDTD methods. *Proceedings of the 28th European Microwave Conference, Amsterdam*, 1:30–35, 1998.
183. Luca Pierantoni, Stefan Lindenmeier, and Peter Russer. Efficient analysis of microstrip radiation by the TLM integral equation (TLMIE) method. In *1998 International Microwave Symposium Digest, Baltimore, ML, USA*, pages 1267–1270, 1998.
184. Stefan Lindenmeier, Luca Pierantoni, and Peter Russer. Time domain modeling of E.M. coupling between microwave circuit structures. In *1999 International Microwave Symposium Digest, Anaheim, CA, USA*, volume 4, pages 1569–1572, June 1999.
185. Luca Pierantoni, Graziano Cerri, Stefan Lindenmeier, and Peter Russer. Theoretical and numerical aspects of the hybrid MoM-FDTD, TLM-IE and ARB methods for the efficient modelling of EMC problems. In *Proceedings of the 29th European Microwave Conference, Munich*, pages 313–316, 1999.
186. Stefan Lindenmeier. *Methoden zur Analyse elektromagnetischer Kopplungen*. Habilitationsschrift, Technische Universität München, München, 1999.
187. Rachid Khlifi and Peter Russer. A novel efficient hybrid TLM/TDMOM method for numerical modeling of transient interference. In *Proceedings of the 22th Annual Review of Progress in Applied Computational Electromagnetics ACES 2006, Miami, FL, USA*, pages 182–187, March 2006.
188. Rachid Khlifi and Peter Russer. A hybrid method combining TLM and mom method for efficient analysis of scattering problems. In *2006 International Microwave Symposium Digest, San Francisco, CA, USA*, pages 161–164, 2006.
189. Rachid Khlifi and Peter Russer. Hybrid space-discretizing method—method of moments for the analysis of transient interference. *IEEE Transactions on Microwave Theory and Techniques*, 54(12):4440–4447, 2006.
190. Rachid Khlifi and Peter Russer. Analysis of transient radiated interferences by the hybrid TLM-IE/MOM algorithm. In *Proceedings of the 37th European Microwave Conference, Munich*, pages 1389–1392, 2007.

191. Martin Aidam and Peter Russer. Derivation of the transmission line matrix method by finite integration. *AEÜ International Journal of Electron. Commun.*, 51:35–39, January 1997.
192. Martin Aidam and Peter Russer. Application of biorthogonal B-spline wavelets to telegrapher's equations. In *Proceedings of the 14th Annual Review of Progress in Applied Computational Electromagnetics ACES, Monterey*, pages 983–990, Monterey, CA, USA, 16–20 March 1998.
193. Martin Aidam and Peter Russer. Comparison of finite difference and wavelet-galerkin methods for the solution of telegraph equations. In *Proceedings of the 28th European Microwave Conference, Amsterdam*, pages 712–717, Amsterdam, 1998.
194. Martin Aidam. *Wavelet-Galerkin Methoden zur Berechnung elektromagnetischer Felder im Zeitbereich*. Dissertation, Technische Universität München, München, 1999.
195. Peter Russer, Mario Righi, Channabasappa Eswarappa, and Wofgang J.R. Hoefer. Lumped element equivalent circuit parameter extraction of distributed microwave circuits via TLM simulation. In *1994 International Microwave Symposium Digest, San Diego, CA, USA*, pages 887–890, 1994.
196. Mario Righi, Channabasappa Eswarappa, Wofgang J.R. Hoefer, and Peter Russer. An alternative way of computing s–parameters via impulsive TLM analysis without using absorbing boundary conditions. *1995 International Microwave Symposium Digest, Orlando, FL, USA*, pages 1203–1206, May 1995.
197. Tobias Mangold and Peter Russer. Modeling of multichip module interconnections by the TLM method and system identification. In *Microwave Conference and Exhibition, 1997 27th European*, pages 538–543, Jerusalem, Sep. 1997.
198. Tobias Mangold, J. Wolf, M. Töpper, H. Reichl, and Peter Russer. Multilayer multichip modules for microwave and millimeterwave integration. *Proceedings of the 28th European Microwave Conference, Amsterdam*, 2:443–448, October 1998.
199. Tobias Mangold and Peter Russer. Full-wave modeling and automatic equivalent-circuit generation of millimeter-wave planar and multilayer structures. *IEEE Transactions on Microwave Theory and Techniques*, 47(6):851–858, June 1999.
200. Tobias Mangold. *Feldmodellierung von verteilten Mehrtorschaltungen und systematische Extraktion von Ersatzschaltungen aus konzentrierten Elementen*. PhD thesis, Technische Universität München, München, 2001.
201. Vitali Chtchekatourov, Larissa Vietzorreck, Walter Fisch, and Peter Russer. Time-domain system identification modeling for microwave structures. In *MMET 2000. International Conference on Mathematical Methods in Electromagnetic Theory*, pages 137–139, 2000.
202. Vitali Chtchekatourov, Fabio Coccetti, and Peter Russer. Full-wave analysis and model-based parameter estimation approaches for y-matrix computation of microwave distributed rf circuits. In *Microwave Symposium Digest, 2001 IEEE MTT-S International*, pages 1037–1040, 2001.
203. Vitali Chtchekatourov, Fabio Coccetti, and Peter Russer. Direct Y–parameters estimation of microwave structures using TLM simulation and prony's method. In *Proceedings of the 17th Annual Review of Progress in Applied Computational Electromagnetics ACES, Monterey*, pages 580–586, May 2001.
204. Fabio Coccetti, Vitali Chtchekatourov, and Peter Russer. Time-domain analysis of RF structures by means of TLM and system identification methods. In *European Microwave Conference, 2001. 31st*, pages 1–4, 2001.
205. Fabio Coccetti and Peter Russer. A Prony's model based signal prediction (PSP) algorithm for systematic extraction of Y-parameters from TD transient responses of electromagnetic structures. In *Proceedings of the 15th International Conference on Microwaves, Radar & Wireless Communications, MIKON 2004*, pages 791–794, 2004.
206. Fabio Coccetti. *Application of System Identification (SI) to Full-Wave Time Domain Characterization of Microwave and Millimeter Wave Passive Structures*. Dissertation, Technische Universität München, München, 2004.
207. Yury Kuznetsov, Farooq Mukhtar, Fabio Coccetti, and Peter Russer. The ultra wideband transfer function representation of complex three-dimensional electromagnetic structures. In

34th European Microwave Conference, Amsterdam, The Netherlands, 11.-15.10.2004, pages 455–458, October 2004.

208. Yury Kuznetsov, Farooq Mukhtar, Timophey Shevgunov, Michael Zedler, and Peter Russer. Transfer function representation of passive electromagnetic structures. In *2005 International Microwave Symposium Digest, Long Beach, CA, USA*, page 4 pp., 2005.
209. Yury Kuznetsov, Farooq Mukhtar, Timophey Shevgunov, Petr Lorenz, and Peter Russer. Application of the stability criterion to the passive electromagnetic structures modeling. In *Microwave Conference, 2006. 36th European*, pages 13–16, 2006.
210. Timophey Shevgunov, Farooq Mukhtar, Yury Kuznetsov, and Peter Russer. Improved system identification scheme for the linear representation of the passive electromagnetic structures. In *Microwaves, Radar & Wireless Communications, 2006. MIKON 2006. International Conference on*, pages 988–991, 2006.
211. Yury Kuznetsov, Farooq Mukhtar, Petr Lorenz, and Peter Russer. Network oriented modeling of passive microwave structures. In *EUROCON, 2007. The International Conference on "Computer as a Tool"*, pages 10–14, 2007.
212. Nikolaus Fichtner, Uwe Siart, Yury Kuznetsov, Farooq Mukhtar, and Peter Russer. TLM modeling and system identification of optimized antenna structures. In *Kleinheubacher Tagung*, Miltenberg, Germany, September 2007.
213. Uwe Siart, Klaus Fichtner, Yury Kuznetsov, Farooq Mukhtar, and Peter Russer. TLM modeling and system identification of distributed microwave circuits and antennas. In *ICEAA 2007, International Conference on Electromagnetics in Advanced Applications*, pages 352–355, Torino, Italy, September 17th–21st, 2007.
214. Timophey Shevgunov, Farooq Mukhtar, Yury Kuznetsov, and Peter Russer. Lumped element network synthesis for one-port passive microwave structures. In *Proceedings of the 17th International Conference on Microwaves, Radar & Wireless Communications, MIKON 2008*, pages 1–4, 2008.
215. Farooq Mukhtar, Yury Kuznetsov, and Peter Russer. Network modelling with brune's synthesis. In *URSI Conference Kleinheubach*, Miltenberg, Germany, October 4th–6th, 2010.
216. Johannes A. Russer, Farooq Mukhtar, Andrey Baev, Yury Kuznetsov, and Peter Russer. Combined lumped element network and transmission line synthesis for passive microwave structure. In *URSI Conference Kleinheubach*, Miltenberg, Germany, October 4th–6th, 2010.
217. Dzianis Lukashevich, Andreas Cangellaris, and Peter Russer. Transmission line matrix method reduced order modeling. In *2003 International Microwave Symposium Digest, Philadelphia, PA, USA*, pages 1125–1128, 2003.
218. Dzianis Lukashevich, Andreas Cangellaris, and Peter Russer. Model order reduction by Krylov space methods applied to TLM electromagnetic field simulation. In *IEEE MTT-S International Microwave Symposium*, pages 200–205, June 2004.
219. Dzianis Lukashevich. *Model Order Reduction (MOR) in Transmission Line Matrix (TLM) Method*. Dissertation, Technische Universität München, München, 2004.
220. Dzianis Lukashevich, Andreas Cangellaris, and Peter Russer. Two-step reduction approach based on the scattering-symmetric lanczos algorithm for TLM-rom. In *Wireless Communications and Applied Computational Electromagnetics, 2005. IEEE/ACES International Conference on*, pages 698–705, 2005.
221. Dzianis Lukashevich, Andreas Cangellaris, and Peter Russer. Broadband electromagnetic analysis of interconnects by means of TLM and Krylov model order reduction. In *Electrical Performance of Electronic Packaging, 2005. IEEE 14th Topical Meeting on*, pages 355–358, 2005.
222. Dzianis Lukashevich and Peter Russer. Oblique-oblique projection in TLM-mor for high-q structures. In *35th European Microwave Conference, Paris, France, 3.-7.10.2005*, pages 849–852, October 2005.
223. Dzianis Lukashevich, Andreas C. Cangellaris, and Peter Russer. Oblique–oblique projection in TLM-MOR for high-$q$structures. *IEEE Transactions on Microwave Theory and Techniques*, 54(10):3712–3720, 2006.

224. Dzianis Lukashevich, Fabio Coccetti, and Peter Russer. System identification and model order reduction for TLM analysis of microwave components. In *Computational Electromagnetics in Time-Domain, 2005. CEM-TD 2005. Workshop on*, pages 64–67, 2005.
225. Dzianis Lukashevich, Özgür Tuncer, and Peter Russer. Fast multipole method based model order reduction for large scattering problems. In *2006 International Microwave Symposium Digest, San Francisco, CA, USA*, pages 1057–1060, 2006.
226. Dzianis Lukashevich, Fabio Coccetti, and Peter Russer. System identification and model order reduction for TLM analysis. *International Journal of Numerical Modelling, Electronic Networks, Devices and Fields*, 20(1–2):75–92, January 2007.
227. Petr Lorenz, José Vagner Vital, Bruno Biscontini, and Peter Russer. A grid-enabled time domain transmission line matrix (TLM-G) system for the analysis of complex electromagnetic structures. In *Computational Electromagnetics in Time-Domain, 2005. CEM-TD 2005. Workshop on*, pages 48–51, 2005.
228. Petr Lorenz, José Vagner Vital, Bruno Biscontini, and Peter Russer. High-throughput transmission line matrix (TLM) system in grid environment for the analysis of complex electromagnetic structures. In *Proceedings of the 21st Annual Review of Progress in Applied Computational Electromagnetics*, pages 706–710, April 2005.
229. Peter Russer, Bruno Biscontini, and Petr Lorenz. Grid-Enabled transmission line matrix (TLM) modelling of electromagnetic structures. In Luciano Tarricone and Alessandra Esposito, editors, *Advances in Information Technologies for Electromagnetics*, pages 399–431. Springer, Heidelberg, 2006.
230. Jürgen Rebel, Martin Aidam, and Peter Russer. A numerical study on the accuracy of TLM-scn formulations for the solution of initial value. In *Proceedings of the 15th Annual Review of Progress in Applied Computational Electromagnetics ACES, Monterey*, pages 628–635, Monterey, CA, USA, 15-20 March 1999.
231. Jürgen N. Rebel, Martin Aidam, and Peter Russer. On the convergence of the classical symmetrical condensed node-TLM scheme. *IEEE Transactions on Microwave Theory and Techniques*, 49(5):954–963, 2001.
232. Jürgen N. Rebel. *On the Foundations of the Transmission Line Matrix Method.* Dissertation, Technische Universität München, München, 2000.
233. Marcelo N. de Sousa, José Vagner Vital, Leonardo R.A.X. de Menezes, and Peter Russer. Evaluation of UWB system coverage with the 2D parflow method. *Proceedings of the 28th General Assembly of the International Union of Radio Science, URSI, Delhi, India*, 2005.
234. Marcelo N. de Sousa, José Vagner Vital, Leonardo R.A.X. de Menezes, and Peter Russer. UWB system coverage using the complex envoltory in 2D TLM power flow (TLMPF). In *2006 International Microwave Symposium Digest, San Francisco, CA, USA*, pages 276–279, 2006.
235. Uwe Siart, Susanne Hofmann, Nikolaus Fichtner, and Peter Russer. Coverage prediction in large scenarios based on the TLM method. In *2008 IEEE Antennas and Propagation Society International Symposium Digest*, pages 1–4, 2008.
236. Uwe Siart, Susanne Hofmann, Nikolaus Fichtner, and Peter Russer. Computation of frequency average power density based on the TLM method. In *European Microwave Conference*, page accepted for publication, October 2008.
237. Petr Lorenz and Peter Russer. Discrete and modal source modeling with connection networks for the transmission line matrix (TLM) method. In *2007 IEEE MTT-S International Microwave Symposium Digest, Honolulu, HI, USA*, pages 1975–1978, 2007.
238. Fabio Coccetti, Larissa Vietzorreck, Vitali Chtchekatourov, and Peter Russer. A numerical study of MEMS capacitive switches using TLM. In *Proceedings of the 16th Annual Review of Progress in Applied Computational Electromagnetics ACES, Monterey*, pages 580–586, Monterey, CA, March 2000.
239. Larissa Vietzorreck, Fabio Coccetti, Vitali Chtchekatourov, and Peter Russer. Numerical methods for the high-frequency analysis of MEMS capacitive switches. *2000 Topical Meeting on Silicon Monolithic Integrated Circuits in RF Systems Digest, Garmisch, 26-28 April 2000*, pages 123–124, April 2000.

240. Larissa Vietzorreck, Fabio Coccetti, Vitali Chtchekatourov, and Peter Russer. Modeling of MEMS capacitive switches by TLM. *2000 International Microwave Symposium Digest, Boston, MA, USA*, pages 1221–1223, June 2000.
241. Larissa Vietzorreck and Peter Russer. Numerical investigation of micromachined structures for thin layers. In *Proceedings of the 30th European Microwave Conference, Paris*, pages 1–4, 2000.
242. Luca Pierantoni, Marco Farina, Tullio Rozzi, Fabio Coccetti, Wolfgang Dressel, and Peter Russer. Comparison of the efficiency of electromagnetic solvers in the time- and frequency-domain for the accurate modeling of planar circuits and mems. In *2002 International Microwave Symposium Digest, Seattle, WA, USA*, pages 891–894, 2002.
243. Fabio Coccetti, Wolfgang Dressel, M. Burger, J. Hasch, and Peter Russer. Analysis of soi cavity resonator by means of a fully automatic time-domain response prediction algorithm. In *Proceedings of the 34th European Microwave Conference, Amsterdam*, pages 265–268, 2004.
244. Peter Russer, Damienne Bajon, Sidina Wane, and Nikolaus Fichtner. Overview and status of numerical electromagnetic field simulation methods applied to integrated circuits. In *IEEE Topical Meeting on Silicon Monolithic Integrated Circuits in RF Systems, SiRF'09. Orlando, FL, USA*, pages 1–8, 2009.
245. Nikolaus Fichtner, Sidina Wane, Damienne Bajon, and Peter Russer. Interfacing the TLM and the TWF method using a diakoptics approach. In *2008 IEEE MTT-S International Microwave Symposium Digest, Atlanta, GA, USA*, pages 57–60, 2008.
246. Nikolaus Fichtner, Sidina Wane, Damienne Bajon, and Peter Russer. Network based hybridization of the TLM and the TWF method. In *ICEAA 2009, International Conference on on Electromagnetics in Advanced Applications*, pages 101–104, 2009.
247. Victor Veselago, Leonid Braginsky, Valery Shklover, and Christian Hafner. Negative refractive index materials. *Journal of Computational and Theoretical Nanoscience*, 3(2):189–218, 2006.
248. Michael Zedler and Peter Russer. Investigation on the dispersion relation of a 3D LC - based metamaterial with an omnidirectional left - handed frequency band. In *2006 International Microwave Symposium Digest, San Francisco, CA, USA*, pages 1477–1479, June 11–14 2006.
249. Michael Zedler and Peter Russer. Three-dimensional CRLH metamaterials for microwave applications. *Proceedings of the European Microwave Association*, pages 151–162, June 2007.
250. Michael Zedler and Peter Russer. Circuit theory approach to the design of metamaterials. In *ICEAA 2009, International Conference on on Electromagnetics in Advanced Applications*, pages 299–302, Torino, Italy, September 14th–18th, 2009.
251. Michael Zedler, Uwe Siart, and Peter Russer. Circuit theory unifying description for metamaterials. *Proceedings of the 29th General Assembly of the International Union of Radio Science, URSI, Chicago*, 2008.
252. Michael Zedler, Christophe Caloz, and Peter Russer. 3D composite right-left handed metamaterials with Lorentz-type dispersive elements. In *Signals, Systems and Electronics, 2007. ISSSE '07. International Symposium on*, pages 217–221, 2007.
253. Michael Zedler, Christophe Caloz, and Peter Russer. Analysis of a planarized 3D isotropic LH metamaterial based on the rotated TLM scheme. In *Proceedings of the 37th European Microwave Conference, Munich*, pages 624–627, Munich, Germany, oct 2007.
254. Michael Zedler, Christophe Caloz, and Peter Russer. A 3-D isotropic left-handed metamaterial based on the rotated transmission-line matrix (TLM) scheme. *IEEE Transactions on Microwave Theory and Techniques*, 55(12):2930–2941, 2007.
255. Michael Zedler, Christophe Caloz, and Peter Russer. Circuital and experimental demonstration of a 3D isotropic LH metamaterial based on the rotated TLM scheme. In *2007 IEEE MTT-S International Microwave Symposium Digest, Honolulu, HI, USA*, pages 1827–1830, 2007.
256. Michael Zedler, George V. Eleftheriades, and Peter Russer. Three-dimensional isotropic scalar metamaterial with drude dispersion for the permittivity and permeability. In *2009 IEEE MTT-S International Microwave Symposium Digest, June 7th–12th, Boston, MA, USA*, pages 149–152, 2009.

257. Ali Eren Culhaoglu, Michael Zedler, Wolfgang J.R. Hoefer, Andrey Osipov, and Peter Russer. Full wave numerical simulation of a finite 3D metamaterial lens. In *Proceedings of the 24th Annual Review of Progress in Applied Computational Electromagnetics, Niagara Falls, Canada*, pages 989–994, Niagara Falls, Canada, March 30th–April 4th, 2008.
258. Ali Eren Culhaoglu, Andrey Osipov, and Peter Russer. Determination of spectral focusing features of a metamaterial slab. In *Proceedings of the 25th Annual Review of Progress in Applied Computational Electromagnetics, ACES, Monterey, CA*, pages 320–325, Monterey, California, USA, 8–12 March 2009.
259. Johannes A. Russer and Wolfgang J.R. Hoefer. A TLM algorithm simulator for the visualization of time discrete electromagnetic processes. In *Proceedings of the Second International Conference on Computation in Electromagnetics*, pages 120–122, London, 1994.
260. Stefan J. R. Müller. *Rausch- und Empfindlichkeitsanalyse linearer Mikrowellennetzwerke*. Dissertation, Technische Universität München, München, 1994.
261. Franz X. Kaertner. Determination of the correlation spectrum of oscillators with low noise. *IEEE Transactions on Microwave Theory and Techniques*, 37(1):90–101, 1989.
262. Franz X. Kärtner. *Untersuchung des Rauschverhaltens von Oszillatoren*. Dissertation, Technische Universität München, München, 1989.
263. Martin H. Schwab. Determination of the steady state of an oscillator by a combined time-frequency method. *IEEE Transactions on Microwave Theory and Techniques*, 39(8):1391–1402, 1991.
264. Martin Schwab. *Ein kombiniertes Zeit–Frequenzbereichsverfahren zur Berechnung periodischer Schwingungen von Oszillatoren*. Dissertation, Technische Universität München, München, 1992.
265. Werner Anzill and Peter Russer. A general method to simulate noise in oscillators based on frequency domain techniques. *IEEE Transactions on Microwave Theory and Techniques*, 41(12):2256–2263, 1993.
266. Werner Anzill, Oskar von Stryk, Roland Bulirsch, and Peter Russer. Phase noise minimization of microwave oscillators by optimal design. In *1995 International Microwave Symposium Digest, Orlando, FL, USA*, pages 1565–1568, 1995.
267. Werner Anzill. *Berechnung und Optimierung des Phasenrauschens von Oszillatoren*. Dissertation, Technische Universität München, München, 1995.
268. Marion Filleböck, Martin Schwab, and Peter Russer. Automatic generation of starting values for the simulation of microwave oscillators by frequency domain techniques. *IEEE Transactions on Microwave Theory and Techniques*, 41(5):809–813, 1993.
269. Marion Filleböck and Peter Russer. Robust continuation method for tuning characteristics computation and global stability analysis of microwave oscillators. In *European Microwave Conference, 1995. 25th*, pages 1225–1229, 1995.
270. Marion Filleböck. *Kombinierte Zeit–Frequenzbereichsmethoden zum Entwurf von Mikrowellenoszillatoren*. Dissertation, Technische Universität München, München, 1996.
271. Josef Hausner, Gerhard R. Olbrich, Peter Russer, and Alejandro A. Valenzuela. Nonlinear approach for the optimization of a dro at 10.4GHz. In *European Microwave Conference, 1988. 18th*, pages 268–273, 1988.
272. L. Eichinger, B. Fleischmann, Peter Russer, and Robert Weigel. A 2 GHz surface transverse wave oscillator with low phase noise. *IEEE Transactions on Microwave Theory and Techniques*, 36(12):1677–1684, 1988.
273. B. Fleischmann, A. Roth, Peter Russer, and Robert Weigel. Low noise phase locked vco at 2.5 GHz for optical transmission networks using fifth harmonic stw delay line. In *European Microwave Conference, 1990. 20th*, pages 1696–1701, 1990.
274. Ralf Klieber, Roland Ramisch, Alejandro A. Valenzuela, Robert Weigel, and Peter Russer. A coplanar transmission line high-T_c superconductive oscillator at 6.5 GHz on a single substrate. *Microwave and Guided Wave Letters, IEEE*, 2(1):22–24, 1992.
275. Volker Güngerich, Martin Schwab, and Peter Russer. Nonlinear design and experimental results of a low-noise varactor tunable oscillator using a coupled microstrip resonator. In *Microwave Symposium Digest, 1992., IEEE MTT-S International*, pages 549–552, 1992.

276. Volker Güngerich, Franz Zinkler, Werner Anzill, and Peter Russer. Reduced phase noise of a varactor tunable oscillator: numerical calculations and experimental results. In *1993 International Microwave Symposium Digest, Atlanta, GA, USA*, pages 561–564, 1993.
277. Volker Güngerich. *Untersuchung breitbandig abstimmbarer rauscharmer integrierter GaAs-MESFET-Mikrowellenoszillatoren*. Dissertation, Technische Universität München, München, 1993.
278. Volker Güngerich, B. Janke, Franz Zinkler, Wolfgang Heinrich, and Peter Russer. MMIC oscillator simulation considering bias-voltage dependence. In *1994 International Microwave Symposium Digest, San Diego, CA, USA*, pages 989–992, 1994.
279. Volker Gungerich, Franz Zinkler, Werner Anzill, and Peter Russer. Noise calculations and experimental results of varactor tunable oscillators with significantly reduced phase noise. *IEEE Transactions on Microwave Theory and Techniques*, 43(2):278–285, 1995.
280. Josef Hausner and Peter Russer. A broadband tunable distributed feedback resonator. In *1991 International Microwave Symposium Digest, Chicago, IL, USA*, pages 739–742, 1991.
281. Josef Hausner. *Mikrowellenoszillator mit abstimmbarem Bragg–Resonator*. Dissertation, Technische Universität München, München, 1991.
282. Jung Han Choi. *A Si Schottky Diode Demultiplexer Circuit for High Bit Rate Fiber Optical Receivers*. PhD thesis, Technische Universität München, München, 2004.
283. Jung Han Choi, Gerhard Olbrich, and Peter Russer. An si schottky diode demultiplexer circuit for high bit-rate optical receivers. *IEEE Transactions on Microwave Theory and Techniques*, 53(6):2033–2042, 2005.
284. Jung Han Choi and Peter Russer. The picosecond pulse transmission on the conductor-backed coplanar waveguide with via holes. *Microwave and Wireless Components Letters, IEEE*, 16(7):419–421, 2006.
285. Mahmoud Al Ahmad, Ruth Maenner, Richard Matz, and Peter Russer. Wide piezoelectric tuning of LTCC bandpass filters. In *2005 International Microwave Symposium Digest, Long Beach, CA, USA*, page 4 pp., 2005.
286. Mahmoud Al-Ahmad. *Wide Piezoelectric Tuning of LTCC Bandpass Filters*. Dissertation, Technische Universität München, München, 2006.
287. Mahmoud Al Ahmad, Richard Matz, and Peter Russer. 0.8 GHz to 2.4 GHz tunable ceramic microwave bandpass filters. In *2007 IEEE MTT-S International Microwave Symposium Digest, Honolulu, HI, USA*, pages 1615–1618, 2007.
288. K.G. Riedel, S.T. Schaal, Karl-Heinz Türkner, and Peter Russer. Thermoradiotherapie bei malignem aderhautmelanom: Neuentwicklung eines mikrowellenhyperthermiesystems. *Fortschritte der Ophthalmologie*, 87(6):543–550, 1990.
289. Peter Russer, Karl-Heinz Türkner, K. Riedel, and S. T. Schaal. Hyperthermia system for treatment of malignant eye tumors. In *Proc. Microwaves and Optronics Conference (MIOP) 1989, Sindelfingen*, February 28th–March 2nd, 1989.
290. Adalbert Bandemer, Peter Russer, and Karl-Heinz Türkner. Acoustooptic time and frequency domain signal analyzer. In *Proc. of the International Symposium on Electromagnetic Compatibility*, pages 428–431, Tokyo, Japan, October 16th–18th, 1984.
291. Adalbert Bandemer. *Ein optischer Hochfrequenzspektrograph zur Zeit-Frequenz-Darstellung nichtstationärer Signale*. Dissertation, Technische Universität München, München, 1988.
292. Robert Weigel. Planar acoustooptic interactions in lithium niobate. In *Proceedings of the International Conference on Nonlinear Optics*, pages 124–136, Ashford Castle, Kong, Ireland, 3–6 May 1988.
293. Kimon Anemogiannis, Peter Russer, and Robert Weigel. Wide-band nonlinear chirp transducers for planar acoustooptic deflectors. In *1989 International Microwave Symposium Digest, New York, NJ, USA*, pages 269–272, 13–15 June 1989.
294. Erwin Biebl, Peter Russer, and Kimon Anemogiannis. SAW propagation on proton-exchanged lithium niobate. In *Ultrasonics Symposium, 1989. Proceedings., IEEE 1989*, pages 281–284, 1989.
295. Erwin Biebl and Peter Russer. Elastic properties of proton exchanged lithium niobate. *Ultrasonics, Ferroelectrics and Frequency Control, IEEE Transactions on*, 39(3):330–334, 1992.

296. Adalbert Bandemer, F. Heiss, and Robert Weigel. Non-linearities in single-mode fibers and calculations of Raman cross talk in a wavelength-multiplexing system. In *Proceedings of the International Conference on Nonlinear Optics*, pages 137–145, Ashford Castle, Kong, Ireland, 3–6 May 1988.
297. Robert Osborne. All-fibre, Nd-YAG-pumped, subpicosecond raman ring laser. In *Proceedings of the International Conference on Nonlinear Optics*, pages 153–158, Ashford Castle, Kong, Ireland, 3–6 May 1988.
298. Robert Osborne. Raman pulse walk-off in single-mode fibers: an exact analysis. *Journal of the Optical Society of America B*, 6(9):1726–1731, 1989.
299. Robert Osborne. *Nonlinear Pulse Propagation in Single-Mode Optical Fibre*. Dissertation, Technische Universität München, München, 1992.
300. Gerd Scholl, Andreas Christ, Hans-Peter Grassl, Werner Ruile, Peter Russer, and Robert Weigel. Efficient design tool for SAW-resonator filters. In *Ultrasonics Symposium, 1989. Proceedings., IEEE 1989*, pages 135–140, 1989.
301. Gerd Scholl, Andreas Christ, Werner Ruile, Peter Russer, and Robert Weigel. Efficient analysis tool for coupled-SAW-resonator filters. *Ultrasonics, Ferroelectrics and Frequency Control, IEEE Transactions on*, 38(3):243–251, 1991.
302. Gerd Scholl, Werner Ruile, and Peter Russer. P-matrix modeling of transverse-mode coupled resonator filters. In *Ultrasonics Symposium, 1993. Proceedings., IEEE 1993*, pages 41–46, 1993.
303. Kimon Anemogiannis, C. Beck, A. Roth, Peter Russer, and Robert Weigel. A 900 MHz SAW microstrip antenna-duplexer for mobile radio. In *1990 International Microwave Symposium Digest, Long Beach, CA, USA*, pages 729–732, 1990.
304. Kimon Anemogiannis, Peter Russer, Robert Weigel, and C. Zimmermann. SAW microstrip front-end for mobile communication systems in the GHz range. In *1991 International Microwave Symposium Digest, Chicago, IL, USA*, pages 973–976, 1991.
305. Erwin Biebl, Kimon Anemogiannis, Robert Weigel, and Peter Russer. High performance mobile communication front-ends in the GHz range using low loss SAW-filters. In *Proceedings of the IEEE Ultrasonics Symposium, 1991*, pages 55–58, 1991.
306. Hans Meier, Robert Weigel, Kimon Anemogiannis, and Peter Russer. SAW microstrip antenna-duplexer for radio communication transceivers in the GHz range. In *Proceedings of the 21st European Microwave Conference, Stuttgart*, pages 398–403, 1991.
307. Hans Meier and Peter Russer. Analysis of leaky surface acoustic waves on litao3 substrate. In *Frequency Control Symposium, 1992. 46th., Proceedings of the 1992 IEEE*, pages 378–383, 1992.
308. Hans Meier and Peter Russer. Analysis of leaky surface acoustic wave reflections. In *Ultrasonics Symposium, 1993. Proceedings., IEEE 1993*, pages 201–204, 1993.
309. Ulrike Rösler, D. Cohrs, A. Dietz, Gerhard Fischerauer, Werner Ruile, Peter Russer, and Robert Weigel. Determination of leaky SAW propagation, reflection and coupling on litao3. In *Ultrasonics Symposium, 1995. Proceedings., 1995 IEEE*, pages 247–250, 1995.
310. Robert Weigel, Andreas Holm, Peter Russer, Werner Ruile, and G. Sölkner. Accurate optical measurement of surface acoustic wave phase velocity. In *Ultrasonics Symposium, 1993. Proceedings., IEEE 1993*, pages 319–322, 1993.
311. Robert Weigel, Andreas Holm, Gerald Soelkner, Werner Ruile, Peter Russer, and Richard Scheps. Laser probing system for the accurate detection of surface acoustic wave phase velocities. In *Visible and UV Lasers*, volume 2115, pages 108–115, Los Angeles, CA, USA, June 1994. SPIE.
312. Andreas Holm, Robert Weigel, Peter Russer, and Werner Ruile. A laser probing system for characterization of SAW propagation on $LiNbO_3$, $LiTaO_3$, and quartz. In *1996 International Microwave Symposium Digest, San Francisco, CA, USA*, pages 1541–1544, 1996.
313. Arye Rosen, Martin Caulton, Paul Stabile, Anna M. Gombar, Walter M. Janton, Chung P. Wu, John F. Corboy, and Charles W. Magee. Silicon as a millimeter-wave monolithically integrated substrate-A new look. *RCA Review*, 42:633–660, December 1981.

314. Arye Rosen, Martin Caulton, Paul Stabile, Anna M. Gombar, Walter M. Janton, Chung P. Wu, John F. Corboy, and Charles W. Magee. Millimeter-wave device technology. *IEEE Transactions on Microwave Theory and Techniques*, 30(1):47–55, January 1982.
315. Josef Büchler, Erich Kasper, Peter Russer, and Karl M. Strohm. Silicon high–resistivity–substrate millimeter–wave technology. *IEEE Transactions on Microwave Theory and Techniques*, 34:1516–1521, December 1986.
316. Karl M. Strohm, Josef Büchler, Peter Russer, and Erich Kasper. Silicon high resistivity substrate millimeter–wave technology. In *1986 International Microwave Symposium Digest, Baltimore, ML, USA*, pages 93–97, June 4th–6th, 1986.
317. K.M. Strohm, Josef Buechler, Erich Kasper, Johann-Friedrich Luy, and Peter Russer. Millimeter wave transmitter and receiver circuits on high resistivity silicon. In *Microwave and Millimetre Wave Monolithic Integrated Circuits, IEE Colloquium on*, pages 11/1–11/4, 1988.
318. Josef Buechler, Erich Kasper, Johann-Friedrich Luy, Peter Russer, and Karl M. Strohm. Planar wband receiver and oscillator. In *European Microwave Conference, 1988. 18th*, pages 364–369, 1988.
319. Josef Buechler, Karl M. Strohm, Johann-Friedrich Luy, Toni Goeller, Sebastian Sattler, and Peter Russer. Coplanar monolithic silicon IMPATT transmitter. In *Proceedings of the 21st European Microwave Conference, Stuttgart*, pages 352–357, 1991.
320. Josef Büchler. *Integrierte Millimeterwellenschaltungen auf Silizium*. Dissertation, Technische Universität München, München, 1990.
321. Josef Buechler. Silicon millimeter–wave integrated circuits. In J.-F. Luy and Peter Russer, editors, *Silicon–Based Millimeter–Wave Devices*, number 32 in Springer Series in Electronics and Photonics, pages 149–192. Springer, Berlin, 1994.
322. Johann Friedrich Luy and Peter Russer. *Silicon-Based Millimeter-Wave Devices*, volume 32 of *Springer Series in Electronics and Photonics*. Springer, Berlin, 1994.
323. Peter Russer and Erwin Biebl. Fundamentals. In Johann Friedrich Luy and Peter Russer, editors, *Silicon-Based Millimeter-Wave Devices*, number 32 in Springer Series in Electronics and Photonics, pages 149–192. Springer, Berlin, 1994.
324. Peter Russer. Si and SiGe millimeter-wave integrated circuits. *IEEE Transactions on Microwave Theory and Techniques*, 46:590–603, May 1998.
325. Erich Kasper, Dietmar Kissinger, Peter Russer, and Robert Weigel. High speeds in a single chip. *IEEE Microwave Magazine*, 10(7):28–33, 2009.
326. Robert Wanner, Martin Pfost, Rudolf Lachner, and Gerhard R. Olbrich. A 47 GHz monolithically integrated sige push-push oscillator. In *IEEE Topical Meeting on Silicon Monolithic Integrated Circuits in RF Systems, September 8th–10th, 2003, Atlanta, GA, USA*, pages 9–12, September 2004.
327. Robert Wanner, H. Schäfer, Rudolf Lachner, Gerhard Olbrich, and Peter Russer. A fully integrated 70 GHz sige low phase noise push-push oscillator. In *2005 International Microwave Symposium Digest, Long Beach, CA, USA*, page 4 pp., 2005.
328. Robert Wanner, H. Schäfer, Rudolf Lachner, Gerhard Olbrich, and Peter Russer. A fully integrated sige low phase noise push-push vco for 82 GHz. In *Gallium Arsenide and Other Semiconductor Application Symposium, 2005. EGAAS 2005. European*, pages 249–252, 2005.
329. Robert Wanner, Rudolf Lachner, Gerhard R. Olbrich, and Peter Russer. A sige monolithically integrated 278 GHz push-push oscillator. In *2007 IEEE MTT-S International Microwave Symposium Digest, Honolulu, HI, USA*, pages 333–336, 2007.
330. Robert Wanner. *Low Phase Noise SiGe Push–Push Oscillators for Millimeter Wave Frequencies*. Dissertation, Technische Universität München, München, 2007.
331. Robert Wanner, Gerhard Olbrich, H. Jorke, Johann-Friedrich Luy, S. Heim, Erich Kasper, and Peter Russer. Experimental verification of the resonance phase transistor concept. In *2004 International Microwave Symposium Digest, Fort Worth, TX, USA*, pages 991–993, June 6th–11th, 2004.

332. Robert Wanner and Peter Russer. The resonance phase transistor cascode circuit. In *IEEE Topical Meeting on Silicon Monolithic Integrated Circuits in RF Systems, September 8th–10th, 2003, Atlanta, GA, USA*, pages 286–289, September 2004.
333. Hristomir Yordanov and Peter Russer. Computation of the electrostatic parameters of a multiconductor digital bus. In *Electromagnetics in Advanced Applications, 2007. ICEAA 2007. International Conference on*, pages 856–859, 2007.
334. Hristomir Yordanov, Michel T. Ivrlač, Josef A. Nossek, and Peter Russer. Field modelling of a multiconductor digital bus. In *Microwave Conference, 2007. European. 37th European*, pages 1377–1380, 2007.
335. Hristomir Yordanov and Peter Russer. Chip-to-chip interconnects using integrated antennas. In *Proceedings of the 38th European Microwave Conference, EuMC 2008*, pages 777–780, Amsterdam, The Netherlands, October 2008.
336. Hristomir Yordanov and Peter Russer. Wireless inter-chip and intra-chip communication. In *Proceedings of the 39th European Microwave Conference, EuMC 2009*, pages 145–148, Rome, Italy, September 2009.
337. Hristomir Yordanov and Peter Russer. Integrated on-chip antennas using CMOS ground planes. In *Proceedings of the 10th Topical Meeting on Silicon Monolithic Integrated Circuits in RF Systems*, pages 53–56, New Orleans, LA, January 2010.
338. Hristomir Yordanov and Peter Russer. Area-efficient integrated antennas for inter-chip communication. In *Proceedings of the 40th European Microwave Conference, Paris*, Paris, France, September 2010.
339. Hristomir Yordanov and Peter Russer. Antennas embedded in CMOS integrated circuits. *Facta universitatis-series: Electronics and Energetics*, 23(2):169–177, 2010.
340. Josef Büchler and Martin Rieger. Analytical calculation of the inductance of Josephson junctions. In H.D. Hahlbohm et al, editor, *SQUID 85, Superconducting Quantum Interference Devices and their Applications*, number 6 in Berlin, pages 89–93. Walter de Gruyter & Co, 1985.
341. Josef Büchler and Martin Rieger. Frequency conversion coefficients of Josephson junctions. *AEU. Archiv für Elektronik und Übertragungstechnik*, 39(6):347–350, 1985.
342. Martin Rieger. *Mikrowellen-Detektion mit Josephson-Elementen*. Dissertation, Technische Universität München, München, 1988.
343. Johannes G. Bednorz and Karl A. Müller. Possible highT_c superconductivity in the Ba- La- Cu- O system. *Zeitschrift für Physik B Condensed Matter*, 64(2):189–193, 1986.
344. Alejandro A. Valenzuela and Peter Russer. High Q coplanar transmission line resonator of $YBa_2Cu_30_{7-x}$ on Mg0. *Applied Physics Letters*, 5:1029–1031, 1989.
345. W. Rauch, Erich Gornik, Alejandro A. Valenzuela, G. Sölkner, F. Fox, H. Behner, G. Gieres, and Peter Russer. Planar transmission line resonators from $YBa_2Cu_3O_{7-x}$ thin films and epitaxial SIS multilayers. *IEEE Transactions on Applied Superconductivity*, 3(1):1110–1113, March 1993.
346. Roland Ramisch, Gerhard R. Olbrich, and Peter Russer. A tapped-delay-line superconductive chirp filter in shielded microstrip. *IEEE Transactions on Microwave Theory and Techniques*, 39(9):1575–1581, 1991.
347. Ralf Klieber, Roland Ramisch, Robert Weigel, Martin Schwab, Roland Dill, Alejandro A. Valenzuela, and Peter Russer. High-temperature superconducting resonator-stabilized coplanar hybrid-integrated oscillator at 6.5 GHz. In *Electron Devices Meeting, 1991. IEDM '91. Technical Digest., International*, pages 923–926, 1991.
348. Ralf Klieber, Roland Ramisch, Robert Weigel, Martin Schwab, Alejandro A. Valenzuela, Roland Dill, and Peter Russer. Single-substrate high-T_c superconducting coplanar oscillator at 6.5 GHz. In *Proceedings of the 1992 Asia-Pacific Microwave Conference, APMC '92.*, pages 127–130, 1992.
349. Christoph Ullrich, Karl F. Warnick, and Peter Russer. Radiation from a monopole antenna in an aperture backed by an absorbing body using a hybrid mom/utd approach. In *2008 IEEE Antennas and Propagation Society International Symposium Digest*, pages 1–4, 2008.
350. Christoph Ullrich. *Effiziente Simulationsmethoden für die Optimierung von komplexen Fahrzeugantennensystemen*. Dissertation, Technische Universität München, München, 2009.

351. Libo Huang, Werner L. Schroeder, and Peter Russer. Estimation of maximum attainable antenna bandwidth in electrically small mobile terminals. In *Proceedings of the 36th European Microwave Conference, Manchester*, pages 630–633, 2006.
352. Libo Huang, Werner L. Schroeder, and Peter Russer. Coexistence of an electrically tunable DVB-H antenna with the GSM transmitter in a mobile phone. In *2007 IEEE MTT-S International Microwave Symposium Digest, Honolulu, HI, USA*, pages 255–258, 2007.
353. Libo Huang and Peter Russer. Tunable antenna design procedure and harmonics suppression methods of the tunable DVB-H antenna for mobile applications. In *Proceedings of the 37th European Microwave Conference, Munich*, pages 1062–1065, Munich, Germany, October 2007.
354. Stefan Lindenmeier, J. F. Luy, and Peter Russer. A multifunctional antenna for terrestrial and satelite radio applications. In *2001 International Microwave Symposium Digest, Phoenix, AR, USA*, May 2001.
355. Stefan Lindenmeier, Gerhard R. Olbrich, Johann-Friedrich Luy, and Peter Russer. A Five-Band antenna for terrestrial and satellite radio services. *Proceedings of the 17th URSI General Assembly 2002, 17.-24. August 2002*, 2002.
356. Robert Wanner, M.I. Sobhy, and Peter Russer. Bidirectional field compensated active antenna. In *Radar Conference, 2006. EuRAD 2006. 3rd European*, pages 65–67, 2006.
357. Tuan Do-Hong, Franz Demmel, and Peter Russer. A method for wideband direction-of-arrival estimation using frequency-domain frequency-invariant beamformers. In *Antennas and Propagation Society International Symposium, 2003. IEEE*, pages 244–247, 2003.
358. Tuan-Do-Hong and Peter Russer. Signal processing for wideband smart antenna array applications. *Microwave Magazine, IEEE*, 5(1):57–67, 2004.
359. Tuan Do-Hong, Franz Demmel, and Peter Russer. Wideband direction-of-arrival estimation using frequency-domain frequency-invariant beamformers: an analysis of performance. *Microwave and Wireless Components Letters, IEEE*, 14(8):383–385, 2004.
360. Karl F. Warnick, Bert Woestenburg, Leonid Belostotski, and Peter Russer. Minimizing the noise penalty due to mutual coupling for a receiving array. *Antennas and Propagation, IEEE Transactions on*, 57(6):1634–1644, 2009.
361. Karl F. Warnick and Peter Russer. Quantifying the noise penalty for a mutually coupled array. In *2008 IEEE Antennas and Propagation Society International Symposium Digest*, pages 1–4, 2008.
362. Hristomir Yordanov, Michel T. Ivrlač, Peter Russer, and Josef A. Nossek. Arrays of isotropic radiators–a field-theoretic justification. In *Proc. ITG/IEEE Workshop on Smart Antennas, WSA 2009*, Berlin, Germany, 30 March–4 April 2009.
363. Florian Krug and Peter Russer. Ultra-fast broadband EMI measurement in time-domain using FFT and periodogram. In *Proceedings of the 2002 IEEE International Symposium on Electromagnetic Compatibility*, pages 577–582, 2002.
364. Florian Krug and Peter Russer. Ultra-fast broadband EMI measurement in time domain using classical spectral estimation. In *2002 International Microwave Symposium Digest, Seattle, WA, USA*, pages 2237–2240, 2002.
365. Florian Krug and Peter Russer. Signal processing methods for time domain EMI measurements. In *Proceedings of the 2003 IEEE International Symposium on Electromagnetic Compatibility*, pages 1289–1292, 2003.
366. Florian Krug and Peter Russer. A new short-time spectral estimation technique for precompliance measurements. In *ICEAA 2003, International Conference on Electromagnetics in Advanced Applications*, pages 247–250, Torino, Italy, September 8th–13th, 2003.
367. Florian Krug and Peter Russer. Statistical evaluations of time-domain EMI measurements. In *Proceedings of the 2003 IEEE International Symposium on Electromagnetic Compatibility*, pages 1265–1268, 2003.
368. Stephan Braun, Florian Krug, and Peter Russer. A novel automatic digital quasi-peak detector for a time domain measurement system. In *Proceedings of the 2004 IEEE International Symposium on Electromagnetic Compatibility*, pages 919–924, 2004.

369. Stephan Braun and Peter Russer. A FPGA based time-domain EMI measurement system for quasi-peak detection and disturbance analysis. In *German Microwave Conference Gemic 2005, April 5-7, Ulm, Germany*, pages 213–216, 2005.
370. Stephan Braun and Peter Russer. The dynamic range of a Time-Domain EMI measurement system using several parallel analog to digital converters. In *16th International Zurich Symposium on Electromagnetic Compatibility*, pages 203–208, 2005.
371. Stephan Braun, Andreas Alt, and Peter Russer. A novel multiresolution high-dynamic ultra-broadband time-domain EMI measurement system. In *2005 International Microwave Symposium Digest, Long Beach, CA, USA*, page 4 pp., 2005.
372. Stephan Braun and Peter Russer. A low-noise multiresolution high-dynamic ultra-broadband time-domain EMI measurement system. *IEEE Transactions on Microwave Theory and Techniques*, 53(11):3354–3363, 2005.
373. Stephan Braun, Martin Aidam, and Peter Russer. Development of a multiresolution time domain EMI measurement system that fulfills CISPR 16-1. In *Proceedings of the 2005 IEEE International Symposium on Electromagnetic Compatibility*, pages 388–393, August 8th–12th, 2005.
374. Stephan Braun and Peter Russer. Taking time-domain EMI measurements according to international EMC standards. *Compliance Engineering Journal*, XXIII 2006 Annual Reference Guide(1):45–54, March 2006.
375. Stephan Braun and Peter Russer. Measurements of spurious emission with a time-domain EMI measurement system using multi-sampling techniques. In *Proceedings of the 17th International Zurich Symposium on Electromagnetic Compatibility, 2006, EMC Zurich 2006*, volume 3, pages 792–795, Singapore, February 2006.
376. Stephan Braun, Stoyan Iliev, Mohammed Al-Qedra, and Peter Russer. A real-time multiresolution time-domain EMI measurement system based on ultra-fast high resolution Analog-to-Digital converters. In *Proceedings of the 16th International Conference on Microwaves, Radar & Wireless Communications, MIKON 2006*, pages 665–668, 2006.
377. Stephan Braun and Peter Russer. Uncertainty analysis and novel test procedures performed with a realtime time-domain EMI measurement system. In *Proceedings of the 2007 IEEE International Symposium on Electromagnetic Compatibility*, pages 1–4, 2007.
378. Stephan Braun, Martin Aidam, and Peter Russer. Development and evaluation of a realtime Time-Domain EMI measurement system for automotive testing. In *Proceedings of the 2007 IEEE International Symposium on Electromagnetic Compatibility*, pages 1–4, Honolulu, HI, USA, July 9th–13th, 2007.
379. Stephan Braun, Arnd Frech, and Peter Russer. A low-noise realtime time-domain EMI measurement system. In *Proceedings of the 18th International Zurich Symposium on Electromagnetic Compatibility, 2007, EMC Zurich 2007*, pages 381–384, 2007.
380. Stephan Braun, Thomas Donauer, and Peter Russer. A real-time time-domain EMI measurement system for full-compliance measurements according to CISPR 16-1-1. *IEEE Transactions on Electromagnetic Compatibility*, 50(2):259–267, 2008.
381. Stephan Braun. *Theorie und Anwendung von Zeitbereichsverfahren zur nonkonformen EMV-Emissionsmessung*. Dissertation, Technische Universität München, München, 2007.
382. Carl Friedrich Gauss. Theoria interpolationis methodo nova tractata. In *Gauss' collected works*, pages 265–330. Goettingen State and University Library, Göttingen, Germany, 1886.
383. Arnd Frech, A. Zakaria, Stephan Braun, and Peter Russer. Ambient noise cancelation with a time-domain EMI measurement system using adaptive filtering. In *Proceedings of the Asia-Pacific Symposium on Electromagnetic Compatibility and 19th International Zurich Symposium on Electromagnetic Compatibility, 2008. APEMC 2008*, pages 534–537, 2008.
384. Arnd Frech, Stephan Braun, and Peter Russer. Time-domain EMI measurements in the presence of ambient noise. In *Proceedings of the 2008 IEEE International Symposium on Electromagnetic Compatibility*, pages 139–142, 2009.
385. Friedrich Hund. *Geschichte der Quantentheorie*. Bibliographisches Institut, Mannheim, 1969.
386. Bernard d'Espagnat. *On Physics and Philosophy*. Princeton University Press, September 2006.

387. Rainer Waser. *Nanoelectronics and Information Technology: Materials, Processes, Devices*. Wiley-VCH, Weinheim, 2nd edition, 2005.
388. Edward L. Wolf. *Quantum Nanoelectronics: An Introduction to Electronic Nanotechnology and Quantum Computing*. Wiley-VCH, Weinheim, March 2009.
389. Peter Russer and Nikolaus Fichtner. Nanoelectronics in Radio-Frequency technology. *IEEE Microwave Magazine*, 11(3):119–135, May 2010.
390. Peter Russer and Franz X. Kaertner. Squeezed-state generation by a DC pumped degenerate Josephson parametric amplifier. *AEÜ Archiv der Elektrischen Übertragung*, 44(3):216–224, March 1990.
391. Franz X. Kaertner and Peter Russer. Generation of squeezed microwave states by a dc-pumped degenerate parametric Josephson junction oscillator. *Physical Review A*, 42(9):5601–5612, November 1990.
392. Horace P. Yuen. Two-photon stimulated emission and pulse amplification. *Physical Review Letters*, 26(9):505–507, June 1975.
393. Horace P. Yuen. Two-photon coherent states of the radiation field. *Physical Review A*, 13(6):2226–2243, June 1976.
394. Mauro Paternostro, Giuseppe Falci, Myungshik Kim, and G. Massimo Palma. Entanglement between two superconducting QUBITs via interaction with nonclassical radiation. *Physical Review B*, 69:214502, June 2004.
395. Jozef Gruska. *Quantum Computing*. McGraw-Hill, New York, 1999.
396. Michael A. Nielssen and Isaac L. Chuang. *Quantum Computation and Quantum Information*. Cambridge University Press, Cambridge, 2000.
397. Mika Hirvensalo. *Quantum Computing*. Springer, Berlin, 2004.
398. Richard P. Feynman. Simulating physics with computers. *International Journal of Theoretical Physics*, 21(6/7):467–488, 1982.
399. Richard P. Feynman. *Feynman Lectures on Computation*. Addison Wesley, Reading, 1996.
400. David Deutsch. Physics and computation. *Quantum Theory, the Church-Turing Principle and the Universal Quantum Computer*, A 400(1818):97–117, July 1985.
401. Mark F. Bocko, Andrea M. Herr, and Marc J. Feldman. Prospects for quantum coherent computation using superconducting electronics. *IEEE Transactions on Applied Superconductivity*, 7(2):3638–3641, June 1997.
402. D.V. Averin. Quantum computing and quantum measurement with mesoscopic Josephson junctions. *Fortschritte der Physik*, 48(9-11):1055–1074, 2000.
403. Yu. Makhlin, G. Schön, and A. Shnirman. Condensed-matter physics: The QUBIT and the cavity. *Nature*, 431:138–139, September 9th, 2004.
404. I. Chiorescu, P. Bertet, K. Semba, Y. Nakamura, C. J. P. M. Harmans, and J. E. Mooij. Coherent dynamics of a flux QUBIT coupled to a harmonic oscillator. *Nature*, 431:159–162, September 9th, 2004.
405. A. Wallraff, D. I. Schuster, A. Blais, L. Frunzio, R.-S. Huang, J. Majer, S. Kumar, S. M. Girvin, and R. J. Schoelkopf. Strong coupling of a single photon to a superconducting QUBIT using circuit quantum electrodynamics. *Nature*, 431:162–167, September 9th, 2004.
406. Albert Einstein, Boris Podolsky, and Nathan Rosen. Can quantum-mechanical description of physical reality be considered complete? *Physical Review*, 47(10):777–780, 15 May 1935.
407. Ryszard Horodecki, Paweł Horodecki, Michał Horodecki, and Karol Horodecki. Quantum entanglement. *Reviews of Modern Physics*, 81(2):865–942, 2009.
408. Siddhartha Sinha and Peter Russer. Quantum computing algorithm for electromagnetic field simulation. *Quantum Information Processing*, 9(3):385–404, 2009.
409. Daniel S. Abrams and Seth Lloyd. Simulation of many-body fermi systems on a universal quantum computer. *Physical Review Letters*, 79(13):2586–2589, September 1997.
410. Peter Russer. Time-domain network methods for electromagnetic field modeling. In Zhizhang Chen and Poman So, editors, *IEEE MTT-S International Microwave Symposium Workshop on New Theories, Applications and Practices of Electromagnetic Field Simulators*, volume WFF, Anaheim, CA, USA, May 28th, 2010.

411. Peter Russer, Nikolaus Fichtner, Paolo Lugli, Wolfgang Porod, and Hristomir Yordanov. Monolithic integrated antennas and nanoantennas for wireless sensors and for wireless intrachip and interchip communication. In *Proceedings of the 40th European Microwave Conference, Paris*, Paris, France, September 2010.
412. Peter Russer, Nikolaus Fichtner, Paolo Lugli, Wolfgang Porod, Johannes A. Russer, and Hristomir Yordanov. Nanoelectronics based monolithic integrated antennas for electromagnetic sensors and for wireless communications. *IEEE Microwave Magazine*, 11(7):58–71, December 2010.
413. Peter Russer. Superconducting nanoelectronic devices. In Luca Pierantoni, Fabio Coccetti, Christophe Caloz, and George W. Hanson, editors, *IEEE MTT-S International Microwave Symposium Workshop on New Microwave Devices and Materials Based on Nanotechnology*, volume WMD, Anaheim, CA, USA, May 24th, 2010.
414. Peter Russer. Superconducting nanoelectronic devices. In *URSI Conference Kleinheubach*, Miltenberg, Germany, October 4th–6th, 2010.
415. Nikolaus Fichtner and Peter Russer. An accelerated hybrid TLM-IE method for the investigation of shielding effectiveness. *Advances in Radio Science*, 8:13–18, 2010.
416. Nikolaus Fichtner and Peter Russer. A hybrid TLM-integral equation method using time-domain plane-waves for shielding effectiveness computations. In *26th Annual Review of Progress in Applied Computational Electromagnetics (ACES)*, Tampere, Finland, April 26th–29th, 2010.
417. Nikolaus Fichtner and Peter Russer. Investigation of a UWB antenna link using the hybrid TD–IE/TLM technique. In *Proceedings of the 40th European Microwave Conference, Paris*, Paris, France, September 2010.
418. Christian Hoffmann, Stephan Braun, and Peter Russer. A broadband time-domain EMI measurement system for measurements up to 18 GHz. In *Proceedings of the European Conference on Electromagnetic Compatibility, 2007, EMC Europe 2010*, Wroclaw, Poland, September 13th–17th, 2010.
419. Christian Hoffmann and Peter Russer. A low-noise high-dynamic range time-domain EMI measurement system for CISPR band E. In *URSI Conference Kleinheubach*, Miltenberg, Germany, October 4th–6th, 2010.
420. Christian Hoffmann and Peter Russer. Measuring electromagnetic interference above 1 GHz in time-domain. In *Proceedings of the European Conference on Electromagnetic Compatibility, 2007, EMC Europe 2010*, Wroclaw, Poland, September 13th–17th, 2010.
421. Stephan Braun, Arnd Frech, Hassan H. Slim, and Peter Russer. Automation of radiated emission measurements with an ultra-fast time-domain EMI measurement system. In *Proceedings of the Asia-Pacific Symposium on Electromagnetic Compatibility and 19th International Zurich Symposium on Electromagnetic Compatibility, 2008. APEMC 2008*, pages 303–306, 2008.
422. Farooq Mukhtar, Hristomir Yordanov, and Peter Russer. Network model of on-chip antennas. In *URSI Conference Kleinheubach*, Miltenberg, Germany, October 4th–6th, 2010.
423. Johannes A. Russer, A. Ramachandran, Andreas C. Cangellaris, and Peter Russer. Phenomenological modeling of passive intermodulation (pim) due to electron tunneling at metallic contacts. In *2006 International Microwave Symposium Digest, San Francisco, CA, USA*, pages 1129–1132, 2006.
424. Johannes A. Russer, Prasad S. Sumant, and Andreas C. Cangellaris. A lagrangian approach for the handling of curved boundaries in the finite-difference time-domain method. In *2007 IEEE MTT-S International Microwave Symposium Digest, Honolulu, HI, USA*, pages 717–720, 2007.
425. Johannes A. Russer, Andreas C. Cangellaris, and Peter Russer. A reciprocity-based methodology for the expedient and accurate prediction of electromagnetic field coupling to multi-conductor transmission lines. In *Proceedings of the 2006 IEEE International Symposium on Electromagnetic Compatibility*, pages 99–101, August 2006.
426. Johannes A. Russer, Andreas C. Cangellaris, and Peter Russer. Electromagnetic field interaction with a transmission line. In I. C. Göknar and L. Sevgi, editors, *Complex Computing-*

Networks Brain-like and Wave-oriented Electrodynamic Algorithms, pages 13–26. Springer, Berlin, 2006.

427. Johannes A. Russer, Prasad S. Sumant, and Andreas C. Cangellaris. A lagrangian approach for the handling of curved boundaries in the finite-difference time-domain method. In *2007 IEEE MTT-S International Microwave Symposium Digest, Honolulu, HI, USA*, pages 717–720, 2007.
428. Johannes A. Russer and Andreas C. Cangellaris. An efficient methodology for the modeling of electromagnetic wave phenomena in domains with moving boundaries. In *2008 IEEE MTT-S International Microwave Symposium Digest, Atlanta, GA, USA*, pages 157–160, 2008.
429. Johannes A. Russer and Andreas C. Cangellaris. Method for enhancing the efficiency of numerical solution of time–periodic transmission line problems with highly disparate time scales. In *Proceedings of the 40th European Microwave Conference, Paris*, Paris, France, September 2010.
430. Johannes A. Russer. *Methodologies for electromagnetic field modeling for computer aided analysis of multi-domain physical interactions*. PhD thesis, Graduate College of the University of Illinois at Urbana-Champaign, Urbana, Illinois, 2010.
431. Johann Wolfgang v. Goethe. *Gedenkausgabe der Werke, Briefe und Gespräche*. Artemis, Zürich, 28 Aug. 1949.
432. Leopold B. Felsen. Lectio Magistralis. In Peter Russer and Mauro Mongiardo, editors, *Fields, Networks, Methods, and Systems in Modern Electrodynamics – A Tribute to Leopold B. Felsen*, pages XIX–XXIX. Springer, Berlin, 2004.
433. Benedictus de Spinoza. *The Ethics; Treatise on the Emendation of the Intellect*. Hackett Publishing Company, 2nd edition, 1991.
434. Thomas Nagel. Conceiving the impossible and the mind-body problem. *Philosophy*, 73(285):337–352, 1998.
435. Colin McGinn. Can we solve the mind-body problem? *Mind*, 98(391):349–366, July 1989.
436. Arthur Schopenhauer. *Kleinere Schriften*. Haffmans, Zürich, 1988.
437. Arthur Schopenhauer. *On the Fourfold Root of the Principle of Sufficient Reasons and On the Will in Nature*. George Bell & Sons, London, 1903.
438. Hugo von Hofmannsthal. *Brief des Lord Chandos: Poetologische Schriften, Reden und erfundene Gespräche*. Insel Verlag, 2000.
439. Erwin Schrödinger. *Mein Leben, meine Weltansicht*. Diogenes, Zürich, 1989.
440. Francis Harold Cook. *Hua-Yen Buddhism: The Jewel Net of Indra*. Pennsylvania State University Press, May 1977.
441. Friedrich Nietzsche. *Also sprach Zarathustra*. 1885.
442. William Shakespeare. *The Tempest*. Act IV, Scene 1, 1610–1611.

MIX
Papier aus verantwortungsvollen Quellen
Paper from responsible sources
FSC® C105338

If you have any concerns about our products,
you can contact us on
ProductSafety@springernature.com

In case Publisher is established outside the EU,
the EU authorized representative is:
Springer Nature Customer Service Center GmbH
Europaplatz 3, 69115 Heidelberg, Germany

Printed by Libri Plureos GmbH
in Hamburg, Germany